MODERN HUMANS

Modern Humans

THEIR AFRICAN ORIGIN AND GLOBAL DISPERSAL

John F. Hoffecker

Columbia University Press
New York

Columbia University Press
Publishers Since 1893
New York   Chichester, West Sussex
cup.columbia.edu

Library of Congress Cataloging-in-Publication Data
Names: Hoffecker, John F., author.
Title: Modern humans : their African origin and global dispersal / John F. Hoffecker.
Description: New York : Columbia University Press, 2017. |
Includes bibliographical references and index.
Identifiers: LCCN 2017025425 (print) | LCCN 2016058573 (ebook) |
ISBN 9780231160766 (cloth : alk. paper) | ISBN 9780231543743 (e-book)
Subjects: LCSH: Human beings—Origin. | Human
beings—Migrations. | Human evolution.
Classification: LCC GN281 .H625 2017 (ebook) | LCC GN281 (print) |
DDC 599.93/8—dc23
LC record available at https://lccn.loc.gov/2017025425

Columbia University Press books are printed on permanent and durable acid-free paper.
Printed in the United States of America

Cover design: Milenda Nan Ok Lee

Cover photo: © Martin Schoeller / Art + Commerce Image Archive

In Memoriam

KARL W. BUTZER

(1934–2016)

CONTENTS

MODERN HUMANS

Chapter One

INFORMATION, COMPLEXITY, AND HUMAN EVOLUTION

Why may we not say that all *Automata* (Engines that move themselves by springs and wheeles as doth a watch) have an artificiall life?

—THOMAS HOBBES (1651)

As Charles Darwin observed in 1871, humans are "the most dominant animal that has ever appeared on the earth."[1] Their dominance is measured not by physical size or by numbers or total biomass, although the latter two are impressive. Rather, it is measured by their control of the environment.

To begin with, humans have occupied virtually every terrestrial habitat on Earth, which is remarkable for a single species, especially because it was largely accomplished more than 45,000 years ago, when the direct ancestors of living humans first spread across much of the land surface of the planet. Moreover, by manipulating both its biotic and its abiotic components, humans have radically altered the environment to suit their own needs, exponentially increasing their numbers and suppressing or eliminating other life-forms along the way. Here, too, the process began more than 45,000 years ago, long before the emergence of the earliest civilizations.

Humans' control of the environment is based on their ability to manipulate objects and materials in complex ways (that is, to make and use complex artifacts). There is nothing remarkable about the capacity to move or modify physical objects and materials, which is widespread in the animal kingdom, but the level of complexity that underlies human technology is unique. Humans can translate a large body of information stored in the brain into a hierarchically organized "artifact" such as a set of winter clothing or a sailboat. The artifacts of humans may be designed with an autonomous function—a *machine*, such as a self-acting rabbit snare or a mechanical clock—that Thomas Hobbes described as "artificiall life" in 1651.[2]

Evolution: The Major Transitions

The evolution of life on Earth, from its beginning more than 3 billion years ago, provides a perspective on the unique ability of humans to translate information from the brain to structure—including functioning structure—in the form of a complex artifact. As John Maynard Smith and Eörs Szathmáry observed, the evolution of living systems has been characterized by a series of "major transitions." Each reflects a fundamental change in *how information is stored, transmitted, and translated*. An example is the transition from single-celled to multicellular organisms, which took place more than 1 billion years ago. And each transition represents a quantum jump to a new level of complexity with *emergent properties*.[3]

Maynard Smith and Szathmáry included humans on their list of major transitions in evolution.[4] Humans store, transmit, and translate information in novel ways; moreover, the changes in information represented by humans pertain to information in the brain (or *neuronal information*) rather than information in the genes. Humans store and transmit neuronal information in the form of language. In fact, the units of language, which include sounds made with the vocal tract, as well as the imagined sounds of the vocal tract reproduced in the brain, are translated from neuronal information. They, too, may be considered a "complex artifact," even when they are not being rendered in material form (for example, written words).[5] And the units of language may be manipulated in hierarchically organized structures with many levels and subcomponents, analogous to a machine (or what Daniel Dennett described as a "virtual machine" in the head).[6]

The capacity of humans to translate neuronal information into material structure, including functioning structure, represents a major transition in evolution, and it parallels the translation of genetic information into functioning structure (living organisms). Some suggest that the earliest life-forms were both information and structure or, more specifically, RNA acting as an enzyme (or a ribozyme).[7] Even the simplest cells (prokaryotes) require the translation of genetic information into proteins and other functioning structures. True cells, or eukaryotes, which may have evolved

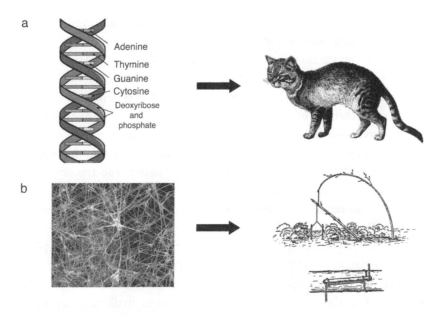

FIGURE 1.1 The translation of information into structure: (*a*) the translation of genetic information into a functioning organism in the form of a cat; (*b*) the translation of information stored in the human brain into functioning structure in the form of a two-state machine, or automaton (rabbit snare made by the Tanaina [see box figure 4.3]). ([*b*] From Cornelius Osgood, *The Ethnography of the Tanaina* [New Haven, Conn.: Yale University Press, 1937], 93, fig. 20. Courtesy of Yale University Publications in Anthropology)

as early as 2 billion years ago, store and transmit a much larger quantity of genetic information than do prokaryotes. They became the basis for multicellular organisms and the metazoa, including vertebrates, which undergo a protracted process of development from a single fertilized eukaryotic cell that entails the translation of massive amounts of genetic information into functioning structure (figure 1.1).[8]

Evolution as Computation

One way to look at evolution is as a form of *computation*. The evolutionary process contains all the basic elements of a computation: "input" (random changes and/or recombination of genetic information that is translated into organisms), "operations" or "defining functions" (natural selection of organisms), and "output" (changes in the genetic information and the organisms).[9] At least some of the changes produce an organism that is better adapted (or exhibits a better "fit") to its environment. And because the input reflects random events, evolution is a "nondeterministic"—even *creative*—computational process, constantly yielding what Darwin described as "endless forms most beautiful."[10] The immense variety is a consequence of the many hierarchical levels of living systems.

The computations of the evolutionary process take place on several levels, however, and with more than one form of information.[11] Each living species is a product of computation with genetic information that occurs on the level of the evolving *lineage*: the information and the organism evolve together—neither can evolve without the other—over time. Thanks to recent and remarkable developments in the recovery and analysis of ancient DNA (aDNA), we have a partial "fossil" record of both the information and the organismal structure to which it is translated (fossil plant and animal remains) in the course of development (figure 1.2*a*).

Long ago, prokaryotes evolved a form of computation on the level of the organism that allows an individual to respond to unpredictable variations in its environment. Proteins in the cell wall of a prokaryote transmit chemical signals—a simple non-genetic form of information—about the presence of potential "food" sources or threats, and the organism responds by moving toward or away from whatever triggered the signal. Unlike the computations of an evolving lineage, the process is a deterministic one with a predictable outcome.[12]

The metazoa evolved a much more complex form of computation on the level of the organism, with a new type of information generated at the cellular level. Specialized eukaryotic nerve cells, or *neurons*, transmit and store information in the electrochemically charged structures (*synapses*) that connect one neuron with others. And, like the mutation and recombination of genetic information, metazoan development includes a randomizing process that renders computation with neuronal information nondeterministic and potentially creative (figure 1.2*b*).

Humans are constantly performing creative computations with neuronal information, but they also use their power to translate neuronal information into material structure to create an entirely new level of evolutionary computation. By making machines, or *automata*, that function independently of a living organism, humans create nonliving structures that act like a simple organism and perform their own computations. A machine also contains all the basic elements of a computation: input (materials or

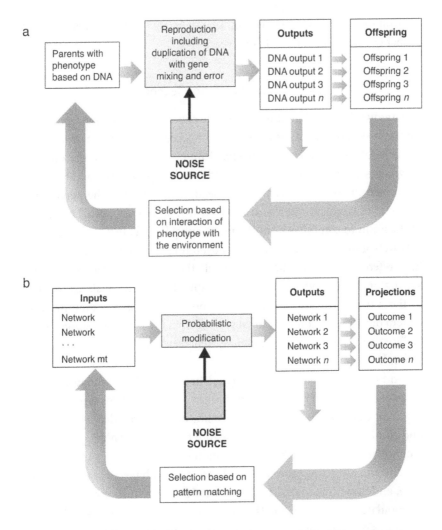

FIGURE 1.2 Evolutionary computation on two levels: (*a*) computation on the level of the evolving lineage; (*b*) computation performed by the metazoan brain on the level of the individual organism. Humans have developed a new level of evolutionary computation in the form of automata. (Redrawn from John E. Mayfield, *The Engine of Complexity: Evolution as Computation* [New York: Columbia University Press, 2013], 143, fig. 5.3 [*a*], 253, fig. 10.1 [*b*])

information), operations or "transition functions," and output (transformed materials or information).[13] Unlike evolving lineages and metazoan brains, however, many machines do not compute with information, but with matter and energy, and most machines perform deterministic computations with predictable output. Machines are not confined to industrial or early civilizations, but have their origin among hunter-gatherer societies and probably date to more than 45,000 years ago.

Language as Information Technology

The unique ability of humans to translate neuronal information into structure also supports another novel form of computation both at and above the level of the organism: language. Humans compute with the units of language—spoken words or imagined words replayed in the brain—in a manner analogous to moving the beads of an abacus. The perceptible "objects" of language (words) are moved around in the "virtual machine" of the brain to compute the problems of everyday life, including the complex web of social relationships that most humans inhabit.[14] Because words have referents that potentially lie outside the immediate spatial-temporal setting of the individual brain, the computations may be about the past or the future.[15] Human language functions not only as a communication system but also as a computational system or, in the words of George Boole, an "instrument of reason."[16] Language represents the original *information technology*.

And if language represents a virtual "technology" for computing with units of information in the form of spoken or imagined vocal sounds, it is apparent that humans also compute with the materials that are modified and combined into their artifacts. Before humans began designing their technology on paper with written symbols and diagrams, their artifacts reflected an unending process of trial and error with pieces of wood, stone, hide, and other materials. The units of technology—raw or cooked—were manipulated in the brain and the hands in a manner analogous to the computations of language. As a result, human technology acquired a character similar to that of language, with a potentially infinite variety of hierarchically structured combinations and recombinations of its material units (box 1.1).

The Rise of the Super-Brain

A major consequence of developing a system of computation from a system of communication is that two or more brains may compute the same problem (that is, may function as a *super-brain*).[17] Humans collectively

BOX 1.1
Modern Humans and "Syntactic Technology"

The technology of *Homo sapiens* is unique within the living world, and it differs in fundamental ways from the technology of all other animals. Because it is a product of the evolution of living systems, *technology* is logically defined in the context of evolutionary biology. It falls into two major categories: (1) appropriated "traits" and (2) redesigned environments. Many animals appropriate pieces of their surrounding environment to function as a trait they otherwise lack, and some animals modify these pieces of the environment to make them functional (for example, a chimpanzee stripping leaves off a twig to function as a termite-fishing stick).[1] Many animals alter their environment to enhance their fitness in some way, such as a bird's nest or a rodent's burrow (and some animals, including humans, do both).

Technology developed among the metazoa and, more specifically, among metazoans such as the cephalopods, insects, and chordates, which evolved both external functioning structures such as jaws and limbs, and a central nervous system to coordinate their functions. Humans are somewhat unusual in having evolved a specialized extremity—the paired hands, which are almost entirely devoted to manipulating pieces of the environment. This appears to be a consequence of a highly unusual locomotor adaptation among mammals—bipedalism—that probably evolved for reasons other than technological function (chapter 3). Humans also evolved a central nervous system with considerable memory-storage capacity and computational powers (which also probably evolved for reasons other than technological function).

The features of modern human technology that render it unique lie in the role that *information* plays in both "appropriated traits" and redesigned environments. In both categories, humans evolved the capacity to translate information stored in the neuronal networks of the brain into the design of technological structures (artifacts and features). This capacity is analogous to the translation of genetic information into the functioning structure of an organism during its development, although the process of translation is very different. There are some examples of the translation of genetic information into technological structure. Some species of ant construct complex underground nests with hierarchically organized networks of tunnels and chambers.[2] And there may be rare examples of the translation of neuronal information into simple technological structure. Modern humans routinely translate neuronal information into complex, hierarchically organized artifacts and features, including those that function independently of the organism (that is, machines or automata).

The technology of modern humans exhibits the same fundamental properties of *syntactic language*, which allows the rapid recombination of a set of finite elements (speech sounds) into a potentially infinite variety of hierarchically organized structures (sentences).[3] This property of language has been labeled "discrete infinity" by Noam Chomsky.[4] Modern humans apply it to the recombination of material objects to generate a wide variety of artifacts and features. They recursively tinker with artifact and feature designs, creating novel combinations of the elements and innovative variations of both. A number of archaeologists have noted the similarities between the properties of syntactic language and the making of artifacts.[5]

It is apparent, moreover, that humans actually *compute* with the components of artifacts and features in a manner analogous to that of computation with words or

(continued)

(continued)

numbers. Innovative variations and novel combinations often are the result of creative or even random manipulations of raw materials (and/or modified pieces of the same) with the hands—and visually and tactilely coordinated feedback with neuronal networks in the brain. This is another unique feature of human technology; no other animal is known to compute with objects and materials as if they were symbols or a form of information.

It also is apparent that—as in the case of language—many individual brains separated by space and time compute the same problems with the materials and objects (for example, how to construct a house). The computations are therefore *collective*, and the solutions arrived at by one group of people in space and time are almost certain to be modified by their descendants and neighbors. The collective character of human technological computation ensures a massive quantity of input and immense computational power (that is, super-brain function).[6]

The construction of machines or automata is another unique feature of human technology (and one that is confined to modern humans). The structural and functional complexity of such artifacts or devices is possible only with "syntactic technology," which allows the translation of a complex design into a hierarchically organized structure with functions, such as a small-mammal trap. (The hierarchically structured products of language may provide a model for the machine.)[7] Making an artifact that functions independently of the organism reflects a major step in the evolution of living systems. Machines or automata perform their own computations (that is, they possess all the elements of a computation) and thus represent a new level of evolutionary computation (beyond the lineage and individual organism).[8] Because of the collective character of human technological computation, machine functions occur on a unique human social-cultural level that transcends individual organisms in space and time.

1. William McGrew, *Chimpanzee Material Culture: Implications for Human Evolution* (Cambridge: Cambridge University Press, 1992).
2. Bert Hölldobler and E. O. Wilson, *The Super-Organism: The Beauty, Elegance, and Strangeness of Insect Societies* (New York: Norton, 2009), 338–339.
3. See, for example, Ray Jackendoff, *Foundations of Language: Brain, Meaning, Grammar, Evolution* (Oxford: Oxford University Press, 2002); and Ray Jackendoff and Eva Wittenberg. "What You Can Say Without Syntax: A Hierarchy of Grammatical Complexity," in *Measuring Grammatical Complexity*, ed. Frederick J. Newmeyer and Laurel B. Preston (Oxford: Oxford University Press, 2014), 65–82.
4. Marc Hauser, Noam Chomsky, and W. Tecumseh Fitch, "The Faculty of Language: What Is It, Who Has It, and How Did It Evolve?" *Science* 298 (2002): 1569–1579.
5. See, for example, André Leroi-Gourhan, *Le Geste et la parole*, vol. 1, *Technique et langage* (Paris: Albin Michel, 1964); and James Deetz, *Invitation to Archaeology* (Garden City, N.Y.: Natural History Press, 1967), 81–101.
6. John F. Hoffecker, "The Information Animal and the Super-brain," *Journal of Archaeological Method and Theory* 20 (2013): 18–41.
7. Daniel C. Dennett, *Consciousness Explained* (Boston: Little, Brown, 1991), 210; Andy Clark, *Being There: Putting Brain, Body, and World Together Again* (Cambridge, Mass.: MIT Press, 1998), and *Supersizing the Mind: Embodiment, Action, and Cognitive Extension* (New York: Oxford University Press, 2011), 44–60.
8. John E. Mayfield, *The Engine of Complexity: Evolution as Computation* (New York: Columbia University Press, 2013).

compute most of their technology, and it is difficult to think of an example—in any temporal or geographic context—of an artifact that is not the product of more than one human brain. In his book *The Social Context of Innovation*, Anthony F. C. Wallace described the contributions of multiple individuals to the development of the steam engine in seventeenth-century England.[18] For a prehistoric setting, Grahame Clark reviewed changes in bow design during the Neolithic of northwestern Europe.[19]

Humans are not the only metazoans that evolved a super-brain (although they are the only metazoans that compute collectively over multiple generations). As the entomologist Thomas Seeley has shown, a honeybee colony collectively computes a new nest site by pooling and comparing information gathered by many individual bees.[20] Collective computation among humans differs significantly from that among honeybees, however, because it is not based on kin selection (a high degree of genetic relatedness among the members of the colony). Collective computation among the members of a human social group is more likely to be competitive (even if the competition is subtle and masked with assurances that the computation is being performed for the common good).

The social integration of the already enlarged human brain—each brain storing and transmitting billions of bits of information—underlies the immense computational power of even the smallest of socioeconomic formations (such as hunter-gatherer bands of 25 to 30 people). There is an analogy to the quantum jump in genome size in the transition from prokaryotes to eukaryotes, which significantly increased the computational power of eukaryotic lineages. The computations of the super-brain are nondeterministic and creative (with much random input). The output is stored among the individual brains and in the structures to which it is translated, as well as in a variety of other forms of external memory storage.

The Evolution of Humans

Humans evolved two specialized organs for the translation of information stored (and manipulated) in the brain into complex, hierarchically organized structures: the paired *hands* and the *vocal tract*. With the vocal tract, the neuronal information is translated into another form of information based on the "structures" of real or imagined sounds. (All forms of information have a physical basis: either matter or energy.)[21] Both the specialized human hand and the vocal tract have deep roots in primate evolution.

The early primates evolved nails in place of claws, reflecting the use of the extremities for grasping objects (including the branches of trees), and most primates employ their forelimbs for manipulation as well as locomotion. What the anatomist John Napier described as the "true hand" is confined to apes and humans, however.[22] The true hand functions primarily as a manipulative organ. The extreme versatility of the human hand almost certainly is an (indirect?) consequence of bipedalism in early humans. The fully modern hand probably was present by 1.5 million years ago (that is, by the time humans were making hand axes).[23]

The vocal tract of living humans is characterized by the migration of the *larynx* (voice box) during infancy to a position in the neck that is low in comparison with that in apes. In this position, the larynx expands the size of the *pharynx* (vocal chamber), allowing humans to generate a wide range of sounds. But reconstructing the evolution of the vocal tract is difficult because it is composed of soft parts and cartilage, which do not preserve in the fossil record. Instead, skeletal features associated with vocal-tract anatomy and function are used to make inferences about the evolution of the larynx in early humans. Examples include the degree of flexure at the base of the cranium and the length of the neck (that is, size and number of cervical vertebrae).[24]

Also helpful is the comparative anatomy and behavior of living primates, which exhibit a wide range of vocal signals. Especially significant are the warning calls of various primates. Among vervet monkeys, each of three different calls has a specific predator referent (that is, the sounds are matched with meanings).[25] The alarm calls of vervet monkeys suggest an evolutionary source for human language, but determining its presence or absence in the fossil record is problematic. In fact, given the uncertain relationship between the anatomy of the vocal tract and language (either might have evolved without the other), clues pertaining to other aspects of the language faculty ultimately may be more useful. One of them is the presence of Broca's area in the endocast of an early *Homo* skull dated to about 1.7 million years ago.[26] Another is evidence that the "critical learning period" for language acquisition in children[27] may be a product of the delayed maturation process in more recent forms of *Homo*.[28]

The special functions of the hands and the vocal tract with respect to the translation of neuronal information into structure (and functioning structure) are closely linked to the evolution of the human brain. Here, too, the roots in primate biology run deep. The primates evolved a "visual brain," with a reduced olfactory bulb and overlapping fields of vision. The catarrhines (Old World monkeys and apes) evolved a cortical area unique among mammals for the visual coordination of motor tasks.[29] Brain size in early humans (australopithecines) remained comparable to that of apes, but, after roughly 2 million years ago, it began to grow steadily—until the dispersal of modern humans.[30]

The pattern and causes of brain growth in *Homo* is one of the most important issues in human evolution. What triggered the process of continual growth after 2 million years ago, and why did it stop with modern humans? Given the high energy demands of a very large brain, the pattern of growth is inevitably tied to human ecology and diet.[31] Significantly, there are changes in the archaeological record at 2 million years ago that suggest a major shift in foraging strategy and diet (and a major expansion of geographic range and habitat).

The continual nature of the enlargement of the human brain suggests what has been termed an "unstable" evolutionary strategy or phenotype between 2 million years ago and the emergence of modern humans.[32] Robin Dunbar argues that continuous selection pressure for the cognitive skills that underlie the management of social relationships was the driving force behind brain growth (social brain hypothesis).[33] Selection for added brain tissue to support such skills conceivably includes increased information-storage capacity and enhanced computational abilities. The "social brain" also is likely related to changes in foraging and diet (the onset of social foraging), and the unstable phenotype may reflect an evolutionary "arms race" among the members of foraging groups that came to a halt only when the brain could expand no further (possibly due to several factors).[34]

Equally important, humans evolved significant changes in brain development that have become apparent only in recent years. The growth and development of the brain in early *Homo* was comparable to that in apes (rapid growth and early maturation).[35] The modern pattern of significantly delayed maturation seems to have evolved quite recently—it may even have been absent among the Neanderthals.[36] Moreover, modern humans appear to have evolved changes in the pattern of brain development that entailed early growth of the parietal lobes and the cerebellum (areas implicated in both language and hand function). The expansion of these parts of the brain during the first year of life produces the characteristic "globular" shape of the modern human cranium.[37]

The challenge to students of human evolution (or paleoanthropologists) is to identify the time and place of important changes in anatomy and behavior in the fossil record within the larger context of more fundamental advances in how humans store, transmit, and translate information. The latter are reflected, above all, in the ways in which living humans translate neuronal information into complex structure, including functioning structure and structure that functions as information (for example, spoken language). The more fundamental changes all seem to be tied to the evolution of the brain, hand, and vocal tract, and these, in turn, appear to be linked to developments in foraging and diet (including bipedal locomotion and social foraging).

Modern Humans as a "Major Transition"

Despite the deep evolutionary roots of language and toolmaking, there is reason to believe that much of what we regard as unique in humans evolved less than 250,000 years ago. Most evidence for the making of complex artifacts with multiple hierarchically organized levels—including machines—postdates the appearance and spread of modern human anatomy. And although it has been impossible to demonstrate the presence or absence of language in a prehistoric context, a phonemic analysis of 500 languages reveals a pattern similar to that of the genetics of living people—relatively recent origin in Africa.[38] Also potentially significant is that compelling evidence for visual art postdates the appearance of modern human anatomy. Visual art exhibits strong parallels with syntactic language in living humans; it represents the translation of neuronal information into another form of information, but in analog rather than in digital form. In sum, much of what Maynard Smith and Szathmáry identified as a "major transition" in humans may be attributed to the origin and spread of *Homo sapiens*.

Modern humans (*H. sapiens*) evolved in Africa between 300,000 and 150,000 years ago (figure 1.3). The modern human anatomical pattern, which comprises several features of the cranium and mandible, as well as

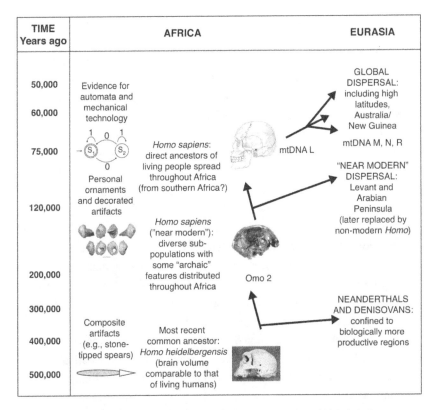

FIGURE 1.3 Major events in human evolution during the past 500,000 years, which include the appearance of anatomically modern (or "near modern") humans in Africa (300,000–150,000 years ago) and their limited dispersal out of Africa (ca. 120,000–90,000 years ago), followed by the spread of the lineages of living people throughout the continent and the beginning of the global dispersal (75,000–60,000 years ago).

some postcranial traits (chapter 2), may be recognized in a set of skeletal remains from Africa, including North Africa, that dates to at least 150,000 years ago.[39] There is one more increase in brain size (subsequently reversed), but the more significant changes in anatomy probably lie in developmental neurobiology—alterations in the timing and pattern of brain growth.[40] Estimated coalescence ages for the maternal and paternal lineages of all living humans are consistent with the dated skeletal remains.[41]

Roughly 120,000 to 90,000 years ago, one or more of the African regional populations of *Homo sapiens* expanded into Eurasia.[42] Their skeletal remains and associated artifacts are well documented in the Levant, but both human fossils and archaeological data suggest that the expansion may have reached much farther into southern Asia, including Southeast Asia (chapter 4). Significantly, it did not extend to Australia, which would have required watercraft, or into higher latitudes, which would have required complex technology to solve the problems of scarce resources and low winter temperatures. The Levantine fossils that represent this early migration out of Africa exhibit most of, but not all, the anatomical features of living modern humans and have been described as "near modern."[43]

At some point between 75,000 and 60,000 years ago, modern humans began another expansion out of Africa. Genetic data indicate that this later migration was undertaken by people who are directly ancestral to all living humans, and, for the most part, their skeletal remains lack the archaic features found in earlier "near modern" human fossils.[44] They appear to have originated in southern Africa.[45] The second expansion ultimately became global in scope, including the occupation of Australia no later than 50,000 years ago and of northern Eurasia before 45,000 years ago (chapters 5 and 6). Retreating glaciers at the end of the Pleistocene opened access to the Americas, which were occupied rapidly after 15,000 years ago (chapter 7).

Paleoanthropologists currently are engaged in a lively debate about the relationship between (1) the cognitive faculties underlying language and complex artifacts and (2) the appearance of modern human anatomy between 300,000 and 150,000 years ago in Africa. Some argue that the two are not necessarily correlated and that many of these faculties are reflected in the archaeological record of non-modern forms of *Homo*, such as the Neanderthals.[46] The most important piece of pertinent archaeology is the evidence for making hafted implements comprising at least three components (wooden handle or shaft, adhesive, and stone blade or point), which is documented among Neanderthals and as early as 500,000 years ago in Africa.[47] The complex design of—and multiple hierarchically organized steps required to produce—a composite tool or weapon is unknown outside the genus *Homo*. Also pertinent is the presence of a gene (FOXP2) in the Neanderthal genome that is associated with language in living people.[48]

Other paleoanthropologists argue that significant advances in human cognitive faculties took place in *Homo sapiens* tens of thousands of years after the appearance of the modern anatomical pattern.[49] They point to the ambiguous archaeological record of "near modern" humans in Africa before 75,000 years ago and their dispersal out of Africa, which yields limited and problematic evidence for the expression of the cognitive faculties found in living people. There are hints of complex pyro-technology in the sub-Saharan sites as early as 164,000 years ago (*Pinnacle Point* [South Africa]) and examples of barbed bone points that may date to 90,000 years ago (*Katanda* [Democratic Republic of Congo]).[50] Shell beads are associated with the expansion of "near modern" humans into the Levant between 120,000 and 90,000 years ago.[51]

It is not until the second dispersal of modern humans out of Africa, however, that unambiguous expressions of the cognitive faculties of living people are evident in the archaeological record. They include indirect evidence for automata in the form of snares by around 65,000 to 62,000 years ago (*Sibudu Cave* [KwaZulu-Natal]) and evidence for mechanical technology (bow and arrow) no later than roughly 44,000 to 42,000 years ago (*Border Cave* [South Africa]).[52] The belated development of these faculties conceivably accounts for the delay between the appearance of *Homo sapiens* in Africa 300,000 to 150,000 years ago and the beginning of the global dispersal 75,000 to 60,000 years ago—especially the invasion of colder regions in northern Eurasia, which had eluded all other forms of *Homo*.[53] The ability of modern humans to design complex technology with three or more hierarchically structured levels (for example, insulated clothing) and mechanical technology, including automata (for example, snares and traps), was probably critical to survival in these regions (chapter 6).

Winter temperatures in many parts of northern Eurasia were below –10°C and thus a special problem for people recently derived from the equatorial zone (who retained the warm-climate anatomy of their immediate ancestors).[54] Low plant and animal productivity (combined with the high caloric demands of a cold-climate setting) required the harvesting of smaller vertebrates—small mammals, birds, and fish—to supplement the low biomass of large mammals (especially in places where few digestible plant foods were available).[55] The archaeological record of modern human dispersal in northern Eurasia yields supporting evidence for insulated clothing and the harvesting of small vertebrates. It also reveals traces of visual art, which may be a proxy for syntactic language.

The seeming disconnect between modern human anatomy and cognition remains a problem in the study of modern human origins and dispersal, and one unlikely to be resolved with available data and analyses. The simplest explanation is that manifestations of modern cognitive faculties have low visibility in the African archaeological record before 75,000 years ago (and in the "near modern" human archaeological record in Eurasia 120,000 to 90,000 years ago) and that some other variable (for example, climate or disease) prevented a successful dispersal out of Africa at this time. Competition with other forms of *Homo*, such as the Neanderthals, may have been a factor in the failure of the dispersal of "near modern" humans, but it also was a problem faced by the people who began the global dispersal after 75,000 years ago.

Regardless of whether or not the neurological basis of modern human cognition evolved with the skeletal features used to define *Homo sapiens*, it is apparent that a significant change in brain function took place at some point between roughly 300,000 and 75,000 to 60,000 years ago. While the composite implements produced in Africa 500,000 years ago (as well as by the Neanderthals in Eurasia) are more complex than any known nonhuman "artifacts," they are simple in comparison with the winter clothing, mechanical devices, and other technologies of modern human hunter-gatherers—especially those of the higher latitudes. These technologies reflect the translation of large quantities of hierarchically organized neuronal information into structures, including self-acting structures with functions (automata). Their production often requires many hierarchically organized steps (that is, reflects a high degree of *computational complexity*). It was only with the ability to generate such complex technology—effectively redesigning themselves and their environment—that humans could inhabit previously uninhabitable places and became what Darwin described as "the most dominant animal that has ever appeared on the earth."

And although not demonstrable in the archaeological record, humans may have developed, during the same interval of time, a comparable degree of computational complexity in their ability to manipulate informational units in the form of spoken and imagined vocal sounds. As noted earlier, this faculty also entails the translation of neuronal information into structure (that is, "information technology"). As a computational system, syntactic language exhibits the same features as modern human technology—nondeterministic with potentially infinite variations on multiple hierarchical levels. The appearance of visual art, which exhibits the same features applied to artifacts (composed of information in analog form), only after the appearance of modern anatomy may reflect the absence of syntactic language before 300,000 to 75,000 years ago. Syntactic language conceivably was preceded by a less complex, non-syntactic form of language (analogous to the composite implements of 500,000 ago).[56] Because the computational system of syntactic language doubles as a communication system, the computations are inevitably collective, creating a super-brain. The super-brain, in turn, generates a phenomenon that may be described as a "neuronal genome" (or simply a *culture*).

This book is about the origin and dispersal of anatomically modern humans. A central theme is that the evolution of *Homo sapiens* represents what Maynard Smith and Szathmáry termed a "major transition" in evolution, entailing fundamental changes in the storage, transmission, and translation of information. Unlike other major transitions, modern humans evolved changes in neuronal information (which, in turn, yielded other forms of non-genetic information). The evolution of modern humans apparently generated increased complexity (both hierarchical and computational). At present, the scope of the changes is revealed only with their global dispersal.

The transition process may have been protracted, however, extending more than 200,000 years (that is, between the appearance of modern human anatomy in Africa and the second out-of-Africa dispersal of *H. sapiens*). Modern humans are derived, moreover, from an earlier form of *Homo* that already had evolved a brain volume comparable to that of living people, as well as the specialized hands and vocal tract that function to translate neuronal information into structure. The predecessors and non-modern contemporaries of *Homo sapiens* already were producing uniquely complex technology (composite implements) and probably speaking some form of language.

Another major theme in this book is that the evolution and spread of modern humans can be understand only within a theoretical framework expanded beyond the neo-Darwinian, or "modern" evolutionary, synthesis to include information and complexity theory. This is because the major transition represented by modern humans entailed changes in non-genetic information that lie outside the synthesis of natural selection theory and population genetics in the mid-twentieth century. And, more generally, the neo-Darwinian synthesis preceded an understanding of the central role of information in evolution. The remainder of this chapter is devoted to a brief history of paleoanthropology and the rise of information and complexity theory.

CHARLES DARWIN AND HUMAN ORIGINS

The scientific study of human origins began with Charles Darwin (1809–1882), who conceived and wrote the two founding documents of paleoanthropology. Darwin established the idea that all living forms are descended from a common ancestor (common descent) and he provided a theory of how evolution works. He explained the relationship between humans and the natural world. He also offered a thoughtful application of his theory to specific problems in human evolution.

On the Origin of Species

At the core of Darwin's theory of evolutionary change was what Ernst Mayr described as "an entirely new and revolutionary principle."[57] It was the idea that nonrandom selection acts on random variation in populations of organisms. In *On the Origin of Species by Means of Natural Selection* (1859), Darwin carefully laid out his argument, beginning with the *variation* that is so obviously present in domesticated species, such as dogs, and that is exaggerated by selective breeding (conscious or unconscious). In the second chapter, he turned his attention to wild plants and animals, acknowledging that variation among them is less obvious than it is among domesticated species, but nonetheless undeniable and to a surprising degree: "I should never have expected that the branching of the main nerves close to the great central ganglion of an insect would have been variable in the same species. . . . [Y]et quite recently Mr. Lubbock has shown a degree of variability in these main nerves in *Coccus*, which may almost be compared to the irregular branching of the stem of a tree."[58] The chapters on variation were followed by discussions of the potential for rapid growth in populations—in even the slowest-breeding organisms—and of natural selection.

Neither Darwin nor his contemporaries understood the "laws of variation" or the general rules of inheritance.[59] The principles of genetics were worked out in the early decades of the twentieth century and synthesized with natural selection theory in the late 1930s and the 1940s. Among the leading geneticists of the period, Sewall Wright (1889–1988) in particular emphasized the stochastic processes underlying variation (or what Jacques Monod later referred to as "a source of noise").[60] But the random character of "mutation" came to be more fully appreciated after the DNA code was deciphered in the 1960s.[61] This revealed the simple digital informational base of evolutionary genetics: four interchangeable nucleic acids that code for the building blocks of protein enzymes and are subject to random copying error during replication.

The Descent of Man

Twelve years after the appearance of *On the Origin of Species*, Darwin published his book on human evolution. The title, *The Descent of Man and Selection in Relation to Sex* (1871), reflected both his decision to use the book as a vehicle for addressing sexual selection (only briefly discussed in *Origin*) and his belief that it was a major factor in human evolution. Despite the lack of fossil evidence for human ancestors in 1871 (besides the discovery of Neanderthal remains in 1856), Darwin's book is filled with careful observations and thoughtful insights, and it continues to contribute to paleoanthropology in the early twenty-first century.[62]

Darwin correctly identified Africa as the likely place of origin of humans and speculated that they are descended from an African ape closely related to gorillas and chimpanzees.[63] Because several naturalists had already discussed the comparative anatomy of apes and humans, Darwin focused his attention on other issues, especially the evolution of human cognitive faculties ("mental powers," as he called them). He emphasized the development of fully bipedal locomotion—which he tied to an early shift from arboreality to open-landscape foraging—and its effect on "endless other structures," especially the hand.[64] Darwin compared "the structure of the hand . . . with that of the vocal organs," drawing an analogy between the making of artifacts and language.[65]

In his discussion of language, Darwin stressed computational function as much as communication, noting that "a complex train of thought can no more be carried on without the aid of words, whether spoken or silent, than a long calculation without the use of figures or algebra."[66] He proposed an explanation for the origin of language that—while largely ignored for more than century—has been revisited by several authors in recent years.[67] Darwin suggested that language was likely derived from the vocalizations and gestures of prehuman ancestors, and he specifically mentioned courtship displays and alarm calls:

[P]rimeval man, or rather some early progenitor of man, probably first used his voice largely as does one of the gibbon-apes at the present day, in producing true musical cadences, that is in singing; we may conclude from a widely-spread analogy, that this power would have been especially exerted during the courtship of the sexes.... The imitation by articulate sounds of musical cries might have given rise to words expressive of various complex emotions.... As monkeys ... utter signal-cries of danger to their fellows, it does not appear altogether incredible that some unusually wise ape-like animal should have thought of imitating the growl of a beast of prey, so as to indicate to his fellow monkeys the nature of the expected danger. And this would have been a first step in the formation of a language.[68]

He further proposed that the subsequent coevolution of language and the brain "has no doubt been far more important," because the use of the former "would have reacted on the mind by enabling and encouraging it to carry on long trains of thought."[69]

Darwin also devoted significant attention to what he termed "the so-called races of man," writing in the introduction that the articulation of his views on the subject was one of the three principal aims of the book.[70] He noted that a number of scholars believed the major living races to represent separate species, independently evolved from earlier human forms (polygenism). Darwin thought this highly improbable, observing that there was no consensus on the definition of races. He pointed out that the features commonly used to define races, such as skin pigmentation, are gradational and that living humans are capable of interbreeding with all other living humans—neither of which is consistent with the differentiation of species in other animals. He also noted that the mental faculties of living humans—including language and the capacity for technological innovation—had clearly evolved before people "first spread over the earth." Darwin concluded that, despite appearances to the contrary, the biological differences among living humans are relatively superficial. Although some of them can be accounted for by local environmental factors (for example, resistance to local pathogens), he argued that most are likely the result of sexual selection.[71]

At this point, Darwin brought the discussion of human evolution to a close and launched into the second (and lengthier) part of his book, on the broader subject of sexual selection in animals. He returned to humans, however, in the final chapter, emphasizing the role of marriage (that is, long-term pair-bonding) as a factor in sexual selection. Among other features, Darwin attributed the scarcity of body hair to sexual selection, but he also attributed most racial variation to the effects of sexual selection on local populations following the global dispersal of modern humans.[72]

Before proceeding further, it should be noted that much of what follows in this book about modern humans was argued or at least suggested by Darwin in 1871, despite the paucity of fossil evidence. Although he repeatedly stressed the lack of fundamental differences between humans and their closest nonhuman relatives, which seems at variance with the notion of modern humans as a "major transition" in evolution, this simply reflects his effort to persuade his readers that common descent applies to all life-forms and that language and toolmaking have clear evolutionary roots in nonhuman animals. Darwin's approach to the evolution of the cognitive faculties unique to modern humans is largely reiterated in this book, with the benefit of a mass of anatomical, genetic, and archaeological data that has accumulated in the past 150 years (as well as some theory that has built on the idea of natural selection acting on random variation).

Darwin's observations on racial variation and its implications for human origins were especially prescient. A century later, a similar pattern of global variation was found with proxy genetic data from living humans.[73] The implication for the evolution of modern humans was clear: they had evolved relatively recently in one region and subsequently spread to other parts of Earth, employing their unique cognitive faculties to adapt to a wide variety of climates and habitats. What Darwin could not know in 1871—and what took paleoanthropology more than a century to work out— was that *Homo sapiens* had been preceded by several earlier forms of *Homo* (each of which represented the "modern humans" of their day), who also dispersed out of Africa, leaving their scattered artifacts and occasional fossil remains across Eurasia.

THE REIGN OF THE ANATOMISTS

The historical irony of paleoanthropology is that as the fossil evidence of early humans began to accumulate, the understanding of human origins and evolution seemed to decline. From the final decades of the nineteenth century through the middle of the twentieth century, paleoanthropology was dominated by specialists in human anatomy, often trained for medicine. With a growing but tiny sample of fossil remains, the anatomists reconstructed human evolution on the basis of selected skeletal traits. The overall pattern was widely viewed not as a Darwinian tree with diverging branches, but as a sequence of stages or grades through which earlier humans had slowly but steadily progressed. Asia was often thought to be the place of human origin, and some leading anatomists believed that the living races of *Homo* had evolved independently from earlier humans in different parts of the world. The observations and insights of Darwin were largely ignored or forgotten.[74]

One reason why Darwin's ideas probably had limited influence on students of human evolution during these decades was that the theory of natural selection was under relentless assault in North America and Europe. Although a majority of biologists had come to accept the idea of evolution by the final decades of the nineteenth century, few agreed that natural selection was the primary mechanism of change and adaptation. Belief in the inheritance of acquired characteristics was widespread. After the development of genetics at the beginning of the twentieth century, many biologists decided that mutation was the driving force of evolutionary change.[75]

The Human Fossil Record

The focus on Asia was stimulated by the first discovery of early human fossils on the island of Java in 1891. A skull fragment with an estimated brain volume (850–900 cubic centimeters [cc]) roughly intermediate between those of modern humans and apes was accorded its own genus and species, but later included in the broader category of *Homo erectus*. In 1923, the first of many finds in a cave system near Beijing in northern China (*Zhoukoudian*) was reported. This material, which eventually yielded a cranial capacity comparable to that of the Java fossils, also was given its own genus and species designation, but later lumped with *Homo erectus* (table 1.1).[76]

Despite favorable conditions for discovery, fossil evidence for human evolution was slow to emerge from Europe. Although more Neanderthal remains were found in the final years of the nineteenth century, their evolutionary relationship to modern humans was unclear. In 1907, a mandible representing a more primitive-looking early human was discovered near Heidelberg. It was classified as a separate species of *Homo* (*H. heidelbergensis*). A human

TABLE 1.1
Human Fossil Discoveries, 1886–1941

Year	Location	Description	Investigator
1886	Belgium (Spy)	*Homo neanderthalensis*	Julien Fraipont
1889–1906	Croatia (Krapina)	*Homo neanderthalensis*	D. Gorjanović-Kramberger
1891	Java (Trinil)	Mandible and skullcap later assigned to *Homo erectus*	Eugene Dubois
1907	Germany (Heidelberg)	Mandible assigned to *Homo heidelbergensis*	Otto von Schoetensack
1908	Germany (Ehringsdorf)	*Homo neanderthalensis*	Franz Weidenreich
1911–1912	England (Piltdown)	Skull assigned to *Eoanthropus dawsoni*, later exposed as fraud	Charles Dawson
1921	South Africa (Broken Hill/ Kabwe)	Cranium and postcranial bones later assigned by some to *Homo heidelbergensis*	Arthur Smith Woodward
1924–1925	South Africa (Taung)	Partial skull and endocast assigned to *Australopithecus africanus*	Raymond Dart
1924–1926	Crimea (Kiik-Koba)	Foot bones assigned to *Homo neanderthalensis*	G. A. Bonch-Osmolovskii
1927	China (Zhoukoudian)	Teeth, mandibles, and skull fragments later assigned to *Homo erectus*	Davidson Black
1933	Germany (Steinheim)	Cranium assigned to *Homo heidelbergensis* or (early) *Homo neanderthalensis*	Fritz Berckhemer
1935–1936	England (Swanscombe)	Cranial fragments assigned to *Homo heidelbergensis* or (early) *Homo neanderthalensis*	A. T. Marston
1936–1939	South Africa (Sterkfontein, other sites)	Crania, mandibles, teeth, and postcranial bones assigned to *Australopithecus*	Robert Broom
1936–1941	Java (Mojokerto, Sangiran region)	Cartial crania later assigned to *Homo erectus*	G. H. R. von Koenigswald
1938	Uzbekistan (Teshik-Tash)	Cranium and postcranial remains (juvenile) assigned to *Homo neanderthalensis*	A. P. Okladnikov

fossil reported in 1911/1912 from Piltdown in southern England, exhibiting a brain volume comparable to that of modern humans but an ape-like jaw, was later exposed as a fraud.[77] From 1933 to 1936, ancient skulls were found in both England and Germany, but eventually were perceived as early Neanderthals. More evidence of pre-Neanderthal humans turned up after the Second World War.[78]

Fossil evidence of human ancestors in Africa did not appear until after 1920, and the earlier history of paleoanthropology in Africa is especially revealing about the state of the field during the first half of the twentieth century. In 1921, a complete cranium and other remains were found at *Broken Hill (Kabwe)* in what is now Zambia. The cranium exhibited some primitive features and a relatively low brain volume (1,280 cc), but was thought to be late—contemporaneous with modern humans outside Africa.[79] Many decades later, it was determined that the opposite was more likely; the cranium was old, with a relatively large brain volume.[80]

In 1925, Raymond Dart (1893–1988) reported the discovery of a partial skull and endocast in South Africa, which he placed into a separate genus (*Australopithecus*) ancestral to *Homo*.[81] Both the dentition and the forward position of the foramen magnum were human-like, but the brain volume was comparable to that of living apes (it was the reverse of the Piltdown find). Dart's interpretation was disputed by leading anatomists of the time such as Sir Arthur Keith (1866–1955), who argued that *Australopithecus* was an extinct ape unrelated to modern humans. This view seemed to be supported by the Piltdown forgery, which was not exposed until after the Second World War. Additional australopith remains were recovered in South Africa by Robert Broom (1866–1951) in the late 1930s, but the controversy continued.[82]

More useful to the study of the human past during the early years of anthropology were the detailed descriptions of living hunter-gatherers (who had almost all disappeared by the end of the twentieth century). Beginning in the final years of the nineteenth century, ethnographers recorded precious information about the economy, diet, technology, language, and religious and social life of foraging peoples around the world. Although there were naïve attempts to apply the ethnography of peoples employing relatively simple technology to the study of earlier forms of humans,[83] the knowledge gained about foraging society and ecology, as well as material culture, were—and remain—an essential frame of reference for modern human origins and dispersal. It is now apparent that at least some of the hunter-gatherers found in southern Africa and southern Asia, as well as Australia, had direct genetic and cultural links to the modern human groups that occupied these areas more than 50,000 years ago (chapters 4 and 5).

The Problem with Paleoanthropology

A major problem in paleoanthropology before 1950 was the lack of reliable chronometric methods for dating the remains of early humans. The time-scale of Earth history, from the Precambrian era to the beginning of recorded history, was still based on guesswork. Although the relative ages of many fossils could be assessed by the correlation of associated faunal remains, even approximations of the absolute dates were unknown.[84] In the absence of absolute dates, paleoanthropologists often ordered the fossils chronologically in accordance with their assumptions about the overall pattern of human evolution. For example, the rather primitive-looking fossils from Ngandong on Java (discovered in 1931 to 1933) were believed to be significantly older than the earliest modern humans. In recent years, they have been dated to 53,000 to 27,000 years ago and apparently represent a late survival of non-modern humans in Southeast Asia.[85]

Another serious problem was the lack of a systematic approach to the analysis of skeletal morphology. In retrospect, the anatomists were too subjective in their selection and measurement of skeletal traits used to classify fossils—and especially to determine phylogenetic, or evolutionary, relationships among fossil humans.[86] As Michael Day observes, the eminent specialists of the 1920s who critiqued Dart's interpretation of *Australopithecus* ignored the dentition, which exhibits a human pattern.[87] In order to make the case that modern human races had evolved locally from earlier human forms, paleoanthropologists such as Franz Weidenreich (1973–1948) selected those features in the fossils that suggest continuity. There were other explanations, however, for why the same trait might show up in two fossils.[88]

And, finally, there was a pervasive racial bias in the approach to human evolution during this period.[89] To what extent this influenced the interpretation of the fossil data cannot be determined, and, in any case, a racial bias—often stated clearly by the researchers themselves—does not by itself render the analyses and conclusions invalid. But given the subjective nature of the analyses and the uncertain age of the fossils, it does not engender confidence in the research results. Darwin had concluded that the differences among the races are relatively superficial and due primarily to sexual selection. Many of the anatomists who dominated paleoanthropology between the publication of *The Descent of Man* and the Second World War reached the opposite conclusion. They saw the differences among the living races as profound—and the explanation for European dominance over colonial peoples.[90] Some of them, such as Keith, applied this conclusion to the origin of modern humans, estimating that several hundred thousand years of independent evolution were necessary to account for the anatomical differences.[91]

In the meantime, the precious sample of human fossil remains suffered a major loss with the disappearance of the Zhoukoudian teeth and bones. Several years after the beginning of the Japanese invasion of China, the fossils were packed into boxes for shipment to the United States for safekeeping. The boxes were taken from Beijing to the port of Tientsin, where they were last seen on December 7, 1941.[92]

THE TRANSFORMATION OF PALEOANTHROPOLOGY

In the years following the Second World War, the study of human evolution underwent a slow but steady transformation. The predominance of skeletal anatomy gave way to a broader approach reflecting the integration of many disciplines. By the end of the twentieth century, molecular phylogenetics had become central to the study of human origins, especially the origin and dispersal of modern humans.

The causes of this transformation were numerous and varied, but one is easily identified. A new generation of paleoanthropologists became active in the late 1940s and the 1950s, and some of them pushed the field firmly in new directions. Sherwood Washburn (1911–2000) promoted field studies of nonhuman primates and played a major role in bringing genetics into the study of human variation and evolution. His student F. Clark Howell (1925–2007) published influential papers on the Neanderthal problem in the 1950s, arguing that the Neanderthals were more likely to represent a geographically isolated species of *Homo* than a stage in human evolution. And Howell, who later organized large-scale interdisciplinary research projects in Europe and Africa, brought the term "paleoanthropology" into common use.[93]

Methodological advances in geology and the life sciences had enormous impact on the field after the Second World War. Radiometric-dating techniques and other chronometric methods, such as tephrochronology (dating of volcanic ashes) and paleomagnetism, provided an absolute timescale for human evolution.[94] Radiometric dating was critical to the debate over the origin and dispersal of modern humans that began in the 1980s. If anything, however, methodological advances in biology had an even greater effect than those in the earth sciences. The analysis of blood-serum proteins as a proxy for genes began even before the decoding of DNA. Applications of the technique were used in the 1960s to address a fundamental issue in paleoanthropology—the evolutionary relationship between humans and apes. The polymerase chain reaction (PCR) technique (1983) and later advances in DNA sequencing (including software for large-scale data analysis) had far-reaching effects on the study of modern human origins and dispersal. Behind the methodological advances lay broader developments in theory, beginning with the "modern evolutionary synthesis" (1936–1947) and continuing with later progress in evolutionary genetics and ecology.

Wider social and political events related to the Second World War also had an impact on the study of human origins, although the direct effects of some of them are impossible to gauge. The use of racial anthropology by the Nazis thoroughly discredited the subject, while the end of colonialism, which was largely triggered by the war, and the emergence of new nation-states in Asia and Africa altered views on race around the world. In the United States, the war became a catalyst for the civil rights movement. African Americans left the South in large numbers to work in war factories in the North and eventually became a political force outside the South (where they were allowed to vote). Segregation of the armed services became an embarrassment after the war, and the Truman administration began to integrate them in 1948.[95] By the early 1960s, the politics of race were having a direct impact on the academic world, especially anthropology.

In the fall of 1957, the first radiometric date was obtained on volcanic rock from Bed I at *Olduvai Gorge* (Tanzania). The estimated age of the volcanic event was 1.75 million years—much older than had been assumed up to that point.[96] Two years later, Louis (1903–1972) and Mary Leakey (1913–1996) reported an australopith fossil (*Zinjanthropus*) from this level, apparently associated with stone tools. In 1964, they published the discovery of remains classified as *Homo habilis* from the same level.[97] More radiometric dates were obtained on the lengthy stratigraphic sequence at Olduvai Gorge, and the chronology was further strengthened with the identification of a paleomagnetic event (Olduvai Event), also dated at other localities. The new fossil discoveries, although controversial, helped cement the shift back to Africa as the likely place of human origin (which had begun with widespread recognition after the Second World War of the australopiths as human ancestors).[98] The geologic studies at Olduvai Gorge inspired similar interdisciplinary research at other sites, and a major project—led by Howell—got under way in the Turkana Basin (Ethiopia) in 1967.[99]

The impact of racial politics on American anthropology became evident in 1962 with the publication of *The Origin of Races* by Carleton Coon (1904–1981). Despite the allusion to Darwin in the title, Coon interpreted the living human races in terms similar to those of Keith and other prewar anatomists (that is, significant biological differences among the geographic races). Accordingly, while acknowledging Africa as the likely place of human origins, he envisioned an early expansion out of Africa and a lengthy period of regional evolution for each of the major races.[100] Coon's book was severely criticized by Washburn, as well as by other prominent anthropologists and the evolutionary geneticist Theodosius Dobzhansky (1900–1975), for its perceived racial undertones and implications. A few months before Washburn addressed the American Anthropological Association on the subject, Coon angrily resigned his post as president of the American Association of Physical Anthropologists over related issues.[101]

Molecular Phylogenetics

The year 1962 also marked the introduction of molecular phylogenetics into the study of human evolution, which until that point had been wholly dominated by anatomy. At a meeting of the New York Academy of Sciences, Morris Goodman (1925–2010) proposed—on the basis of comparative analyses of blood-serum proteins in humans and apes—that humans should be grouped in the same family as chimpanzees and gorillas as a subset of the Asian and African apes. This conclusion was consistent with what Darwin and Thomas Henry Huxley (1825–1895) had written in the nineteenth century (on the basis of comparative anatomy), but was at variance with the primate taxonomy of the time, which placed humans in a family (Hominidae) separate from that of all the apes (Pongidae).[102]

Goodman's conclusion also suggested a relatively late divergence of humans and apes, which was at variance with the paleoanthropology of the time. Fossils had been unearthed in both Africa and southern Asia that were thought to represent the earliest known human. They consisted of teeth and jaw fragments that date as far back as 14 million years ago and were assigned to the genus *Ramapithecus*.[103] Goodman did not dispute the hominid status of these remains, however, and instead argued that rates of protein evolution were slower in apes and humans than in other primates.[104]

Skeptical of the proposed slowdown in protein evolution, Washburn encouraged his student Vincent Sarich (1934–2012) to develop an approach to estimating when the human line had diverged from the ape lineage. Together with biochemist Allan Wilson (1934–1991), Sarich worked out a technique for estimating divergence times among living taxa based on the measurement of immunological distance in serum-albumin proteins. In a groundbreaking paper published in 1967, they estimated the time of divergence between apes and humans, with a "molecular clock," at roughly 5 million years ago.[105] This eventually triggered a major battle with leading paleoanthropologists, including Louis Leakey, who accepted the interpretation of *Ramapithecus* and insisted on the primacy of the fossil data. The battle dragged on until the end of the 1970s and was resolved only when the fossils themselves were reevaluated and reinterpreted—by specialists in anatomy—as a nonhuman ancestor of the orangutan (*Sivapithecus*).[106]

The fall of *Ramapithecus* was a turning point in the transformation of paleoanthropology, although many years passed before Goodman's proposed changes to primate taxonomy were formally adopted. Sarich and Wilson offered an alternative to the unsystematic approach of the anatomists to the fossil data. Although their immunological-distance technique was primitive by current standards and provided only an indirect measurement of the genes, it was objective and quantitative. They set a precedent that would play a role in the coming debate over the origin and dispersal of modern humans.

Part of the problem of integrating genetics into the study of human evolution in the 1970s was that evolutionary genetics was undergoing its own transformation. The decoding of DNA in the 1960s and the development of methods for reading gene sequences revealed many surprises and forced a reassessment of how the evolutionary process works. As early as 1966, analyses of humans and fruit flies yielded greater than expected genetic variability in both taxa.[107] Much non-coding DNA (or "junk DNA") was discovered, while coding DNA was found to be redundant. These discoveries led to the seemingly radical "neutral theory" of evolution (sometimes described as "non-Darwinian evolution"), proposed in 1968 by Motoo Kimura (1924–1994). They also helped explain the basis of the "molecular clock," which had aroused the skepticism of many paleoanthropologists.[108]

The emerging techniques of genetic analysis also were applied to the problem of variation in living humans, although the implications for human evolution were largely ignored at the time. In 1972, Dobzhansky's former student Richard Lewontin (b. 1929) examined variation in a number of genes, including those that code for serum proteins and red blood cell enzymes, with a global sample of humans. He found that most variation (85.4 percent) was *within* rather than between populations and that, when an additional 8.3 percent of variation between groups within the widely recognized racial groups was accounted for, only 6.3 percent of variation remained between them (for example, Caucasians versus Native Americans).[109] Lewontin questioned the biological basis for racial classification.

Once again, molecular analyses were used to provide a quantitative measure of a phenomenon that had been measured rather subjectively and influenced—to say the least—by investigator bias. And once again, Darwin's observations and insights of 1871 had been confirmed. Unlike the relevance of Sarich and Wilson's research, the relevance of Lewontin's work for paleoanthropology was not appreciated in the early 1970s.[110] This probably was due to the fact that the problem of modern human origins and dispersal had yet to be framed in terms that would render genetic diversity in living human populations a highly relevant issue. Nevertheless, Lewontin noticed something that later would become an important part of the debate. Observing that genetic diversity was significantly lower in Native Americans, Aborigines, and Oceanians than in other groups, he concluded that this "must reflect something about their early history rather than their general breeding structure."[111] Reduced diversity in these populations apparently is a consequence of recent dispersal out of Africa (that is, they are genetic subsets of a more diverse parent population in Eurasia, which is a subset of the more diverse African parent).

Epigenetics and the Evo Devo Revolution

The publication of the human genome in 2001 was another major event with implications for—and applications in—paleoanthropology.[112] The small percentage of the genome that codes for proteins (roughly 20,000 genes, or less than 2 percent) underscores the importance of other portions of the genome.[113] During the past decade, increasing attention has been focused on the role of "junk DNA" in regulating gene expression through RNA "machines." Although protein-coding DNA dominates the prokaryotic genome, more complex forms of life, including humans, exhibit an increasingly high percentage of non-protein-coding DNA.[114]

As Sean Carroll noted a few years ago, *developmental biology* (or embryology) was omitted from the modern evolutionary synthesis of 1936 to 1947.[115] It was only after later discoveries, such as the role of *Hox* genes in animal development (1980s), that the critical evolutionary role of regulatory genes came to be more fully appreciated (the *"evo devo* [evolutionary development] revolution"). Related discoveries concerning the function of regulatory genes and their interaction with environmental variables drove another revolution in *epigenetics*.[116]

Both epigenetics and developmental biology are important to the study of human evolution, and perhaps especially to the question of modern human origins. The genetic substitutions that underlie the modern human anatomical pattern, and the uniquely modern cognitive faculties that accompanied or followed the changes in skeletal anatomy, probably took place in the regulatory component of the genome—not in the highly conserved protein-coding sequences. The changes in the timing and pattern of brain growth identified in the fossil record appear to have been changes in developmental biology. And, more generally speaking, the translation of neuronal information into structure that characterizes the transition to *Homo sapiens* parallels the development of the organism translated from genetic information.

INFORMATION, COMPLEXITY, AND HUMAN EVOLUTION

The discovery that living organisms are based on a set of digital information that codes for the constituents of proteins and enzymes had a profound effect on evolutionary biology. Information was not part of Darwin's theory of natural selection, and it was omitted from the modern evolutionary synthesis. But in the years following the discovery of the genetic code in 1953, it became apparent that information plays a fundamental role in the evolutionary process. Moreover, as Maynard Smith observed, it provides an opportunity to quantify that process.[117]

What Is Information?

The concept of *information* has been defined in two ways. Several years after the end of the Second World War, electrical engineer Claude Shannon (1916–2001) defined information—in the context of a message transmitted from source to receiver—as the reduction of uncertainty (or *entropy*). The uncertainty of the message increases as a function of alternative possible messages. If the message has no possible alternatives (the contents of the message are known), the message has no uncertainty and contains no information. If there are many possible alternative messages, both the uncertainty about and the quantity of information in the message increase. Invoking the term "entropy" as a synonym for "uncertainty," Shannon drew an explicit parallel between his definition of information (sometimes labeled "Shannon information") and statistical mechanics (box 1.2).[118]

The other definition of information concerns the contents of a message, rather than its relationship to possible alternative messages: *semantic* or *intentional* information.[119] A formal definition was offered by the American philosopher Fred Dretske (1932–2013), who described it in 1981 as "that commodity capable of yielding knowledge" and emphasized that "false information and *misinformation* are not kinds of information—any more than decoy ducks or rubber ducks are kinds of ducks."[120] Semantic information is composed of *symbols* and clearly applies to human language. The arbitrary relationship between specific DNA base-pair triplets (codons) and the amino acids for which they code means that genetic information also is composed of symbols (although much of DNA is non-coding).[121]

Maynard Smith and others concluded that semantic information is more appropriate for biology than is Shannon information, but the two concepts may be used together to explain the evolution of living systems.[122] Regardless of how it is defined, information exhibits two fundamental properties. First, it can exist only in some physical form, either matter or energy. Conceivably, information may take any physical form, and it may

BOX 1.2
Information Theory and Its Applications

The beginning of *information theory* lies in the solution to a problem in thermodynamics proposed by the physicist Leo Szilard (1898–1964) in the 1920s. The problem had been posed decades earlier by James Clerk Maxwell (1831–1879), who imagined that a tiny intelligent being (later dubbed "Maxwell's demon") could violate the second law of thermodynamics (the entropy of an isolated system increases over time) by sorting fast from slow particles of gas into separate chambers, one becoming hot and the other cold.[1] Maxwell's thought experiment reflected the perceived stochastic, rather than Newtonian mechanical, behavior of the particles. Szilard solved the problem by observing that acquiring information about the speed of individual particles would impose an energy cost on the demon, thus increasing entropy and preserving the second law.[2]

In 1948, Claude E. Shannon, a researcher at Bell Telephone Laboratories, published a paper that became the founding document of the field of information theory.[3] Shannon was addressing practical problems faced by the communications industry: What is the most efficient (cost-effective) way to transmit a message, and how can it be done in the context of background noise? The question of efficiency arose from the recognition that spoken and written language is highly redundant: a message in any natural language contains more symbols than are absolutely necessary to communicate the information from one person to others. The problem of "thermal noise" in electric communication systems—a source of information loss—was a long-standing one for the industry.

In order to address these problems systematically, Shannon constructed a simple theoretical model of a communication system and formulated a mathematical definition of the substance (information) being transmitted from source to destination. He defined information not in terms of content or meaning, but as the measurable extent to which it reduces the uncertainty (or entropy) in a message among "a set of possible messages." In mathematical terms, information became the logarithm of the number of possible choices,

$$H = -\Sigma\, p_i \log p_i$$

where H is the information and p is the probability that the ith message will be chosen. Shannon quantified units of information not in terms of words, but as digits in binary form or *bits* (a contraction of "binary digits") (box figure 1.2).[4]

Despite its seemingly narrow focus on problems in communications technology, Shannon's paper had a broad impact on the idea of information.[5] At its core was Szilard's revolutionary concept of information as a quantifiable material entity, which had implications for biology and anthropology, as well as for other social sciences and the concept of mind. A few years later, biologists discovered that living systems are based on digital-information units—life evolves through the operation of selection on the random recombination (entropy) of these units—providing a basis for the quantification of the (nondeterministic) evolutionary process.[6]

The birth of information theory coincided with a major push to develop a programmable digital computer. Shannon's thinking about information had been influenced by his wartime work in cryptography.[7] As recounted by James Gleick in his history of

(continued)

(continued)

Information
source

Transmitter

Message

Signal

Received
signal

NOISE
SOURCE

Receiver

Destination

Message

BOX FIGURE 1.2 Claude Shannon's simple model of a communications system. (Redrawn from Claude E. Shannon, "A Mathematical Theory of Communication," *Bell System Technical Journal*, July 1948, 379–423)

information, fellow cryptographer Alan Turing, who cracked the Germans' Enigma code (chapter 3), spent some time at Bell Labs during the war and often met with Shannon to discuss matters of mutual interest—especially the design of a computer.[8] Turing had described the basic concept behind a programmable computer (or "universal machine") in 1936.[9]

The war and its aftermath provided the impetus and resources for translating the concept into a working machine. In 1941, researchers at Bell Labs designed an artillery fire-control system that was a form of analog computer, and the U.S. Army funded the development of the first digital programmable computer (ENIAC) to calculate shell trajectories.[10] In the early 1950s, the army also funded the development of a computer (MANIAC) to calculate the effects of thermonuclear blasts. The basic design was developed by John von Neumann (1903–1957) and his colleagues after the war at the Institute for Advanced Study in Princeton.[11]

Computer engineering had an even greater impact on the idea of information and its implications for the concept of mind than Shannon's work on communication. Shannon had quantified the process of transmitting thought from one brain to another, but Turing had developed a model for computation inside the brain. Turing's real goal was the design of a machine that could *think*,[12] and von Neumann drew explicit comparisons between the brain and the computer with respect to the storage and processing of information. Von Neumann stressed the difference between information in continuous form (analog) and information in discrete or discontinuous form (digital).[13] Before his premature death, he was working on a theory of "self-reproducing automata," which blurred the distinction between living systems and complex technology.[14]

1. See, for example, Melanie Mitchell, *Complexity: A Guided Tour* (Oxford: Oxford University Press, 2009), 43–47.
2. Leo Szilard, "Über die Entropieverminderung in Einem Thermodynamischen System bei Eingriffen Intelligenter Wesen," *Zeitschrift für Physik* 53 (1929): 840–856.
3. Claude E. Shannon, "A Mathematical Theory of Communication," *Bell System Technical Journal* 27 (July 1948), 379–423. See also John R. Pierce, *An Introduction to Information Theory:*

Symbols, Signals and Noise, 2nd ed. (New York: Dover, 1980); and Luciano Floridi, *Information: A Very Short Introduction* (New York: Oxford University Press, 2010).

4. Shannon's paper was preceded by (and made reference to) papers that had been published in the same journal two decades earlier: H. Nyquist, "Certain Factors Affecting Telegraph Speed," *Bell System Technical Journal,* April 1924, 324; and R. V. L. Hartley, "Transmission of Information," *Bell System Technical Journal,* July 1928, 535. The term "bit" had been invented by John W. Tukey in 1945. See also Jon Gertner, *The Idea Factory: Bell Labs and the Great Age of American Innovation* (New York: Penguin, 2012).

5. Claude E. Shannon and Warren Weaver, *The Mathematical Theory of Communication* (Urbana: University of Illinois Press, 1949).

6. Christoph Adami, "The Use of Information Theory in Evolutionary Biology," *Annals of the New York Academy of Sciences* 1256 (2012): 49–65.

7. Shannon's "Mathematical Theory of Communication" was an expansion of his classified report "A Mathematical Theory of Cryptography" (memorandum MM 45-110-02, September 1, 1945, Bell Laboratories).

8. James Gleick, *The Information: A History, a Theory, a Flood* (New York: Pantheon, 2011), 204–268. Fundamental ideas about the computer were articulated in Alan M. Turing, "On Computable Numbers, with an Application to the *Entscheidungsproblem,*" *Proceedings of the London Mathematical Society* 42 (1936): 230–265.

9. Turing, "On Computable Numbers."

10. Paul E. Ceruzzi, *Computing: A Concise History* (Cambridge, Mass.: MIT Press, 2012), 38–48.

11. George Dyson, *Turing's Cathedral: The Origins of the Digital Universe* (New York: Pantheon, 2012).

12. Ibid., 225–242.

13. John von Neumann, *The Computer and the Brain,* 3rd ed. (New Haven, Conn.: Yale University Press, 2012).

14. John von Neumann, *Theory of Self-Reproducing Automata,* ed. Arthur W. Burks (Urbana: University of Illinois Press, 1966). See also William Poundstone, *The Recursive Universe: Cosmic Complexity and the Limits of Scientific Knowledge,* rev. ed. (Mineola, N.Y.: Dover, 2013).

comprise discrete, or discontinuous, units (*digital* information) or continuous units (*analog* information) or a combination of both.[123] Second, information must be "about something" or possess a context. Even Shannon information, which may take the form of a single unit of information (bit), must refer to something outside the message, while semantic information, by definition, must correspond to another entity.[124]

Although there has been some discussion of information as "a third primitive component of reality" along with mass and energy, information may be understood more simply in terms of its role in the evolution of complex systems.[125] Information is an essential component or feature of very complex systems, such as living organisms, which must evolve as lineages or become extinct.[126] A lineage can exist only with a replicating and heritable informational base, and can evolve only through adaptive changes in the informational base as a result of natural selection on random variation in the information.

How Is Complexity Measured?

Complexity also has been defined in more than one way, but it may be more useful to ask how it should be measured. In the early 1960s, economist Herbert Simon (1916–2001) proposed that the complexity of a system (living or nonliving) could be measured with respect to the number of hierarchical levels and subsystems (*hierarchical complexity*).[127] A system comprising multiple subsystems may exhibit emergent properties not found in the subsystems (for example, a multicellular organism versus a unicellular organism).[128]

Most other definitions of complexity make reference to information.[129] Complexity has been defined as the shortest possible description of an entity (Kolmogorov complexity). For example, the expressions "$10 \times 10 \times 10$" and "10^3" are equally complex, even though the former is composed of more symbols than the latter. Measures of algorithmic or computational complexity may be applied to living and nonliving systems. Such measures include quantity of input and number of steps required to produce output.[130] Another definition, applied specifically to a living system, is that complexity corresponds to—and may be measured in term of—the quantity of accumulated genetic information about its environment.[131]

As in the case of information, the varying definitions of complexity are not incompatible with one another and may be applied together in the context of evolutionary biology. Simon's concept of hierarchical complexity, which is simple to measure, may be the most useful for the evolution of living systems. Discussions of complexity also have touched on whether or not it represents a set of fundamental laws (complexity theory). The issue seems to overlap with a continuing debate on the underlying principles of living systems.[132]

Information, Complexity, and Evolutionary Biology

The discovery that life is based on digital information yielded an immediate benefit—*in silico* research, or computer modeling of the evolutionary process, which is difficult to observe and measure in a natural setting (and within a reasonable interval of time). Despite the recognition of its central role in evolution by some leading theorists, however, there was little formal application of information theory in biology in the decades that followed the decoding of DNA. The chief reason probably was the belated development of techniques for fast sequencing of DNA, which began to yield vast quantities of data in the early twenty-first century and gave birth to the science of *bioinformatics*.[133]

As Sewall Wright observed, the evolutionary process is neither wholly deterministic nor probabilistic, but a combination of both.[134] The probabilistic, or nondeterministic, component of evolution lies in the random mutation of genetic information, which provides the ultimate source of phenotypic variation. In sexually reproducing organisms, the random recombination of genetic information provides additional variation. The evolutionary process has harnessed a physical phenomenon (entropy) that is essential for adaptation to unpredictable changes in the environment, underscoring the central role of information in the process. (It should be noted that natural random mutation rates are too high and a constant threat to the fitness of an organism, as indicated by the evolved redundancy of the genetic code [multiple codons specify the same amino acid] and enzyme "proofreading" of replicated DNA sequences).[135]

The evolution of living systems may be characterized as a process of nondeterministic computation with genetic information, where DNA represents both the input and the output.[136] Natural selection provides the deterministic component. The process operates above the level of the individual organism. Both the information and the organisms evolve, but the latter only as lineages and not as individuals (see figure 1.2).

As a consequence of evolutionary computation, each genome accumulates information about its environment. The information is not a representation of the environment, but a digital program for an organism (including a *super-organism*, such as a honeybee colony) designed to reproduce in its environment.[137] Despite the potential for mutation, DNA is a relatively stable medium for storage; even after the death of an organism, much of the genome (especially the mitochondrial genome [mitogenome]) may remain intact for thousands of years under favorable conditions.[138] The storage capacity of the genome is large, especially in eukaryotes, which contain an average of 10^7 to 10^{11} base-pairs. Genetic information is *transmitted* during mitosis and meiosis, and *translated* during protein synthesis.

The fossil record indicates a general trend in evolution toward increased complexity. The trend is evident not only in the proliferation of diverse life-forms, but also in the greater complexity of individual organisms (for example, multicellular organisms are preceded by unicellular life). However, many lineages exhibit no significant increase in complexity over lengthy periods of time, suggesting that changes in complexity occur only under certain circumstances.[139]

As noted at the beginning of the chapter, Maynard Smith and Szathmáry proposed a series of eight "major transitions" in evolution:

- Replicating molecules → populations of molecules in compartments
- Unlinked replicators → chromosomes
- RNA as gene and enzyme → DNA and protein (origin of genetic code)
- Prokaryotes → eukaryotes
- Asexual clones → sexual populations
- Protists → animals, plants, and fungi (multicellular life)
- Solitary individuals → colonies (super-organisms)
- Primate societies → human societies (language)

Recently, Szathmáry revised the list, condensing the first three into two, adding plastids (photosynthesis), and subtracting sex as major transitions. To these transitions may be added the immune and nervous systems.[140] Each transition represents a change in *how information is stored, transmitted, and/or translated* (to which may be added manipulated or computed). And each transition is completed when "entities that were capable of independent replication before the transition can replicate only as part of a larger whole after it," suggesting that each represents an increase in hierarchical complexity.[141]

The general trend toward increased complexity in evolution may be measured by other approaches mentioned earlier. In addition to the accumulation of information about the environment through the process of evolutionary computation, evolution appears to yield increased computational complexity. There are significant differences in the quantity of input (DNA) and the number of steps required to translate information into structure and function (that is, a living organism) between, for example, prokaryotes and eukaryotes. The pattern also may be illustrated by graph theory (increased network complexity = increased computational complexity).[142]

Eukaryote Origins and Multicellular Life

After the origin of life (represented by the first three "major transitions"), the evolution of "true cells" was the most consequential of the identified transitions. Eukaryotes store and transmit a vastly increased quantity of genetic information (roughly 250-fold increase in genome size over prokaryotes). The larger quantity of information—which requires a significantly increased source of energy to translate—provided the necessary basis for the emergence of multicellular life.[143]

The transition to eukaryotes is closely linked to the origin of sex and complex life-forms (metazoa) and represents major changes in the role of information in the evolutionary process. In addition to the exponential increase in the information that they store and transmit, multicellular organisms reproduce in an entirely different way from that of prokaryotes. Instead of subdividing following the duplication of genetic information, the multicellular organism develops from an embryonic eukaryote that translates information into structure. In most eukaryotes, the developing organism is constructed from a variety of differentiated and specialized cells. The developmental process may be protracted and entail multiple stages and organismal forms. Most of the increase in the eukaryote genome lies in the non-coding component, reflecting the greatly increased role of regulatory genes and epigenetic processes in the developmental biology of complex life.[144]

In sexually reproducing eukaryotes, the genetic information in the embryonic cell is randomly recombined from two parents, each of which contributes half of the information base of the new organism. Sexual reproduction generates much higher variability in the population and yields a more dynamic process of evolutionary change and adaptation than does asexual reproduction. It also creates a new type of selection—*sexual selection*—whereby organisms replace environmental variables and information interacts with information in the evolutionary process.[145] In addition to intraspecific sexual selection, complex species were forced to adapt to other complex species (for example, evolutionary "arms race"), as the biological environment became increasingly complicated and unpredictable (complexity begat more complexity).[146]

The metazoa evolved a new form of "temporary" information (neuronal information) and a specialized organ for computing with it (the brain). Neuronal information, which is stored in both digital and analog form as electrochemical signals in the synaptic connections among neurons, exists at an organizational level above genetic information (that is, above the cellular level). Its properties are strikingly different from those of genetic information—it is stored, transmitted, translated, and manipulated (or computed) in different ways. Neuronal information has higher energy costs and storage space requirements per bit than genetic information. It is stored in an unstable medium, subject to loss unless reinforced ("use or lose"). Unlike genetic information, neuronal information is not necessarily transmitted from parent to offspring, and it is not translated into structure (except among humans). The evolution of neuronal information contributed to a more complex world in which spatial and temporal variations in the environment (including the social environment) became more unpredictable.[147]

By any measure, the transition from prokaryotes to eukaryotes and multicellular organisms marked a major increase in complexity. Eukaryotes are widely believed to represent an evolved symbiosis of prokaryotes, and the evolution of multicellular organisms produced more organizational levels with emergent properties.[148] The changes in the role of information rendered the process of evolutionary computation more complex in terms of quantity of input and number of steps. Moreover, sexual reproduction introduced a new source of randomness into the nondeterministic computations of the process—once again at a higher level of organizational complexity (that is, one organism → two organisms).

Large-brained metazoans developed another source of random input to evolutionary computation with neuronal information. Among some groups, the formation of synaptic connections during early brain development is highly random.[149] The ability to perform nondeterministic computations provides a potential for generating unpredictable moves in predator–prey relationships and socially competitive game playing (as well as in anticipating an opponent's move by recursively generating possible alternatives). It presumably is the source of creativity in the large metazoan brain, which is manifest in the novel variations in the songs of male humpback whales, as well as in the creative use of language, visual art, and other media among humans.[150]

Information, Complexity, and Human Evolution

It is difficult, if not impossible, to explain the origin and dispersal of *Homo sapiens* within the framework of the modern evolutionary synthesis of 1936 to 1947. Modern humans are inextricably tied to non-genetic forms of information—including structures translated from non-genetic information—that evolve in accordance with processes not described in the synthesis of natural selection and population genetics. A wider evolutionary framework that encompasses multiple forms of information (with varying properties) and both living and nonliving complex systems (that is, with a broader definition of life) is required.

The most important development in human evolution is the unique human capacity for translating information in the brain into structure, including other forms of information based on such structure. It underlies the two most striking features of the genus *Homo*: syntactic language and complex artifacts, including automata. As both a communication and a computational system, the digital-information units of language are based on "structures" in the form of vocal sounds, translated from the neuronal networks of the brain. The complex artifacts based on information in the brain often comprise many hierarchically organized subcomponents and may function with moving parts (analogous to an organ) or even independently (analogous to an organism). By translating information into complex artifacts, humans contrive to design their own adaptive traits with non-genetic information. These "traits" are not subject to the material constraints of structures translated from genetic information; they may be fashioned from nonorganic matter.

As noted earlier, random input generated by developmental neurobiology (and the environment) provides some creative power to the vertebrate brain, analogous to random processes that underlie innovation in evolutionary computation with genetic information.[151] But only in humans is this creative potential expressed through the translation into complex structures and forms of information (for example, words) that may be transmitted outside the brain.[152] As a result, humans are innovative with technology, incorporating novel materials and procedures, modifying and rearranging components, and adding new subsystems. The same creativity is applied to the manipulation of information in digital form (for example, syntactic language) and in analog form (for example, visual art).

By evolving a novel computational system with units of information borrowed from their system of vocal communication, humans found themselves with a means of collective computation. Information sharing and computation among multiple brains (super-brain) represents a higher level of organizational complexity, but not one achieved through the process of evolving a super-organism (for example, honeybee colony).[153] The distinction is important because it explains the underlying—often subtle—competitive character of collective computation among human social groups. Combined with the creative potential of the individual brain, this perpetual struggle invests the manipulation of non-genetic information among groups with a dynamism that it would not otherwise possess. It parallels the process of evolutionary computation with genetic information among competing individuals in a population.

Collective computation with non-genetic information is neatly illustrated by human technology.[154] Despite the potential for innovation and random variation, humans typically make and use artifacts in accordance with a *technological algorithm* developed over many generations (and often with contributions from other groups). Most human technology is too complex for each individual to reinvent (or recompute). The algorithms, which may comprise multiple materials and many hierarchically organized steps, are learned during childhood and adolescence. In hunter-gatherer societies, they are often gender-specific; in more complex societies, many algorithms are transmitted within groups of technical specialists (for example, metallurgists). There is a parallel between a genetic program for the development of an organism and a non-genetic technological algorithm in a human social group.

Information and complexity theory have other useful applications to human technology, especially in the context of modern human origins and dispersal. As noted, the hierarchical organization of technology—number of levels and subcomponents—may be the most important measure of its *structural complexity*. Equally significant is the relationship between structural and *functional complexity*, and the latter has been formally defined in the same terms as information—that is, the reduction of uncertainty.[155] The increasing structural complexity of human technology reflects the increasing complexity of the problems that it is designed to solve through a series of hierarchically organized functions, each of which addresses a subproblem. Ultimately, humans establish an entirely new level of evolutionary computation by creating self-acting artifacts, or automata, that perform their own computations. (The measurement of technological complexity is discussed further in chapter 2.)

The traditional culture concept, originally conceived by nineteenth-century ethnologists, has fallen into disfavor among many anthropologists in recent years.[156] But if culture is defined as a body of non-genetic information—and structure based on non-genetic information—shared among the members of a social group, it is as real a phenomenon as a genome. As already noted, all types of information, including the seemingly ephemeral thoughts in the conscious brain, exist as some form of matter or energy. Culture may be redefined as a *neuronal genome* or even a "neuronal lineage," and its emergence in the genus *Homo* introduced new processes of change—and new structures in the form of complex artifacts—to evolution (box 1.3).

By the time modern humans dispersed out of Africa 75,000 to 60,000 years ago, they had acquired a capacity for translating information in the brain into complex structure. "Artifacts" with three or more hierarchical levels (for example, winter clothing) are represented or may be inferred from the archaeological record of the dispersal and its aftermath. The underlying computational complexity of such technology was a prerequisite for the successful colonization of habitats that other forms of *Homo* had been unable to occupy. The nondeterministic character of the computations was essential for the innovations that supported rapid dispersal into these variable habitats (for example, desert margins, boreal woodland, and shrub tundra).[157]

The capacity for syntactic language—with a comparable level of nondeterministic computational complexity—may be inferred from the global distribution of languages among living peoples, which are unlikely to have evolved independently among two or more populations). It, too, may have been critical to the occupation of habitats with scarce plant and animal resources, by facilitating the creation of widespread alliance networks based on the complex computations that underlie long-term relationships with a large number of in-laws and nonrelatives.

Modern humans produced at least two new forms of information at the time of their (second) dispersal out of Africa: visual art, which is a uniquely human type of information in analog form, and digital notation, which presumably was used to enhance computation. They built the first

BOX 1.3
Information Theory and the Culture Concept

"These days," wrote Adam Kuper in 1999, "anthropologists get remarkably nervous when they discuss culture."[1] The development is an unexpected one, analogous to evolutionary biologists becoming nervous about natural selection. The concept of culture is central to anthropology. It was defined in 1871 by British ethnologist Edward B. Tylor (1832–1917) as "that complex whole which includes knowledge, belief, art, morals, law, custom, and any other capabilities and habits acquired by man as a member of society."[2]

Franz Boas (1858–1942) and his students emphasized the historical (and unpredictable) quality of culture in their argument with the speculative "cultural evolutionists" of the late nineteenth century (whose ranks, ironically, included Tylor).[3] The concept was further elaborated in the early twentieth century by Boas's students Clyde Kluckhohn (1905–1960) and Alfred Kroeber (1876–1960), who famously described culture as a "super-organic" phenomenon. Both Kluckhohn and Kroeber stressed the importance of symbols in culture (over social structure and artifacts).[4]

Within academic anthropology, the principal critique of the culture concept came from another of Boas's students, Edward Sapir (1884–1939), who argued that Kroeber and others had failed to appreciate "the role of the individual in history."[5] Sapir accused them of having "reified" an abstraction. Other anthropologists also decried the "reification" of culture,[6] and it became a criticism of the concept from outside anthropology as well.

Although Margaret Mead (1901–1978) was unimpressed with the potential contribution of information theory (or *cybernetics*) to anthropology in 1950,[7] her encounter with Claude Shannon preceded developments in genetics that offered a unifying concept for biology and culture. The subsequent discovery that life is based on a set of digital symbols (and that the underlying computations of an evolving lineage are as nondeterministic and even "creative" as the workings of the human brain) offered a parallel with language and other aspects of culture, as broadly defined.

If culture is defined as a set of symbols or information (in both digital and analog form) shared among the members of a social group, it is not an abstraction but a physical entity represented by various forms of matter and energy that are the basis for the information (including the electrochemical signals in the neuronal networks of individual brains).[8] The information takes many forms developed by humans, including spoken words (vocal sounds), visual art, and written symbols. And, following Tylor, it is reasonable to include the nonsymbolic structures based on translated information shared among a social group—such as a pair of pliers—as part of "culture."

Both the "historical particularist" approach of Boas and the "materialist" approach of *cultural ecologists* like Julian Steward (1902–1972) are reconcilable with the notion that culture represents a set of information (or symbols) and structures based on information shared by members of a society. The multiple sources of random input in the human brain provide the basis for continual nondeterministic (and unpredictable) computational output with neuronal units of information, as well as the many forms of structure and information based on the translation of neuronal information. Much of the output is generated in the course of (often subtly) competitive social interactions, but some of it is applied to obtaining food, combating predators and pathogens, and solving other problems of adaptation to the physical and biological environment. In his influential critique of *cultural materialism*, Marshall Sahlins drew a false dichotomy between "materialism" and "idealism," failing to understand that the creative manipulation of symbols is as fundamental to the evolution of living systems as it was to Picasso and T. S. Eliot.[9]

1. Adam Kuper, *Culture: The Anthropologist's Account* (Cambridge, Mass.: Harvard University Press, 1999), 226. Kuper suggested that it is best "to avoid the hyper-referential word altogether, and to talk more precisely of knowledge, or belief, or art, or technology, or tradition, or even of ideology" (x).
2. E. B. Tylor, *Primitive Culture* (London: Murray, 1871), 1. Tylor's book was published in the same year as Charles Darwin's *The Descent of Man and Selection in Relation to Sex*.
3. George W. Stocking, Jr., *Race, Culture, and Evolution: Essays in the History of Anthropology* (Chicago: University of Chicago Press, 1968). There is some parallel between the pseudo-Darwinian approach of the anatomists who reconstructed human evolution as a progression of stages and the cultural "evolutionists" who, although ostensibly applying Darwin's ideas to society and culture, proposed a progression of social and cultural stages through which all peoples either had or would "evolve."
4. Kuper, *Culture*, 56–58.
5. Regna Darnell, "The Anthropological Concept of Culture at the End of the Boasian Century," *Social Analysis* 41 (1997): 42–54.
6. According to A. R. Radcliffe-Brown, "We do not observe a 'culture,' since that word denotes, not any concrete reality, but an abstraction, and as it is commonly used a vague abstraction" ("On Social Structure," *Journal of the Royal Anthropological Institute* 70 [1940]: 2, quoted in Kuper, *Culture*, xiv).
7. James Gleick, *The Information: A History, a Theory, a Flood* (New York: Pantheon, 2011), 242–249. "Cybernetics" was another term for information theory, adopted by Norbert Wiener, *Cybernetics: Or Control and Communication in the Animal and the Machine* (Cambridge, Mass.: MIT Press, 1948).
8. Rolfe Landauer, "Information Is Physical," *Physics Today* 44 (1991): 23–29.
9. Marshall Sahlins, *Culture and Practical Reason* (Chicago: University of Chicago Press, 1976).

mechanical artifacts, and they probably included self-acting machines, or automata (for example, small mammal trap), which may be described as a simple form of artificial life. (Machines also perform an evolutionary computation, although it is typically a deterministic one.) Eventually, modern humans produced many other novel forms of information, including writing, written mathematics, and, ultimately, mechanical devices and automata for computation.

Syntactic language may have provided the model for both visual art and automata. Paleoanthropologists and archaeologists have long pondered the relationship between language and tools (or, more generally, language and artifacts).[158] The dating of the oldest stone tools—more than 3 million years ago and associated with the australopiths (chapter 3)—suggests that artifacts must precede sentences. However, the indirect evidence of the fossil record with respect to the timing and pattern of development (chapter 4) indicates that syntactic language probably evolved with anatomically modern humans (or "near modern" humans) more than 150,000 years ago, while archaeological evidence for visual art and automata seem to be more recent and associated with the global dispersal of the living lineages out of Africa after 75,000 years ago.

For paleoanthropologists, the prime directive is an explanation of how and why humans evolved their capacity for translating information in the brain into structures, including other forms of information and functioning structures. The task is an interdisciplinary one, requiring the integration of anatomical, genetic, and archaeological data within a spatially and temporally varying environmental context. The fossil anatomical data provide an essential framework for the evolution of brain volume and organization, as well as the evolution of the human hand and vocal tract. They also offer important information on the evolution of human development (evo devo), including developmental neurobiology. The rapidly accelerating study of genetics in living and fossil humans soon may identify the key substitutions that define modern humans through comparative analysis with non-modern genomes. Analyses of the genetics of living people and aDNA provide information on the timing and routes of the dispersal within and out of Africa. They also yield insights into population density and human paleoecology (both for modern humans leaving Africa and for their non-modern cousins in Eurasia). Finally, only the archaeological data offer evidence for the translation of neuronal information into structure, including other forms of information.

Chapter Two

MODERN HUMAN ORIGINS AND DISPERSAL

The Synthesis

[I]n a genetic sense, everyone on this planet looks like an African.
—SVANTE PÄÄBO (1995)

Before attempting to explain the evolution of *Homo sapiens* (chapters 3 and 4), it is necessary to establish the facts of palaeoanthropology as they pertain to the origin and dispersal of modern humans. What is the basis for concluding that the modern human anatomical pattern was present initially in sub-Saharan Africa and only later in other parts of the world? What are the genetic data for interbreeding between *Homo sapiens* and various non-modern humans outside Africa? Where and when do we see evidence of major technological innovations related to the dispersal out of Africa?

Our current understanding of modern human evolution is a product of the postwar transformation of paleoanthropology described in chapter 1. It began with the discovery of fossils of anatomically modern humans in Africa that yielded radiometric dates significantly older than those of fossils of modern humans outside Africa. The comparative age of the fossils became the basis of the Recent African Origin (RAO) model for *Homo sapiens*. In the late 1980s, the analysis of mitochondrial DNA (mtDNA) from a broad sample of living people indicated Africa as the source of maternal lineages outside Africa and provided an estimate for the time of divergence. This was followed by many other genetic studies and, eventually, analysis of some ancient DNA (aDNA) from dated skeletal remains in the Northern Hemisphere. New archaeological finds yielded evidence for the emergence of modern human cognitive faculties in Africa and offered clues to their role in the dispersal out of Africa.

THE ORIGIN AND DISPERSAL OF THE RAO MODEL

In 1967, a fossil-hunting team led by Richard Leakey recovered human skull fragments and other remains from the Kibish Formation on the banks of the Omo River in Ethiopia (figure 2.1).[1] The *Omo-Kibish I* skull was recognized as a robust-looking modern human, but it was dated by Karl W. Butzer to roughly 130,000 years ago—earlier than modern humans were thought to be present anywhere.[2] A few years later, the age of the skull found at *Broken Hill* (*Kabwe* [Zambia]) was reassessed as Middle Pleistocene (more than 130,000 years old); today, it is thought to be more than ten times older than had been believed in the 1960s.[3] During the 1970s and 1980s, more modern human skeletal remains from East and South Africa were dated to time ranges that antedated the appearance of modern humans outside Africa, including *Border Cave* (South Africa), *Florisbad* (South Africa), *Eliye Springs* (Ethiopia), and *Klasies River Mouth* (South Africa).[4]

At the International Congress of Human Paleontology, held in Nice (France) in October 1982, arguments for a recent African origin of modern humans were presented by the British paleoanthropologists Michael Day and Christopher Stringer and by the German anatomist Günter Bräuer.[5] Based on the anatomy and estimated age of fossils like Omo-Kibish I, their reasoning was simple—modern humans appeared first in Africa and must

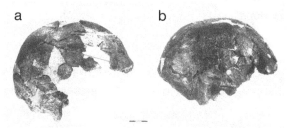

a   b

FIGURE 2.1 The skull fragments recovered in 1967 near the Omo River by Richard Leakey, which document the presence of anatomically modern humans in sub-Saharan Africa before 100,000 years ago. The anatomical differences between (*a*) Omo-Kibish I and (*b*) Omo-Kibish II underscore the high degree of variability among *Homo sapiens* during this period (chapter 4). (From "New Clues Add 40,000 Years to Age of Human Species," February 16, 2005, National Science Foundation, https://www.nsf.gov/news/news_summ.jsp?cntn_id=102968)

therefore have spread from Africa to other parts of the world. Bräuer also argued for some genetic admixture between the spreading African population and local non-modern populations in Eurasia.

At the same conference, Day and Stringer proposed a set of anatomical traits—many of them metric—for defining modern humans.[6] Most of the traits on their list are features of the cranium, including frontal angle (less than 130°), parietal angle (less than 138°), and occipital angle (greater than 114°). Collectively, they describe a cranial vault with a distinctive globular shape, which reflects the derived developmental biology of *Homo sapiens* (chapter 4).[7] Day and Stringer's effort to define modern humans more systematically and quantitatively represented a conscious departure from the subjective approaches to the human fossil record typical of earlier years (chapter 1). Since 1982, other paleoanthropologists have built on their list, adding some other traits.[8]

By the 1980s, a new generation of fossil anatomists had arisen to argue for a modified version of the multiple-origins (or "multi-regional") model originally advocated by Franz Weidenreich and others. As before, this alternative view of the origin of modern humans rested primarily on observed continuities in skeletal features between local non-modern and modern people in various parts of the world. Its proponents saw particularly strong local continuity in the Far East and Australasia. In contrast to some versions of the earlier model (for example, that of Carleton Coon), they emphasized the role of *gene flow*—rather than parallel evolution—in producing common (modern human) features among different regions: "The pattern of regional variation was maintained throughout most of the Pleistocene by a balance between the local forces promoting homogeneity within populations and regional distinctions between them (selection, drift), and multidirectional gene flow which . . . need not have been particularly high."[9] In this respect, there was overlap with the RAO model proposed by Bräuer. Nevertheless, they disputed the evidence for early modern humans in Africa.[10]

Mitochondrial Eve and the Middle Stone Age

The debate entered a new phase in January 1987 with the publication of a study of mtDNA in living human populations by Rebecca Cann, Mark Stoneking, and Allan Wilson.[11] The mitochondrion contains a small loop of DNA (comprising about 16,500 base-pairs) that is nonrecombining, partly nonfunctional (that is, does not code for proteins), and abundant (each cell contains hundreds or thousands of mtDNA loops). Portions of it mutate—through copying errors—at a comparatively rapid rate and thus are particularly well suited to the study of short-term evolutionary events. Because mtDNA is transmitted only from the female to her offspring, it reflects events only in the maternal lineage (box 2.1).

Cann, Stoneking, and Wilson analyzed mtDNA differences among a sample of 147 females representing lineages from various regions of the world. Along the same lines as Richard Lewontin's study in 1972,[12] they noted that overall variation was low—a limited number of mutations had accumulated since the global dispersal of modern humans. They found that mtDNA mutations were higher in the African lineages (since mutations had been accumulating for a longer period of time), and—based on the construction of a tree of mtDNA relationships around the world—they concluded that modern human female lineages were derived from an ancestor who lived in Africa roughly 200,000 years ago.

The publication of the paper by Cann and her colleagues generated a media sensation; the hypothesized female ancestor was subsequently dubbed "African Eve" or "mitochondrial Eve." Both the original paper and a follow-up publication in 1991 were subjected to intense scrutiny and criticism. The latter was focused specifically on the mtDNA sample and the statistics used to generate the genetic tree.[13] Within a few years, however, new studies were published by other researchers, confirming the results of the original analysis. In 1995, Satoshi Horai and his colleagues reported comprehensive sequencing of mtDNA bases from three individuals derived from Africa, Europe, and East Asia, respectively, yielding a slightly younger estimate of divergence time from an African source (roughly 145,000 years).[14]

The 1980s also saw the redating of the African Middle Stone Age (MSA) to a significantly older time range. For many decades, the MSA had been viewed as roughly contemporaneous with the Upper Paleolithic of Europe (less than 40,000 years old) and—like the Broken Hill skull—an indication of stagnation and backwardness in sub-Saharan Africa. Although some older dates on MSA sites were reported in the 1960s and 1970s, the critical threshold seems to have been reached in the 1980s with dates of more than 100,000 years at sites such as *Mumba Rock Shelter* (Tanzania), *Prospect Farm* (Kenya), Florisbad (South Africa), and Klasies River Mouth (South

BOX 2.1
The Origin of Modern Humans and Mitochondrial DNA

The infusion of data and analyses of mtDNA transformed the debate over the RAO model, which until 1987 had been argued on the basis of human fossil anatomy.[1] Although the initial study was critiqued on methodological grounds, the conclusions were supported by many subsequent analyses of mtDNA, and the origin and spread of modern humans is still discussed primarily in terms of the divergence and diaspora of maternal lineages.[2] Early mtDNA studies were based on defining mutations, but as DNA-sequencing techniques improved, the whole mitogenome was analyzed. The significantly larger non-recombining portions of the Y chromosome also were analyzed to collect information on the paternal lineages, and, eventually, researchers began to shift the emphasis to whole genomes. Nevertheless, mtDNA remains an important part of the study of the evolution and dispersal of modern humans.

Each cell in the human body contains organelles known as *mitochondria*, which generate power (through the production of *adenosine triphosphate* [ATP]) for cellular functions. Some types of cells have more than 1,000 mitochondria. Every mitochondrion includes a circular DNA molecule that contains the only genetic information stored outside the cell nucleus. According to the symbiotic theory of eukaryote origins, mitochondria evolved from prokaryotes that became incorporated into true cells.[3]

Each loop of mtDNA is a haploid genome comprising about 16,500 base-pairs, inherited through only the female in humans and most other animals. Most of the mtDNA base-pairs code for proteins or RNA, but two relatively small segments (each containing fewer than 600 nucleotides), known as *hypervariable region 1* and

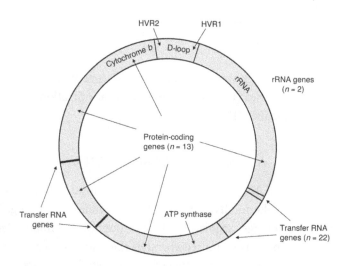

BOX FIGURE 2.1 The human mitochondrial genome: a circular mtDNA molecule, containing about 16,570 nucleic acid base-pairs, showing the location of HVR1 and HVR2 as well as regions that code for RNA and proteins. (Redrawn from Toomas Kivisild, "Maternal Ancestry and Population History from Whole Mitochondrial Genomes," *Investigative Genetics* 6 [2015]: 3, fig. 1)

(*continued*)

(continued)

hypervariable region 2 (HVR1 and HVR2), are non-coding. Because they have no effect on fitness, mutation rates are highest in hypervariable segments (box figure 2.1).

If the weakness of mtDNA as a source of information lies in its small size (compared with the roughly 58 million base-pairs on the non-recombining Y chromosome or the approximately 3.2 billion base-pairs in the nuclear genome as a whole), its advantage lies in its abundance. In the early days of DNA-sequencing techniques, mtDNA was the primary source of genetic information about modern human lineages. Commercial firms continue to use a few key marker substitutions in HVR1 and HVR2 to assign individuals to specific lineages (or mtDNA haplogroups), although some firms now sample more widely from nuclear DNA.[4] Sometimes, only the mtDNA is sufficiently well preserved in skeletal remains for analysis.[5]

1. Rebecca L. Cann, Mark Stoneking, and Allan C. Wilson, "Mitochondrial DNA and Human Evolution," *Nature* 325 (1987): 31–36.
2. See, for example, Stephen Oppenheimer, *The Real Eve: Modern Man's Journey Out of Africa* (New York: Basic Books, 2003); Spencer Wells, *Deep Ancestry: Inside the Genographic Project* (Washington, D.C.: National Geographic Society, 2007).
3. John Maynard Smith and Eörs Szathmáry, *The Origins of Life: From the Birth of Life to the Origin of Language* (Oxford: Oxford University Press, 1999), 70–77.
4. For example, AncestryDNA, www.ancestry.com/DNA.
5. For example, S. Benazzi et al., "The Makers of the Protoaurignacian and Implications for Neandertal Extinction," *Science* 348 (2015): 793–796.

Africa).[15] The new dates were obtained from a variety of novel methods such as uranium series (U-series), amino acid racemization (AAR), and electron spin resonance (ESR).

Y-Chromosome and Ancient DNA

In 1995, the first analysis of variations in Y-chromosome DNA (Y-DNA) was published.[16] Like mtDNA, most Y-DNA is nonrecombining, but it comprises a much longer sequence of bases (roughly 58 million). Because the Y chromosome is transmitted through the male lineage, it complements the mtDNA sequences. A major Y-DNA study reported in 2000 confirmed a recent African origin for male lineages and yielded an estimated divergence time of only 59,000 years.[17] Two years later, another new aspect of the research on human evolutionary genetics appeared with the first analysis of aDNA extracted from a human fossil—the original Neanderthal find from *Feldhofer Cave* (Germany). Svante Pääbo and other genetics researchers had finally overcome the technical challenge of avoiding contamination of aDNA with modern DNA and providing a new means of ground-truthing the RAO model.[18]

Proponents of the "multi-regional" model of modern human origins continued to emphasize evidence for continuity (that is, genetic admixture) between local non-modern and modern human populations. They identified Neanderthal traits in skeletal remains of *Homo sapiens* recovered from Portugal (*Lagar Velho*) and Romania (*Peştera cu Oase*).[19] An analogous argument had been put forward by archaeologists who attributed to cultural assimilation the mixture of artifacts typically produced by Neanderthals with those assumed to have been made by modern humans.[20]

During the 1990s, new archaeological discoveries in Africa provided evidence of behavioral changes to complement the emergence of modern human anatomy in sub-Saharan Africa. They included personal ornaments (ostrich-eggshell beads) from *Enkapune ya Muto* (Kenya) that date to more than 40,000 years ago, and barbed bone points from *Katanda* (Congo) that date to about 90,000 years ago.[21] Richard G. Klein integrated the archaeological data with the human fossils and argued that neuro-anatomical changes in *Homo sapiens* had triggered modern behavior—including syntactic language—and the global dispersal ("neural hypothesis").[22] More evidence of modern behavior later surfaced at *Blombos Cave* (South Africa), including lumps of red ocher engraved with geometric designs that date to around 75,000 years ago.[23]

By the beginning of the twenty-first century, however, genetics seemed to dominate the debate over modern human origins. Much of the research now focused on the global dispersal of modern humans—using comparative analysis of DNA sequences from various living populations to identify specific movements of people from Africa to parts of Eurasia, Australia, and the Western Hemisphere. By identifying genetic markers or *haplogroups* (a set of mutations characteristic of a specific clade or lineage) and calculating their divergence times, geneticists could reconstruct elements of the dispersal.[24] For example, Y-DNA haplogroup C-M130 apparently was carried to Australia by the first male lineage to occupy the island continent, and the time of its divergence from its parent marker currently is estimated at more than 70,000 years ago.[25] Coalescence estimates for the major Native American haplogroups indicate that dispersal in the Western Hemisphere began 15,000 to 14,000 years ago.[26]

After 2000, more analyses of aDNA recovered from the skeletal remains of both Neanderthals and modern humans in Eurasia were published. These provided an opportunity to ground-truth elements of the global-dispersal model based on the genetics of living humans by identifying specific genetic markers in dated *Homo sapiens* remains. For example, mtDNA extracted from a 37,000-year-old skeleton from the central East European Plain in 2010 was matched with a marker (mtDNA haplogroup U) that is widespread in Eurasia and thought to have been carried into eastern Europe by way of the Caucasus Mountains after 50,000 years ago.[27] In 2013, the same mtDNA haplogroup was identified from remains from southern Siberia that date to about 25,000 years ago.[28]

The following year, a 45,000-year-old bone from western Siberia was assigned to mtDNA haplogroup R and Y-DNA haplogroup K, both with close links to African lineages, while the *Anzick* skeleton from Montana (ca. 12,700–12,500 years old) was assigned to mtDNA D and Y-DNA Q-L54, both with links to Northeast Asian lineages.[29] In early 2016, Cosimo Posth and his colleagues reported the discovery of the oldest non-African maternal lineages (mtDNA M and N) in several European specimens that date to around 40,000 to 34,000 years ago (these lineages are absent in living Europeans).[30] Ancient DNA recently has been used to improve the calibration of the "molecular clock" for the mitochondrial genome (or "mitogenome") (figure 2.2).[31]

Studies of aDNA also provided an opportunity to examine the idea that some genic exchange took place between modern and non-modern populations during the global dispersal of the former. To the surprise of many, evidence for gene flow between *Homo sapiens* and two non-modern contemporaries was reported in separate analyses published in 2010 (box 2.2).[32] A comparison of the genome of Neanderthals (based on aDNA) with that of living humans revealed that the non-African population (including Native Americans) carries roughly 2.5 percent of the Neanderthal genome.[33] Recent analyses of aDNA from early *Homo sapiens* remains in Europe found that the percentage of Neanderthal DNA was originally higher (as much as 10 percent in a 40,000-year-old specimen from Romania) but has steadily decreased.[34]

Equally surprising, living Native Australians also carry as much as 6 percent of the genome of another close relative—the Denisovans—documented initially on the basis of aDNA extracted from a finger-bone

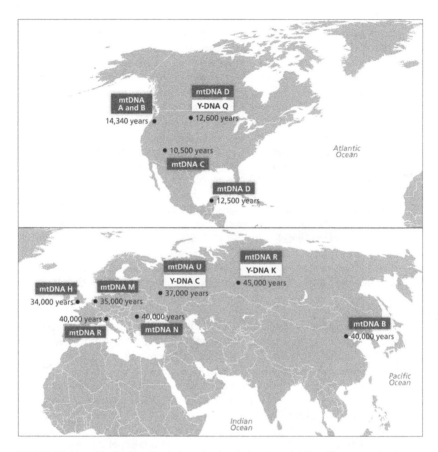

FIGURE 2.2 Map of the Northern Hemisphere, showing the location and dating of human remains that have yielded aDNA, as well as the identified mtDNA and Y-DNA haplogroups, or maternal and paternal lineages, respectively. (Base map adapted from en.wikipedia.org/wiki/File: BlankMap-World6.svg)

fragment from *Denisova Cave* in southwestern Siberia.[35] Recently, additional fossil and genetic data have been reported.[36] Some peoples in neighboring parts of the southwestern Pacific carry smaller amounts of Denisovan DNA. At the start of the second decade of the new millennium, paleoanthropology achieved a milestone as the anatomists who had critiqued the RAO model and the interpretation of genetic data happily embraced the latter.

The analyses of both non-modern and modern human genomes in 2010 reflected continuing advances in DNA-sequencing techniques. These, which included the first commercially available "next-generation sequencing"

BOX 2.2
Ménage à Trois: Modern Humans, Neanderthals, and Denisovans

A few years ago, the question of interbreeding between modern humans and Nean-derthals was debated on the basis of anatomical traits identified in the fossils. In 2003, for example, Erik Trinkaus and his colleagues noted the presence of a Neanderthal characteristic (*horizontal-oval type mandibular foramen*) in the modern human jaw from Oase in Romania.[1] Now the question is investigated primarily by genetics, studying aDNA from Neanderthals and modern humans, as well as by analyses of living people. The discovery in 2010 of another non-modern human species in northern Eurasia, from Denisova Cave in the Altai Mountains (southwestern Siberia), was quickly followed by recognition that many living humans carry some genes from both Neanderthals and Denisovans.[2] The pace of research has been brisk, and the topic has become a major issue in paleoanthropology.

Neanderthals (*Homo neanderthalensis*) and modern humans are descended from a common ancestor that lived in Africa roughly 500,000 years ago and probably is represented by the Broken Hill (Kabwe) cranium from Zambia. This large-brained form of *Homo* migrated out of Africa several hundred thousand years ago (chapter 4), gradually evolving the distinctive features of the Neanderthals in Europe and pro-ducing a closely related Asian species in the form of the Denisovans (the date of the Neanderthal–Denisovan split currently is estimated at between 473,000 and 381,000 years ago).[3] When modern humans expanded out of Africa after 75,000 years ago, they encountered both taxa in Eurasia.

The original paper on the draft sequence of the Neanderthal genome concluded that while only non-Africans carry some Neanderthal genes (1–4 percent), Neander-thals were spread across Eurasia, suggesting that interbreeding took place when modern humans first migrated out of Africa.[4] Recent research indicates that the pic-ture is more complicated—that modern humans mated with both Neanderthals and Denisovans at different times and places. Neanderthals also interbred with Deniso-vans. And although gene flow between Neanderthals and modern humans was initially thought to be unidirectional (Neanderthals → modern humans), new aDNA evidence reveals the presence of some modern human DNA in a Neanderthal specimen.[5]

By early 2016, at least five pulses of gene flow had been identified among the three taxa: (1) an early event (soon after modern humans migrated out of Africa) that accounts for a small percentage of Neanderthal genes among all living non-Africans, and (2) later Neanderthal–modern human admixture represented in only living Euro-peans and South and East Asians, followed by (3) interbreeding between Neander-thals and East Asians.[6] Living Melanesians (including Native Australians) carry evidence of (4) later admixture with Denisovans (but not the later mating with Neanderthals), and there is evidence of (5) very late interbreeding in Europe based on the analysis of aDNA from Oase, which exhibits a comparatively high percentage of Neanderthal DNA and is thought to be only several generations removed from a Neanderthal ancestor (box figure 2.2).[7]

What was the impact of gene transfer (or *introgression*) from Neanderthals and Denisovans to modern humans? Much of the non-modern DNA acquired by *Homo sapiens* apparently was deleterious and selected out of the modern genome. New analyses of aDNA in modern human specimens from Europe, dating to more than 20,000 years ago, indicates a steady decline in the percentage of Neanderthal

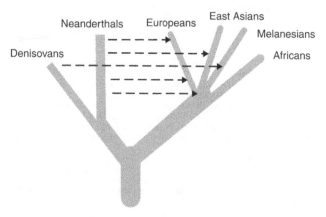

Neandertals   Europeans   East Asians

Melanesians

Denisovans   Africans

Common ancestor

BOX FIGURE 2.2 Gene flow among Denisovans, Neanderthals, and modern humans between roughly 100,000 and 50,000 years ago in Eurasia. (Modified from Ann Gibbons, "Five Matings for Moderns, Neandertals," *Science* 351 [2016]: 1251, unnumbered fig.)

genes.[8] Among those that remain among non-Africans, some are associated with increased risk for various diseases, including diabetes and lupus. Others are associated with upper-respiratory infections, mood disorders, and obesity.[9] The pronounced scarcity of both Neanderthal and Denisovan genes in the X chromosome may reflect detrimental effects on male fertility.[10] It has been suggested that Neanderthal DNA continues to reduce the fitness of non-Africans by 0.5 percent or more.[11]

At least a few of the Neanderthal and Denisovan genes acquired by modern humans, however, seem to have helped them adapt to local conditions. Michael Dannemann, Aida M. Andrés, and Janet Kelso recently identified a cluster of genes apparently derived from both Neanderthals and Denisovans that provide immunity against bacteria, fungi, and parasites.[12] Neanderthal genes associated with blood coagulation may have conferred selective advantage in the European population, while a genetic adaptation to high altitude among modern Tibetans is thought to be tied to Denisovan DNA.[13]

It should be noted, though, that few, if any, of the anatomical adaptations to cold climate (for example, short forelimbs and thick chest), long recognized among the European Neanderthals, were acquired by the modern humans who occupied the coldest parts of Eurasia. It also should be noted that the absence of Neanderthal and Denisovan genes among Africans suggests that the cognitive faculties of modern humans were largely or wholly unaffected by their interbreeding with other forms of *Homo* outside Africa.[14] The growing body of data on genetic admixture between modern and non-modern populations during the dispersal of modern humans finally allows a quantitative test of the "multi-regional model" of the origin of *Homo sapiens*.

1. Erik Trinkaus et al., "An Early Modern Human from the Peștera cu Oase, Romania," *Proceedings of the National Academy of Sciences* 100 (2003): 11231–11236.
2. David Reich et al., "Genetic History of an Archaic Hominin Group from Denisova Cave in Siberia," *Nature* 468 (2010): 1053–1060; David Reich et al., "Denisova Admixture and the

(continued)

First Modern Human Dispersals into Southeast Asia and Oceania," *American Journal of Human Genetics* 89 (2011): 516–528.

3. Matthias Meyer et al., "Nuclear DNA Sequences from the Middle Pleistocene Sima de los Huesos Hominins," *Nature* 531 (2016): 504–507. The Neanderthals and Denisovans are more closely related genetically to each other than either is to modern humans.

4. Richard E. Green et al., "A Draft Sequence of the Neandertal Genome," *Science* 328 (2010): 710–722.

5. Ann Gibbons, "Five Matings for Moderns, Neandertals," *Science* 351 (2016): 1250–1251. One issue that researchers recently have confronted is the difficulty of sorting out Neanderthal versus Denisovan contributions to the genome of non-African modern humans.

6. Benjamin Vernot et al., "Excavating Neandertal and Denisovan DNA from the Genomes of Melanesian Individuals," *Science* 352 (2016): 235–239.

7. Ann Gibbons, "Ancient DNA Pinpoints Paleolithic Liaison in Europe," *Science* 348 (2015): 847.

8. Qiaomei Fu et al., "The Genetic History of Ice Age Europe," *Nature* 534 (2016): 200–205.

9. Corinne N. Simonti et al., "The Phenotypic Legacy of Admixture Between Modern Humans and Neandertals," *Science* 351 (2016): 737–741.

10. Sriram Sankararaman et al., "The Combined Landscape of Denisovan and Neanderthal Ancestry in Present-Day Humans," *Current Biology* 26 (2016): 1241–1247.

11. Kelley Harris and Rasmus Nielsen, "The Genetic Cost of Neanderthal Introgression," *Genetics* 203 (2016): 881–891.

12. Michael Dannemann, Aida M. Andrés, and Janet Kelso, "Introgression of Neandertal- and Denisovan-like Haplotypes Contributes to Adaptive Variation in Human Toll-like Receptors," *American Journal of Human Genetics* 98 (2016): 22–33.

13. Simonti et al., "Phenotypic Legacy of Admixture Between Modern Humans and Neandertals," 739–740; Emilia Huerta-Sánchez et al., "Altitude Adaptation in Tibetans Caused by Introgression of Denisovan-like DNA," *Nature* 512 (2014): 194–197.

14. Although traces of Neanderthal/Denisovan ancestry have been found in isolated African populations (the Luhya and Gambians), they are believed to represent recent admixture with non-African populations, according to Vernot et al., "Excavating Neandertal and Denisovan DNA from the Genomes of Melanesian Individuals," 237.

method in 2000 (*massively parallel signature sequencing* [MPSS]), provide increasingly fast and inexpensive means for decoding complete genomes.[37] Although medical, forensic, and commercial applications drove the rapid development of these techniques, they have had a significant secondary impact on the study of modern human origins and dispersal. Whole-genome analyses of living populations and—where possible—of aDNA extracted from dated human remains have increasingly replaced the earlier studies of mtDNA and Y-DNA. These analyses have increased the size of the database by orders of magnitude (while the mitogenome contains fewer than 17,000 base-pairs, the whole human genome contains roughly 3.6 billion base-pairs). The immense body of data provides a firmer basis for assessing the relationships among the lineages.

The reconstruction of the Neanderthal and Denisovan genomes also allowed for whole-genome comparisons between modern humans and their closest relatives (estimated divergence of roughly 500,000 years).[38] It now is possible to identify the base-pair substitutions that differentiate modern humans from the two other descendants of their immediate ancestor (often classified as *Homo heidelbergensis*); the non-modern forms presumably lack the genetic basis for the cognitive faculties unique to *Homo sapiens*. Growing recognition of the critical role of the non-protein-coding portions of the genome (most of which was formerly classified as "junk DNA") suggests that the critical substitutions will be found among the regulatory genes (chapter 1).[39] More specifically, the striking differences in developmental biology between modern humans and their closest relatives suggest that the key substitutions should be found in parts of the genome that regulate brain growth.

Finally, the reconstruction of non-modern human genomes may provide a means of testing the multi-regional model with genetic data rather than fossil anatomical features. As described earlier, the model attributes the geographic variation observed in living people ("racial" variation) to genic exchange between the modern humans who evolved in Africa 300,000 to 150,000 years ago and regional non-modern populations (for example, European Neanderthals). New research has yielded evidence of multiple times and places in Eurasia where interbreeding among modern humans, Neanderthals, and Denisovans took place (see box 2.2). If the multi-regional model is correct, there should be significant correlation between the non-modern human genes carried by living people and the "geographic clusters" identified as genetic markers of racial variation in modern humans (for example, the 6 percent Denisovan introgression in Melanesians should overlap conspicuously with the East Asian and/or Oceanian clusters).[40]

OUT OF AFRICA: THE GLOBAL DISPERSAL

Charles Darwin chose the slow-breeding elephant to illustrate the explosive potential of population growth in *On the Origin of Species*. A pair of elephants, he noted, could produce roughly 15 million descendants within 500 years.[41] All populations of living organisms will grow exponentially until limited by available food or another factor, and all populations eventually will expand into neighboring habitats unless checked by the same factors. Thus it is not necessary to explain why modern humans (or earlier forms of *Homo*) expanded out of Africa, but only to understand what conditions otherwise constrained movement into Eurasia. The natural rate of increase represents a potentially constant source of pressure for population growth beyond the carrying capacity of the environment and expansion into adjoining habitats.

What factors are likely to have governed or influenced the dispersal of modern humans out of Africa? Climate change and its effects on plant and animal life probably were major variables.[42] Between 130,000 and 50,000 years ago, significant fluctuations in temperature and moisture occurred in both the Northern and Southern Hemispheres, affecting local and regional patterns of weather. Areas adjoining northeastern Africa that could have provided land access to Eurasia, such as the Arabian Peninsula, may have been uninhabitable during intervals of drier climate.[43]

Later intervals of warmth probably influenced the migrations of modern humans in northern Eurasia. Both human fossil and archaeological evidence suggest a major expansion roughly 48,000 to 45,000 years ago during a warm phase that corresponds to Greenland Interstadial 12 in the ice-core record (chapter 6). The effects of warmer and wetter climates on human movements may have been twofold: increased habitability of previously unoccupied places, such as central Siberia, and accelerated population growth (that is, increased population pressure) in more productive areas, such as the Levant, already occupied by modern humans.

One indirect effect of climate change that had a considerable impact on the global dispersal was the coalescence of the Laurentide and Cordilleran glaciers in North America about 55,000 years ago—presumably a result of slightly cooler temperatures at this time.[44] The merged ice sheets apparently blocked access to most of the Americas until about 15,000 years ago. Even though modern humans colonized Beringia no later than 32,000 years ago (and possibly much earlier), they were unable to occupy most of North America and South America until warming climates removed the physical barrier to the Western Hemisphere (chapter 7).[45]

Another indirect effect of climate change was the expansion of terrestrial habitat in areas exposed by the drop in the global sea level during cooler periods (as a result of billions of gallons of evaporated ocean water converted to glaciers on land). Most of this land was created along coastlines around the world, but in some places—such as the area between Australia and New Guinea and that between Chukotka and Alaska—lower sea levels created an extensive lowland plain. The presence of the plain between northeastern Eurasia and northwestern North America was especially significant, because wetter conditions in the southern portions of the Bering Land Bridge generated comparatively high plant productivity at latitude 60° North (chapter 7).[46]

The global dispersal of modern humans, which began 75,000 to 60,000 years ago, took several forms, including the migration of people into areas already inhabited by other forms of *Homo* and the colonization of areas not previously occupied by humans. There is little information about either in history or ethnography, and it is clear that each form of dispersal created a very different fossil, genetic, and archaeological record.[47]

The Migration into Eurasia

Modern humans in Africa expanded into two major regions occupied by other forms of *Homo*: southern Eurasia and parts of northern Eurasia (Europe, Central Asia, and southwestern Siberia). In these regions, in-migrating groups of *Homo sapiens* not only had to adapt to new habitat, but also had to compete for resources with a congeneric taxon (since *H. sapiens* was an invasive species). The distribution of non-modern human DNA in living populations suggests that modern humans first encountered the Neanderthals in western Eurasia and later mixed with the Denisovans in East Asia (probably Southeast Asia) (see box 2.2).[48]

Despite the evidence for their interbreeding with both non-modern taxa, it appears that modern humans effectively replaced the Neanderthals and Denisovans, driving them to extinction. Estimates of effective population size (the number of breeding individuals) based on analysis of aDNA (see box 7.1) suggest that the Neanderthal population was small and subject to periodic stress; it may have been especially vulnerable to competition from another species (further exacerbated by reduced resource availability during cold-climate intervals).[49] Modern humans acquired some genes from the non-modern Eurasian populations that probably helped them adapt to local conditions (especially local pathogens); this portion of the Neanderthal/Denisovan genetic legacy may, in fact, have been critical to the occupation of Eurasia by modern humans. But most of the non-modern introgression was deleterious and has been weeded out by selection.[50]

With rare and isolated exceptions, the African populations that migrated into Eurasia after 75,000 years ago retained the diagnostic anatomical features of *Homo sapiens* (that is, there is little evidence for hybridization in modern human skeletal remains), although fossil remains of the earliest modern humans in Eurasia are scarce.[51] The absence of Neanderthal and Denisovan DNA in African populations (see box 2.2) suggests that interbreeding with non-modern humans had little or no lasting impact on the cognitive faculties of modern humans. In sum, living non-African populations (including Native Americans) are modern humans who carry a genetic legacy of interbreeding with Neanderthals and/or Denisovans, but are not a hybridized form of *Homo*.

There are two reasons why modern humans may have had a competitive advantage over local non-modern humans in Eurasia, despite their status as an invasive or a colonizing species. Both reasons are related to the unique human capacity for translating non-genetic information into structure, as discussed in chapter 1, and for executing more complex "evolutionary computations" with these structures (that is, complex technology and syntactic language as "information technology").

First, in-migrating modern humans seem to have had the ability to design novel technologies for extracting resources—and expanding their dietary breadth and ecological niche—that were inaccessible to the local non-modern population.[52] The pattern is represented most clearly in northern Eurasia, where there is evidence for large-scale harvesting of small mammals and aquatic resources as early as 45,000 years ago (the technologies themselves have no current archaeological visibility in this time range) (chapter 6). It also could have been a factor in southern Eurasia, where there is indirect evidence of watercraft before 50,000 years ago. Modern humans may have developed new technologies for deep-water fishing along the South Asian coast or imported them from Africa (chapter 5). Both Neanderthals and Denisovans probably lacked the complex technologies for harvesting small vertebrates (including mechanical instruments and self-acting devices, or *automata*) found among recent hunter-gatherers.

Second, in-migrating modern human groups apparently possessed the same capacity for organizing and maintaining large social networks found among recent hunter-gatherers.[53] Here, again, the pattern is clearly evident in northern Eurasia, where the transportation of raw materials over hundreds of kilometers from their source is documented before 40,000 years ago—and represents one of the most striking contrasts between the archae-ological record of modern humans and that of other forms of *Homo*.[54] While long-distance transport in regions characterized by low plant and animal productivity (for example, East European Plain) probably reflects wide foraging ranges, it also likely represents the movement of materials within social networks maintained by complex matrices of kinship, marriage ties, and alliances. If Neanderthals and Denisovans lacked a syntactic language (perhaps speaking a simpler non-syntactic form of language), only mod-ern humans may have possessed the computational power to establish and sustain webs of long-term relationships among hundreds of individuals.

Despite modern humans' expansion of dietary breadth to resources unavailable to the Neanderthals and Denisovans, it is apparent that they also competed effectively with the non-modern populations for the same resources, especially large mammals in northern Eurasia. A comparative analysis of faunal remains in modern human and Neanderthal sites sug-gests significant niche overlap between the two taxa.[55] Some of the com-plex technologies developed by modern humans (for example, mechanical spear-thrower) may have given them a competitive advantage in exploit-ing common prey.[56] The same weaponry may also have provided an edge in direct conflicts between modern humans and local non-modern popu-lations, combined with a potential to organize larger groups through their alliance networks (organized aggression or warfare was not endemic, but was present among recent hunter-gatherers).[57]

If the Neanderthals are exceptionally well known among non-modern forms of *Homo* on the basis of recent aDNA analyses and a formidable col-lection of fossils and archaeological sites, the Denisovans remain a shad-owy presence in Asia before and after the arrival of *Homo sapiens*. Despite their fossil identification with the Altai Mountains in southwestern Sibe-ria, traces of their admixture among living people (Melanesians) indicate that the Denisovans were concentrated in Southeast Asia or perhaps more broadly across South Asia.[58] The remains in Denisova Cave probably rep-resent a northern outlier—just as the Neanderthal remains at this site rep-resent an eastern outlier of a population centered in the more biologically productive areas of western Eurasia. There is a large body of archaeologi-cal remains and at least some skeletal material (especially isolated teeth) in South and Southeast Asia, including southern China, that likely belong to the Denisovans (chapter 5).

The Colonization of Australia, Northeast Asia, and the Americas

Modern humans also invaded places never previously inhabited by the genus *Homo*. These newly occupied lands fall into two categories: (1) entire continents previously isolated from earlier human settlement by natural barriers, including Australia and the Americas; and (2) large areas of northern Eurasia where environmental factors (low plant and animal productivity, cold winters, and/or others) precluded earlier human settlement. The "colonization" (a term from evolutionary ecology) of these vast domains also is likely related to the unique human capacity for translating nongenetic information into structure and for performing complex evolutionary computations with this ability. The absence of boats or rafts suitable for navigation across more than 70 kilometers of ocean between Southeast Asia and Australia apparently kept earlier humans out of Australia and New Guinea (chapter 5). And the inability to design complex clothing and shelter technology to cope with extreme low temperatures, as well as the mechanical technology (including automata) required to harvest small vertebrates, likely prevented Neanderthals and Denisovans from occupying Northeast Asia—which, in turn, kept them out of the Americas (later blocked by coalescing glaciers) (chapters 6 and 7). The capacity to organize large alliance networks across the landscape also may have been critical to the successful occupation of places like Northeast Asia and the desert interior of Australia.

The boating technology that provided modern humans with access to Australia and New Guinea (conjoined by lower sea levels at the time and often referred to collectively as the continent of *Sahul*) may have been intimately connected to fishing and a South/Southeast Asian or even an African coastal marine adaptation. All this technology possesses very low visibility in the archaeological record, however. The conclusion that boats and/or rafts brought modern humans to Sahul is based on sea-level chronology—there was no other way to get there. There is archaeological evidence, though, for the occupation of dry interior zones of Australia before 40,000 years ago (although the development of desert adaptations similar to those of recent hunter-gatherers, such as the *Aranda*, probably occurred thousands of years after the initial colonization of Sahul).[59]

A combination of low temperatures and low plant and animal productivity probably excluded all human settlement from the coldest and driest portions of Northeast Asia until the arrival of modern humans (box 2.3), who may have first occupied this part of the world during a warm interval that began about 48,000 years ago. And although it has long been assumed that Neanderthals inhabited the East European Plain, some recent research suggests that their sites may have been confined to the southern margins of the plain (chapter 6). Most evidence for complex clothing and shelter technology, as well as direct or indirect evidence for mechanical devices and automata, before 30,000 years ago has been recovered from Northeast Asia and eastern Europe. The oldest eyed needles (evidence for tailored clothing), for example, were found in the northern Caucasus Mountains and Altai region.[60] Evidence for harvesting small mammals (possible use of traps and snares) and eating freshwater aquatic foods was found on the central East European Plain and in Northeast Asia.[61]

As discussed in chapter 1, it is the direct and indirect evidence for technology that was both structurally and computationally more complex than that known among any other animal, including the Neanderthals and Denisovans, that provides the basis for concluding that *Homo sapiens* represents a major transition in evolution related to the translation of nongenetic information into structure. The capacity for designing such complex technology may have developed along with modern human anatomy in Africa between 300,000 and 150,000 years ago, but currently is documented only after the second dispersal out of Africa after 75,000 years ago—and specifically in conjunction with the colonization of places where environmental variables appear to have excluded all other forms of *Homo*. Early evidence for long-distance transport of materials (with implications for computational complexity associated with syntactic language) also has emerged from these regions.[62]

Despite the presence of modern humans in Beringia before 30,000 years ago, a large body of evidence (fossil, genetic, and archaeological) indicates that the colonization of midlatitude North America and South America took place after 15,000 years ago.[63] The timing almost certainly reflects the retreat of ice sheets in coastal and interior areas of northwestern North America that had sealed off most of the Western Hemisphere from all the terrestrial mammals for the preceding 40,000 years.[64] Because the dispersal of modern humans in the Americas was comparatively late—three or four times more recent than that in Eurasia and Australia—it exhibits high archaeological visibility. The relatively rich record of the initial dispersal,

BOX 2.3
Net Primary Productivity, Temperature, and the Dispersal of Modern Humans

Modern humans invaded a wide range of habitats and climate zones during their dispersal out of Africa, including habitats never previously occupied by humans. These were places characterized by very low plant and animal productivity—where resources were thinly distributed across the landscape—and extremely low winter temperatures. They included much of Northeast Asia and all of the Arctic. The ecology of recent hunter-gatherers provides insight into understanding how modern humans adapted to these habitats with an array of technological and organizational strategies (chapter 6).[1]

Biological productivity is divided into primary (plants) and secondary (herbivores), and *net primary productivity* (NPP) is the amount of biomass formed by plants (chiefly through photosynthesis) per unit area and time. NPP often is measured in terms of total dry organic matter (grams) per square meter per year ($g/m^2/yr$), but sometimes is measured only in terms of total generated carbon ($g/C/m^2/yr$), which is about 45 percent of total dry organic matter.[2] NPP varies significantly by terrestrial habitat around the world (box figure 2.3). It is highest in wet tropical zones (such as sub-Saharan Africa, southern Asia, and lowland South America) and lowest in very cold and/or dry regions (such as North Africa and the Canadian Arctic). NPP for tropical rain forest, for example, ranges between 1,000 and 3,500 $g/m^2/yr$, while those for temperate grassland ranges between 200 and 1,500 $g/m^2/yr$ and arctic tundra between only 100 and 400 $g/m^2/yr$.[3] Terrestrial NPP is influenced heavily by available moisture, although it also is affected by temperature (cold air has a lower capacity to hold moisture).[4]

Although plants vary substantially in the amount of generated organic matter available to herbivores—some plants invest heavily in anti-herbivore defenses—NPP correlates well with both herbivore biomass and human population density.[5] The latter includes *hunter-gatherer population density*, as reconstructed from ethnographic data, although it does not apply to hunter-gatherers in coastal areas, who exploited rich marine foods unrelated to local terrestrial-plant production. For example, NPP for Nunivak Island (southwestern Alaska) is only about 90 $g/C/m^2/yr$, but its *Yupik* inhabitants maintained a population density of 23 persons per 100 square kilometers.[6] For hunter-gatherers in interior areas, there is a correlation between NPP and population density. The *Yukaghir* of northeastern Asia, who inhabited arctic tundra with an NPP of only 40 $g/C/m^2/yr$, supported an estimated 0.5 person per 100 square kilometers, while the *Hadza* of northern Tanzania, who occupy a tropical woodland with an NPP of about 680 $g/C/m^2/yr$, maintain 15 to 24 persons per 100 square kilometers. As might be expected, NPP correlates inversely with mobility among interior hunter-gatherers: people who live in areas with low NPP travel greater distances in their annual cycle of movements.[7]

The *technological complexity* of hunter-gatherers, however, correlates inversely with temperature better than with NPP. Among recent hunter-gatherers, the most complex technology is found in the Arctic, as well as other places where continental climates produce extreme winter lows (for example, subarctic Siberia).[8] In this context, the pattern is found in both interior and coastal areas; in fact, the most complex hunter-gatherer technology is associated with arctic coastal economies with an emphasis on marine-mammal hunting. The need for complex technology in these places lies in the significantly higher caloric demands of a cold-climate setting,

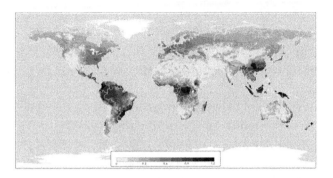

BOX FIGURE 2.3 World map of net primary productivity, or plant production, measured in grams of carbon per square meter/per year. (From https://www.uni-klu.ac.at/socec/bilder/ NPPt_600.png)

where available plant foods are scarce. Hunter-gatherers who live in the Arctic require as much as 48 percent more energy from their diet (almost entirely derived from mammals, birds, and/or fish) than people who inhabit the tropics (and consume primarily plant foods).[9]

Recent hunter-gatherers in cold interior habitats employed complex technology to harvest fish and other smaller vertebrates (for example, trap-lines and weirs) because terrestrial large-mammal resources were insufficient for year-round food supply, while arctic coastal foragers developed complex technologies for marine-mammal hunting (including deep-water whaling). And hunter-gatherers in both cold-climate interior and coastal settings required complex clothing and shelter technology for extreme low winter temperatures (the January mean in the region occupied by the Yukaghir is about –30°C).[10]

1. Robert L. Kelly, *The Lifeways of Hunter-Gatherers: The Foraging Spectrum* (Cambridge: Cambridge University Press, 2013).
2. O. W. Archibold, *Ecology of World Vegetation* (London: Chapman & Hall, 1995).
3. Helmut Lieth, "Primary Production of Major Vegetation Units of the World," in *Primary Productivity of the Biosphere*, ed. Helmut Lieth and Robert H. Whittaker (New York: Springer, 1975), 205, table 10-1.
4. Robert H. Whittaker, *Communities and Ecosystems*, 2nd ed. (New York: Macmillan, 1975), 201–206.
5. S. J. McNaughton et al., "Ecosystem-Level Patterns of Primary Productivity and Herbivory in Terrestrial Habitats," *Nature* 341 (1989): 142–144; Gary W. Luck, "The Relationships Between Net Primary Productivity, Human Population Density and Species Conservation," *Journal of Biogeography* 34 (2007): 201–212.
6. Kelly, *Lifeways of Hunter-Gatherers*, 178–184, table 7-3.
7. Ibid., 77–113.
8. Wendell H. Oswalt, *An Anthropological Analysis of Food-Getting Technology* (New York: Wiley, 1976); Robin Torrence, "Time Budgeting and Hunter-Gatherer Technology," in *Hunter-Gatherer Economy in Prehistory: A European Perspective*, ed. Geoff Bailey (Cambridge: Cambridge University Press, 1983), 11–22.
9. G. A. Harrison et al., *Human Biology: An Introduction to Human Evolution, Variation, Growth, and Adaptability*, 3rd ed. (Oxford: Oxford University Press, 1988), 479–484.
10. Wendell H. Oswalt, "Technological Complexity: The Polar Eskimos and the Tareumiut," *Arctic Anthropology* 24 (1987): 82–98.

supported by genetic analyses and aDNA, provides a unique window on the process of hunter-gatherer colonization of an uninhabited landscape (chapter 7).

Both the genetics of living Native Americans and the chronology of the archaeological sites indicate an extraordinarily rapid dispersal throughout North and South America. The entire Western Hemisphere appears to have been settled in less than 2,000 years by a rapidly growing population that initially spread thinly over large areas.[65] Many of the earliest dated sites, especially in North America, represent small, short-term occupations containing few artifacts. One factor that may have contributed to the rapid growth and spread of the Native American population is that it probably was derived from a large and diverse source population in Beringia that— according to the genetically based *Beringian Standstill Hypothesis* (or *Beringian Incubation Model*)—occupied a refugium between Northeast Asia and Alaska for thousands of years before the migration into North and South America.[66] Other factors likely included the warming climates after 15,000 years ago and their effect on plant and animal productivity, especially on the rich coastal resources along the Northwest Pacific and farther south.

Centers of Secondary Dispersal

A new analysis of global Y-DNA sequences published in 2016 reported evidence for a major diversification of paternal lineages in South and Southeast Asia roughly 55,000 to 50,000 years ago.[67] This observation is consistent with an earlier analysis of modern human teeth that identified Southeast Asia as an early center of modern human dispersal.[68] It also is supported by analyses of the maternal lineages that indicate a parallel expansion of subclades in mtDNA haplogroup M between 55,000 and 44,000 years ago.[69] Southeast Asia is characterized by extremely high levels of net primary productivity and correspondingly high levels of human population density.[70]

Southeast Asia appears to have been the first of multiple centers of secondary dispersal that became local sources of population expansion after modern humans migrated out of Africa. They are important for two reasons. First, they played a major role in the later stages of the global dispersal of modern humans—specifically the colonization of the Americas and Oceania. Second, they often obscured the earlier stages of the global dispersal by burying traces of it in the genetics of living people, although archaeological data (as well as aDNA) sometimes are helpful in excavating the human genetic stratigraphy.

Southwestern Europe became another center of secondary dispersal after 40,000 years ago, as subclades in mtDNA haplogroup U spread into eastern Europe and beyond.[71] The timing of this event appears to have coincided with an interval of pronounced warmth roughly 38,000 to 36,000 years ago (following a period of intense cold) that likely triggered a burst of human population growth in the biologically richest area of northern Eurasia. It seems to have obscured the earlier arrival in Europe of mtDNA haplogroup M (corresponding to the initial movement out of Africa, but not found today among Europeans).[72] In eastern Europe, this secondary dispersal is represented by aDNA and associated archaeological remains assigned to the Aurignacian industry.[73]

Perhaps the most surprising center of secondary dispersal is Beringia, which apparently supported relatively high levels of NPP during cold intervals of the past 50,000 years as a result of milder temperatures and moist air circulating around the North Pacific Ocean.[74] The human refugium population postulated by the Beringian Standstill Hypothesis is thought to have expanded after 15,000 years ago both eastward into North America and westward into Northeast Asia.[75] The latter is reflected in linguistic data, but apparently not in archaeological data.[76] Beringia thus seems to have played a significant part in the peopling of the Americas, while another population expansion in Southeast Asia probably helped promote the last major phase of the global dispersal—the settlement of Oceania.[77]

THE SYNTHESIS

The recent African origin and global dispersal of *Homo sapiens* have been reconstructed on the basis of three primary lines of evidence: (1) dated fossil remains of modern humans (and their non-modern contemporaries); (2) genetic analysis of living humans and study of aDNA in both modern and non-modern human fossils; and (3) archaeological data (artifacts, features, and associated remains). Although each has its strengths and weaknesses, they collectively provide a formidable body of evidence for the RAO model and the global dispersal of modern humans, including the routes and timing of specific events.

- *Fossil data.* The classification and dating of fossils of modern humans provided the original basis for the RAO model and remains the most fundamental set of evidence for the global dispersal. However, the quantity of dated remains, from both Africa and other parts of the world, is small, especially in comparison with the rapidly accumulating mass of genetic data. The starkness of the contrast is illustrated by the reconstruction of a species of *Homo*—the whole genome comprising billions of bits of information—initially from aDNA extracted from a finger-bone fragment found in Denisova Cave.[78] Or by the wealth of information obtained through a whole-genome analysis of a 45,000-year-old modern human from a leg-bone fragment discovered in western Siberia (chapter 6).[79] Modern human skeletal remains are particularly scarce in sites occupied during the early stages of the dispersal outside Africa; when present, they are typically confined to small fragments or isolated teeth that often are difficult to assign to a specific taxon.[80]

The principal weakness of the fossil anatomical data is that the phenotypes identified from the fossils may or may not reflect phylogenetic relationships, although phylogenetic relationships often were inferred by paleoanthropologists, especially in earlier years. A classic example is shovel-shaping of incisor teeth, which was identified in *Homo erectus* fossils from the Far East and is widespread among modern humans in East Asia. It was one of several traits selected by Weidenreich and others from the East Asian fossils that were believed to show that living Asian peoples had evolved directly from a local non-modern human population. The same character appears in other times and places among living and fossil humans, however, and does not necessarily indicate a direct genetic link.[81] More recently, initial studies of the *Kennewick* skeleton, found on the Columbia River in northwestern North America and dated to around 8,500 years ago, identified some "Caucasoid" features that encouraged speculation about its relationship to living Native Americans. A whole-genome analysis of aDNA from the skeleton, published in 2015, confirmed its status as a Native American (chapter 7).[82]

Human skeletal remains provide important information on the evolution of growth and development, however, which appears to be a critical factor in modern human origins. Growth rates are determined primarily on the basis of incremental features of the teeth.[83] The comparative measurement of modern and non-modern human fossils reveals that the pattern of delayed maturation in living people evolved late in *Homo*. Fossil anatomical data also offer information on the derived pattern of brain growth in modern humans that produces the characteristic globular cranium (box 2.4).[84]

• *Genetic data.* The second body of evidence is the genetic analysis of living humans, combined with the study of aDNA recovered from skeletal remains. As a consequence of the application of new techniques for rapid sequencing of DNA, genetic data on the origin and dispersal of modern humans have exploded and now dominate the discussion. The data are in *digital* form and are more amenable to quantitative analysis than are the phenotypes (anatomical traits in analog form), especially in very large numbers. Genetic data have become the principal basis for reconstructing the broad pattern of the global dispersal, yielding a general model that may be tested in the field with the other sources of data (fossils, artifacts, and aDNA). Combined with aDNA analyses, genetic data on living peoples have become the primary means of investigating the degree of assimilation or interbreeding between modern humans and their non-modern contemporaries (although analysis of anatomical traits observed in the fossils still plays a role in the debate [see box 2.2]).[85]

The most significant constraint on the genetic data remains uncertainty about mutation rates, especially with respect to mtDNA.[86] Not only do rates of mutation (or substitution) vary among the different categories of DNA, but they also vary over time and within those categories at different sites (that is, locations on the DNA sequence). Because different substitution rates have been applied to the data, the analyses have yielded varying estimates of the time of divergence of specific haplogroups. Within the past few years, however, a sufficient amount of ancient mtDNA from reliably dated contexts has been recovered to begin improving the calibration of the human mitochondrial clock.[87]

Another problem is that the divergence of and mutations in lineages are unlikely to coincide, and one usually takes place without the other ("incomplete lineage sorting"). As a result, genetic analyses of two or more groups may not yield accurate information about their phylogenetic relationships.[88] The new DNA-sequencing techniques and whole-genome analysis apparently have mitigated the problem with vastly larger samples.[89] For example, the initial mtDNA study of the original Denisovan specimen failed to reveal its close phylogenetic relationship to the Neanderthals and

BOX 2.4
Measuring Growth and Development in Human Fossil Remains

The pattern of growth and development in *Homo sapiens*, especially the development of the brain, is critical to the unique cognitive faculties of living humans. In comparison with the African apes, modern humans exhibit a pattern of significantly increased postnatal growth and delayed maturation, and it seems to be tied to language acquisition. The development and application of rigorous quantitative methods for determining the timing of growth and development in human fossils is therefore important to the study of the origin of modern humans.

Estimates of growth rates in fossil humans have consistently been based on teeth. In 2001, Christopher Dean and his colleagues used the incremental markings in tooth enamel (*perikymata*) to determine rates of growth in early humans. They found that the perikymata were widely spaced among australopithecines and early *Homo*, indicating faster growth.[1] In 2007, Tanya Smith and her colleagues applied a noninvasive measurement technique (*x-ray synchrotron microtomography*) to teeth in a 300,000-year-old juvenile mandible from Morocco (*Jebel Irhoud*) (box figure 2.4).[2] Their results established that the pattern of slow growth and late development found among living humans was present in early representatives of *Homo sapiens*. A followup study found evidence for faster growth rates in the Neanderthals.[3]

In 2004, *computer tomography (CT) scanning* was applied to an early *Homo* skull from Java (*Mojokerto*) to determine the age at death on the basis of various cranial

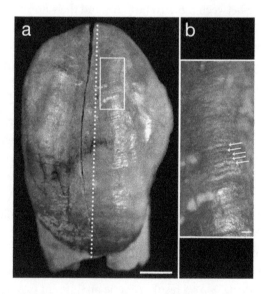

BOX FIGURE 2.4 The unerupted lower-left canine germ of the juvenile from Jebel Irhoud: (*a*) stereo-microscope overview, with the position of the area enlarged in *b*; (*b*) perikymata (*arrows*), surface manifestations of long-period Retzius lines, were counted from the cusp tip to the cervix on the original tooth. (From Tanya M. Smith et al., "Earliest Evidence of Modern Human Life History in North African Early *Homo sapiens*," *Proceedings of the National Academy of Sciences* 104 [2007]: 6129, fig. 1. Copyright 2007 National Academy of Sciences)

developmental features. Although the estimated age of the individual was only about 1 to 1.5 years, the brain was already 72 to 84 percent of its total adult size, reflecting very rapid growth.[4] More recently, the analysis of developmental patterns in Neanderthal crania was reconstructed with the aid of three-dimensional geometric morphometrics. The results revealed that the process of early development that produces the "globular" shape of the modern human cranium (*globularization phase*) is unique to *Homo sapiens* (and not seen in the Neanderthals).[5]

1. Christopher Dean et al., "Growth Processes in Teeth Distinguish Modern Humans from *Homo erectus* and Earlier Hominins," *Nature* 414 (2001): 628–631.
2. Tanya M. Smith et al., "Earliest Evidence of Modern Human Life History in North African Early *Homo sapiens*," *Proceedings of the National Academy of Sciences* 104 (2007): 6128–6133.
3. Tanya M. Smith et al., "Dental Evidence for Ontogenetic Differences Between Modern Humans and Neanderthals," *Proceedings of the National Academy of Sciences* 107 (2010): 20923–20928.
4. H. Coqueugniot et al., "Early Brain Growth in *Homo erectus* and Implications for Cognitive Ability," *Nature* 431 (2004): 299–302.
5. Philipp Gunz et al., "A Uniquely Modern Human Pattern of Endocranial Development: Insights from a New Cranial Reconstruction of the Neandertal Newborn from Mezmaiskaya," *Journal of Human Evolution* 62 (2012): 300–313.

modern humans—later clarified with whole-genome analysis. And whole-genome analysis of a 37,000-year-old skeleton from the central East European Plain yielded information about Neanderthal gene flow not recovered from the earlier mtDNA research (figure 2.3).[90]

As already noted, aDNA has become an important means of testing the predictions of the genetic model; the haplogroups defined through comparative analysis of mtDNA and Y-DNA among living humans may actually be identified in skeletal remains that date to the time of the global dispersal. A major concern in aDNA analysis has been the threat of contamination from people who handle the sample. This concern seems to have been justified in some cases,[91] and the development of effective laboratory protocols to prevent contamination from the DNA of living people has been critical to the success of this research. Another constraint on aDNA work is the influence of climate on the long-term preservation of DNA, which survives best in higher latitudes and cooler temperatures.[92] Until recently, aDNA has been confined to samples from latitudes above 40° North,[93] but researchers working in a warm-climate setting (for example, Caribbean islands) now report recovery of aDNA from contexts as old as 2,000 years.[94]

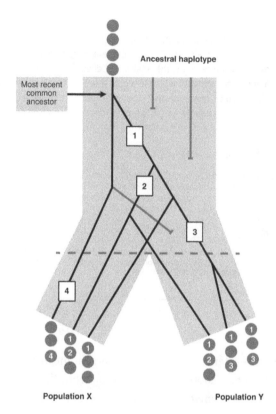

FIGURE 2.3 Incomplete lineage sorting. (Redrawn from Ryosuke Kimura, "Human Migrations and Adaptations in Asia Inferred from Genomic Diversity," in *Emergence and Diversity of Modern Human Behavior in Paleolithic Asia*, ed. Yousuke Kaifu et al. [College Station: Texas A&M University Press, 2015], 36, fig. 3.2)

- *Archaeological data.* Another line of evidence for both the RAO model and global dispersal is archaeological data. Artifacts and associated debris recovered from sites in Africa (and the Levant) that are 75,000 years old or older provide some evidence for the cognitive changes that are associated with the evolution of modern humans—they complement the fossil anatomical evidence for a recent African origin. The earlier archaeological record of *Homo sapiens* in Africa (and the Levant) is especially notable for the appearance of personal ornaments (as early as 135,000 years ago) and examples of natural objects decorated with simple geometric designs.[95] These artifacts presumably are related to social interactions and may reflect an intensification and/or expansion of alliance networks (that is, enhanced computational abilities applied to social life).

There also is evidence in sub-Saharan Africa (mostly southern Africa) for new types of tools and weapons, including the mass-production of small stone bladelets and geometric microliths (ca. 80,000 years ago), bone points and awls (ca. 75,000 years ago), and barbed bone harpoons (possibly as early as 90,000 years ago).[96] In addition to the appearance of new categories of artifacts, the archaeological record yields evidence for material objects and features that are structurally more complex (for example, exhibit three or more hierarchical levels of organization) and/or reflect more complex underlying computations than those made by other forms of *Homo*. Sites that date to before 75,000 years ago in Africa (specifically southern Africa), have yielded evidence for computationally complex applications of heat treatment to stone (ca. 164,000 years ago) and processing of ocher (ca. 100,000 years ago).[97] The pattern conceivably reflects enhanced computational abilities applied to materials.

Evidence for significantly more complex material computation and structurally complex artifacts and features coincides with the dispersal of modern humans out of Africa. Sometimes the presence of computationally and structurally/functionally complex technology may be inferred from more commonly preserved related artifacts (for example, eyed needles = sewn clothing) or indirectly from other sources of evidence (for example, high values for $\delta^{15}N$ in human bone = fishing technology?). The most direct evidence for structurally complex artifacts (in terms of hierarchical complexity) is found in the category of visual art, but these data currently are confined to less than 50,000 years ago, and the earliest known examples have been documented in sites outside Africa (box 2.5).

The artifacts, features, and associated debris recovered from datable archaeological sites constitute the overwhelming majority of material remains left behind by modern humans as they dispersed out of Africa. Archaeological data offer a potential proxy for groups of modern humans, and in most places they provide the earliest dated evidence for the arrival of *Homo sapiens*—given the scarcity of human skeletal remains. In the future, it may be possible to correlate specific types of artifacts and/or groups of artifact assemblages with particular haplogroups.[98]

The use of archaeological remains as a proxy for modern humans is limited by the problematic relationship between many of the artifacts and specific human taxa. Although modern humans created items that their non-modern contemporaries probably never designed or produced—such

BOX 2.5
Measuring Technological Complexity

Modern human technology is uniquely complex—organizationally, functionally, and computationally—and reflects a unique capacity to translate non-genetic information into structure in complex, hierarchically organized form. Like other forms of evolutionary computation, human technology is the product of *nondeterministic computation* and reflects a source of random variation and input. When humans design or redesign an artifact, they engage the same cognitive faculties that support potentially infinite creative variation in language and art—and express it in material form. The unique complexity and innovative character of modern human technology is a product of evolution, or, more precisely, of evolutionary computation with non-genetic information. Modern humans applied their capacity for complex technology to create an entirely new level of evolutionary computation in the form of machines, or automata (see box 4.4).

The complexity of human technology is a major issue in paleoanthropology and a long-standing basis for inferences about the cognitive faculties of early humans.[1] Accordingly, paleoanthropologists have considered various ways to measure technological complexity in the archaeological record. The most common approach in archaeology entails the reconstruction of the "procedural steps" required to transform one or more raw materials into an artifact (also known as the *chaîne opératoire*), which is a complete or partial measure of computational (or *algorithmic*) complexity.[2] It has been applied widely to lithic technology (for example, Levallois core reduction and biface manufacture) and occasionally to non-lithic technology, revealing greater complexity in the technology of early humans than previously believed.[3]

A general measure of complexity for preindustrial technology was proposed in the 1970s by Wendell H. Oswalt, who subdivided instruments (for example, bow and arrow) and facilities (for example, deadfall trap) into their irreducible components ("technounits") and calculated the number of parts for each artifact or feature. Parts not functionally differentiated from one another (such as the more or less identical wooden stakes used to construct a fish weir) were not counted as separate technounits.[4] The *technological complexity* of each instrument and facility was determined on the basis of the total number of its technounits.[5]

Oswalt counted technounits for food-getting technology ("subsistants") among 36 hunter-gatherer and horticulturalist groups, drawing on earlier ethnographic publications for detailed descriptions of specific instruments and facilities. For example, a spear made by the Aranda (Australia) is broken down as follows:

spear, used with throwing-board: wood point + wood barb + sinew, point-barb binder + wood foreshaft + wood shaft + sinew, point-foreshaft binder + resin, point-foreshaft binder + sinew, foreshaft-shaft binder + resin, foreshaft-shaft binder = 9 technounits

The spear was determined to be more complex than, for example, a fish trap (6 technounits) made by the *Klamath* (northwestern North America), but less complex than a salmon drag gill net (12 technounits) made by the *Deg Hit'an* (Alaska).[6] Oswalt classified all subsistants with moving parts (mechanical technology) as "complex," however.

Oswalt's analysis revealed some interesting patterns among recent hunter-gatherers,[7] but his approach has not been applied widely with archaeological data because most of the instruments and facilities in his sample were made from perishable materials. Moreover, his methodology provides no measure of structural or functional complexity, both of which are potentially more significant than technounit counts with respect to the evolution of human cognitive faculties. Among complexity theorists, structure often has been measured in terms of the number of hierarchical levels and the organization of embedded components and subsystems.[8] Although there are isolated examples of nonhuman artifacts or features with three hierarchical levels of organization (for example, the underground nest complexes of some ant species), none of them is based on computation with neuronal information; only humans create technology with three or more hierarchical levels based on the manipulation of non-genetic information.[9]

A formal definition of functional complexity is based on an application of information theory (see box 1.1) to technology: the "functional design complexity of an artifact" is defined as a measure of its probability of reducing entropy (that is, a measure of its *information content*) in the context of the problem that it is designed to solve. There is a relationship, moreover, between functional and hierarchical complexity. As the complexity of problems increased (that is, more steps are required to compute a solution), living systems adapted by *decomposing* the problems into sets of smaller subproblems, each of which could be solved independently.[10]

In the context of hunter-gatherer technology, for example, many groups that inhabit interior areas in northern North America harvest large quantities of riverine fish by installing a combination of weirs and traps at selected times and places designed to divert the fish into the traps. Beginning with its strategic placement in a stream channel at a specific time in the annual cycle, the weir–trap complex "performs" a sequence of functions that redirect the fish toward and into the traps, from which they cannot escape. The functional complexity of the weir–trap complex (neither a mechanical device nor an automaton) is reflected in its hierarchical design, which ensures a high probability of success (and a corresponding step-by-step reduction of uncertainty) (box figure 2.5).

As with Oswalt's technounit counts, measuring the structural and/or functional complexity of technology in the archaeological record is difficult because most hunter-gatherer technology—especially the more complex instruments and facilities—is made of perishable materials and has limited visibility. Complex technology in the archaeological record is inferred from isolated components (for example, arrow-point = bow and arrow), related artifacts (for instance, eyed needle = tailored clothing), or indirect sources (for example, concentration of small-mammal remains = snares and/or traps).[11] As described in chapters 4, 5, and 6, there is such evidence for structurally and functionally complex technology in the archaeological record of modern humans in and outside Africa after the beginning of the global dispersal of modern humans between 75,000 and 60,000 years ago.

1. See, for example, Stanley H. Ambrose, "Paleolithic Technology and Human Evolution," *Science* 291 (2001): 1748-1753, and "Coevolution of Composite-Tool Technology, Constructive Memory, and Language: Implications for the Evolution of Modern Human Behavior," *Current Anthropology* 51 (2010): S135–S147; Nathan Schlanger, "Understanding Levallois: Lithic Technology and Cognitive Archaeology," *Cambridge Archaeological Journal* 6 (1996): 231–254; and Lyn Wadley, "Recognizing Complex Cognition Through Innovative Technology in Stone Age and Palaeolithic Sites," *Cambridge Archaeological Journal* 23 (2013): 163–183. See also Thomas Wynn and Frederick L. Coolidge, "Archeological Insights into Hominin Cognitive Evolution," *Evolutionary Anthropology* 25 (2016): 200-213.

(continued)

(continued)

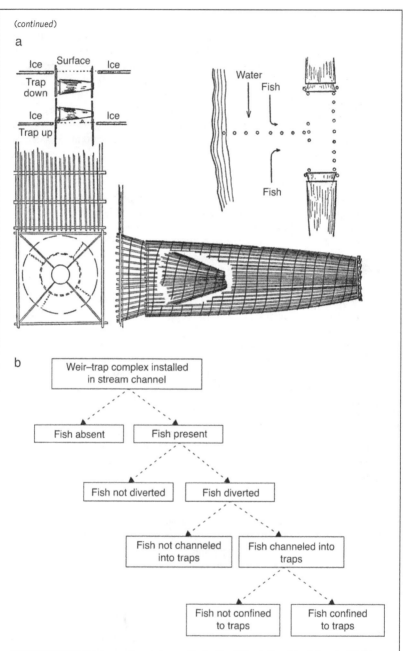

a

b

BOX FIGURE 2.5 Fish trap-weir complex used by the Deg Hit'an in the winter (a), and a "decision tree" for a trap-weir complex (b), illustrating how it addresses the decomposed problem of fish capture by solving a series of smaller sub-problems with multiple "functions" (although it lacks moving parts). The structural complexity of the artifact reflects its functional complexity (high probability of reducing uncertainty or "entropy"). ([a] From Cornelius Osgood, *Ingalik Material Culture* [New Haven, Conn.: Yale University Press, 1940], 229–230. Courtesy of Yale University Publications in Anthropology)

2. Nathan Schlanger, "The Chaîne Opératoire," in *Archaeology: The Key Concepts*, ed. Colin Renfrew and Paul Bahn (Abington: Routledge, 2005), 25-31.

3. Miriam Noël Haidle, "How to Think a Simple Spear," in *Cognitive Archaeology and Human Evolution*, ed. Sophie A. de Beaune, Frank L. Coolidge, and Thomas Wynn (Cambridge: Cambridge University Press, 2009), 57-73; Marlize Lombard and Miriam Noël Haidle, "Thinking a Bow-and-Arrow Set: Cognitive Implications of Middle Stone Age Bow and Stone-Tipped Arrow Technology," *Cambridge Archaeological Journal* 22 (2012): 237-264; Charles Perreault et al., "Measuring the Complexity of Lithic Technology," *Current Anthropology* 54 (2013): S397-S406.

4. Along similar lines, the mathematician A. N. Kolmogorov (1903-1987) proposed that the complexity of an object or a system should be measured by the shortest possible description of it, in "On Tables of Random Numbers," *Sankhyā: The Indian Journal of Statistics*, ser. A, 25 (1963): 369-375.

5. Wendell H. Oswalt, *Habitat and Technology: The Evolution of Hunting* (New York: Holt, Rinehart and Winston, 1973), and *An Anthropological Analysis of Food-Getting Technology* (New York: Wiley, 1976).

6. Oswalt, *Anthropological Analysis of Food-Getting Technology*, 264-283.

7. Robin Torrence, "Time Budgeting and Hunter-Gatherer Technology," in *Hunter-Gatherer Economy in Prehistory: A European Perspective*, ed. Geoff Bailey (Cambridge: Cambridge University Press, 1983), 11-22; John F. Hoffecker, *Desolate Landscapes: Ice-Age Settlement in Eastern Europe* (New Brunswick, N.J.: Rutgers University Press, 2002); Mark Collard, Michael Kemery, and Samantha Banks, "Causes of Toolkit Variation Among Hunter-Gatherers: A Test of Four Competing Hypotheses," *Canadian Journal of Archaeology* 29 (2005): 1-19.

8. Herbert A. Simon, "The Architecture of Complexity," *Proceedings of the American Philosophical Society* 106 (1962): 467-482.

9. Bert Hölldobler and E. O. Wilson, *The Super-Organism: The Beauty, Elegance, and Strangeness of Insect Societies* (New York: Norton, 2009), 338-339.

10. Dan Braha and Oded Maimon, "The Measurement of a Design Structural and Functional Complexity," *IEEE Transactions on Systems, Man, and Cybernetics Part A: Systems and Humans* 28 (1998): 527-535; Francis Heylighen, "The Growth of Structural and Functional Complexity During Evolution," in *The Evolution of Complexity: The Violet Book of "Einstein Meets Magritte,"* ed. Francis Heylighen, Johan Bollen, and Alexander Riegler (Dordrecht: Kluwer, 1999), 17-44.

11. Ian Gilligan, "The Prehistoric Development of Clothing: Archaeological Implications of a Thermal Model," *Journal of Archaeological Theory and Method* 17 (2010): 15-80; Lyn Wadley, "Were Snares and Traps Used in the Middle Stone Age and Does It Matter? A Review and a Case Study from Sibudu, South Africa," *Journal of Human Evolution* 58 (2010): 179-192; Paola Villa et al., "Border Cave and the Beginning of the Later Stone Age in South Africa," *Proceedings of the National Academy of Sciences* 109 (2012): 13208-13213.

as objects of art—they often are absent in sites occupied during (and after) the global dispersal. The most common category of artifacts—stone tools—comprises a range of types produced by modern and non-modern humans alike. The problem is illustrated by the current controversy over the authorship of the stone blades that underlie the ash from the Toba supervolcanic eruption (ca. 74,000 years ago) at *Jwalapuram* (India), and whether they represent evidence of an early dispersal of modern humans into South Asia.[99] On the other side of the coin is *Byzovaya* (northern Russia), where bifaces and flake tools have been cited as evidence for a late survival of Neanderthals in northeastern Europe.[100] The only stone-tool types that may be diagnostic of *Homo sapiens* are microblades and the small cores that produced them.

Modern humans continued to make simple stone tools, such as flake scrapers and pebble choppers, that were indistinguishable from those made by their non-modern contemporaries (and their ancient predecessors). These tools often were made, used, and discarded at places where animals were killed and butchered, for example; sometimes they were fashioned expediently from local raw materials of poor quality that were convenient, contributing to their archaic appearance. In the absence of artifacts that are diagnostic of *Homo sapiens*, it is often unclear who made these tools, and a number of Eurasian archaeological sites that date to the general period of the global dispersal cannot be firmly assigned to a specific human taxon. There also are sites that contain a mixture of forms typically made by modern and non-modern humans; until recently, many of them were attributed to the latter, but their status is problematic.[101]

Assuming that they can be reliably assigned to *Homo sapiens*, artifacts and features sometimes provide vital information about the global dispersal that cannot be obtained from the other categories of data. Archaeological remains yield evidence for organizational adaptations to environments where resources were widely dispersed (for example, movement of raw materials over great distances). And only the archaeological data offer clues about how modern humans adapted to local conditions through innovation in technology. An example is the fragment of an eyed needle that dates to about 40,000 years ago from a cave in the northern Caucasus region that indicates tailored clothing (technological adaptation to low temperature).[102] The weakness of this category of data is the comparatively poor preservation of non-stone artifacts—which typically are the source of information about the complex technology of modern humans that reflects their capacity for innovation—and the uncertainties inherent in reconstructing or inferring such technology from isolated fragments and other clues.

The Fossil Evidence

The fossil evidence for the recent African origin and global dispersal of modern humans comprises a set of bones and teeth that may be assigned to *Homo sapiens* on the basis of a list of diagnostic anatomical traits and dated to at least 150,000 years ago in Africa, but to less than 75,000 years ago in most other parts of the world. There is no compelling evidence for modern human dispersal in the Western Hemisphere until about 15,000 years ago, and the settlement of the Americas—with the possible exception of Alaska/Yukon—represents a later phase of the global dispersal (as does the occupation of much of Oceania).

The skeletal anatomical traits considered diagnostic of modern humans include the following cranial features: (1) vertical frontal bone, (2) typically bipartite rather than continuous browridge, (3) high cranial vault with outward bulging of parietal bones ("globular"), (4) rounded occipital bone without transverse torus, (5) highly flexed cranial base, (6) relatively flat and retracted facial region (not projecting outward), (7) presence of canine fossa or pronounced hollowing of bone below each orbit, and (8) presence of chin (or mental eminence) on the mandible. As noted at the beginning of this chapter, Day and Stringer added metric values to many of these features (figure 2.4).[103]

Post-cranial traits include the following: (1) barrel-shaped thorax, (2) pelvis lacks lateral flare and exhibits more vertical iliac blades, (3) proximal margin of scapula typically unisulcate or with a single groove (occasionally bisulcate),and (4) distal phalanx of thumb usually two-thirds as long as proximal phalanx.[104] Although individually these cranial and post-cranial traits do not always indicate modern humans, when taken together or combined into indices, they are a reliable means of classification.

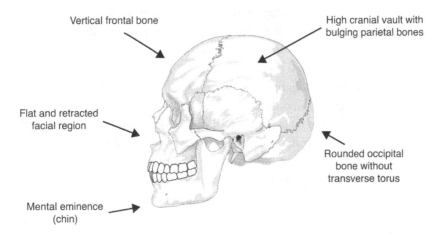

FIGURE 2.4 Anatomical traits of the skull considered diagnostic of *Homo sapiens*. (Adapted from http://images.all-free-download.com/images/graphiclarge/human_ skull_side_ view_ clip_art_15525.jpg)

Some of the diagnostic traits nevertheless are noticeably absent in specimens from Africa that antedate the global dispersal after 75,000 years ago (and specimens from the Levant that date to 120,000–90,000 years ago), many of which have been characterized as "archaic" or "near modern" (the assignment of some [for example, *Ngaloba* or Laetoli H.18 from Tanzania] to modern humans has been questioned).[105] The "mosaic" pattern probably reflects a gradual evolutionary transition to *Homo sapiens* in Africa[106] and the likelihood that living people are directly descended from only one of the modern human subpopulations in Africa (and the Levant) before 75,000 years ago (chapter 4).

Teeth preserve better than bone in the fossil record; although small (crown diameter typically less than 1 centimeter), isolated human teeth are easily identified—in contrast to most small bone fragments—in an archaeological context and contain a number of diagnostic features (that is, they carry a large amount of information relative to their size).[107] Isolated teeth sometimes represent the earliest dated skeletal remains of modern humans in regions outside Africa (for example, eastern Europe). They also serve as population markers, analogous to the mtDNA and Y-DNA haplogroups, and provide information about the timing and pattern of the global dispersal (box 2.6).

The most significant exception to the pattern of an expansion out of Africa after 75,000 years ago is the presence of "near modern" human remains in the Levant that date to between 120,000 and 90,000 years ago.[108] These fossils apparently reflect a movement into a small area of Southwest Asia that adjoins Northeast Africa during a warm phase (MIS 5). The subsequent reappearance of Neanderthals in the Levant suggests that they displaced the *Homo sapiens* population in the region, possibly during a cold phase that occurred roughly 75,000 years ago (MIS 4).[109] Recent discoveries of teeth assigned to modern humans in Southeast Asia that date to more than 75,000 years ago have encouraged speculation that the early dispersal into Eurasia extended significantly farther east and south, but this possibility remains to be confirmed (chapter 5).[110]

The remaining modern human fossils reflect the global dispersal out of Africa, which probably began between 75,000 and 60,000 years ago. They include a limb bone from *Tabon Cave* (Philippines) that may be as old as 58,000 years and the recently reported cranium from *Manot Cave* (Israel) that dates to about 55,000 years ago.[111] The dating of an older skull from

BOX 2.6
Dental Morphology and the Origin and Dispersal of Modern Humans
G. RICHARD SCOTT

Most living humans are unaware that the human dentition exhibits extensive morpho-logical variation within and between groups. The majority of traits are found on the occlusal and lingual surfaces of the teeth and are not readily visible to the casual observer. When people smile, they reveal only the facial or labial (near the lips) sur-faces of the upper and lower anterior teeth (incisors and canines), which exhibit little variation. The hidden traits are extra cusps, ridges, fissures, tubercles, and roots expressed on the chewing surfaces of the teeth (occlusal) or, as with root traits, hid-den below the gum line. Anthropologists and paleontologists have known about this variation for a century, but until recently the literature on dental morphology was beset by problems of trait definition and observational error.

In the 1980s, population studies of dental morphology began to change. Taking a cue from the classic work of Albert A. Dahlberg (1909–1993),[1] Christy G. Turner II and his students defined more than 30 crown and root traits and developed a widely adopted system of classification.[2] Turner examined thousands of skeletons through-out the world—with special emphasis on Native Americans, East and Southeast Asians, Australians, Pacific Islanders, and Europeans—while two of his students examined large samples from Africa and India. Influenced by the modern evolutionary synthesis and rapid advances in genetics after 1950, Turner's initial hope was that morphological traits like Carabelli's cusp had simple modes of inheritance. If that proved valid, researchers could reduce phenotypic observations to gene frequencies and address variation in terms of gene flow, genetic drift, and natural selection.

After some initial successes with pedigree studies, it became apparent that dental traits are not simple Mendelian autosomal dominant or recessive traits, but threshold traits with polygenic modes of inheritance. This conclusion did not, however, diminish their utility in population studies. As with other threshold traits, phenotype frequencies reflect underlying genetic variation (environmental factors have little impact on the development of morphological traits). Once other issues were resolved, such as sex-ual dimorphism and inter-trait association, dental morphology was employed to address issues of affinity at the local, regional, continental, and global levels—analogous to studies of nuclear genetic markers.

The first major application of tooth morphology to a broad anthropological issue was the problem of Native American origins. In 1971, Turner developed a three-wave model for the settlement of the Americas based on a single root trait: three-rooted lower first molars.[3] This trait is common in Eskimo–Aleuts (30–50 percent) and rare in American Indians (5 percent). The dichotomy had long been recognized for other bio-logical traits (for example, blood groups and anthropometrics), but Turner identified a third group—Na-Dené speakers in Alaska, Canada, and the American Southwest—that is intermediate between the two extremes. Accordingly, he proposed three waves con-sisting of Macro-Indians (all American Indians of North and South America), Na-Dené-Athapaskans (possibly including Northwest Coast populations), and Eskimo-Aleuts.

Turner also applied dental morphology to issues in Asian prehistory. After a study of Jōmon, Ainu, Japanese, and Chinese dentitions, he concluded that the Jōmon were ancestral to the Ainu and not to modern-day Japanese. Tooth morphology indicated

(continued)

(continued)

that the modern population of Japan is descended from mainland Asian groups that invaded the Japanese archipelago during the Yayoi period (ca. 2,200 years ago).[4] Turner later concluded that there are two distinct dental patterns in Asia: Sinodont (North and East Asian) and Sundadont (Southeast Asian). Eight traits, in particular, set the two groups apart; in general, Sinodont teeth exhibit complex specialized crowns, while Sundadont dentitions are less sculpted and more generalized. (The Australian dentition did not fit neatly into either, although Turner thought that it exhibits "proto-Sundadonty.")[5] One ramification of the dental dichotomy was that Asia became a springboard for dispersal into the last major regions colonized by human populations— the Western Hemisphere and the Pacific. All Western Hemisphere populations are descended from Sinodont groups, while Polynesians (and some Melanesians and Micronesians) are descendants of Southeast Asian Sundadonts.

On a global scale, Turner used dental morphological data to estimate relative degrees of similarity among recent human populations. He focused on what he identified as 29 key crown and root traits. As some traits are consistently rare or almost invariant, the analysis performed on 16 major geographic groups was based on 22 variables (for example, upper central incisor winging, upper central incisor shoveling, upper first molar Carabelli's trait, three-rooted lower first molar). The mean measures of divergence (pairwise distances) calculated among these groups are illustrated by the dendrogram in box figure 2.6a, which shows four major divisions of living humankind. The most highly divergent group is composed of five Native American samples. The next cluster to diverge includes Europe, New Guinea, South Asia (India), and sub-Saharan Africa. The final two groups are represented by Australia and the Pacific, and East, Southeast, and Central Asia.

For the most part, the clusters make sense in terms of historical relationships, but to be consistent with the genetic data, the first group to branch off should be African,

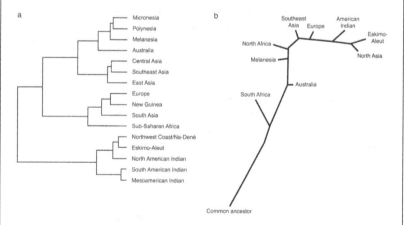

BOX FIGURE 2.6 Mean measures of divergence: (a) calculated among 16 recent human populations based on 22 dental variables (crown and root traits), showing four major divisions; (b) between a hypothetical dental ancestor and nine major geographic groups.

not Native American. What are possible explanations for this unexpected finding? First, many early dental anthropologists focused on Native American teeth because they provided such a contrast to European teeth, which are known for their simplicity and lack of morphological complexity. The traits selected by Turner (and before him, Dahlberg and P. O. Pedersen) have a strong Asia-centric bias (for example, shovel-shaped incisors, incisor winging). Additionally, the traits that characterize Northeast Asian and Native American populations often involve enhanced sculpting of the tooth crowns, and it is possible that natural selection played a role in producing more complex teeth. The inclusion of more Afridont traits (including upper central incisor labial convexity, the Bushman canine, and mid-line diastema) would produce a different dendrogram. Unfortunately, these data are not currently available.

An alternative to using Turner's key crown and root traits, with their potential bias toward Asian and Asian-derived populations, is to adopt a cladistic approach and focus on ancestral and derived crown and root traits. To arrive at an ancestral pattern for dental morphology, all that is required is to evaluate hominoid and early hominin teeth. For example, in 1926, William King Gregory and Milo Hellman noted that Miocene hominoid lower molars had five cusps that showed consistent contact at the central occlusal fossa between the metaconid (cusp 2) and the hypoconid (cusp 3).[6] This produced the well-known Y-5 pattern, which characterizes not only extinct and extant hominoids but also early hominins. Other ancestral states include four-cusped upper molars, two-rooted upper first premolars, three-rooted upper second molars, two-rooted lower first and second molars, and no absence or reduction of the third molar. All told, 13 traits were selected that attained a frequency of either 100 percent or 0 percent, as representing the ancestral state of the earlier hominin dentition. Polymorphic traits such as shovel-shaped incisors and Carabelli's trait were not included in the analysis because an ancestral state cannot be recognized.

Mean measures of divergence were calculated between a hypothetical dental ancestor and nine major geographic groups. The populations that were least derived should be in closest proximity to the ancestor. Groups with the most-derived dentitions would be the farthest removed. For this analysis, an unrooted tree is more appropriate than the classic tree diagram. The unrooted tree, in which all groups emanate from a common ancestor, is shown in box figure 2.6b. Although all modern human populations are far removed from the hypothetical ancestral state, the least-derived group is African, while the second-least-derived group is Australian. The most-derived groups, not surprisingly, are the highly sculpted dentitions of East Asians and Native Americans. Melanesia, North Africa, Southeast Asia, and Europe fall between these two extremes. Although more work on assessing dental morphological patterns among world populations is warranted, these results are consistent with the predictions of the RAO model.

1. Albert A. Dahlberg, "Materials for the Establishment of Standards for Classification of Tooth Characters, Attributes, and Techniques in Morphological Studies of the Dentition" (Zollar Laboratory of Dental Anthropology, University of Chicago, mimeo, 1956).
2. Christy G. Turner II, Christian R. Nichol, and G. Richard Scott, "Scoring Procedures for Key Morphological Traits of the Permanent Dentition: The Arizona State University Dental Anthropology System," in *Advances in Dental Anthropology*, ed. Mark A. Kelley and Clark Spencer Larsen (New York: Wiley-Liss, 1990), 13–31; G. Richard Scott and Joel D. Irish, *Tooth*

(continued)

(continued)

Crown and Root Morphology: The Arizona State University Dental Anthropology System (Cambridge: Cambridge University Press, 2017).

3. Christy G. Turner II, "Advances in the Dental Search for Native American Origins," Acta Anthropogenetica 8 (1984): 23–78; "Dental Evidence for the Peopling of the Americas," National Geographic Society Research Reports 19 (1985): 573–596; and "The First Americans: The Dental Evidence." National Geographic Research 2 (1986): 37–46.

4. Christy G. Turner II, "Dental Evidence on the Origins of the Ainu and Japanese," Science 193 (1976): 911–913, and "Teeth and Prehistory in Asia," Scientific American, February 1989, 88–96.

5. Christy G. Turner II, "Major Features of Sundadonty and Sinodonty, Including Suggestions About East Asian Microevolution, Population History, and Late Pleistocene Relationships with Australian Aboriginals," American Journal of Physical Anthropology 82 (1990): 295–317.

6. William King Gregory and Milo Hellman, The Dentition of Dryopithecus and the Origin of Man, Anthropological Papers of the American Museum of Natural History 28 (New York: American Museum of Natural History, 1926).

Liujiang (southern China) (ca. 68,000–61,000 years ago) is considered problematic.[112] However, a metatarsal bone reported in 2010 from Callao Cave (Philippines)—which indicates an early water crossing—is dated to about 67,000 years ago.[113]

Younger remains are known from northern Eurasia, such as the isolated teeth from Grotta del Cavallo (Italy) and the recently reported leg bone from Ust'-Ishim (western Siberia), both dated to about 45,000 years ago.[114] Isolated teeth from Kostenki (Russia) indicate the presence of modern humans on the central East European Plain before 40,000 years ago.[115] The oldest dated fossils from Australia (Lake Mungo) are about 40,000 years old, but archaeological evidence indicates that modern humans had arrived there at least 10,000 years earlier.[116] All human remains in the Americas postdate 14,000 years (figure 2.5).[117]

Some human remains from southern China that yield dates older than 75,000 years may indicate an earlier Homo sapiens dispersal into tropical Asia. They include isolated teeth from Luna Cave (ca. 127,000–70,000 years old) and a mandible from Zhiren Cave (116,000–106,000 years old).[118] These fossils conceivably represent a Southeast Asian extension of the movement of "near modern" humans into the Levant about 120,000 years ago. The dating and/or taxonomic status of the southern China remains is uncertain, however, and there is no supporting genetic evidence for modern humans in East Asia before 75,000 years ago.[119] An alternative explanation for the fossils is that they represent the local non-modern human population (presumably, the Southeast Asian Denisovans), who were almost certainly

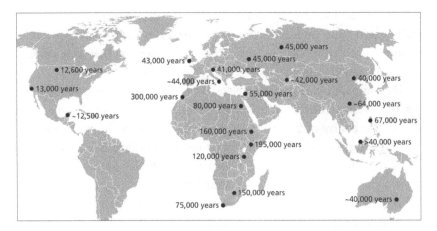

FIGURE 2.5 Map of the world, showing the location and dating of human remains assigned to *Homo sapiens*. (Base map adapted from en.wikipedia.org/wiki/File:BlankMap-World6.svg)

present in this region before the arrival of modern humans, but are known thus far only on the basis of aDNA and isolated skeletal parts from Northeast Asia (that is, no other fossil remains have been assigned to this taxon). Given the close phylogenetic relationship, the skeletal anatomy of the southern Denisovans is likely to be similar to that of *Homo sapiens.*

In addition to providing evidence for the broad pattern and timing of the global dispersal, skeletal materials of modern humans reveal a phenotypic expression of the pattern of variability observed earlier in the genetic data. An analysis of cranial and dental traits among modern humans shows that overall variability is less among people outside Africa, suggesting that—as in the case of the genetics—the non-African population is a subset of the African population.[120]

The skeletal materials also yield evidence for adaptations to local environmental conditions during and after the dispersal. Stable-isotope analysis of *Homo sapiens* bones provides indirect clues to diet, and the isotope values obtained on bones from several sites in northern Eurasia indicate heavy consumption of freshwater foods as early as 45,000 years ago, reinforcing other lines of evidence for an expansion of the ecological niche that reflects innovations in complex technology.[121] The skeletal remains also reveal that—as late as 35,000 years ago—the modern human occupants of northern Europe retained anatomical adaptations to warm climates (for example, high brachial indices) that were presumably inherited from their recent African ancestors.[122] As noted earlier, the dissonance between genetically based traits and local climates underscores the importance of the technological "traits" (for example, insulated clothing) that were created from non-genetic computation.

The comparative analysis of skeletal traits provided the original basis for inferences about interbreeding between modern and non-modern humans during the course of the dispersal, and continues to yield some evidence for this pattern (see box 2.2). The most notable example is a Neanderthal anatomical feature (*horizontal-oval type mandibular foramen*) reported from a modern human mandible found in the Peştera cu Oase (Romania).[123] The mandible recently yielded aDNA confirming its taxonomic status as *Homo sapiens* (and representing one of the basal Eurasian mtDNA lineages), but revealing an exceptionally high amount of Neanderthal DNA (estimated at 6–9 percent of the genome).[124]

The Genetic Evidence

The genetic data offer potentially critical information on the evolution of modern human cognitive faculties in sub-Saharan Africa. While the skeletal remains provide an important clue—evidence that the slow growth and development characteristic of living humans evolved very late in *Homo*—only the genetic data can confirm that the appearance of modern human anatomy coincides with changes in the genome related to the derived and peculiar pattern of brain growth. The genetic changes, in turn, are plausibly tied to syntactic language because it can be acquired only during the extended critical learning period—which apparently did not exist in early *Homo* and may have evolved only with modern human anatomy.

This research was inaugurated quite recently and only limited results are available, but it holds clear potential to address central questions about the evolution of cognition in modern humans: syntactic language and (what appears to be a closely linked faculty for) complex material computation. The key lies in comparative analyses of the genome of modern humans with those of the Neanderthals and Denisovans, because the latter, as noted earlier, presumably lacked the substitutions that underlie the unique pattern of growth and development in *Homo sapiens*.[125] Ironically, the non-modern genome is better known than the modern genome for the time period before 75,000 years ago (before modern human dispersed out of

Africa) because it is based on aDNA from Europe and Siberia.[126] The modern genome is based on DNA from living and recent people, although it now is being supplemented with aDNA from modern humans who colonized higher latitudes after 50,000 years ago.

Analysis of the Neanderthal genome already has revealed the presence of a gene implicated in speech function (FOXP2) that was the subject of earlier speculation about the evolution of cognition in modern humans.[127] A potentially significant set of base substitutions in modern humans that are involved in the organization of the neocortex (organization of neuronal networks) has been identified in *human accelerated region 1* (HAR1). Other genes related to modern human brain development exhibit substitutions or mutations in the recent past. Substitutions in *human accelerated region 2* (HAR2) are related to the developmental biology of the hand, potentially related to material computational complexity.[128]

It was genetic data—the mtDNA study by Cann, Stoneking, and Wilson in 1987—that lifted the debate over the RAO model out of the familiar terrain of arguments based on the interpretation of skeletal traits. In the almost three decades since their publication, a vast body of genetic evidence pertaining to the African origin and global spread of modern humans has accumulated. Much of the recent data are in the form of whole-genome analyses, which represent an exponential increase in the quantitative database. And it was genetic data that lifted the debate over Native American origins out of the familiar terrain of arguments based on archaeological remains (and, to a lesser extent, on linguistics and the interpretation of skeletal traits).

According to current estimates, the most recent common ancestors of all living males and females existed about 190,000 years ago and 180,000 to 112,000 years ago, respectively.[129] The global coalescence estimates significantly antedate the firmly documented presence of modern human anatomy outside Africa. The genetic data also support the RAO model with respect to the overall pattern of diversity in living humans. The pattern of decreasing diversity as a function of distance from Africa is observable in mtDNA, Y-DNA, and autosomal DNA, and it indicates that the non-African populations are a genetic subset of the African groups. The non-African groups are younger and have had less time to diversify (table 2.1).[130]

The genetics of living humans provide a general model for the global dispersal, including the dispersal in Africa.[131] The movements of various

TABLE 2.1

Estimated TMRCA for Maternal and Paternal Lineages and Their Inferred Relationship to Major Events in the Origin and Dispersal of Modern Humans

Maternal Lineages		Paternal Lineages		Event
mtDNA	Estimated Coalescence Age	Y-DNA	Estimated Coalescence Age	
Modern human/ Neanderthal lineages	389,000 years ago	Modern human and Neanderthal lineages	Ca. 588,000 years ago	Divergence of modern humans and Neanderthals/ Denisovans
All lineages	143,000 years ago	All lineages	Ca. 190,000 years ago	Origin of modern humans in Africa
Haplogroup L3	72,000 years ago	Haplogroups DE and CF	Ca. 76,000 years ago	Out-of-Africa migration of modern humans
Haplogroup P	56,770 years ago	Haplogroup C-M130	Ca. 75,000 years ago	Settlement of Sahul (Australia/ New Guinea)
Haplogroup N	Ca. 51,000 years ago			Early dispersal in Eurasia
		Subclade HO	Ca. 51,000 years ago	Lineage diversification in South Asia
Haplogroups U5 and U6	50,380 years ago			Early population expansion in Europe
Haplogroup M	Ca. 49,000 years ago			Early dispersal in Eurasia
Subclades A2, C1, and D1	Ca. 31,000 years ago			Divergence of Native American founding lineages from Asian parent populations
Subclades H1 and H3	Ca. 18,000 years ago			Post–Last Glacial Maximum reoccupation of Europe

Maternal Lineages		Paternal Lineages		Event
mtDNA	Estimated Coalescence Age	Y-DNA	Estimated Coalescence Age	
Subclades B4a1a1a B4a1a1a1	15,830 years ago			Settlement of Oceania
		Haplogroup Q1a-M3	Ca. 15,000 years ago	Initial dispersal in the Americas

Sources: Based on Qiaomei Fu et al., "A Revised Timescale for Human Evolution Based on Ancient Mitochondrial Genomes," *Current Biology* 23 (2013): 555, table 3; Adrien Rieux et al., "Improved Calibration of the Human Mitochondrial Clock Using Ancient Genomes," *Molecular Biology and Evolution* 31 (2014): 2787, table 2; Fernando L. Mendez et al., "The Divergence of Neandertal and Modern Human Y Chromosomes," *American Journal of Human Genetics* 98 (2016): 728–734; Cosimo Posth et al., "Pleistocene Mitochondrial Genomes Suggest a Single Major Dispersal of Non-Africans and a Late Glacial Population Turnover in Europe," *Current Biology* 26 (2016): 827–833; G. David Poznik et al., "Punctuated Bursts in Human Male Demography Inferred from 1,244 Worldwide Y-Chromosome Sequences," *Nature Genetics* (2016): 3, fig. 2.

modern human groups (represented by genetic markers or haplogroups) have been reconstructed on the basis of estimated divergence times from their parent groups (calculated from substitutions in mtDNA and Y-DNA) and their recent geographic distribution. For the maternal lineages, the starting point is mtDNA haplogroup L0, whose living representatives, the *Khoisan* hunter-gatherers of southern Africa, exhibit the highest genetic diversity in the world population. The three major African subgroups (L1–L3) are derived from L0 and include the *Mbuti* (pygmies) of Central Africa (L1), much of the *Bantu*-speaking population of sub-Saharan Africa (L2), and the youngest subgroup, which is also represented in the Middle East (L3) (figure 2.6).

The coalescence date (or "time to the most recent common ancestor" [TMRCA]) of mtDNA haplogroup L3 recently was calculated (with improved calibration based on dated aDNA) at about 72,000 years ago (93,000–54,000 years ago). This is the best current estimate for the beginning of the dispersal out of Africa (and divergence of the three basal Eurasian maternal lineages: mtDNA haplogroups M, N, and R).[132] For the paternal lineages, the basal haplogroup is Y-DNA A ("African Adam"), and the ancestral Eurasian lineage is Y-DNA CF-P143, with an estimated TMRCA of around 76,000 years ago.[133]

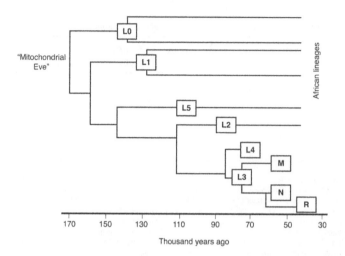

FIGURE 2.6 The primary maternal lineages (mtDNA haplogroups), beginning with "Mitochondrial Eve" and the early African lineages (L0–L5), and including the source group (L3) for the out-of-Africa dispersal of lineages between 75,000 and 60,000 years ago and the earliest Eurasian lineages (M, N, and R).

The initial migration into Eurasia seems to have been eastward across southern Asia and rapidly into Australia (mtDNA M, N, and R) before 50,000 years ago.[134] The movement into Europe seems to have taken place about 50,000 years ago, and recent aDNA research reveals that the same maternal lineages (mtDNA M, N, and R) were present before 40,000 years ago (although mtDNA M is absent among living Europeans).[135] Daughter groups of haplogroups N and R (haplogroup F and haplogroups A and B) moved into East Asia at roughly the same time, and A and B are among the lineages that probably first occupied Beringia, but only later dispersed throughout the Americas (ca. 15,000 years ago).[136] The dispersal pattern indicated by Y-DNA is similar, although not identical, to that based on mtDNA data.[137]

Analysis of aDNA extracted from dated human skeletal remains offers the most important means of ground-truthing the global dispersal, especially in northern Eurasia and North America, where cooler climates are more likely to have preserved DNA molecules. The potential of even very small fragments of bone to generate a complete genome from aDNA is encouraging; during the next decade, most analyses probably will be performed on isolated fragments that may not even be assignable to *Homo sapiens* on the basis of anatomy. The recovery of nonhuman aDNA from sediment—rather than from bones and teeth—suggests that human sedimentary ancient DNA (sedaDNA) may be found in the same context, especially in permafrost regions.[138] An important factor in dating aDNA related to the dispersal is the recent application of new pretreatment techniques for radiocarbon dating of old (more than 30,000 years) samples of wood charcoal and bone, as well as the development of reliable calibration curves.[139]

The number of modern human specimens analyzed for aDNA is growing rapidly, and includes recently reported remains from Italy (ca. 41,100–38,500 years ago) and western Siberia (ca. 45,000 years ago), both assigned to mtDNA haplogroup R.[140] Both mtDNA haplogroups M and N have been identified in aDNA from Europe that dates to between 40,000 and 35,000 years ago.[141] Whole-genome analyses have been performed on samples from Europe and Siberia. A recently reported study of remains from northern China (42,000–39,000 years ago) documents the appearance of mtDNA B.[142] Within the next few years, the dispersal of modern humans in the Northern Hemisphere probably will be based primarily on dated aDNA, rather than on the model generated from the analysis of DNA in living people.

In North America, aDNA has been recovered from all the major Native American mtDNA and Y-DNA haplogroups in remains that date to 10,000 years ago or earlier. Maternal lineages A and B have been reported from human coprolites in Oregon (ca. 14,340 years ago), while skeletal remains from a site in Nevada (ca. 10,560–10,250 years ago) were assigned to mtDNA haplogroup C.[143] The analysis of the Anzick skeleton from Montana (12,700–12,560 years ago) indicates that it represents maternal lineage D (and paternal lineage Q-L54), while the lesser known mtDNA haplogroup X has been identified from slightly younger remains in Washington (ca. 8,500 years ago).[144]

The genetic data also are yielding information on population size and growth rates, which are important variables in the global dispersal. Estimates of population size—which have been generated from mtDNA, Y-DNA, X-chromosome DNA (X-DNA), and other autosomal DNA—consistently yield very low numbers for effective population size (breeding population) at the time of the dispersal (see box 7.1). By one estimate, the global dispersal was launched by an effective population of only about 1,000 individuals.[145] If accurate, such estimates help account for the overall low genetic variability in the global population that Lewontin first noted in

1972. They also help account for the striking phenotypic variations that evolved in local populations during the postglacial epoch (that is, modern geographic or racial variation). Although some of this variation reflects the long-term influence of local environmental factors, such as solar-radiation intensity and average winter temperatures, much of it probably is a consequence of genetic founder's effect (possibly reinforced by sexual selection) intensified by the rapid spread of a small population.[146]

Finally, as discussed earlier, the genetic data provide definitive evidence of interbreeding and genic exchange among modern humans, Neanderthals, and Denisovans during the dispersal of *Homo sapiens* (see box 2.2). The issue was formerly addressed primarily on the basis of skeletal anatomy and was controversial, reflecting the uncertainties of interpreting the anatomical evidence. In addition to shedding light on the role of interbreeding with non-modern humans during the global dispersal, traces of genetic admixture in living people and aDNA samples are—as Mark Stoneking observed—a "useful marker of population relationships."[147] Like the genetic markers for the maternal and paternal lineages, the introgression from Neanderthals and Denisovans helps reconstruct the timing and routes of the dispersal.[148] It also may provide a means for evaluating the genetic contribution of non-modern groups to geographic or "racial" variation of among recent and living people (box 2.7).

The Archaeological Evidence

Archaeological data are critical to understanding the origins and spread of *Homo sapiens*. Only artifacts and features provide clues about the cognitive faculties unique to modern humans. As with the changes in skeletal anatomy, this evidence emerged first in Africa. And only archaeological data shed light on the organizational and technological innovations that allowed modern humans to occupy most of the available terrestrial habitats on Earth. More generally, the facts of the archaeological record support the thesis that the evolution and dispersal of *Homo sapiens* may be understood as a "major transition" in evolution that entailed fundamental changes in the storage, transmission, and translation of (non-genetic) information. The transition also manifests increased complexity of computation—both with non-genetic information and with materials. Modern humans created new forms of non-genetic information.

BOX 2.7
Recent African Origin, Global Dispersal, and the Problem of Human
Geographic "Races"

The recent African origin and global dispersal of *Homo sapiens* is highly relevant to long-standing controversies about the emergence of geographic "races" among living people—and even to the question of their existence.[1] Charles Darwin (who referred skeptically to "the so-called races of man") viewed biological variation among living people as relatively superficial, and in 1871 he explicitly tied this observation to the idea that humans evolved their defining characteristics in one place before they "spread over the earth" (chapter 1).[2]

A century later, the evolutionary geneticist Richard Lewontin reported that most genetic variation in living people is *within* rather than between geographic groups, and he questioned the value of "race" as a biological concept.[3] Lewontin's work was challenged in 2003 by A. W. F. Edwards, who noted that a correlation matrix and principal components analysis—as opposed to a locus-by-locus analysis of the data—yielded groups that correspond to traditional "racial" classifications.[4] The critique did not dispute Lewontin's central conclusion—that most variation is within, not between, populations—but underscored that some genetic variation exhibits a geographic pattern. The sequencing of DNA samples (with a focus on short tandem repeats) reveals the same pattern, yielding five continental clusters (Africa, Europe, East Asia, Australasia, and the Americas).[5] What is the source of this variation, and what does it mean? Does it warrant the subdivision of the global population into geographic "races"?

The genetic data indicate that all living humans are descended from one female and male who lived in Africa less than 200,000 years ago (that is, less than 8,000 generations ago). After 75,000 to 60,000 years ago, a subset of the African population spread rapidly across Eurasia and Australia, and subsets of this population later dispersed in the Americas and Oceania. Both the scope and the speed of the dispersal were a function of an unprecedented capacity for rapid adaptation to novel habitats. Like a Russian *matroshka* doll, each successive phase of the dispersal produced a less diverse genetic sample or subset of its source population. The impact on the human genome was a pronounced founder's effect on local subpopulations (that is, random variations in gene frequencies as a result of "sampling error"). Many of the local populations became isolated or partially isolated by natural barriers (for example, mountain ranges or bodies of water) and, eventually, language and cultural barriers.

The presence or absence of many selectively neutral traits in living human subpopulations (for example, epicanthic fold) is likely to reflect random variation in the original dispersing group, rather than genetic adaptation to local conditions.[6] Variations in the size and shape of the nose, which do not appear to possess any adaptive value in the context of specific habitats or climates, but is a common feature in differentiating "races," is another example. Sexual selection, which Darwin suspected to have been a significant factor in human geographic variation, most probably has played a role in increasing or decreasing some of these traits in regional populations.

Over sometimes extended periods of time, regional populations evolved genetic adaptations to local conditions, including pathogens, ultraviolet radiation, and effective temperature. These adaptations, especially to ultraviolet radiation (skin pigmentation) and temperature (body size and shape), have played a major role in perceived

(continued)

racial differences, but they are clinal—primarily latitudinal variation. The discontinuities are obvious only where subpopulations from different latitudinal zones are brought together by historical circumstances, and latitudinal gradients in skin pigmentation are found in Asia and Africa, irrespective of the pattern of continental genetic clusters. These traits sometimes contribute to the pattern of geographic variation (for example, Africa versus Europe) but are controlled by independent variables.

Finally, new data on the Neanderthal and Denisovan genomes indicate that—despite the evidence of interbreeding between these taxa and *Homo sapiens*—genetic traits in local non-modern populations are unlikely to have been a significant source of "racial" variation in living humans. Although an exception to this may be several genotypes from both Neanderthals and Denisovans that provide immunity against bacteria, fungi, and parasites, the distribution of these traits does not correspond to that of the major geographic "races."[7] In cases where both local non-modern and living human populations share traits, such as short distal extremities and red hair in Europe, there is evidence that they evolved independently in modern humans (see box 2.2).[8]

In sum, the pattern of genetic variation in living people that corresponds to the traditionally defined major "races" is tied in part to the founder's effect on local subpopulations of rapid global dispersal. The random variations in some of these gene frequencies likely were altered by sexual selection in local marriage networks. Adaptation to pathogens and climate conditions contributed to geographic variation, but most of the variation was clinal and often cut across the other patterns of regional variation. The contribution of local non-modern populations was minor (and also cut across traditional "racial" boundaries). Even though regional or continental clusters of traits correspond to traditional "racial" classifications, Darwin's skepticism about the significance of geographic variation in modern humans was justified.

1. Ian Tattersall and Rob DeSalle, *Race? Debunking a Scientific Myth* (College Station: Texas A&M University Press, 2011), 101–143.
2. Charles Darwin, *The Descent of Man and Selection in Relation to Sex* (London: Murray, 1871), 1:240–250.
3. Richard C. Lewontin wrote that "racial classification is now seen to be of virtually no genetic or taxonomic significance" ("The Apportionment of Human Diversity," *Evolutionary Biology* 6 [1972]: 397).
4. A. W. F. Edwards, "Human Genetic Diversity: Lewontin's Fallacy," *BioEssays* 25 (2003): 798–801.
5. See, for example, Noah Rosenberg et al., "Genetic Structure of Human Populations," *Science* 298 (2002): 2381–2385. See also Eran Elhaik et al., "Geographic Population Structure Analysis of Worldwide Human Populations Infers Their Biogeographical Origins," *Nature Communications* 5 (2014): 3513; and Nicholas Wade, *A Troublesome Inheritance: Genes, Race, and Human History* (New York: Penguin Press, 2014), 95–122.
6. Tattersall and DeSalle, *Race?*, 151.
7. Michael Dannemann, Aida M. Andrés, and Janet Kelso, "Introgression of Neandertal- and Denisovan-like Haplotypes Contributes to Adaptive Variation in Human Toll-like Receptors," *American Journal of Human Genetics* 98 (2016): 22–33. The genotypes associated with defense against local pathogens are widespread in non-African populations (that is, all but one of the defined geographic "races" of living people).
8. Modern humans in Europe before 30,000 years ago exhibited the longer distal extremities found in low-latitude populations (see, for example, John F. Hoffecker, *Desolate Landscapes: Ice-Age Settlement in Eastern Europe* [New Brunswick, N.J.: Rutgers University Press, 2002], 153–158), while the gene for red hair in living Europeans is not the same gene for this feature found in Neanderthal aDNA (Carles Lalueza-Fox et al., "A Melanocortin 1 Receptor Allele Suggests Varying Pigmentation Among Neanderthals," *Science* 318 [2007]: 1453–1455).

The skeletal anatomy of modern humans is associated with a fundamental change in the archaeological record. The emergence and spread of modern humans were accompanied by the appearance of entirely *new categories of artifacts*: (1) objects for material social display (for example, personal ornaments and decorated items), (2) visual art (three-dimensional sculptures, two-dimensional engravings, drawings, and paintings), (3) objects or features exhibiting digital notational information (engraved or drawn/painted), and (4) musical instruments (initially simple wind instruments, such as pipes). The artifacts of social display may have been preceded by earlier forms, which conceivably include the large stone bifaces (*hand axes*) produced by several non-modern species of *Homo*, although this interpretation remains controversial and seemingly untestable.[149] The regional stylistic variation in small bifaces and points in Africa after 300,000 years ago probably reflects the injection of social display into stone-tool production (these objects served economic and social ends) (table 2.2).[150]

The appearance of *new forms of information* is the most striking aspect of the archaeological record of modern humans. Beyond the "dance language" of the honeybee (which translates information into body movements) and the predator-specific alarm calls of some nonhuman primates

TABLE 2.2
Archaeological Record of the Genus *Homo*, Before 30,000 Years Ago

Species	Tools	Display	Visual Art	Music	Notation
Homo sapiens	Mechanical tools and weapons (plus automata)	Personal ornaments and decorated objects	Sculpture, paintings, and engravings	Wind instruments	Digital notation (engraved)
Homo neanderthalensis	Composite tools and weapons	Personal ornaments (?)			
Homo heidelbergensis	Composite tools and weapons	Large bifaces (?)			
Homo ergaster/erectus	Large bifaces	Large bifaces (?)			
Homo habilis/rudolfensis	Choppers and flake tools				

(which match sound and meaning), only the hypothesized early—and presumably non-syntactic—language of early *Homo* represents a form of information in the biological world that is neither genetic nor neuronal. The arrival of *syntactic language* (a more complex version of non-syntactic language) is unlikely to have postdated the TMRCA for all living human male and female lineages (ca. 190,000–145,000 years ago) and may have coincided with the evolution of delayed growth and maturation in *Homo sapiens*. The sudden appearance of *visual art* (analog) and digital *notation*—neither of which have any parallel among other animals—is another major event in evolution. But in postglacial times and quite recently, modern humans have created other novel forms of information, including written, electric, and electronic symbols.

The artifacts of modern humans exhibit an unprecedented level of computational and structural/functional complexity (see box 2.5). The pattern is manifest especially in visual art and technology, but eventually is evident in all categories of artifacts and features. By 35,000 years ago, both two- and three-dimensional art reflected a complex structure (up to three or more hierarchical levels) (figure 2.7). Comparable complexity in technology is evident by 40,000 years ago, with the appearance in northern Eurasian

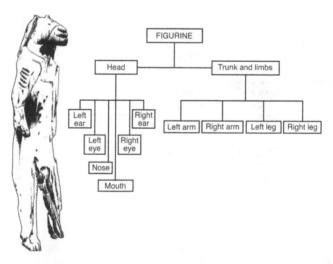

FIGURE 2.7 The reassembled *Löwenmensch* figurine from the site of *Hohlenstein-Stadel* (Germany), dating to roughly 35,000 years ago (*left*), and its hierarchical organization, which exhibits more than three levels (*right*). Visual art yields the earliest direct evidence of complex hierarchical organization (three or more levels) in the artifacts and features of modern humans.

archaeological sites of eyed needles, which indicate tailored insulated clothing (also implied by the presence of modern humans in areas where mean January temperatures would have fallen far below 0°C).[151] Tailored insulated clothing among recent hunter-gatherers exhibits a complex hierarchical structure with three or more organizational levels. Before modern humans, the most complex known examples of technology are the composite tools and weapons of *Homo heidelbergensis* and the Neanderthals, which comprised at least three components (each of a different material), two hierarchical levels, and several subassemblies (chapter 4).

The technology of modern humans is unique among that of all animals, including apparently all earlier humans, because it contains examples of *mechanical technology*, or artifacts and devices with moving components. Mechanical technology performs part of the material computation during its use or operation, and modern humans also created devices to operate or "compute" materials automatically or without the movement of the human body (machines or automata) (see box 4.4). The earliest evidence for automata is faunal remains from *Sibudu Cave* (KwaZulu-Natal), dating to around 65,000 to 62,000 years ago, that suggest the possible use of snares.[152] There is evidence for bow and arrow from another South African site (Border Cave) by about 44,000 to 42,000 years ago, and fire drills probably were made at the time that modern humans first invaded northern Eurasia (before 45,000 years ago).[153]

The earlier archaeological record of modern humans reveals that the new artifact categories and evidence of increased complexity in the manipulation of information and materials did not appear immediately and concurrently, but gradually and sequentially. Examples of personal ornaments in the form of perforated marine-shell beads were present as early as around 135,000 years ago in North Africa (and the adjoining Levant, briefly occupied by *Homo sapiens* at the time).[154] The ornaments also do not necessarily reflect more complex material computation—they may have been hierarchically organized as multiple necklace strings with other attachments—but they may indicate an intensification of alliance networking. If so, this conceivably was tied to an enhanced capacity for computation related to social interactions—the complex manipulation of relationships.

Natural objects decorated with engraved geometric designs include lumps of red ocher and fragments of ostrich eggshell from sites in South

Africa that date to 75,000 and 60,000 years ago, respectively.[155] They conceivably represent a new form of semantic information, but are more parsimoniously interpreted as examples of decorative design, analogous to the trim on a building or an automobile, devoid of any meaning or symbolism. As such, they may be lumped into the same broad category of social-display items with the personal ornaments. They are a non-genetically based version of the patterns of body display that have evolved in many animals (for example, stripes on a tiger), and, in addition to artifacts, it may be assumed that modern humans decorated their bodies with tattoos and designs painted with mineral pigment, charcoal, and clay before 60,000 years ago.

Archaeological evidence for new types of tools and weapons also emerged first in Africa. Barbed bone harpoons that date to about 90,000 years ago were reported from East Africa, although the accuracy of the dates has been questioned. Bone awls and points were reported from the same site in South Africa that produced the 75,000-year-old pieces of red ocher decorated with geometric designs. And—in addition to the indirect evidence for snares (ca. 65,000–62,000 years ago)—the decorated fragments of ostrich eggshell from South Africa (ca. 60,000 years old) are assumed to be shards of water containers or canteens.[156]

There are other indications of technological innovation and complexity in a comparatively early context. They include geometric microliths—small stone flakes or blades retouched into the shapes of triangles, crescents, and trapezes—dating to as early as 80,000 years ago in South Africa. Moreover, these artifacts represent a relatively brief and localized phenomenon, suggesting a sudden burst of innovation that is not characteristic of nonmodern toolmaking. The geometric pieces probably were parts of composite tools or weapons—possibly even arrows or throwing-dart points.[157] There is evidence of freshwater fishing (before 100,000 years ago), which suggests that some new technology had been developed to acquire resources that had been inaccessible, as well as indirect evidence (composition of fauna) for snares from South Africa.[158]

Examples of visual art and musical instruments, as well as notation, do not materialize in the archaeological record until after 50,000 years ago. A carved piece of ivory that dates to about 45,000 years ago from Russia (Kostenki 14) probably is a fragment from a small human figurine, while rock art—recently reported from Southeast Asia (*Sulawesi*)—is dated to roughly 40,000 to 35,000 years ago.[159] Several well-known human figurines from western Europe, including the famous *Löwenmensch* sculpture, date to about 35,000 years ago (see figure 2.7). The earliest known examples of notation and musical instruments (also from western Europe) are from the same time period.

The gradual and sequential appearance of the various categories of artifacts and features may indicate a pattern of development or expression of modern human cognition. Alternatively, it may reflect—at least in part—variations in the archaeological visibility of the artifact categories. These variations are likely to represent, in turn, differences in both materials and setting. More than 75,000 years ago (before modern humans dispersed out of Africa), visual art may have been created entirely from wood and thus has not been preserved, for example. Most of the artifacts and features just mentioned were discovered only recently (since 2000), and new finds could alter the picture significantly in the near future.

Conversely, the early appearance of artifacts related to social display and alliance networking and/or the seemingly late development of mechanical technology may reveal something about the evolution of modern human cognition. It also may reflect the "major transition" in evolution and fundamental changes in information discussed in chapter 1. If, as argued here, increased computational complexity in *Homo* was driven primarily by the cognitive challenges of manipulating multiple social relationships to individual advantage, artifacts related to social display are a logical, if not predictable, material manifestation of enhanced computational complexity in an animal already producing a variety of artifacts. The use of materials for social display conceivably preceded and stimulated broader applications of material computation.

If the application of increased computational complexity to technology was gradual and delayed, it could explain the success (and speed) of the global dispersal out of Africa, which began 75,000 to 60,000 years ago, compared with the limited and ultimately unsuccessful dispersal of "near modern" humans that took place between 120,000 and 90,000 years ago. The "near modern" humans may have lacked the material computational faculty to develop the complex technologies—including mechanical tools and weapons and automata—required for colonizing environments where temperatures were low and resources scarce.

Archaeological data shed much light on the global dispersal of *Homo sapiens*, although uncertainties persist in assigning specific remains to modern or non-modern humans in various parts of the world. In some region, modern humans were the first hominin species to occupy the landscape, and artifacts and other debris are invariably the earliest indications of their presence (given the scarcity of human skeletal remains). Thus, for example, the oldest known archaeological sites in Australia/New Guinea (ca. 50,000 years old) and South America (ca. 14,000 years old) represent the initial dispersal into these regions. In both cases, the dating of the sites is consistent with the estimated time of arrival based on the calculated divergence time of mtDNA and Y-DNA haplogroups (box 2.8).[160]

In parts of northern Eurasia, including both eastern Europe and southern Siberia, the initial movement of modern humans into various regions may be represented by an archaeological proxy that has yet to yield diagnostic skeletal remains (that is, has yet to be confirmed as a proxy for *Homo sapiens*). In parts of central and eastern Europe, archaeological sites that date to around 45,000 years ago contain stone points and blades that were struck from prepared or Levallois cores that are similar to an industry found in the Levant at this time and suspected to be the product of modern humans.[161] However, there are no diagnostic human skeletal remains, and the artifacts are similar to some made by non-modern groups in an earlier context. The same pattern is apparent in southern Siberia: these sites are plausible archaeological proxies for dispersing groups of modern humans, but they have yet to be confirmed as such.

Throughout Eurasia, there are archaeological sites that contain artifacts diagnostic of anatomically and cognitively modern humans in the form of ornaments, visual-art objects, and complex technologies. Stone bladelets and microblades may be added to this list; although their production is seemingly within the technological capacity of various non-modern groups, they appear confined to modern human sites in both Africa and Eurasia and may be considered diagnostic of *Homo sapiens*. However, sites containing diagnostic artifacts are rarely the earliest known manifestations of modern humans, and, in many cases, they seem to postdate their arrival.

The archaeological data also are widely believed to yield information on "cultural" assimilation between modern humans and their non-modern contemporaries, especially the Neanderthals. The pattern offers a parallel to the evidence of interbreeding provided by genetic data (see box 2.2). In theory, evidence of cultural assimilation would consist of artifact assemblages that contain a mixture or combination of forms that are diagnostic

BOX 2.8
Dating the Global Dispersal of Modern Humans

Karl W. Butzer's application of *uranium-thorium* (U/Th) dating to the crania from Omo-Kibish provided the basis for the Recent African Origin model.[1] U/Th dating is one of a number of radiometric (or isotopic) dating methods applied to fossils of modern humans from Africa that antedate the global dispersal. Radiometric dating is based on the measurement of radioactive decay of naturally occurring unstable isotopes of certain elements (either into other isotopes of the same element or into another element). Because radioactive decay occurs at a fixed rate, the amount of decay is a function of the time elapsed since the entity or deposit containing the unstable isotope was formed. The elapsed time is estimated from the ratio of the unstable parent isotope that remains to its daughter isotope products.

In the case of U/Th, the unstable thorium isotope ^{230}Th decays into ^{234}U over about 500,000 years. The decay of two unstable uranium isotopes (^{238}U and ^{235}U) also can be measured, and all three methods are included in the category of *uranium-series* (U-series) dating. U-series dating can be applied to marine and lake deposits, as well as to aquatic shells. Another category of radiometric methods applied to relatively old fossils of modern and non-modern humans (or their depositional context) is *potassium/argon* dating, which includes measurement of the decay of the unstable potassium isotope ^{40}K into the stable argon isotope ^{40}Ar and measurement of the ratio of ^{40}Ar to ^{39}Ar to estimate the K/Ar ratio (*argon/argon* dating). Potassium/argon dating has a time range that exceeds the age of Earth and can be applied to volcanic rock (therefore useful for dating volcanic ashes associated with human fossils).[2]

The radiometric method applied most widely to the dating of the dispersal of modern humans out of Africa is *radiocarbon*, which measures the decay of the unstable carbon isotope ^{14}C to the stable isotope ^{12}C. If the organic fraction of bone is preserved, human skeletal remains may be dated directly with radiocarbon. Associated nonhuman bone or shell, wood charcoal (for example, from a hearth), and the organic component of associated soil also may be dated with this method. The principal limitation of radiocarbon dating is that ^{14}C decays relatively quickly, and samples retain very little to measure after 40,000 years (and may be contaminated by minute quantities of younger carbon). Accordingly, dating the early phases of the dispersal into Eurasia and Australia/New Guinea with this method is problematic.

In recent years, improvements in radiocarbon methods have extended its effective range to about 50,000 years, allowing dating of all but the earliest phases of the global dispersal.[3] More rigorous pretreatment in the lab has reduced the potential for contamination from younger carbon in older samples. Wood-charcoal samples may be prepared with the *acid-base oxidation-stepped combustion* (ABOx-SC) protocol, which was used to date samples of known age from the tephra that resulted from the *Campanian Ignimbrite* (CI) volcanic eruption.[4] For the dating of bone, which is especially susceptible to contamination from younger carbon, the collagen fraction—if adequately preserved—may be subjected to *gelatin extraction/ultra-filtration* to remove potential contaminants.[5]

Two other improvements of recent years include the widespread use of *accelerator mass spectrometry* (AMS) for dating radiocarbon samples, and the development and

(continued)

(continued)

application of reliable calibration curves. AMS measurement of samples entails the direct counting of ^{14}C and ^{12}C atoms and is more precise than the earlier method (gas proportional and liquid scintillation counting). The accurate calibration of dates is necessary because past fluctuations in atmospheric radiocarbon affect the initial quantity of ^{14}C in samples. Depending on the time period, the discrepancy between radiocarbon years and calendar years may be several millennia.[6]

Even the most effective techniques of pretreatment and measurement are useless on radiocarbon samples older than 50,000 years (the ^{14}C has completely decayed), and the earliest dates on the migration of modern humans out of Africa have been obtained by other methods. These procedures are less precise than radiocarbon dating and suffer from other potential uncertainties. Luminescence dating, which includes both *thermoluminescence* (TL) and *optically stimulated luminescence* (OSL), measures the number of electrons that accumulate in the absence of sunlight as a function of time in various minerals such as quartz and feldspar. It has been applied to sediments in many archaeological sites older than 50,000 years.[7] Another method is *electron spin resonance* (ESR), which also measures the accumulation of electrons in crystals and is most often applied to tooth enamel.[8]

Widespread *chrono-stratigraphic markers* have helped date the dispersal of modern humans. The CI tephra, which is firmly dated to about 40,000 years ago, has been used to calculate some of the earlier archaeological traces of modern humans in Italy, the

BOX FIGURE 2.8 The stratigraphic profile of Shlyakh, located in southern Russia near the Don River and excavated by P. E. Nehoroshev and L. B. Vishnyatsky, showing radiocarbon dates on ultra-filtered collagen and soil organics (humate fraction), as well as the position of the Laschamp Paleomagnetic Excursion. (Photograph by the author [2013])

Balkans, and eastern Europe (reflecting the massive ash plume that spread eastward from its source near the Bay of Naples).[9] In southern Asia, the volcanic ash from the Toba eruption, which occurred about 74,000 years ago, has been used (in conjunction with OSL) to date artifacts that some archaeologists attribute to modern humans.[10] Paleomagnetic stratigraphy (or *paleomagnetism*), which is based on past fluctuations in the intensity and direction of the magnetic field of Earth, also contains widespread chrono-stratigraphic markers, including the *Laschamp Paleomagnetic Excursion* (ca. 41,000 years ago).[11]

The site of *Shlyakh* (southern Russia), which contains a stone-artifact industry that may be a proxy for modern humans in northern Eurasia, provides an example of the application of multiple methods to dating the dispersal. Analysis of the sediments at Shlyakh indicated that the artifacts are buried in low-energy stream deposits. No charcoal was found at the site, and large-mammal bones (primarily bison) associated with the artifacts were poorly preserved. However, sufficient collagen was found in several bones for gelatin extraction/ultra-filtration of samples for AMS dating. Two samples of soil organics recovered from a buried soil overlying the main occupation levels also were dated by radiocarbon. Earlier work at Shylakh identified the Laschamp Paleomagnetic Excursion in sediments containing the lowermost artifacts.[12] The main occupation levels appear to date to between 45,000 and 30,000 years ago (box figure 2.8).

1. Karl W. Butzer, "The Mursi, Nkalabong, and Kibish Formations, Lower Omo Basin, Ethiopia," in *Earliest Man and Environments in the Lake Rudolf Basin: Stratigraphy, Paleoecology, and Evolution*, ed. Yves Coppens et al. (Chicago: University of Chicago Press, 1976), 12–23.
2. Radiometric and non-radiometric dating methods applied to the study of human evolution are described in Richard G. Klein, *The Human Career: Human Biological and Cultural Origins*, 3rd ed. (Chicago: University of Chicago Press, 2009), 22–54.
3. Jim Allen and James F. O'Connell, "Both Half Right: Updating the Evidence for Dating First Human Arrivals in Sahul," *Australian Archaeology* 79 (2014): 86–108.
4. M. I. Bird et al., "Radiocarbon Dating of 'Old' Charcoal Using a Wet Oxidation, Stepped-Combustion Procedure," *Radiocarbon* 41 (1999): 127–140; Katerina Douka, Thomas Higham, and Andrey Sinitsyn, "The Influence of Pretreatment Chemistry on the Radiocarbon Dating of Campanian Ignimbrite–Aged Charcoal from Kostenki 14 (Russia)," *Quaternary Research* 73 (2010): 583–587.
5. Will Beaumont et al., "Bone Preparation at the KCCAMS Laboratory," *Nuclear Instruments and Methods B* 268 (2010): 906–909.
6. See, for example, Bernhard Weninger and Olaf Jöris, "A ^{14}C Age Calibration Curve for the Last 60 ka: The Greenland-Hulu U/Th Timescale and Its Impact on Understanding the Middle to Upper Paleolithic Transition in Western Eurasia," *Journal of Human Evolution* 55 (2008): 772–781.
7. Steven L. Forman, James Pierson, and Kenneth Lepper, "Luminescence Geochronology," in *Quaternary Geochronology: Methods and Applications*, ed. Jay Stratton Noller, Janet M. Sowers, and William R. Lettis (Washington, D.C.: American Geophysical Union, 2000), 157–176.
8. B. A. Blackwell, "Electron Spin Resonance Dating," in *Dating Methods for Quaternary Deposits*, ed. N. W. Rutter and N. R. Catto (St. John's, Newfoundland: Geological Association of Canada, 1995), 209–268.
9. Francesco G. Fedele, Biagio Giaccio, and Irka Hajdas, "Timescales and Cultural Process at 40,000 BP in the Light of the Campanian Ignimbrite Eruption, Western Eurasia," *Journal of Human Evolution* 55 (2008): 834–857; John F. Hoffecker et al., "From the Bay of Naples to the River Don: The Campanian Ignimbrite Eruption and the Middle to Upper Paleolithic Transition in Eastern Europe," *Journal of Human Evolution* 55 (2008): 858–870.
10. Michael Petraglia et al., "Middle Paleolithic Assemblages from the Indian Subcontinent Before and After the Toba Super-Eruption," *Science* 317 (2007): 114–116.
11. A. H. L. Voelker et al., "Radiocarbon Levels in the Iceland Sea from 25 to 53 kyr and Their Link to the Earth's Magnetic Field Intensity," *Radiocarbon* 42 (2000): 437–452; R. Traversi et al., "The Laschamp Geomagnetic Excursion Featured in Nitrate Record from EPICA-Dome C Ice Core," *Scientific Reports* 6 (2016): 20235.

(continued)

(continued)

12. P. E. Nehoroshev, "Rezul'taty Datirovaniya Stoyanki Shlyakh," *Rossiiskaya arkheologiya* 3 (2006): 21-30. Shlyakh was excavated by P. E. Nehoroshev and L. B. Vishnyatsky (both Russian Academy of Sciences) in 1990/1991 and from 1998 to 2001. A new stratigraphic profile was recorded in 2013 by Vance T. Holliday (University of Arizona), and analysis of soil-micromorphology thin sections was performed by Paul Goldberg (Boston University). The radiocarbon dates on bone collagen were obtained by John R. Southon (University of California, Irvine) and on soil organics by Scott J. Lehman (University of Colorado, Boulder). A series of OSL dates was obtained on sediments collected by the author from various levels at the site in 2013 by Steven L. Forman (Baylor University), but the age estimates were discordant with the other dating results and have been discounted.

of modern humans and others. It might even consist of individual artifacts that comprise attributes of both. The archaeological evidence for assimilation is problematic, however; in each case, there are plausible alternative explanations for patterns attributed to some form of cultural exchange.

The best-known example is the *Châtelperronian* industry of the Franco-Cantabrian region, which is found in several cave sites and contains both stone tools that are characteristic of assemblages produced by local Neanderthals (side scrapers and denticulates) and artifacts that are diagnostic of modern humans (ornaments and bone tools). At two sites, diagnostic Neanderthal skeletal remains are associated with Châtelperronian assemblages, leading many to conclude that the industry reflects a flow of cultural traits from in-migrating modern humans to local Neanderthals. There is a growing suspicion, however, that the Châtelperronian layers are simply a mechanical mixture of the latest Neanderthal and the earliest modern human occupations during an episode of extreme cold (Heinrich Event 4), when the effects of frost action on cave sediments would have been particularly intense.[162]

A large number of other sites across Europe have yielded a combination of typical Neanderthal and modern human artifacts, and they also often have been interpreted as evidence of cultural assimilation by local Neanderthals at the time that modern humans were spreading into the continent. These sites invariably lack diagnostic human skeletal remains (although some have yielded isolated teeth), but have been assumed to be the product of Neanderthals, based on analogy with the Châtelperronian. New multivariate approaches to the analysis of teeth indicate that modern humans probably made most of or all these assemblages, however, and the simple artifacts thought characteristic of Neanderthal toolmaking (for example, side scrapers) are more parsimoniously explained as expedient tools similar to those produced by later modern humans.[163]

The most interesting contribution of the archaeological data to the global dispersal lies in their potential for revealing the *organizational and technological adaptations* that made it possible, although most of the data are from sites occupied several millennia after the initial arrival of modern humans (figure 2.8). In addition to tailored insulated clothing and fire drills in northern Eurasia, there is archaeological evidence for the expansion of the diet to small vertebrates, which has implications for novel technologies, including snares, traps, nets, throwing darts, and other instruments and devices, most of which are mechanical. The open-air occupation in Russia that yielded possible traces of early figurative art (Kostenki 14)

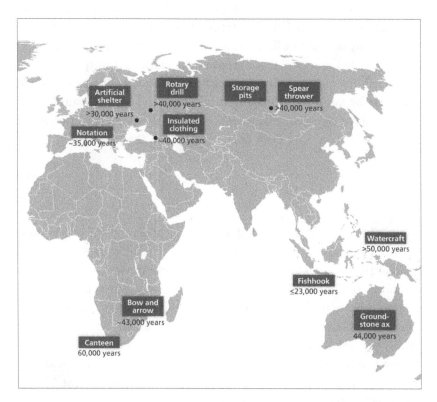

FIGURE 2.8 Map of the Eastern Hemisphere, showing the location and estimated age of some technological innovations of modern humans. (Base map adapted from en.wikipedia.org/wiki/File: BlankMap-World6.svg)

also contains a large concentration of hare remains that indicates harvesting of small mammals.[164]

Other major innovations include watercraft (inferred from the more than 70-kilometer crossing between Southeast Asia and Australia), artificial shelters (documented at open-air sites at least as early as 30,000 years ago), and domesticated dogs (based on new information from sites in Belgium and eastern Europe). There is evidence of fishing technology before 30,000 years ago in Southeast Asia,[165] while the quantity of bird remains in a few sites that are more than 40,000 years old suggest that—as in the case of the small-mammal remains—some technologies for hunting birds (for example, throwing darts) may have been developed.[166]

The establishment of widely dispersed alliance networks over large areas is suggested by the movement of raw materials over distances of hundreds of kilometers from their source. The materials may have been exchanged between groups or carried by one group, but either way, the transport of materials over unprecedented distances indicates social interactions over very large areas. Fossil shells recovered from an approximately 45,000-year-old context on the central East European Plain indicate transport over 500 kilometers from their source—exceeding non-modern human material transport by an order of magnitude.[167] These and other organizational adaptations to environments where resources were widely scattered and unpredictable may have been possible only with the faculty of syntactic language.

Other Evidence

In addition to fossil, genetic, and archaeological data, there is some supporting evidence for RAO model and the global dispersal of modern humans from other sources. One of them is genetic evidence from microorganisms that inhabit both the human body and its technological extension in the form of clothing. The pattern of decreasing DNA diversity as a function of distance from Africa is found in a bacterium (*Helicobacter pylori*) that inhabits the gastrointestinal tract of roughly half the living human population.[168] And comparative DNA analysis of human body lice (*Pediculus humanus*), which inhabit clothing, also reveals a recent African origin (estimated dispersal date of ca. 72,000 years ago).[169]

Supporting evidence from linguistics includes a study of phonemic diversity with a global sample of more than 500 languages. According to the results of the study, *phonemes* (units of sound) in modern human languages exhibit the same pattern observed in DNA—decreasing diversity as a function of distance from Africa.[170] Significantly, *morphemes* (units of meaning or information) in languages do not exhibit this pattern, and the distinction is important to everything that follows in this book.

Chapter Three

AN EVOLUTIONARY CONTEXT FOR
HOMO SAPIENS

God hath made man upright, but they have sought out many inventions.

—ECCLESIASTES 7:29

In the aftermath of the First World War, the German military establishment searched for a way to prevent foreign powers from decoding secret messages, which had been a problem during the war. Within a few years, they decided to purchase and deploy a large number of encryption machines known as "Enigma," which converted information to indecipherable noise with a series of rotating scramblers. The orientation of the scramblers could be reset on a daily basis. By 1926, foreign intelligence services realized that they could not decode intercepted messages sent by the German military.[1]

The Enigma code was cracked by a young British mathematician named Alan Turing (1912–1954) at the beginning of the Second World War. Although Turing had clues (or "cribs") to work with, the key to deciphering the messages coded by the Enigma machine—intercepted each morning—lay in the very rapid production of alternative possibilities with another machine (named "bombe"). Eventually, Turing and his colleagues were running about 50 interconnected bombes to determine the German code for the day.[2]

Turing's solution to the problem of deciphering the Enigma code was fundamentally similar to the workings of the immune system. More generally, it was similar to the process of evolutionary computation (chapter 1) and probably to the way the modern human brain functions. The immune system produces proteins that bind to foreign antigens and block their potentially hostile functions within the organism. In order to bind to the antigen, the protein must possess a chemical structure specific to that antigen. The immune system adapts to the (potentially infinite) variations in antigen structure by simultaneously and recursively generating an immense range of alternative possibilities. When a protein successfully binds to a novel antigen, the system "selects" for the chemical structure of the antigen by producing more individual proteins with the same structure. The process of generating alternative possibilities entails the random recombination of the genetic sequences ("hyper-mutation") that code for the proteins.[3]

In a similar way, living systems evolve by generating random variations in the genetic information base that codes for organism structure and function. The alternative possibilities are potentially infinite. Environmental factors select for specific structures and functions through differential reproductive success. The genetic information adapts to variations in environment, or—as an information theorist would say—accumulates information about the environment through the reduction of uncertainty.[4] The evolutionary process is a form of nondeterministic computation that exploits the random copying errors of genetic transmission to create novel solutions to the problems of adaptation.[5]

The brain of at least some metazoans, including the large-brained vertebrates, apparently functions in a similar way.[6] The process of pattern recognition in the human brain is thought to entail a very rapid search among alternative possibilities.[7] The *massive parallel processing* of data is similar to the mass-production of possible alternative structures of antigens in the immune system, and to the simultaneous operation of Turing's bombe machines to identify elements of the Enigma code.[8] If reversed, the process would be a source of creative recombination (and somewhat analogous to sexual recombination of genetic information in the metazoa).

THE EVOLUTION OF THE BRAIN: COMPUTING WITH NON-GENETIC INFORMATION

The appropriate evolutionary context for humans often is assigned to their order (Primates) within the class Mammalia. This approach is logical, since many characteristically or uniquely human features have evolutionary roots in the early primates, including prosimians.[9] But the ways that humans store, transmit, and translate information—as noted by John Maynard Smith and Eörs Szathmáry—are so fundamentally different from those of their primate and even mammalian ancestors that a wider and deeper evolutionary context is justified.[10]

The significance of the changes that produced modern humans is better appreciated from the perspective of the evolution of all living systems, from their beginnings more than 3 billion years ago to the automata created by *Homo* ("artificial life"). The origin of life and information was briefly discussed in chapter 1, along with the transition to eukaryotes and multicellular life. In this chapter, the evolutionary context of modern humans is approached more narrowly, beginning with the metazoan brain, which represents the earliest form of evolutionary computation with *semantic nongenetic information* (for a definition of "semantic information," see chapter 1). It may be argued that all significant developments in human evolution concern the structure and function of the brain (including the evolution of the human hand and vocal tract). Humans altered the evolutionary process by developing the means by which non-genetic information in the brain can be translated into structure both inside and outside the brain.

The Evolution of the Metazoan Brain

The brain is an organ that receives non-genetic information from the sensory organs, processes this information, and computes responses that often involve motor functions (for example, movement of the limbs), but may include a variety of other body functions (for example, elevated blood flow to the skin). The primary purpose of the brain is to detect spatial-temporal variations in the environment that may affect the fitness of the organism, which may include a significant decline in ambient temperature, the approach of a threatening predator, or the appearance of a potential mate.[11]

The brain evolved among complex animal life (that is, metazoa). In prokaryotes and plants, its primary function is performed by other structures. Bacteria evolved receptors in the form of proteins embedded in their prokaryote cell walls that detect variations in their immediate environment, such as the presence of sugar molecules (energy source) or heavy-metal ions (toxins). The receptors transmit signals to other parts of the cell that trigger an appropriate response—move toward or away from the detected entity, depending on its predictable effect on the fitness of the bacterium. The transmitted signals represent a non-genetic form of information—presumably, the earliest evolved form of non-genetic information. Plants also evolved signaling mechanisms as an adaptation to variations in their environment, but their immobile character negates the need for a central nervous system.[12]

Like other metazoan organs, the brain is composed of an interconnected set of specialized eukaryotic cells. The specialized cells of the brain are nerve cells, or neurons, which store and transmit information in the form of electrochemical signals where they connect with other neurons (synapses). Each neuron may connect with hundreds or thousands of other neurons, and the information-storage capacity of the brain grows exponentially with the total number of neurons.

As noted in chapter 1, the non-genetic information stored and transmitted in neurons ("neuronal information") exhibits a set of properties that is very different from those of genetic information. Most important, neuronal information is a *temporary* form of information received from the environment, rather than from a parent organism, and it may be lost or replaced with new information. Unlike genetic information, which is digital, neuronal information may be either digital or analog. However, like genetic information, it is semantic, and it may possess a complex hierarchical organization.

Although the earliest metazoans (possibly similar to sponges) probably lacked both neurons and brains, one of most primitive surviving phyla (Cnidaria), which contains sea anemones and corals, evolved a diffuse network of neurons ("skin brain") that functions in a manner analogous to that of the receptor proteins of bacteria. A central nervous system (CNS) or brain may have first evolved in an entirely hypothetical taxon known as *Urbilateria*, which is the postulated ancestor of all the bilateral phyla (metazoa other than sponges and cnidarians). Urbilateria would have lived during the Ediacaran (ca. 600 million years ago) and potentially carried much of the genetic information that produced the subsequent "Cambrian explosion" of metazoan life.[13]

As the metazoa evolved during the Paleozoic (542–251 million years ago), some groups developed especially large and complex brains containing millions of neurons and billions of potential synaptic connections. The largest brains and most developed sensory systems evolved independently in the vertebrates (phylum Chordata) and cephalopods (phylum Mollusca). The evolution of larger, more complex brains was driven by various factors, including the cognitive demands of being a predator, navigating complex environments (especially on land), and gaming the social environment. If the last of these became an important factor in the evolution of *Homo*, the first is thought to have been critical to the evolution of early brains, as well as later developments. The earliest bilateral animals with a central nervous system (flatworms) are widely thought to have been predators, and the evolution of their CNS is attributed to the cognitive requirements of detecting and pursuing prey in a marine environment (table 3.1).[14]

The metazoan brain performs *evolutionary computations with non-genetic information* analogous to the computations with genetic information in evolving lineages.[15] Although the responses of bacteria to environmental variations reported by their protein receptors represent a computation with non-genetic information, it is a simple computation (performed with "Shannon information"). The computations of the metazoan brain with neuronal information are complex, performed with semantic information. Their complexity may be measured in terms of both input (quantity of information received from the sensory organs) and defining functions or computational steps (standard measures of computational complexity).

The first vertebrates (jawless fish), which appeared about 470 million years ago, probably also were predators. They evolved more complex olfactory and visual systems as part of their adaptation. They are associated with the development of the *cerebellum*, which coordinates motor responses and played a special role in the evolution of *Homo*. And they evolved—through gene replications—a primitive form of *hemoglobin*. Because of the importance of hemoglobin in supplying a large quantity of oxygen to the brain, its evolution in early vertebrates was critical to the subsequent development of very large brains in higher vertebrates. By contrast, the other animal group that evolved relatively large brains—the cephalopods—never evolved a gene for hemoglobin, and ultimately were constrained in terms of brain size.[16]

The early jawed vertebrates (cartilaginous fishes) evolved the *myelin sheath* to insulate the axon (the elongated appendage of the neuron), which allows faster electrochemical signaling and a more densely interconnected brain. They also evolved *adaptive immunity* against threats from the micro-environment (bacteria and viruses). Although not part of the CNS, the immune system was a major evolutionary development because it represented a novel mechanism for the recombination of genetic information. The process is fundamentally similar to evolution in general, but much faster, and has been compared with the function of the vertebrate brain.[17]

TABLE 3.1

Major Developments in the Evolution of the Metazoan Brain, Including the Vertebrate, Mammalian, and Primate Brains

Time	Taxonomic Group	Feature(s)	Significance
35 million years ago	Cercopithecoidea (Old World monkeys)	Trichromatic color vision; increased size of neocortex (relative to overall brain size)	Color vision expands visual spectrum; increased neocortex expands social network size (?)
55 million years ago	Prosimians (early primates)	Expanded visual-processing area (reduced olfactory area); ventral premotor area	Visual-processing area expands visual representations in brain; ventral premotor area enhances motor control
220 million years ago	Mammalia (mammals)	Neocortex; expanded auditory range	Neocortex integrates other parts of brain and stores information about environment, including visual maps
460 million years ago	Chondrichthyes (cartilaginous fish)	Myelin sheath (insulates axon); four-chain hemoglobin	Myelin sheath allows faster electrochemical signaling and more densely interconnected brain; four-chain hemoglobin binds and releases oxygen more efficiently
470 million years ago	Agnatha (jawless fish [earliest vertebrates])	More complex sensory systems (olfactory and visual); single-chain hemoglobin; cerebellum	Olfactory and visual systems increase sensory input; hemoglobin supplies large quantities of oxygen to brain; cerebellum ("little brain") coordinates sensory input with motor function
More than 500 million years ago	Platyhelminthes (including flatworms)	Central nervous system (CNS) concentrated in head	Earliest known brain (centralized information processing) preceded by diffuse network of neurons ("skin brain") in cnidarians
600 million years ago	Urbilateria (?)	Central nervous system (?)	Hypothesized earliest brain in hypothetical taxon

Some vertebrates became terrestrial roughly 370 million years ago, confronting a more complicated and unpredictable environment that drove brain evolution into new realms. Temperature fluctuations are more pronounced on land than in the sea and required further adaptations. Higher vertebrates later evolved the means of homeostasis, or maintaining a constant (warm) body temperature, which buffered them from the effects of changing temperatures but required significantly greater amounts of energy or food. Predation was a solution to the added food demands and, once again, seems to have been a factor in brain evolution. The earliest mammals (which appeared roughly 220 million years ago) are thought to have been predatory, although parenting behavior—also tied to the high energy demands of homeostasis—played a role in the development of the mammalian CNS as well.

The mammals evolved new features of the sensory system, most notably a capacity for hearing high-frequency sounds, which may have been related to both predation (detecting insects) and parenting (receiving distress calls from infants that were inaudible to reptiles and birds). But their most important innovation in brain anatomy was the *neocortex*. This is a sheet-like, multilayered structure that integrates other parts of the brain and stores information about the environment, including visual maps of the landscape.[18] Long-term storage of neuronal information gave it a new function: it became a permanent or semipermanent component of individual adaptation. Extended parental care increased the importance of gathering and storing neuronal information in mammalian young.[19]

The Prosimian Brain: The Visual Animal

The first "true" primates evolved during the Eocene, roughly 55 million years ago, probably from an extinct group such as the plesiadapiformes. Their appearance may be tied to a general expansion in the number and diversity of mammals following the mass extinctions of the late Cretaceous, about 65 million years ago, which wiped out the last of the dinosaurs and opened new ecological niches for various groups. The early primates were prosimians (represented today by lemurs, lorises, tarsiers, and bushbabies). They were small nocturnal predators, adapted to the widespread tropical forests of the early Tertiary (65–1.8 million years ago). Most of, if not all, their distinctive features are plausibly attributed to the demands of moving through trees and locating and catching small prey (for example, insects).[20]

Fossil prosimians exhibit reduced snouts (relative to their primitive mammalian ancestors) and large front-facing eyes, which provided stereoscopic vision. They also evolved grasping hands with nails (as opposed to claws). The anatomy of their brains reflects a significantly expanded visual-processing area and correspondingly reduced olfactory area. From the comparative anatomy of living prosimians, we have learned that the early primates evolved an increased density of photoreceptors in the center of the retina and significantly expanded visual representations in the brain. Also important from our perspective was the appearance of a unique cortical area for enhanced visual control of muscle movements (*ventral premotor area*).[21]

While increased visual control of limb function is of obvious relevance to later evolutionary developments in *Homo*, such as precision movements of the hands, the overall increase in the detail and complexity of representations in the visual cortex seems equally significant in light of the evolved human faculty for generating hierarchically organized structures of information and artifacts. Visual representations—especially in primates—are the most complex structures of information (in analog form) generated in the brain from sensory data. They are processed in a hierarchical and recursive fashion that has been described as the reverse of the process by which modern humans create the digital structures of language (or the analog structures of complex artifacts).[22]

At a conference held in Britain in 1958 on the "mechanisation of thought processes," computer scientist (and early artificial-intelligence researcher) Oliver Selfridge (1926–2008) proposed a model for pattern recognition that may be applicable to both natural and artificial intelligence. The essence of the model (which Selfridge labeled *pandemonium*) was that many individual cells, or "demons," in a machine or brain may be programmed to identify individual elements of a larger pattern. Collectively, the firing cells could distinguish one pattern from another (figure 3.1).[23] Unless the patterns to be recognized are extremely simple, the system is practical only with massive parallel processing of data with alternative possibilities—analogous to Turing's simultaneous bombes deciphering the Enigma code or the workings of the immune system.

Selfridge's pandemonium model helps explain how the brain generates visual representations and recognizes patterns in data received by other sensory organs.[24] It also may provide insights into how the higher primate—and, more broadly, the mammalian—brain produces creative thought. The recursive construction of sentences (and, potentially, of artifacts) in the modern human brain conceivably engages a similar process of simultaneously generating a massive set of alternative possibilities.[25] The potential

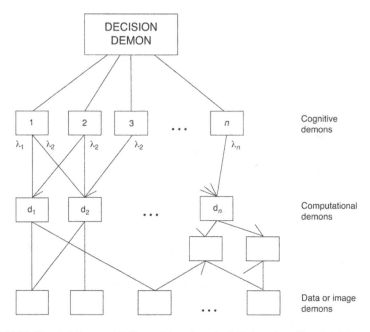

FIGURE 3.1 Oliver Selfridge's model of "amended pandemonium." (Redrawn from Oliver G. Selfridge, "Pandemonium: A Paradigm for Learning," in *Mechanisation of Thought Processes: Proceedings of a Symposium Held at the National Physical Laboratory on 24th, 25th, 26th and 27th November 1958* [London: Her Majesty's Stationary Office, 1959], 517, fig. 3)

for the creative recombination of neuronal information stored in the brain probably extends far beyond living humans, because it is manifest in the creative variations of songs in male humpback whales,[26] and perhaps beyond the mammals to other vertebrates.[27] It is obvious in humans only because they translate large quantities of creatively recombined information to structure outside the brain in the form of spoken and written words, visual art, and other categories of information-based artifacts.

Anthropoid Cognition: Alarm Calls and Tools

Monkeys appear in the African fossil record in the early Oligocene, roughly 35 million years ago. As are prosimians, early monkeys are associated with tropical forests, but not with predation; they appear to have been frugivores.[28] Old World monkeys evolved an important addition to primate visual processing widely believed to reflect their need to identify fruit in a dense forest backdrop of green. As the result of a gene duplication, they developed *trichromatic color vision* based on three retinal cone pigments (rather than two cone types, which is common among other mammals). This, in turn, drove the evolution of specialized neural structures for processing the enriched stream of incoming visual information.[29]

Absolute brain size increased in the monkeys in concert with their larger body size, but—when the energy requirements of the brain are accounted for—the ratio of brain to body size remained similar to that of the prosimians.[30] The relative size of the neocortex increased in the monkeys, however. Because the size of the neocortex correlates well with that of the social group, the pattern has been ascribed to increased social complexity.[31] And while some have questioned this interpretation (noting that neocortex size also correlates with other variables),[32] there is evidence for the development in some monkeys of large cortical areas that control facial expression, and this almost certainly is tied to social behavior.[33]

Several New World and Old World monkeys evolved predator-specific alarm calls in the context of the social group. This feature is relatively widespread among the primates, including some prosimians (lemurs), but is especially well known in vervet monkeys (*Cercopithecus aethiops*). In a classic paper published in 1980, Robert Seyfarth, Dorothy Cheney, and Peter Marler reported that vervet monkeys in Kenya had developed specific vocal warning signals for three different predators (leopards, eagles, and snakes) that elicit responses as appropriate to the threat (figure 3.2).[34] The alarm calls are issued only in a social context and apparently are influenced by the genetic relationship of caller to receiver. Since 1980, alarm calls have been documented in other primates.[35]

The significance of predator-specific alarm calls in vervet monkeys and other primates lies in the fact that they represent a simple form of *categorization*, which is an essential first step in the development of language. Each distinctive vocal sound is assigned to a specific entity in the environment and is broadly analogous to a proper noun. However, the calls are given only in the presence of the predator; they are not used outside the immediate spatial-temporal context of the entity to which they refer. And, in contrast to human language, the calls are inherited (transmitted from parent to offspring by genetic information), although there is evidence that learning plays some role in their development. Their recurring presence in the primates is nevertheless pertinent to the evolution of language in *Homo*.

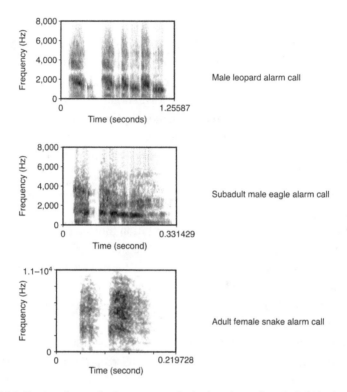

FIGURE 3.2 Vervet monkey vocalizations: sonograms for the three alarm calls, each of which refers to a particular predator (leopard, eagle, and snake). (From "Vervet Monkey Vocalizations," http://www.psych .upenn.edu/~seyfarth/Baboon%20research/vervet%20vox.htm)

Fossil discoveries reported in 2013 include the oldest known ape remains, dating to 25 million years ago, from East Africa.[36] The age of the fossils supports earlier estimates based on molecular phylogenetics.[37] The early apes were small, and their fossils are not easily differentiated from those of monkeys. They are most reliably distinguished by a feature that also separates them from humans. The humerus (upper arm bone) and the heads of both the radius and the ulna (lower arm bones) exhibit characteristic shapes that reflect the function of the elbow joint, which allows considerable flexibility in arm movements. On the basis of this feature, fossils assigned to the genus *Proconsul* were classified as an early ape (with many monkey-like features) dating to the Miocene (ca. 23–5 million years ago).[38]

The apes underwent a major diversification and expansion of geographic range into Eurasia after a land connection was reestablished with Africa around 17 million years ago. Many of the later Miocene apes were comparatively large, and they evolved various ground-dwelling adaptations. Among them is *Dryopithecus*—its fossil remains discovered in the mid-nineteenth century—which exhibits features in common with those of the living African apes, and represents a plausible ancestor for both living apes and humans. A significant development in *Dryopithecus*, though, likely was related to climbing trees rather than moving on the ground.[39]

The anatomist John Napier (1917–1987) concluded that the "true hand" had evolved in apes and that the differences between ape and human hands are superficial.[40] The design of ape and human hands reflects the absence of weight-bearing loads because the extremities of the forelimbs typically do not function as feet. The ape hand has a dual function: it is used for both suspension in trees and manipulation of objects (and modern African apes evolved an eccentric form of walking with their hands, or "knuckle walking"). Humans subsequently evolved an even more specialized hand; it is used for manipulation, with little or no locomotor function (except during infancy).

The visually coordinated control of the hands for tool use by chimpanzees was noted by Charles Darwin in 1871 and was expanded 90 years later by Jane Goodall to include *toolmaking*.[41] Examples include sponges (made by crumpling a leaf), insect-fishing sticks (made by stripping smaller twigs off a branch), and rock hammer and anvil for cracking nuts. These and other tools are examples of the modification and reduction of natural objects, as well as the *combination* of objects.[42] Chimpanzees apparently do not translate a complex design based on neuronal information into a tool or weapon. Significantly, however, the making and using of tools among chimpanzees is transmitted by learning, not by genetic inheritance, which suggests an analogy with the body of non-genetic information that humans transmit from one generation to the next (traditionally termed "culture").

Equally important is evidence for the use of material objects in "language" experiments with captive chimpanzees. In such experiments, chimpanzees—which produce a narrower range of vocal sounds than do humans—have been most successful in learning and using symbols in the form of a physical object or sign. The pygmy chimpanzee, or bonobo (*Pan paniscus*), named Kanzi reportedly mastered more than 200 symbols for various phenomena (including himself) in the form of lexigrams in experiments conducted by Sue Savage-Rumbaugh.[43] Kanzi also learned to flake simple stone tools.[44] Other researchers used plastic tokens in experiments with chimpanzees and a juvenile orangutan. Chimpanzees also have been taught to communicate with hand gestures, and studies of wild chimpanzees reveal that they use several dozen gestures in their natural setting.[45]

Language experiments with apes have been controversial for many years. Much of the criticism has been focused on the interpretation of the results.[46] The consensus at present is that no ape has demonstrated the faculty for syntactic language in any form, even when material symbols or hand gestures are substituted for vocalizations. Some apes have shown the capacity for categorizing objects (including themselves) and actions in their environment, like the predator-specific alarm calls of vervet monkeys, but on a wider scale and with the added ability of linking symbol to referent in the absence of the latter. And some apes have demonstrated the ability to do computations with several symbols, comparable to the proto-linguistic faculty of a young child. (But, as in the case of the tools made by chimpanzees, the computations are severely limited in terms of hierarchical complexity.) The experiments indicate that there is much individual variability in the symbol-using abilities of apes and that material objects are particularly suitable as symbols.

Although chimpanzees and other great apes are not known to use alarm calls in the wild, gibbons (*Hylobates* sp.), classified as lesser apes, produce vocal warning cries. In fact, gibbons share a number of features with humans that are not found in the great apes. They evolved highly specialized forelimbs (for swinging from the branches of trees) and walk bipedally as much as 6 to 7 percent of the time, and this appears to be related to the structure and function of their larynx, which is human-like in some respects.[47] Most important, gibbons establish long-term pair-bonds between males and females (they are the least sexually dimorphic of the apes). As in other species, the bonds are established with the help of elaborate courtship displays between the sexes, and—in addition to alarm calls—gibbons employ their vocal faculties to generate unusually complex courtship songs.

As mentioned in chapter 1, Darwin specifically invoked the courtship songs of gibbons in his speculations about the origins of language and music in humans. The complex hierarchical structure of gibbons' songs, revealed by recent research, has been compared with the syntax of human language.[48] Similar observations have been made about complex courtship

BOX 3.1

The Evolution of the Human Vocal Tract

The vocal tract in *Homo sapiens* is one of two specialized organs that function to translate semantic information in the brain into structures outside the brain, including those that represent other forms of information. Like the hand, the vocal tract is a highly precise instrument, capable of translating complex digital information into coded sounds that may be transmitted to other brains. And, like the hand, the specialized functional anatomy evolved within the past few million years. Reconstructing the evolution of the human vocal tract is more difficult than tracking evolutionary changes in the hand, however.[1]

As in most mammals, the *larynx* (voice box) is positioned relatively high in the neck in the African apes—at the same level as the first three cervical vertebrae. In this location, it can shift position to separate the passageway to the lungs (*trachea* [windpipe]) from the tube to the stomach (*esophagus*). In living humans, the larynx migrates downward during infancy to a location opposite the fourth to seventh cervical vertebrae, where it cannot separate the windpipe from the esophagus.[2] This creates a choking risk—which accounts for several thousand deaths each year—but expands the size of the *pharynx* (vocal chamber). Using the vocal folds of the larynx, living humans generate a wide range of sounds in this enlarged vocal chamber (box figure 3.1).

The key elements of the vocal tract are composed of soft parts, or cartilage, and they do not preserve in the fossil record. Various researchers have attempted to reconstruct the vocal-tract morphology of pre-modern humans on the basis of related skeletal parts that have been preserved. The degree of flexure at the base of the cranium was one suggested indication of a lowered position of the larynx, but the relationship between the two has been questioned. Other suggestions have included the length of the neck, morphology of the *hyoid* bone, and size of the *hypoglossal canal*, but these also have proved problematic.[3]

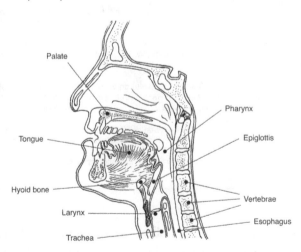

BOX FIGURE 3.1 The human vocal tract. (From John F. Hoffecker, *Landscape of the Mind: Human Evolution and the Archaeology of Thought* [New York: Columbia University Press, 2011], 86, fig. 3.4)

(continued)

(continued)

The size of the thoracic canal (containing the upper spinal cord) appears to be a reliable indicator of the degree of fine motor control of breathing (lung pressure), however, which is as critical to speech as the position of the larynx. The small size of the thoracic canal in the 1.5-million-year-old partial skeleton of *Homo ergaster* from *Nariokotome* (Kenya)—roughly half the size of the canal in modern humans—indicates a more ape-like condition, suggesting a significant difference in speech capabilities.[4] Given the presence of Broca's area in the 1470 endocast, a reasonable inference is that early *Homo* evolved a form of language that was less complex than that of modern humans (for example, a non-syntactic language or "proto-language").[5] By contrast, the Neanderthals exhibit an enlarged thoracic canal similar to that of modern humans (and also possessed the FOXP2 gene, implicated in speech function and once thought unique to *H. sapiens*), suggesting that the speech capability of modern humans evolved no later than their most recent common ancestor (ca. 500,000 years ago).[6]

In any case, the relationship between speech function and language is complicated, and the presence or absence of enhanced breathing control and a modern human vocal tract cannot be equated with the presence or absence of syntactic language (or non-syntactic language). A syntactic language is theoretically possible without a modern human vocal tract (or, for that matter, without the use of *any* vocal tract [for example, sign language]). No language employs the full range of possible sounds, and thus a syntactic language based on a much smaller range of sounds is conceivable.[7]

1. Leslie Aiello and Christopher Dean, *An Introduction to Human Evolutionary Anatomy* (London: Academic Press, 1990), 232–243; W. Tecumseh Fitch, *The Evolution of Language* (Cambridge: Cambridge University Press, 2010), 297–337.
2. Jeffrey Laitman, "The Anatomy of Human Speech," *Natural History*, August 1984, 20–27; Philip Lieberman, *The Biology and Evolution of Language* (Cambridge, Mass.: Harvard University Press, 1984), 271–283.
3. Lieberman, *Biology and Evolution of Language*, 287–313, and *Eve Spoke: Human Language and Human Evolution* (New York: Norton, 1998); W. Tecumseh Fitch, "The Evolution of Speech: A Comparative Review," *Trends in Cognitive Sciences* 4 (2000): 258–267. See also Fitch, *Evolution of Language*, 329–335.
4. A. M. MacLarnon and G. P. Hewitt, "The Evolution of Human Speech: The Role of Enhanced Breathing Control," *American Journal of Physical Anthropology* 109 (1999): 341–363.
5. Alan Walker and Pat Shipman, *The Wisdom of the Bones: In Search of Human Origins* (New York: Knopf, 1996), 261–278.
6. Wolfgang Enard et al., "Molecular Evolution of *FOXP2*, a Gene Involved in Speech and Language," *Nature* 418 (2002): 869–872; Richard E. Green et al., "A Draft Sequence of the Neandertal Genome," *Science* 328 (2010): 710–722.
7. Aiello and Dean, *Introduction to Human Evolutionary Anatomy*, 232.

songs in some bird species by Robert Berwick and his colleagues.[49] The songs do not contain any semantic information (although they may be considered a form of Shannon information), but represent the translation of neuronal information into an aesthetically appealing pattern of sound, analogous to the visual pattern of a peacock's tail. Their potential relevance to language is that the evolved neural pathways and motor responses used to translate information into such hierarchically structured patterns are a conceivable preadaptation to speech function (box 3.1). Along with categorization, the making of tools, and the use of material symbols, they represent part of the evolutionary context of human cognition.

THE HUMAN LINEAGE: AUSTRALOPITHS AND *HOMO*

Between the time that humans diverged from African apes (ca. 5 million years ago) and the appearance of modern humans, three major interrelated developments occurred in the human lineage: (1) the evolution of *bipedalism* and increased specialization of the hands (australopiths); (2) a radical shift in *foraging strategy* associated with expansion into drier and cooler habitats, where resources were more widely dispersed (early *Homo*); and (3) the steady growth of *brain size* after around 2 million years ago (*Homo* sp.).

Bipedalism represents the defining feature of the human lineage and probably triggered the divergence from quadrupedal apes. It is widely believed to reflect increased time spent on the ground (as opposed to a more arboreal life) and may be tied to the shrinkage of forests in sub-Saharan Africa after 8 million years ago.[50] It had little effect on overall brain volume, but undoubtedly was critical to the evolution of an increasingly versatile and sensitive pair of hands, and it now appears likely that at least some of the australopiths were making stone tools.

The shift in foraging strategy, which took place about 2 million years ago, also is conceivably tied to the reduction of arboreal habitat in tropical Africa, although the most striking consequence in the fossil record was a massive range expansion to the temperate zone of Eurasia. The new pattern of foraging furthered the trend toward increased bipedal locomotion on the ground and specialized use of the hands to fashion artifacts, but there were other consequences for human evolution that cannot be overemphasized. Foraging by early *Homo* was collective and centralized—gathered food was brought to a central location—as indicated by the pattern of remains in the archaeological record. There was an increased consumption of meat and probably some cooking (controlled use of fire), and there is fossil anatomical evidence for pair-bonding (reduced sexual dimorphism) and some form of language (Broca's area identified in a fossil endocast).[51]

Both pair-bonding and language (or "proto-language") likely are connected to collective or social foraging.

It probably is no coincidence that the volume of the brain began to increase with the shift to *central-place foraging* and greater meat consumption. Among primates (as well as social carnivores), the size of the neocortex correlates with the size of the social group. And the expansion of the human brain between around 2 million years ago and the appearance of modern humans (300,000–150,000 years ago) is plausibly attributed to selection pressure on the ability to develop *reciprocal alliances* with people, especially people other than siblings and parents or offspring.[52] Managing a long-term pair-bond with the opposite sex requires a similar ability. At the same time, increased meat consumption probably was critical to meeting the high energy costs of larger brains.[53]

If the social network was essential to a collective foraging strategy, selection would have favored individuals who manipulated the reciprocal relationships to their long-term reproductive advantage—as well as individuals who established a wider network of alliances. The result apparently was a perpetual "arms race" for the cognitive abilities that support such social skills (that is, expanded memory capacity and probably increased computational complexity), which drove more or less continual growth in human brain volume until it reached a point at which further growth was constrained by one or more other factors (for example, thermoregulation).[54]

Australopiths: "Bipedal Apes"

The human clade (or tribe Hominini) probably diverged from an African ape lineage between 6 and 4 million years ago. Half a century after Vincent Sarich and Allan Wilson's explosive paper on the age of the first human, molecular phylogenetics remains the primary basis for estimating the time of divergence between the two lineages.[55] The African fossil record for the interval between 8 and 5 million years ago still is sparse and fragmentary. Although there is a 7-million-year-old fossil (*Sahelanthropus*) from western Central Africa, its hominin status remains problematic. The earliest humans may include *Orrorin* (ca. 6 million years ago) and *Ardipithecus kadabba* (ca. 5.7 million years ago), but most members of the human subfamily before the appearance of *Homo* belong to the australopiths, which now comprise four genera (*Australopithecus, Paranthropus,*

Ardipthecus, and *Kenyapithecus*) in sub-Saharan Africa between 4.4 and 1 million years ago.[56]

The australopiths have been characterized as "bipedal apes," and they lacked most of the features that distinguish the genus *Homo* from other hominoids (humans and apes). To begin with, they retained a number of arboreal adaptations that suggest continued tree climbing. These included comparatively long forearms, grasping feet, and curved fingers. The australopiths exhibit pronounced size dimorphism between the sexes (males, on average, were 40 to 50 percent larger than females), which suggests an absence of long-term pair-bonding. And they show relatively little increase in brain volume above the African ape level.[57]

Nevertheless, the australopiths did undergo some significant evolutionary change with respect to the hand. Fossil hand bones of *Australopithecus afarensis* ("Lucy"), dating to 3.2 million years ago, possess some typical ape features (such as slender apical tufts, or fingertips, on the distal phalanges) but also exhibit a relatively long thumb (about 50 percent of the length of the index finger), which is characteristically human and indicates that Lucy's hand was capable of at least some of the wide range of movements and positions of the modern human hand. Further developments in hand function are evident after 2 million years ago in early *Homo*.[58]

Humans eventually evolved a uniquely versatile and precise instrument for translating neuronal information into other forms outside the brain, based on the visually coordinated movements of no fewer than 27 bones and 35 muscles located at each forelimb extremity.[59] The movements and positions of the human hand can be combined and recombined like the sounds produced by the vocal tract to transmit information in a fully syntactic linguistic form (sign language), and the range of artifacts that can be created by translating neuronal information through complex sequential hand movements also is potentially unlimited (box 3.2).[60]

At least some of the australopiths made and used stone tools (and probably made some non-stone implements as well). In 2015, stone cores and flakes were dated to about 3.3 million years ago at the site of *Lomekwi 3* in West Turkana (Kenya). The assemblage, which comprises about 150 artifacts and includes percussors, worked cobbles, and potential anvils,[61] antedates the oldest known *Homo* fossils by 500,000 years and probably was produced by an australopith (*Kenyapithecus platyops* is the only hominin species known from the region during this period). Two later australopiths

BOX 3.2
The Evolution of the Hand

The hand has deep evolutionary roots in the primates, among the few groups of mammals that carry food to their mouths (as opposed to bringing their mouths to food),[1] and as early as the Eocene, the fossil ancestors of modern prosimians had evolved grasping hands with nails instead of claws. Finger–thumb opposition appeared in the catarrhines, which evolved a variety of specialized hand adaptations—all but one of which combined locomotor and manipulative functions.[2]

Fossil hand bones of *Proconsul*, the early Miocene ape from Africa, reveal a hand that—like other aspects of its anatomy—combines typical ape and monkey features.[3] By the later Miocene (ca. 9.5 million years ago), a generic ape pattern is evident in hand bones of *Dryopithecus* from western Eurasia.[4] The African apes subsequently evolved their own specialized hand, with shortened thumb and elongated fingers (for suspending from branches) and other features related to their peculiar knuckle-walking locomotion.[5] The later fossil record of African apes is extremely sparse, however, and the sequence of events that yielded the modern ape hand is unknown.

The early human fossil record also is sparse and fragmentary, and it is not known if changes in the hand took place at the time that the lineage of humans diverged from that of African apes. The earliest indication of change appears 3.2 million years ago in East Africa with "Lucy" (*Australopithecus afarensis*). Among the remains recovered from *Hadar* (Ethiopia) are more than 50 hand bones, including carpals, metacarpals, and phalanges.[6] Analysis of the bones revealed many similarities to the hand bones of living apes, including slender apical tufts (fingertips) on the ends of the phalanges. It also revealed features lacking in the apes, but found among later humans—most notably, a relatively long thumb (50 percent of the length of the middle digit). Lucy probably was capable of some of, but not all, the grips and other hand movements of modern humans (box figure 3.2).[7]

The changes in hand morphology and function that took place 3.2 million years ago were not accompanied by any increase in brain volume or evidence of stone-tool production. But after 2.6 million years ago, there are modest increases in brain size (in both earliest *Homo* and the robust australopithecine *Paranthropus*), and stone tools are found in East Africa. Hand bones of early *Homo* associated with stone tools and dated to at least 1.75 million years ago were recovered in 1960 at *Olduvai Gorge* (Tanzania). The bones were described by John Napier, who noted the expanded apical tufts (broad fingertips) similar to those of modern human fingers. But the Olduvai hand retains some ape-like features (for example, the morphology of the proximal and middle phalanges), and Napier wondered if it had been capable of a full precision grip.[8]

Of comparable age to the finds from Olduvai are hand bones from *Swartkrans* (South Africa), tentatively assigned to the australopithecine *Paranthropus robustus* (also associated with stone tools). These hand bones differ from the earlier australopithecine pattern observed at Hadar and exhibit a number of human-like features, including expanded apical tufts, like those of the Olduvai specimen.[9] It is not certain, however, that these bones belong to *Paranthropus*—some remains of *Homo* also are present in the same deposits).[10]

A fully modern human hand seems to have evolved by roughly 1.5 million years ago in *Homo* (both *H. ergaster* and *H. erectus*), although the supporting fossil evidence remains somewhat elusive. A few hand bones were recovered with the partial

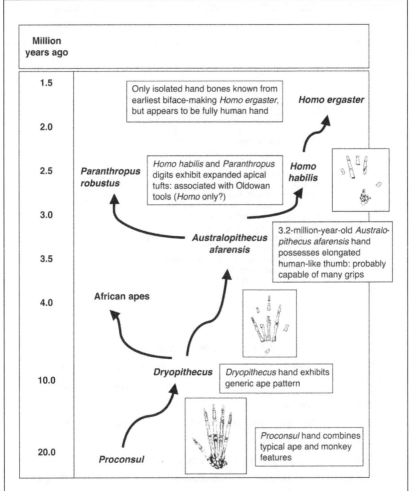

Million years ago

1.5	Only isolated hand bones known from earliest biface-making *Homo ergaster*, but appears to be fully human hand — *Homo ergaster*
2.0	
2.5	*Paranthropus robustus* — *Homo habilis* and *Paranthropus* digits exhibit expanded apical tufts: associated with Oldowan tools (*Homo* only?) — *Homo habilis*
3.0	
3.5	*Australopithecus afarensis* — 3.2-million-year-old *Australopithecus afarensis* hand possesses elongated human-like thumb: probably capable of many grips
4.0	African apes
10.0	*Dryopithecus* — *Dryopithecus* hand exhibits generic ape pattern
20.0	*Proconsul* — *Proconsul* hand combines typical ape and monkey features

BOX FIGURE 3.2 The evolution of the human hand. (From John F. Hoffecker, *Landscape of the Mind: Human Evolution and the Archaeology of Thought* [New York: Columbia University Press, 2011], 51, fig. 2.7)

skeleton of *Homo ergaster* from Nariokotome, which is dated to about 1.5 million years ago.[11] While these isolated specimens provide a less complete portrait of the hand than do the bones from Hadar and Olduvai, they are essentially modern in size and shape—like the Nariokotome arm bones—and probably reflect the arrival of a fully modern human hand.

1. John Napier, *Hands*, rev. Russell H. Tuttle (Princeton, N.J.: Princeton University Press, 1993).
2. Ibid., 57–58.
3. Alan Walker and Pat Shipman, *The Ape in the Tree: An Intellectual and Natural History of Proconsul* (Cambridge, Mass.: Belknap Press of Harvard University Press, 2005), 74–75.
4. Salvador Moyà-Solà and Mieke Köhler, "A *Dryopithecus* Skeleton and the Origin of Great-Ape Locomotion," *Nature* 379 (1996): 156–159.

(continued)

(continued)

5. Napier, *Hands*, 67–77.
6. Michael E. Bush et al., "Hominid Carpal, Metacarpal, and Phalangeal Bones Recovered from the Hadar Formation: 1974–1977 Collections," *American Journal of Physical Anthropology* 57 (1982): 651–677.
7. Mary W. Marzke, "Joint Functions and Grips of the *Australopithecus afarensis* Hand, with Special Reference to the Region of the Capitate," *Journal of Human Evolution* 12 (1983): 197–211. See also Leslie Aiello and Christopher Dean, *An Introduction to Human Evolutionary Anatomy* (London: Academic Press, 1990), 385–388.
8. John Napier, "Fossil Hand Bones from Olduvai Gorge," *Nature* 196 (1962): 409–411, and *Hands*, 88–89; Aiello and Dean, *Introduction to Human Evolutionary Anatomy*, 389–392. The hand bones include three carpals, 11 phalanges, and one metacarpal fragment from Olduvai Hominid 7 assigned to *Homo habilis* from Bed I at Olduvai Gorge.
9. Randall L. Susman, "Hand of *Paranthropus robustus* from Member 1, Swartkrans: Fossil Evidence for Tool Behavior," *Science* 240 (1988): 781–784.
10. Richard G. Klein, *The Human Career: Human Biological and Cultural Origins*, 3rd ed. (Chicago: University of Chicago Press, 2009), 235–237.
11. Alan Walker and Richard Leakey, "The Postcranial Bones," in *The Nariokotome* Homo erectus *Skeleton*, ed. Alan Walker and Richard Leakey (Berlin: Springer, 1993), 136–138. The bones include two first metacarpal shafts (?) and two phalanges.

that overlap with *Homo* also may have been stone-tool makers. The robust *Paranthropus* is associated with stone artifacts, while the remains of another late form, *Australopithecus garhi*, were found with animal bones that exhibit tool cut-marks.[62]

Evidence for the australopiths' diet has been gathered from several sources, including stable-isotope analysis of their bones, microwear analysis of their teeth, and plant phytoliths extracted from dental calculus. Although much variability is revealed among samples from different taxa, places, and time periods, there are repeated indications of an emphasis on plants found in grassy woodland or shrubland, including fruits and leaves.[63] Tubers (or "underground storage organs") may have been a major energy source, as they are among the *Hadza*, modern hunter-gatherers who live in a similar environment (savanna-woodland in equatorial Africa).[64] A recent analysis of the diet of a very late South African species (*A. sediba*) that coexisted with early *Homo* roughly 2 million years ago included traces of wood and bark.[65]

Early *Homo*: Central-Place Foraging

The shift in foraging strategy that took place about 2 million years ago in sub-Saharan Africa built on the earlier pattern of energy-efficient bipedal locomotion and non-locomotor use of the hands for making and using tools. The new strategy was an adaptation to a drier (and potentially cooler) habitat, however, where resources were more widely scattered and probably more difficult to find than in a savanna-woodland. It most likely entailed a significant increase in the consumption of meat—either scavenged or hunted or both. Archaeological data indicate that it was a central-place strategy, which is consistent with the predictions of theoretical ecology. The consequences of the shift on social behavior and the brain were considerable and created the circumstances under which anatomically modern humans later evolved.

The human fossil record in sub-Saharan Africa for 2.8 to 1.8 million years ago is richer than the record for the earlier transition (to bipedalism), but nevertheless is confusing and difficult to interpret. There were at least two—possibly three—species of larger-brained hominins in eastern and southern Africa at this time, and their evolutionary relationships to both the australopiths and later *Homo* are unclear.[66] The oldest *Homo* remains were reported in 2015 from *Ledi-Geraru* in Afar (Ethiopia) and have not yet been assigned to a particular species.[67] There also are archaeological sites containing stone tools and cut-marked animal bones, and it is not clear which hominins occupied these sites.[68] The changes in the fossil (and archaeological) record may be summarized as follows:

- *Skeletal anatomy.* Fossil remains from sites dating to 2.8 to 1.8 million years ago exhibit much variability in the dimensions of the teeth and cranium, but indicate a significant increase in the braincase (both allometric and non-allometric), especially in specimens assigned to *Homo rudolfensis*. Fossils of *Homo rudolfensis* include the famous 1470 skull from East Turkana (Kenya), with an estimated cranial capacity of 720 cubic centimeters (cc), or roughly 40 percent larger than that of *Australopithecus*. The 1470 endocast revealed what Ralph Holloway described as a complex "modern-human-like" third inferior frontal convolution—the area of the brain that contains Broca's area (Brodmann's areas 44 and 45).[69] The rib cage became less flared, and the pelvis narrower than in the australopiths, reflecting a smaller gut.[70] Remains assigned to *Homo* (as well as to *Paranthropus*) reveal new features of the hand, including expanded apical tufts.[71] Finally, it should be noted that sexual dimorphism began to decline after 2.5 million years ago, although it did not reach modern human levels (males, on average, are 8 percent larger than females) until relatively late in human evolution. In early *Homo*, the endocranial volume of males is only about 20 percent larger than that of females (representing as much as a 50 percent reduction in dimorphism from the australopiths).[72]

• *Diet.* Sites occupied or visited by tool-using humans in Africa between 2.4 and 1.8 million years ago contain large-mammal bones that exhibit cut-marks and percussion marks, indicating that bones were being broken and stripped of meat.[73] Faunal assemblages that date to about 2 to 1.8 million years ago reveal the accumulation of complete or nearly complete carcasses of small bovids, which probably reflects hunting (as opposed to scaveng-ing), but more selective retrieval of body parts of medium-size bovids.[74] Stable-carbon-isotope analyses of early *Homo* remains from the Turkana Basin indicate a high ratio of ^{13}C to ^{12}C and an emphasis on C_4 plants (grasses, sedges), which probably reflects the increased consumption of grassland animals.[75] The smaller gut, dominated by the small rather than the large intestine, suggests greater consumption of meat.[76] At the same time, the diet of recent foraging peoples in tropical and subtropical envi-ronments suggests that the consumption of plants remained high.[77]

• *Technology.* Sites occupied or visited by humans in Africa also contain stone tools, most or all of which may be assigned to the Oldowan Industrial Complex. The artifacts are categorized as (1) *hammerstones*, or heavy stones used to strike other pieces of stone; (2) variously shaped *cores*, from which flakes were struck; and (3) *flakes*, which vary in size and shape and some of which were chipped along the edge (*retouched flakes*).[78] Experimental work with a bonobo (Kanzi) suggests that fashioning Oldowan tools lies beyond the capabilities of living apes, probably because of differences in the func-tional anatomy of the hand.[79] Although archaeological evidence for the use of controlled fire before 780,000 years ago is problematic, several lines of evidence point toward the cooking of food before 1.8 million years ago.[80]

Representatives of the genus *Homo* invaded grassland habitat in sub-Saharan Africa roughly 2 million years ago, and the shift is reflected both in the distribution of their sites and in the carbon-isotope analyses men-tioned earlier.[81] Presumably, it was a response to further expansion of C_4 grassland in Africa after 2.5 million years ago.[82] By 1.8 million years ago, *Homo* occupations are found in North Africa and mid-latitude Eurasia as far as 40° North. The contrast with the australopiths, which were confined to southern and equatorial Africa (their remains are never found above latitude 16° North), is stark.[83] It is apparent that some humans had developed the ability to forage successfully in environments that were significantly less productive biologically than the tropical and subtropical grassy woodlands of sub-Saharan Africa.

In the late 1970s, Glynn Isaac (1937–1985) proposed a social-foraging model to explain the pattern of archaeological sites between 2.5 and 1.8 million years ago in Africa. They include (1) a few artifacts found with the remains of a single animal species (for example, hippopotamus), and (2) concentrations of stone artifacts associated with the remains of several animal species. Isaac suggested that the former represent places where an individual animal was killed and/or butchered, while the latter represent central locations to which the meat obtained at multiple kill-butchery places was carried. He further argued that the pattern of "central-place foraging," combined with the use of tools to process animal carcasses and the sharing of hunted and gathered foods among members of the group, constituted a "novel complex of adaptations."[84] The model was controversial because the food-sharing hypothesis was not testable and because it was proposed at a time of increased skepticism about the interpretation of animal bones in archaeological sites.

Although the food-sharing issue was never resolved, discoveries and research during and after the 1980s altered the picture of early *Homo* adaptations and provided support for other elements of Isaac's social-foraging model. Especially important was the discovery of sites in less biologically productive settings—in both the tropics and the middle latitudes—that date to between 2 and 1.8 million years ago. Their presence ties the changes in anatomy, diet, and technology between 2.5 and 1.8 million years ago to foraging in habitats where resources were more widely dispersed across the landscape than they were in the biologically rich settings to which humans had been confined. Also significant is the accumulated body of research on the processes of site formation and the analyses of animal bones in the sites, which support Isaac's contention that (1) meat was being obtained from large mammal carcasses (by either scavenging or hunting or both), and (2) humans were concentrating artifacts and animal bones in a central place.[85]

At *Kanjera South*, on the southern shore of Lake Victoria (western Kenya), stone artifacts and a large number of mammal bones were found in stream and lake deposits that date to roughly 2 to 1.8 million years ago. The analysis of the small-bovid remains, which exhibit cut-marks from tools (and percussion marks from hammerstones), indicates the recovery of complete or nearly complete carcasses and suggests either confrontational scavenging or, more likely, hunting. The medium-size bovid remains, however, hint at more selective retrieval of body parts (for example, a high percentage of head parts), which suggests that they probably were scavenged.[86]

Compelling evidence for central-place foraging recently has emerged from the famous and unique *FLK Zinjanthropus* site at Olduvai Gorge (1.84 million years old), where Manuel Domínguez-Rodrigo and his colleagues found artifacts and large-mammal remains concentrated at a location *near* a spring—but not at the spring itself—indicating that the carcasses probably had been brought to the site by its human occupants.[87] The analysis of the sediments, which included a carbonate bed deposited by the nearby spring at the time of occupation, was integrated with the interpretation of the artifacts and faunal remains (figure 3.3).

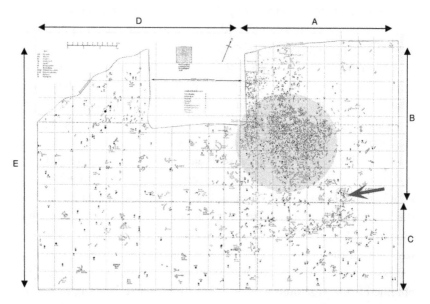

FIGURE 3.3 Map of the floor of the FLK *Zinjanthropus* site, with microfaunal accumulation (*arrow*), including bird bones. (Reproduced from M. Domínguez-Rodrigo et al., "New Excavations at the FLK *Zinjanthropus* Site and Its Surrounding Landscape and Their Behavioral Implications," *Quaternary Research* 74 [2010]: 322, fig. 6, with permission of Elsevier)

Increased meat consumption probably was critical to the occupation of less-productive habitats—there is a strong negative correlation between biological productivity and meat consumption among recent hunter-gatherers—and Oldowan tools likely were an important part of the strategy.[88] Sharp flakes were needed to dismember carcasses and strip meat off bones, while hammerstones were used to sever joints and smash bones open to obtain marrow. And when resources are clumped and unpredictable—such as a large-mammal carcass—optimal foraging models predict a central-place foraging pattern.[89]

A more powerful factor may have been mating strategy and sexual selection. The significant decline in sexual dimorphism in early *Homo* likely indicates that the shift to long-term pair-bonding between males and females had taken place.[90] Pair-bonding evolves when females require a substantive contribution from males to ensure the survival and success of their offspring. In species like humans, where the offspring are particularly dependent on their parents for an extended period of time, the required contribution from males may be extreme.[91] Under these circumstances, selection favors males who invest sufficient time, energy, and/or resources in the care of their offspring, and it will favor females who are adept at identifying and acquiring reliable mates. Pair-bonding would have affected every aspect of human social life. It probably was essential for ensuring that the increased energy demands of a larger brain during child development were met, and for allowing early *Homo* to occupy less-productive habitats. It also would have reinforced central-place foraging by increasing the need for home bases.

As Bernard Chapais observed, pair-bonding was the single most important step in creating modern human society. The second most critical development was the advent of reciprocal exogamous mate exchange—specifically, the exchange of females between male kin groups—which underlies the social networks established and maintained in hunter-gatherer societies.[92] Unlike pair-bonding, there are no markers for reciprocal exogamy in the fossil (or early human archaeological) record and, apparently, no clues as to when this mating pattern evolved.[93]

The parallels between the collective computations of a human social group and those of a honeybee colony (or super-brain) were noted in chapter 1. In both taxa, collective computation is based on a faculty for the transmission of semantic information from one brain to others. And in both taxa, information transmission may have evolved in the context of a social-foraging strategy. As with early humans, honeybees evolved in tropical Africa and dispersed northward into the less-productive habitats of the temperate zone. The colony implements what Thomas Seeley characterized as an "information-center" foraging strategy.[94] Foragers double as scouts and employ their "dance language" to describe the locations of food sources to other foragers when they return to the hive. The strategy allows the colony to focus its attention on high-quality food patches.[95]

The honeybee model offers a possible explanation for *why* humans may have evolved a simple form of language, or "proto-language," between 2.5 and 1.8 million years ago as part of their wider adaptation to foraging in less-productive habitats. Information sharing, which is an important part of foraging among recent hunter-gatherers, could have been especially helpful in the timely identification of animal carcasses (high-quality but unpredictable "patches" with a limited shelf life) in areas too large for the group to cover as a whole. Analogous to Oldowan tools (which probably have their roots in the tools made and used by some African apes), a proto-language could have developed from preexisting gestures and vocalizations similar to those found in the living apes, but perhaps further evolved among some or all of the australopiths. Vocalizations used for categorization—like the alarm calls of vervet monkeys—would seem to be the most likely source for the information units of a proto-language in *Homo*.

Despite its apparent simplicity and presumed similarity to nonhuman primate communication, a proto-language along the lines just described would represent an evolutionary transition with respect to *information*. It would entail the translation of neuronal information into another form of information (variations in sound waves between 20 and 2,000 hertz affected by manipulation of the vocal tract on flowing air generated by the lungs). The transmitted information (in an unstable form) would be immediately converted back to neuronal information by the receiver and presumably stored for a limited period of time in the neocortex. Unlike genetic information, proto-linguistic information could be transmitted from one individual to others irrespective of genetic relationships (outside parent–offspring or sibling ties). Thus proto-language would alter the connection between genetic and neuronal information, decoupling the latter from the former to some extent.

THE EXPANSION OF THE BRAIN IN *HOMO*

Following the estimated 40 percent increase in cranial volume evident in at least some forms of *Homo* between 2.5 and 1.8 million years ago, brain size continued to expand, more or less continually, until the appearance of anatomically modern humans (ca. 300,000 years ago). When endocranial volume is scaled against time for this interval, it reveals a pattern of steady increase—roughly 100 cc for every 500,000 years.[96] Early modern humans possessed a brain that was literally twice the size of that of early *Homo* (ca. 1,500 cc), although the comparison must be adjusted for body size, which increased from an estimated 50 to 60 kilograms (early *Homo*) to roughly 65 to 70 kilograms (mean body weight today for Africa is about 60 kilograms) (figure 3.4).

The number of synaptic connections would have increased exponentially with the number of neurons, moreover. Each neuron is connected to as many as 1,000 others, and while the number of neurons would have doubled in *Homo*, it is the number of synapses that provides the best estimate

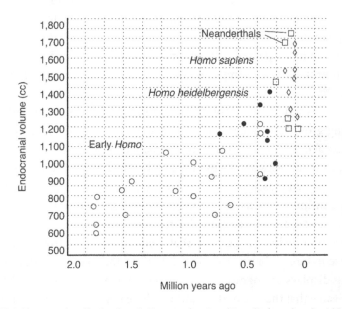

FIGURE 3.4 The expansion of brain volume in *Homo* as a function of time. (Redrawn from Daniel E. Lieberman, *The Evolution of the Human Head* [Cambridge, Mass.: Belknap Press of Harvard University Press, 2011], 566, fig. 13.12*a*)

of information-storage capacity. Today, the human brain contains an estimated 85 billion neurons, including roughly 20 billion neurons in the neocortex, and an estimated 10^{14} to 10^{15} synaptic connections (somewhat reduced in volume from the early *H. sapiens* brain).[97]

What factor or factors drove the expansion of the human brain after the appearance of *Homo*? It was once assumed that the evolution of the large human brain needed no explanation because of its obvious advantages. Now it is easier to measure the energy costs of the brain, which are considerable, and it is clear that the benefits—in terms of enhanced reproductive success—must have been significant to offset those costs. The adult human brain consumes roughly 280 to 420 kilocalories a day, which is about 20 to 25 percent of metabolic energy at rest, and the proportionally larger brain of an infant may consume more than 60 percent of total metabolic energy.[98] By contrast, the energy needs of the chimpanzee brain are only about 100 to 120 kilocalories a day. It is estimated that the early *Homo* brain of around 800 cc would have required about 200 kilocalories a day.[99]

To begin with, it is apparent that the shifts to pair-bonding, central-place foraging, and higher meat consumption were necessary preconditions for the growth in *Homo* brain volume. The use of controlled fire for cooking may have been an important factor. As the primatologist Richard Wrangham noted, research during the past few decades has demonstrated that cooked food releases more energy during digestion due to the effect of heat on proteins (for example, 40 percent increase in the protein value of an egg).[100]

It also would seem—in order to account for continued growth—that foraging groups either gradually increased their per capita energy intake between about 2 million and 300,000 years ago or evolved some form of energy trade-off between the brain and other organs or tissues (for example, liver or selected muscles) or both.[101] In any case, the pattern of continual or nearly continual expansion of the brain suggests constant selection pressure, as opposed to one or more discrete events.

Richard Dawkins suggested that sexual selection drove the steady increase in human brain size. He emphasized the role of the brain in generating displays of cognitive abilities—a "mental peacock's tail."[102] (It is conceivable that the size of the cranium alone may have been an advantage in attracting mates, the increased number of neurons and synapses being a secondary effect.) The shift to pair-bonding associated with early *Homo* lends added credibility to this idea. Pair-bonding would have sharply altered the direction and consequences of sexual selection, as males and females chose mates on the basis of predicted long-term behavior. This, in turn, would have selected for individuals of both sexes with superior computational faculties (and supporting memory-storage capacity) for accurate prediction, along with the ability to manipulate long-term relationships to individual advantage ("Machiavellian intelligence") (box 3.3).[103]

More familiar to paleoanthropologists is the *social brain hypothesis*, proposed by Robin Dunbar, which is based on the observed correlation between neocortex size and social group size in primates (as well as several other mammalian orders). According to this model, overall brain size—specifically, that of the neocortex—grows with the increased information-processing demands of a larger group (this is, more relationships to manage). Again, both computational skills and memory-storage capacity are implicated.[104] If pair-bonding in early *Homo* lends credence to sexual selection as a factor in brain expansion, the development of exogamous mate exchange (which probably took place at some point before 500,000 years ago) is equally critical to the social brain hypothesis, because it ensures a low mean coefficient of genetic relationship among members of a group. It is only the cooperative relationships among nonrelatives that select for "Machiavellian intelligence."

The two explanatory models actually complement each other and represent variations on a theme. In many respects, building and maintaining cooperative or reciprocal relationships among nonrelatives is similar to establishing a long-term pair-bond or reciprocal relationship between a male and a female. Both relationships would logically select for similar traits—the computational skills required to predict behavior in others outside the parent–child or sibling ties and to manipulate the relationship to individual advantage. Both sexual selection and what might be termed "social selection" probably would contribute to brain expansion.

The exponential increase in the size of the *Homo* brain recalls the expansion of the size and complexity of the genome during the transition from prokaryotes to eukaryotes (chapter 1). In each case, the easing of energy constraints on growth was followed by an explosive increase in the amount of information associated with an individual organism. The 250-fold expansion of the genome in eukaryotes crossed a threshold, opening the door not only to more complex cells, but to multicellular and

BOX 3.3
Game Theory and Human Social Networking

Roughly 2 million years ago, humans evolved a radical shift from the foraging strategy of the higher primates by bringing food (including scavenged and probably hunted meat) and other resources back to a home base—that is, central-place foraging, which represents the optimal solution to dispersed and unpredictable resources. Among recent hunter-gatherers, it entails reciprocal cooperation among male–female pairs and their in-laws, who exchange information about resources, sometimes hunt and gather collectively, and may exchange food and other resources.[1]

Like syntactic language and mechanical technology, the organization of living human societies is unique among animals. Recent hunter-gatherer societies comprise a network of exogamous pairs and their offspring, who exchange materials and information, in addition to marriage partners.[2] Because of the exogamous exchange of mates, the mean coefficient of genetic relatedness is low in comparison with that of nonhuman primate groups,[3] and more complex human societies exhibit an even lower coefficient of relatedness. Cooperation among nonrelatives in such societies cannot evolve through kin selection, but only through the evolution of a capacity for establishing and maintaining reciprocal relationships (*reciprocal altruism*).[4]

Noting a correlation between the size of the neocortex and that of the social network in primates (as well as social carnivores), Robin Dunbar proposed that the expansion of the human brain was driven by the demands of social alliance making ("social brain hypothesis").[5] He suggested that those demands included both massive memory-storage capacity—to accommodate a wide range of long-term relationships with other individuals—and social computational skills.

The evolutionary biologist Robert Trivers, who formulated the concept of "reciprocal altruism," also stressed the computational demands of long-term cooperation: "Because human altruism may span huge periods of time, a lifetime even, and because thousands of exchanges may take place, involving many different 'goods' and with many different cost/benefit ratios, the problem of computing the relevant totals, detecting imbalances, and deciding whether they are due to chance or to small-scale cheating is an extremely difficult one."[6]

Trivers proposed the following formula for *subtle cheating*, which he defined as "reciprocating, but always attempting to give less than one was given, or more precisely, to give less than a partner would give if the situation were reversed":[7]

$$\Sigma_{i,j} \, (b_{qi} - c_{qj}) > \Sigma_i \, (b_{qi} - c_{ai}) > \Sigma_{i,j} \, (b_{aj} - c_{ai})$$

where the ith altruistic act performed by the altruist has a cost to her of c_{ai} and a benefit to the subtle cheater of b_{qi} and where the jth altruistic act performed by the subtle cheater has a cost to him of c_{qj} and a benefit to the altruist of b_{aj}.

Both Trivers and John Maynard Smith noted the parallel between managing reciprocal alliances and classic *game theory*.[8] Game theory was developed in the 1940s by John von Neumann (see box 1.2) and Oskar Morgenstern (1902–1977) as an alternative approach to economics.[9] But the game-theory concept that Trivers and Maynard Smith regarded as most relevant to reciprocity and "small-scale cheating" was developed later by John Nash (1928–2015) and other employees of the infamous RAND Corporation. Known as the "prisoner's dilemma," it imagined a game between two players seeking both cooperation and individual advantage.[10]

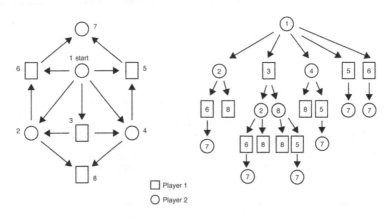

BOX FIGURE 3.3 A "state-space graph" for a simple two-player game (*left*), and a corresponding "game tree" that shows possible alternative moves (*right*), illustrating the potentially complex computations underlying game playing. (Redrawn from Philip C. Jackson Jr., *Introduction to Artificial Intelligence*, 2nd ed. [New York: Dover, 1985], 124, fig. 4-2)

Game theory provides a means of assessing—at least in a general way—the complexity of the computations underlying the management of a long-term reciprocal relationship within a human social network. The choices and decisions of one person may be modeled as a single-player game of strategy with a "game tree," in which the possible alternative "moves" (actions) are represented by *nodes* connected by branches of the tree (box figure 3.3). The deep complexity of modern human computation in this context is measured by the immense quantity of input in the form of stored memories about the actions and perceived motivations of the various people with whom the player interacts, accumulated over many years.[11]

1. Steven J. Mithen, *Thoughtful Foragers: A Study of Prehistoric Decision Making* (Cambridge: Cambridge University Press, 1990); Robert Whallon, William A. Lovis, and Robert K. Hitchcock, eds., *Information and Its Role in Hunter-Gatherer Bands* (Los Angeles: Cotsen Institute of Archaeology Press, 2011); Robert L. Kelly, *The Lifeways of Hunter-Gatherers: The Foraging Spectrum* (Cambridge: Cambridge University Press, 2013), 137–165 (including explicit discussion of "reciprocal altruism" on 146).
2. Bernard Chapais, *Primeval Kinship: How Pair-Bonding Gave Birth to Human Society* (Cambridge, Mass.: Harvard University Press, 2008).
3. John Yellen and Henry Harpending, "Hunter-Gatherer Populations and Archaeological Inference," *World Archaeology* 4 (1972): 244–252; Kim R. Hill et al., "Co-residence Patterns in Hunter-Gatherer Societies Show Unique Human Social Structure," *Science* 331 (2011): 1286–1289.
4. Robert L. Trivers, "The Evolution of Reciprocal Altruism," *Quarterly Review of Biology* 46 (1971): 35–57.
5. Robin I. M. Dunbar, "The Social Brain Hypothesis," *Evolutionary Anthropology* 6 (1998): 178–190.
6. Trivers, "Evolution of Reciprocal Altruism," 46.
7. Ibid.
8. Ibid., 38; John Maynard Smith, *Evolution and the Theory of Games* (Cambridge: Cambridge University Press, 1982), 167–173.
9. John von Neumann and Oskar Morgenstern, *Theory of Games and Economic Behavior* (Princeton, N.J.: Princeton University Press, 1944). For a highly readable account of game theory, intertwined with the life of John von Neumann, see William Poundstone, *Prisoner's Dilemma: John von Neumann, Game Theory, and the Puzzle of the Bomb* (New York: Anchor Books, 1992).

(continued)

(continued)

He described *Theory of Games and Economic Behavior* as "one of the most influential and least read books of the twentieth century" (41).

10. Poundstone, *Prisoner's Dilemma*, 94–131. In "Evolution of Reciprocal Altruism," Trivers cited R. Duncan Luce and Howard Raiffa, *Games and Decisions* (New York: Wiley, 1957). Luce also was a RAND Corporation employee. See also Alexander J. Stewart and Joshua B. Plotkin, "Collapse of Cooperation in Evolving Games," *Proceedings of the National Academy of Sciences* 111 (2014): 17558–17563.

11. Quantity of input is a standard measure of algorithmic or computational complexity. See, for example, Christos H. Papadimitriou and Kenneth Steiglitz, *Combinatorial Optimization: Algorithms and Complexity* (Mineola, N.Y.: Dover, 1998).

metazoan organisms.[105] It also altered the process of evolution, because the eukaryote genome evolves in a way different from that of the prokaryote genome. It is reasonable to ask if the immense capacity of the later *Homo* brain for storage and manipulation of neuronal information (approaching 10^{14}–10^{15} synaptic connections) also crossed a threshold, opening the door to the unique cognitive faculties of modern humans and new forms of information.[106]

If proto-language evolved in early *Homo*, it would have added another dimension to information processing and a new level of complexity with emergent properties. The flow of semantic information among brains within a social group—even if small by modern human standards—would have created a collective form of information storage and computation (or "super-brain"). Collective information storage and manipulation in the honeybee swarm transforms it into what Seeley described as a "cognitive entity." [107] Here, too, there is a parallel with the transition in genetic information from prokaryote to eukaryote, because only the latter can generate a multicellular entity composed of communicating and collaborating—and, eventually, specialized—parts. The combined effects of sexual and social selection would produce a competitive and potentially dynamic evolutionary setting in human society, however.

The Mystery of the "Hand Ax"

Aside from the expansion of brain volume, the most noticeable development in *Homo* between 2 million and 500,000 years ago was the appearance and spread of the *hand ax*. Oldowan technology was fundamentally reductive. Cobbles and rock fragments were broken and chipped to create working surfaces or edges; flakes struck off cobbles or fragments were retouched along their margins. In the 1980s, Nicholas Toth found a close relationship between the shape of the original rock or blank and that of the finished artifact, suggesting that the tool did not necessarily reflect a design in the brain of the toolmaker.[108] Many Oldowan artifacts appear to represent points on a multidimensional continuum, and the tool "types" recognized by archaeologists may exist primarily in their brains.

By 1.76 million years ago, some representatives of *Homo* were chipping large bifacial artifacts out of rock fragments that exhibit an ovate shape in three dimensions.[109] The large bifaces or hand axes also were reductive, but in contrast to the Oldowan tools, they appear to reflect the imposition of a design or a mental template on the rock. Some archaeologists have objected to the term "mental template," noting that the early bifaces are especially crude and that even some later specimens exhibit a relationship between the original blank and the final form.[110] By 1 million years ago, however, there seems little doubt that many bifaces were the product of a design based on information in the brain. Many of the later bifaces were fashioned through a process of three irreversible steps: (1) striking a large stone flake from a core or fragment; (2) roughing out the general ovate form; and (3) trimming the margins to refine the shape, often yielding a highly symmetrical final form (figure 3.5).[111]

A more long-standing debate over hand axes concerns their function. What were these artifacts used for, and how did their often finely rendered shape contribute to their function(s)? Microscopic analysis of the edges indicates that some specimens were used to dismember large-mammal carcasses.[112] In some regions, however, *Homo* butchered carcasses with simple chopping tools and flakes, and it is difficult to see how the ovate shape of the biface contributed to its function as a butchering implement. Moreover, many bifaces were either not used or used only lightly, and it is unclear if they had any intended function as a tool.[113]

An alternative explanation, advanced at the end of the 1990s, is that the large bifaces were a material form of *display* (that is, an artifactual peacock's tail).[114] This idea has been controversial, and much of the criticism has been focused on the suggestion that hand axes represent male sexual displays.[115] But they may indicate general displays of social solidarity in the context of the networks based on reciprocity that humans eventually developed—and

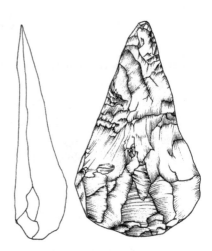

FIGURE 3.5 A large biface or hand ax from the site of *Swanscombe* (England), illustrating the more refined appearance of later bifaces (length = 16 centimeters). (Redrawn from John Wymer, *Lower Palaeolithic Archaeology in Britain as Represented by the Thames Valley* [London: Baker, 1968], 51, fig. 14)

that probably helped drive the expansion of the *Homo* brain during this extended interval of time. Material displays of social solidarity are not uncommon in modern humans (for example, national-flag lapel pin).

We will probably never know why early humans made hand axes, because there is no way to rigorously test and refute alternative explanations for their function. The appearance of material forms of social display after 1.8 million years ago is consistent, however, with the suggestion that the changes in foraging strategy that occurred between 2.8 and 1.8 million years ago placed a premium on social-gaming skills. From this perspective, the hand ax represents another manifestation of the selection pressures that drove brain expansion. It reflects the investment of time and materials in computing social relationships.

Modern Human Origins and the "Social Brain"

In chapter 4, the paleoanthropological record of the origin of modern humans—beginning with their immediate ancestors in Africa about 500,000 years ago—is described and interpreted. As noted in chapter 1, *Homo sapiens* is interpreted here as a "major transition" in evolution, entailing fundamental changes in the storage, transmission, and translation of information, and comparable to earlier such transitions described by John Maynard Smith and Eörs Szathmáry more than two decades ago.[116] Some of the changes are evident in the paleoanthropological record before the advent of the modern human anatomical pattern, however, roughly 300,000 to 150,000 years ago.

The changes in the storage, transmission, and translation of information are interpreted as the result of intense selection pressures created by the competitive social environment (that is, reciprocal alliance network). These selection pressures probably underlay the evolution of enhanced cognitive faculties related to managing a growing set of reciprocal relationships with non-siblings (including male–female pair-bonds), to the long-term reproductive advantage of individuals. They likely drove the steady expansion of the brain after about 2 million years ago (social brain hypothesis)[117] or after the development of a collective foraging pattern that allowed *Homo* to occupy habitats where resources were less plentiful and less predictable. Total brain volume seems to have reached its maximum limit (relative to body size) about 300,000 to 150,000 years ago in sub-Saharan Africa. From this point onward, selection apparently promoted the development of other means of improving the reproductive success of the "social brain." The most important of these was a more complex system of computation, represented by syntactic language.

Chapter Four

RECENT AFRICAN ORIGIN

That language is an instrument of human reason, and not merely a medium
for the expression of thought, is a truth generally admitted.

—GEORGE BOOLE (1854)

In 1797, villagers near Toulouse in southern France captured a boy who had
been living in the woods since the age of five. The boy escaped, was recap-
tured and escaped again, but later emerged from the wild voluntarily.
Although it was apparent that he could hear, the boy (who was now thought
to be 12 years old) was taken to the National Institute of the Deaf in Paris.
He was eventually named Victor of Aveyron and effectively adopted by a
medical student and later physician, Jean-Marc-Gaspard Itard (1775–1838),
who attempted to teach him language and eventually wrote a book about
the experience.[1]

Although able to speak a few words, Victor never learned to use syntactic language and became an important part of the evidence for a "critical
period" in the development of a child with respect to language acquisition.[2] There have been a few other cases of "feral" children, as well as children denied exposure to spoken language by unusual circumstances. The
best-known example of the latter is a girl born in 1957 in California
("Genie"), who was kept in total isolation for many years by an abusive
parent and, while later making substantial progress in nonverbal communication, also was unable to acquire full language ability.[3]

These cases, although fortunately few in number, provide a vital clue
to the evolution of modern humans. They support the hypothesis that
children pass through a critical period—typically between age five and
puberty—for language acquisition, after which it is difficult and perhaps

impossible to acquire syntactic language. (Adults can pick up a second language, but fluency is not easy to attain, especially as they grow older.) They indicate that language faculty is neither simply inherited genetically nor learned by experience, but is the product of a unique and unprecedented integration of brain development and non-genetic information. The neural networks of the developing brain appear to be shaped in part by exposure to language spoken by adults.

The process of language acquisition reflects the high degree of *cerebral plasticity* in modern humans, especially juveniles. Many animal species undergo a critical period in brain development with respect to the visual system or other aspects of cognition, but the influence of environmental stimuli on the organization of the human brain during growth is extreme, and apparently accounts for both the cognitive faculties and certain neurological disorders unique to *Homo sapiens*.[4] The contrast with our closest living relatives is evident in the timing of brain growth in the infant and child. In chimpanzees, the brain of the newborn is already 50 percent of its adult size, which is attained within two years. In living humans, the brain of the newborn is only 25 percent of its adult size, and the period of growth is literally ten times greater: 20 years.[5] Cerebral plasticity seems to be a major variable in the acquisition of syntactic language and the evolution of modern humans.

Language as Computation

Like many other metazoan forms of life, humans perform complex computations with information received and stored in the neural networks of the brain—especially visual information—related to navigation and other functions. Modern humans also perform computations with the units of language (and mathematics), which is both a means of communication and a system of computation. The computational properties of language lie in the generation of virtual objects (or what the philosopher Andy Clark has termed "material symbols") in the form of spoken or imagined words and numbers.[6] As Charles Darwin and others have noted, the mental manipulation of these material symbols—like the physical manipulation of the beads of an abacus—appears to be essential for complex thoughts and calculations, especially when they entail multiple hierarchical levels with embedded components (for example, a compound sentence with subordinate clauses).[7] As words and numbers are moved around in the neural networks of the adult brain, it becomes what Daniel Dennett labeled a "virtual machine."[8] Language is the original *information technology*.

And it is the computational—not the communicative—aspects of language that are likely to have been selected for, because they almost certainly conferred significant reproductive advantages on individuals who acquired this faculty. Despite the enormous impact of syntactic language on human communication, it is difficult to explain the evolution of any form of communication that does not increase the fitness of the individual relative to others.[9] The use of language for communication probably is a secondary effect, or—as Noam Chomsky suggested a few years ago—it "might turn out to be a kind of epiphenomenon."[10]

How is the *computational complexity* of syntactic language defined and measured? Some years ago, Herbert Simon proposed a general definition and measure of "complexity" in terms of hierarchical levels and subcomponents (chapter 1).[11] Computation with syntactic language (and higher mathematics) generates hierarchically structured products with many levels and subcomponents. In a classic paper on language and object manipulation in living humans, Patricia Greenfield described both as examples of "hierarchical complexity."[12] The computational complexity of ordinary language use has been compared with a polynomial function or quadratic equation (both of which exhibit a hierarchical structure).[13]

The combinatorial possibilities of language—given its potentially complex hierarchical structure—are limitless and exhibit what Chomsky labeled the property of "discrete infinity."[14] And because most of the units of information have referents assigned by convention (that is, represent public symbols), the computations of language (and mathematics) may be applied to subjects that lie outside the immediate spatial and temporal setting of the brain in which they are being generated. From an evolutionary perspective, this may be their most important feature because they offer the potential of *predicting the future* (that is, reducing uncertainty), including the possible future computations of other brains and the behavior of the individuals in which they reside. In the complex and competitive environment of human social groups—organized by pair-bonds and reciprocal alliances—this faculty is likely to enhance reproductive fitness (chapter 3).

The combinatorial possibilities of language are not only limitless, but also unpredictable—living humans seem to be capable of generating random thoughts—and the computations of language have been characterized as "nondeterministic."[15] The phenomenon reminds us of the ancient debate over determinism and "free will" and of René Descartes's (1596–1650) famous distinction between the *body* ("a machine made by the hand of God") and the *mind*, with its "freedom peculiar to itself."[16] Our perspective on biology has changed since the seventeenth century, however, and it now is possible to think of the body of an organism as the product of another nondeterministic process—genetic evolution—with its own mechanism for generating random variation (chapter 1). The simplest explanation for the "freedom peculiar" to the human brain is a parallel process of informational computation that incorporates some degree of randomness (that is, "controlled entropy").

There is reason to believe, moreover, that the process of nondeterministic computation in the brain is not unique to humans; that is, the uniqueness of modern human cognition lies elsewhere (chapters 1 and 3). Male humpback whales are capable of generating novel songs, which presumably requires a similar faculty for the creative recombination of neuronal information.[17] The presence of this ability in two mammalian orders—Cetacea and Primates—that diverged more than 50 million years ago suggests that it is widespread among placental mammals. In his "theory of neuronal group selection," Gerald Edelman emphasized the importance of random processes in the brain across the whole phylum Chordata.[18] They conceivably play a role analogous to that of random genetic variation in the nongenetic adaptations of the most complex metazoa. What makes humans unique is the translation of the nondeterministic computations with information in the brain into complex structures outside the brain (including the "material symbols" of spoken language). It gave birth to a new kingdom of artificial creatures.

Modern Human Origins

The problem of the origin of modern humans is framed here in terms of *how* and *why* the computational complexity represented by syntactic language evolved in *Homo sapiens*. The *when* and *where* are established, at least in broad terms: 300,000 to 150,000 years ago in Africa. The evolution of modern human cognition has often been framed with reference to the origin of language (or symbols), but the emphasis has been on the communicative rather than the computational aspects of language. Because the spoken symbols of language have no fossil or archaeological visibility, archaeologists sometimes have directed attention to preserved material symbols, such as visual art or personal ornaments (or what the psychologist Merlin Donald termed "external symbolic storage").[19]

In my view, the long-standing emphasis on symbolism is misplaced. Symbols are universal in living systems in the form of genetic information and also are found in the form of neuronal information among almost all metazoans. In a few cases, such as the honeybee dance and vervet monkey alarm calls, other forms of information and symbolism have evolved. And, as discussed in chapter 3, early humans probably evolved a simple form of language (or non-syntactic proto-language) roughly 2 million years ago. Modern humans evolved their unique system of computation from this older system of communication, although they undoubtedly expanded the lexicon far beyond the original set of words, or "material symbols."

Once modern humans had transformed their brains into virtual computational machines, they applied their faculty for creating complex, hierarchically structured formations to all spheres of life. Personal ornaments, abstract designs, paintings, and sculptures all exhibit the potential for hierarchical complexity and "discrete infinity." They are secondary effects of enhanced computation in *Homo sapiens*, but they also represent new forms of information that had major consequences for human life.

Even more consequential was the impact on technology, which exhibits the same potential for hierarchical organization and design innovation (although it unfolds gradually in the archaeological record after about 150,000 years ago). Modern humans redesigned themselves as organisms and eventually redesigned much of their environment, engineering their rapid dispersal over Earth and subsequent generation of massive population centers and altered landscapes. The technology of modern humans was not a new form of information, but a new form of *structure*—like an organism—based on complex or semantic information.

Computational Complexity and Cerebral Plasticity

The hypothesis presented here for the evolution of modern human cognition ("computational complexity model") may be summarized as follows:

- Modern humans exhibit a unique faculty for computation (the computational aspects of syntactic language) that is characterized by the manipulation of digital units of information borrowed from vocal communication in complex hierarchically organized sequences ("hierarchical complexity").

- The computational complexity of modern humans is based on the translation of neuronal information into higher-order information units in the form of spoken and unspoken vocalizations ("material symbols") that are manipulated in the neural networks as a "virtual machine."

- The computational complexity of modern humans evolved as a feature of brain development, reflecting a high degree of cerebral plasticity, which, in turn, is a function of the massive complexity of the brain in later *Homo* (that is, much of the organization of the brain is derived from environmental input rather than genetic information). Children denied adequate exposure to syntactic language during the critical period of development are unable to acquire the faculty of language (that is, the computational complexity inherent in syntactic language).

- The process of language acquisition in children is closely related to the process through which the child acquires the faculty for manipulating objects in complex, hierarchically structured, sequences ("practical intelligence" or technological competence).

- The computational complexity of modern humans evolved in conjunction with an existing body of information in the form of proto-language (spoken and unspoken sounds with referents) and structure in the form of complex artifacts—both of which play a critical role in the development of the brain.

The close relationship between the acquisition of language and "practical intelligence" (object manipulation or tool use) may offer insight into how the computational complexity of modern humans evolved. The issue has deep roots in the field of developmental psychology, beginning with the stages of child development recognized by Jean Piaget (1896–1980) and the work of L. S. Vygotsky (1896–1934).[20] Drawing on available primate studies for comparison, Vygotsky argued that the acquisition of "technical thinking" in the child precedes—and is critical to—the development of "intelligent speech." Vygotsky also emphasized the computational significance of language and saw parallels with "the use of tools."[21]

The archaeological record now reveals the making of composite artifacts (or "complex artifacts") no later than 500,000 years ago. These weapons and tools eventually included both stone-tipped spears and hafted cutting/scraping implements comprising at least three elements (each of a different raw material) that were brought together in a preconceived hierarchically organized design. Because artifacts of comparable complexity are absent in the earlier archaeological record, the computations underlying the manufacture of composite tools and weapons probably were more complex than those within the ability of early *Homo*. The technological algorithms incorporated more input (materials) and required more steps or operations (chapter 1).

There is some evidence that the pronounced delay in growth and maturation so characteristic of modern humans began at roughly the same time as the appearance of complex artifacts (ca. 500,000 years ago).[22] Because there is a clear relationship between the timing of brain growth and the acquisition of syntactic language in living people, their possible co-occurrence suggests that an earlier jump in computational complexity may have evolved along the same path: delayed maturation and increased cerebral plasticity. The predecessors of modern humans also may have undergone a critical period of early brain development, influenced by a body of information and structure collectively generated by the social group over an extended period of time. The appearance of complex artifacts in the archaeological record may signal the initial phase of a process that eventually led to the computational complexity of modern humans.

Brain Size and the Origin of Modern Humans

The wider evolutionary context for cerebral plasticity in *Homo* lies in the extraordinary growth of cranial volume described in chapter 3. By roughly half a million years ago years ago, brain size had reached a level comparable to that of living people (when adjusted for latitude). Although further growth occurred after 500,000 years ago—and even after 300,000 years ago—cranial volume subsequently returned to the earlier level (1,250–1,299 cubic centimeters [cc] for tropical Africa).[23] The immediate ancestors of modern humans had evolved a brain that was roughly twice the size of that of early *Homo*, with an exponential increase in the number of potential synaptic connections.

The most striking consequence of this massively expanded brain was that much of it was and is constructed with non-genetic information. As John Mayfield observed, the metazoan genome simply lacks the information-storage capacity to provide more than basic design elements for the modern human brain.[24] The metazoan brain—especially in mammals—evolved the ability to store a large quantity of information about the environment, including the social environment. But the *Homo* brain of 500,000 years ago seems to have arrived in a new place altogether. Not only was the organization of the developing brain heavily influenced by non-genetic information, but a significant amount of that information was not derived from the natural environment, but from the social environment—in other words, human groups. The information probably included the "material symbols" of proto-language and certainly included the technological algorithms of complex artifacts.

This chapter begins not with the appearance of *Homo sapiens*, but with the most recent common ancestor of modern humans, Neanderthals, and Denisovans, roughly 500,000 years ago. The taxon often is referred to as *Homo heidelbergensis* (named on the basis of a fossil discovered in Germany in the early twentieth century, despite its African origin) and is associated with the earliest evidence for complex artifacts.[25] The temporal scope reflects the perceived importance of such artifacts in the development of computational complexity in modern humans. The later phase of the evolution of modern humans is represented by the advent of the modern anatomical pattern (*Homo sapiens*) as early as 300,000 years ago in the African fossil record. It is followed by archaeological evidence for discrete infinity and innovative technology, which are interpreted here as secondary effects of enhanced computational complexity (that is, syntactic-language faculty). Also pertinent to the timing of the origin of anatomically modern humans are the combined estimates of time to the most recent common ancestor for all living maternal and paternal lineages, which currently yield a range of around 190,000 to 143,000 years ago (see table 2.1), providing at least a minimum date for *Homo sapiens*.[26]

The wider evolutionary context for *Homo sapiens* also encompasses the growing body of data against which models of modern human origin—including the computational complexity model—may be tested and potentially refuted. Beyond the archaeological data already mentioned, there are comparative anatomical and genetic data from the fossil record on maturation rates and brain development in modern humans and their immediate ancestors. Both the sequence of dental development as a proxy for life history and fossil crania indicate that the delayed maturation and extended brain growth characteristic of living humans was absent in the australopithecines and early *Homo*.[27] The trend toward slow maturation appears to have begun with *Homo heidelbergensis*,[28] and many short DNA sequences related to epigenetic regulation of brain development have been identified in modern human, Neanderthal, and Denisovan genomes (suggesting a common origin in or before *H. heidelbergensis*).[29]

Evidence for a second phase of delayed maturation and an altered trajectory of brain development has emerged from a comparative analysis of Neanderthal and modern human fossils. An early *Homo sapiens* mandible from North Africa exhibits an essentially modern pattern of dental development, while Neanderthal fossils reveal at least subtle life-history differences with modern humans.[30] Equally important is the discovery (based on a comparative analysis of infant crania) that the early "globularization phase" of development is unique to *Homo sapiens* and reflects the rapid expansion of the neocortex and cerebellum during the first year of life.[31] The relative expansion of the cerebellum may be significant because of its dual role in toolmaking and language.[32]

THE MOST RECENT COMMON ANCESTOR

All living people are descended directly from a large-brained species of *Homo* that lived in Africa about 500,000 years ago. Comparative genomics indicates that both Neanderthals and the mysterious Denisovans of southwestern Siberia also are descended from this African species of *Homo*, which represents the most recent common ancestor (MRCA) of all three taxa. Roughly 500,000 years ago (the dating is uncertain), it migrated out of Africa into Eurasia and became the source for at least two *Homo* lineages that were still present when modern humans dispersed out of Africa about 75,000 to 60,000 years ago. In fact, comparative genomics also indicate that some genic exchange took place between modern humans and both Neanderthals and Denisovans.

As already noted, the African ancestor of all three lineages often is referred to as *Homo heidelbergensis*, although there is some disagreement among paleoanthropologists as to how this term should be used. It was first applied to a mandible recovered at *Mauer*, near the city of Heidelberg, Germany, in 1907 at a time when the human fossil record was very sparse (and thus there was little material available for comparison). When the *Broken Hill (Kabwe)* cranium was recovered in Africa in 1921, a connection to the Mauer jaw was not obvious. Therefore, this skull, which was erroneously believed to be very recent, was accorded its own species designation (*Homo rhodesiensis*). In the 1970s, however, anatomical parallels were noted between the Broken Hill skull and a cranium recovered in 1960 from *Petralona* (Greece). Because the Petralona skull appeared to be of comparable age to the Mauer jaw (and both were of comparable age to the now re-dated Broken Hill cranium), all three became linked together, and some researchers placed them all in the taxon *Homo heidelbergensis* (figure 4.1).[33]

Technically, even the earliest Eurasian representatives of this group should be placed in separate lineages, despite their obvious similarities to the African fossils, and other investigators have assigned all European fossils in this clade to the Neanderthals.[34] Over several hundred thousand years, the European population evolved—through genetic drift and adaptation to local conditions—a set of distinctive features that are typical of the classic Neanderthals. The Denisovans (who were more closely related to the Neanderthals than to African *Homo*) presumably evolved their own set of characteristics, although to date the Denisovans are known almost entirely through ancient DNA (aDNA). If the Eurasian fossils are placed in separate lineages (or in one Neanderthal lineage with two branches), it no longer makes sense to apply the species designation *heidelbergensis* to the African fossils.

The term *Homo heidelbergensis* is used here nevertheless because it is widely accepted among paleoanthropologists (it is common in the

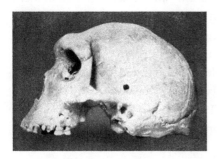

FIGURE 4.1 The skull from Broken Hill (Kabwe), often assigned to *Homo heidelbergensis*. (From *"Homo rhodesiensis,"* Wikipedia, en.wikipedia.org/wiki/Homo_ rhodesiensis)

literature) and because it has heuristic value: it captures a set of anatomical traits and associated archaeological evidence that represent a critical point of departure for modern humans, despite the multiple lineages subsumed among the major fossils (and their associated archaeological remains). In fact, while the later Eurasian fossils are excluded from this group, I believe that they may be considered proxies for *Homo heidelbergensis* in certain ways. Although the Eurasians evolved their own distinctive traits, their archaeological record is fundamentally similar to that of their African ancestors and suggests similar cognitive faculties. More specifically, the Neanderthals and Denisovans seem to have inherited the enhanced computational complexity of *Homo heidelbergensis*—reflected in their manufacture of complex artifacts—but never developed the further enhancement evident in the archaeological record of *Homo sapiens* after 150,000 years ago. Because aDNA is preserved in only the northern fossils, the Neanderthal and Denisovan genomes may provide essential clues to the genetic changes related to brain development that distinguish modern humans from their immediate ancestors, as well as clues to why *Homo heidelbergensis* was different from early *Homo*.[35]

Homo heidelbergensis represents a major shift in human evolution and, in a general sense, may be considered the first "modern human." Our immediate African ancestor exhibits three important—and potentially interrelated—characteristics that set it apart from early *Homo* and tie it more closely to ourselves. The first is brain size, which, as already noted, had reached an essentially modern level by 500,000 years ago, at least for a tropical African setting. Arguably, both the size of the brain and the constraints on further growth were key factors in human evolution from this point onward. Second, several lines of evidence indicate that the rate of maturation (and specifically of brain maturation) was significantly slower in *Homo heidelbergensis* than in early *Homo* and closer to that of modern humans. Delayed growth of the brain is associated with the acquisition of advanced cognitive faculties, including syntactic language, and the advent of this trend seems especially important. And third, *Homo heidelbergensis* was the first human to make complex artifacts, which may indicate an increase in computational complexity in the same way that the archaeological record of *Homo sapiens* after 150,000 years ago may be interpreted as a product of enhanced computation.

Skeletal Anatomy

The major fossils are listed in table 4.1 and include four crania from sub-Saharan Africa and several cranial and postcranial remains from Eurasia. An isolated limb-bone fragment from *Boxgrove* (southern England) documents the northernmost extent of the group, yields interesting information on adaptation to climate, and is associated with an important archaeological site. The small sample from Africa underscores the need to include closely related specimens from Eurasia in the definition of the taxon, especially with respect to cranial capacity (brain size). In general, the selected Eurasian fossils antedate 400,000 years and exhibit few or no derived features that characterize the non-African lineages—most notably the Neanderthals. Although no East Asian specimens are in the table, several have been suggested as possible representatives of *Homo heidelbergensis* (figure 4.2).[36]

Several paleoanthropologists have suggested a set of skeletal traits for *Homo heidelbergensis* that collectively distinguish this species from both early *Homo* and modern humans (and Neanderthals). Among others, they include (1) well-developed and continuous supraorbital torus (brow ridge), (2) increased width of parietal bone (and more rounded) relative to early *Homo*, (3) reduced postorbital constriction relative to early *Homo*, (4) more rounded occipital bone, (5) high arched temporal squama, (6) reduced facial prognathism, and (7) elongated superior pubic ramus.[37] As Daniel Lieberman observes, some of these features—such as the reduced postorbital constriction and more rounded occipital bone—are related to the larger brain; in general, *Homo heidelbergensis* is a robust version of anatomically modern humans.[38]

For the specimens listed in table 4.1, estimated cranial capacity ranges between 1,100 cc (*Lake Ndutu* [Tanzania]) and more than 1,400 cc (*Vértesszöllös* [Hungary]), but these are outliers. Most of the measurable fossils are 1,200 to 1,250 cc (Ralph Holloway and his colleagues suggested a somewhat higher mean of 1,266 cc), which is only slightly below the mean for tropical Africa among living people (1,250–1,299 cc).[39] It should be noted that the dating of all the fossils assigned to *Homo heidelbergensis* is approximate (most are given as ranges of tens or hundreds of thousands of years) and that the actual age of any one specimen could be significantly more or less. Moreover, the precise stratigraphic provenience of some specimens

TABLE 4.1

African and Eurasian Fossils Assigned to the Taxon *Homo heidelbergensis*

Site	Estimated Age (method)	Description	Cranial Capacity (cc)	Comments
Bodo (Ethiopia)	640,000–550,000 years (argon/argon)	Partial skull with face; parietal of another skull; distal humerus shaft	1,250	Cut-marks on one skull fragment
Elandsfontein (South Africa)	600,000 years (biostratigraphy)	Skullcap and mandible fragment	1,225	Associated with artifacts and faunal remains; mandible similar to Mauer
Lake Ndutu (Tanzania)	500,000–300,000 years (archaeological stratigraphy)	Cranium	1,100	Age estimated on the basis of associated artifacts
Broken Hill (Kabwe [Zambia])	300,000–200,000 years (ESR)	Cranium; maxilla fragment; humerus fragment; pelvic bones; femur fragments; tibia fragment	1,325	Discovery of cranium, originally classified as *Homo rhodesiensis* (1921)
Narmada (India)	600,000–400,000 years (biostratigraphy)	Skullcap	More than 1,200	Associated with hand axes
Petralona (Greece)	400,000–200,000 years (U-series)	Cranium	1,230	Similar to Broken Hill and original link between African and Eurasian fossils for this taxon
Mauer (Germany)	500,000 years (biostratigraphy)	Mandible		Discovery of and first fossil classified as *Homo heidelbergensis* (1907)
Boxgrove (England)	500,000 years (biostratigraphy)	Tibia shaft		Associated with hand axes and evidence of large-mammal butchery
Arago (France)	425,000–400,000 years (ESR, U-series)	Cranium; mandible; pelvic bone; isolated teeth; others	1,166	Evidence for intentional removal of a portion of cranium

Site	Estimated Age (method)	Description	Cranial Capacity (cc)	Comments
Vértesszöllös (Hungary)	Ca. 400,000 years (U-series, biostratigraphy)	Occipital bone; deciduous teeth	More than 1,400	Associated with Lower Paleolithic artifacts, but no hand axes

Sources: Based on Michael Day, *Guide to Fossil Man: A Handbook of Human Palaeontology*, 3rd ed. (Chicago: University of Chicago Press, 1977); Günter Bräuer, "The Origin of Modern Anatomy: By Speciation or Intraspecific Evolution?" *Evolutionary Anthropology* 17 (2008): 24, table 1; Richard G. Klein, *The Human Career: Human Biological and Cultural Origins*, 3rd ed. (Chicago: University of Chicago Press, 2009), 290–308; Chris Stringer, *Lone Survivors: How We Came to Be the Only Humans on Earth* (New York: Times Books, 2012), 253–261.

FIGURE 4.2 Map of Africa and Eurasia, showing the location of fossils assigned to *Homo heidelbergensis* or its close relatives, dated to between roughly 600,000 and 400,000 years ago. (Base map adapted from en.wikipedia. org/wiki/File:BlankMap-World6.svg)

(for example, Broken Hill) is uncertain. Overall, brain size exhibits a significant increase (ca. 25 percent) relative to that of early *Homo* and is comparable to that of living humans, when adjusted for latitude and mean annual temperature.[40]

Evidence for maturation rates is derived primarily from fossil teeth, which generate microscopic lines of incremental growth (*perikymata*) in the enamel during maturation. In modern humans, the perikymata are tightly packed together, reflecting a slow rate of growth relative to that of apes and early humans. The growth rate of teeth is a reliable proxy for the overall rate of maturation, and this includes the timing of brain growth. Perikymata packing in early *Homo*, as well as in the australopiths, is comparable to that in modern and fossil African apes.[41] The faster growth rate in early *Homo* is supported by an independent assessment of brain development in a 1.8-million-year-old fossil from Java (see box 2.4).[42]

The evidence for a slower rate of growth in *Homo heidelbergensis* than in early *Homo* is based on measurements of a large sample of teeth from *Atapuerca* (Spain) that probably date to roughly 400,000 years ago and indicate a pattern similar to that of living people.[43] Although these fossils are widely considered early Neanderthals, they are not far removed from the most recent common ancestor of modern humans and Neanderthals. Evidence for delayed maturation in the later Neanderthals also supports a slower growth rate in *Homo heidelbergensis*, because it is more likely that Neanderthals and modern humans inherited the same trait from a common ancestor than that each evolved it independently.[44] There are conflicting estimates of growth rates in Neanderthals, however, and the subject is controversial. It has been suggested that the later Neanderthals evolved faster growth rates than their African ancestors as part of their adaptation to a cold climate.[45]

Diet and Ecology

As in the case of early *Homo*, most of the information about diet and ecology in *Homo heidelbergensis* is derived from a small number of datable archaeological sites that contain well-preserved animal (and, occasionally, plant) remains. The critical question is whether there was a significant change—or even a modest but measurable change—between earlier humans and *Homo heidelbergensis*. The issue is important because diet and ecology are potentially tied to brain size, maturation rate, and technology, and change or constancy in the food that people ate and how they acquired it may help explain the anatomical and behavioral developments that precede the emergence of modern humans.

Overall, there is little evidence for a significant change in diet and ecology with the appearance and spread of *Homo heidelbergensis*, but this may not apply to their northern representatives, who evolved into the Neanderthals and Denisovans. Sites occupied by *Homo heidelbergensis* yield the remains of large mammals that were either hunted or scavenged. The case for large-mammal hunting is strongest in some Eurasian sites. But there is no sign of the mass-kill techniques (for example, driving herds of bison or horses into natural traps) that later became common. Nor is there evidence for the systematic harvesting of small mammals, birds, and fish—all of which require specialized technologies in the form of throwing darts, traps, and other complex devices (and many of which include mechanical components).

The heavy consumption of a diverse array of plant foods, including roots and tubers, is assumed, especially for the lower latitudes, although supporting evidence is rare and often ambiguous. The best example is *Gesher Benot Ya'aqov* (Israel), which produced remains of a variety of nut-bearing plants, as well as pitted stones that probably were used for processing these and other vegetal foods.[46] The tubers likely were extracted from hard surface soils with the use of pointed stones or bifacial picks, and while some of these technologies were specific to plant foods, they were simple and are found today among nonhuman primates.

At the site of *Elandsfontein* (South Africa), which is thought to be at least 600,000 years old and contains skeletal remains assigned to *Homo heidelbergensis* (see table 4.1), the remains of a wide variety of ungulates and carnivores—as well as some elephant and baboon, and a few smaller mammals—were found concentrated around a former marsh or spring. The distribution of skeletal parts is skewed toward axial elements (skulls, vertebrae, and ribs), which cannot be explained by differential preservation, and it may be assumed that limbs were removed from the carcasses. Traces of stone-tool marks on a large sample of ungulate bones were rare (less than 1 percent), however, and less common than the gnaw-marks of carnivores and porcupines. Thus while both skeletal remains and artifacts document the recurring presence of *Homo heidelbergensis* at Elandsfontein, human procurement of large mammals at the site seems to have been limited.[47]

By contrast, several Eurasian sites contain clear evidence for the consumption of large mammals. Horses probably were hunted and butchered at Boxgrove and at *Schöningen* (Germany), thought to date to about 500,000 and 400,000 years ago, respectively.[48] And later European sites occupied by the "classic" Neanderthals yield abundant taphonomic evidence for the hunting and butchering of large animals.[49] The analysis of stable isotopes in Neanderthal bone indicates heavy consumption of meat from various mammalian taxa, including mammoth and rhinoceros, but also shows up to about 20 percent of the Neanderthals' protein as having derived from plants.[50] It is not clear, however, if the emphasis on large mammals at higher latitudes reflects superior hunting abilities in the northern descendants of *Homo heidelbergensis* or simply local environmental conditions, because the same pattern can be observed among recent foraging peoples.[51]

Complex Artifacts

The most significant development in *Homo heidelbergensis*, as stressed in earlier discussions, is the making of *complex artifacts*. The most recent common ancestor of modern humans and Neanderthals produced composite tools and weapons assembled from at least three components, exhibiting a hierarchically structured design (with two organizational levels). Complex artifacts signal a "major transition" in evolution with respect to the storage, transmission, translation, and manipulation of information. Earlier human and nonhuman artifacts are reductive—including the hook tools fashioned from leaves by New Caledonian crows—and although they occasionally involve the use of objects in combination or sequence, no known examples entail the integration of different components into a unified design.[52] Complex artifacts also represent the first "uniquely human" trait in the fossil record. They reflect an increase in material computational complexity, as do the even more complex artifacts that begin to appear after the evolution and spread of *Homo sapiens*.

Not a single composite tool or weapon is known from the archaeological record before 20,000 years ago, but their presence 500,000 years ago and later is reliably inferred from two lines of evidence: (1) traces of hafting wear on stone points and retouched flakes that were inserted into the ends of wooden spear shafts or into lateral grooves on wooden handles, and (2) residue of adhesives, which eventually included a number of materials such as bitumen and pine resin. Additional supporting evidence has emerged in the form of impact fractures on points—confirming their use as projectile tips—and trimming on the blanks apparently applied to improve their fit into the haft.[53]

The technological algorithm involved multiple steps in the making of each component before they were assembled into the composite artifact. The stone blanks were struck off prepared cores—itself a complicated process comprising several irreversible steps—and then modified to improve the fit. The fashioning of the wooden shaft or handle also must have required a sequence of steps, including the cutting of a groove or slot, either at the end or along the side of the shaft or handle, presumably with a sharp stone implement. The adhesive also may have involved some preparation, as well as procurement. Finally, the parts were assembled into an integrated whole (figure 4.3).

Although evidence for the production of composite artifacts as early as 500,000 years ago was reported for the first time in 2012, a link between hafting and *Homo heidelbergensis* was implied by the earlier discovery that both Neanderthals and modern humans were making composite tools and weapons.[54] As in the case of delayed maturation rates, the most parsimonious explanation for the pattern is that the trait was derived from the most recent common ancestor of Neanderthals and modern humans. Another

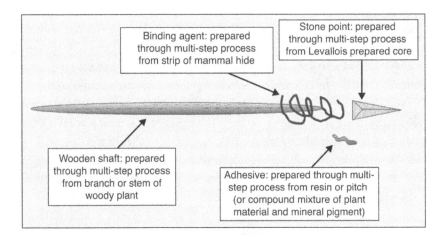

FIGURE 4.3 A complex artifact (stone-tipped spear) assembled from four components, each of which requires a multi-step sequence of preparation from one or more materials. Each component nests within the larger organization of the composite artifact, which exhibits two levels.

clue was offered by the dating (reported some years ago) of prepared cores to about 500,000 years ago.[55] Paleoanthropologists have nevertheless been slow to appreciate the significance of complex artifacts in human evolution.

Two other technologies associated with *Homo heidelbergensis* also may reflect the increase in computational complexity. The analysis of a wooden spear recovered from the 400,000-year-old site of Schöningen by Miriam Haidle revealed a lengthy and complex algorithm or sequence of production steps, including procurement of materials for—and flaking of—the multiple stone tools used to shape the wood.[56] The earliest reliable evidence for the controlled use of fire also is linked to *Homo heidelbergensis*, although, as discussed in chapter 3, it appears increasingly likely that early *Homo* was cooking food. But it seems that only modern humans mastered the technology of *making* fire.[57]

Homo heidelbergensis in Perspective

The appearance of *Homo heidelbergensis* marks the third major shift in human evolution. The first was bipedal locomotion, linked to foraging in relatively open tropical environments. An indirect consequence was a more specialized hand, which is evident in the later australopiths (along with simple stone tools). The second was central-place foraging, linked to a major expansion out of the tropical zone and into the temperate zone, where resources were more widely dispersed and less predictable in space and time. This shift is tied to pair-bonding (indicated by the reduction in sexual dimorphism) and is widely believed to have produced a simple form of language. It also entailed a larger brain, more complex stone-tool technology (and a more specialized hand), and a greater emphasis on meat consumption.

By half a million years ago, the brain had reached an essentially modern volume in its tropical African setting. The changes that took place with the evolution and spread of *Homo heidelbergensis* probably were related to the size of the brain and included delayed maturation (reducing the size of the brain at birth) and complex artifacts. The underlying computations required for the production of complex artifacts may indicate that *Homo heidelbergensis* had evolved a higher faculty for the manipulation of information in the brain (that is, computational complexity). The pattern anticipates and parallels the later changes that are evident with the evolution and spread of *Homo sapiens*. Modern humans also exhibit delayed growth (as well as some unique patterns of brain development) and a major increase in the complexity of their artifacts. The material computations underlying the hierarchically structured artifacts of modern humans are commensurate with those of syntactic language.

The parallel between the two may indicate that both forms of *Homo* evolved through a similar process and that *Homo heidelbergensis* may be considered the first phase of the evolution of modern humans. If living children pass through a critical period for the acquisition of syntactic language, perhaps the children of *Homo heidelbergensis* experienced a similar critical period for the acquisition of the computational faculties that underlay the making of complex artifacts. Modern humans seem to have evolved along a similar pathway, developing a larger brain and further delays in the postnatal growth of the brain—greater computational complexity tied to increased cerebral plasticity during early childhood. As noted earlier, the reciprocal social networks that conceivably drove the evolution of superior computational faculties in modern humans probably were present in *Homo heidelbergensis*.[58]

The intensification of social networking may be indicated by the emergence of regional artifact styles in Africa. After 300,000 years ago, distinct forms become apparent among stone points and bifacially chipped pieces (figure 4.4).[59] Most of the variation may be stylistic rather than functional, and it conceivably reflects an increased investment in material expressions of social identity among local groups, leading to a convergence of particular regional forms. This, in turn, may indicate enhanced computational faculties applied to establishing and maintaining reciprocal alliances and social networks. A similar pattern is evident with the transition to modern humans, which was followed by the appearance of personal ornaments and artifacts decorated with abstract designs, along with indications of wider networks (for example, long-distance movement of materials).

At the other end of the social spectrum, competition within and/or between groups may explain the evidence for violence and possible cannibalism on two *Homo heidelbergensis* crania, from *Bodo* (Ethiopia) and *Arago* (France) (see table 4.1). It may also explain the appearance of complex artifacts because—despite the enhanced power and efficiency of hafted tools and stone-tipped spears—they had no obvious impact on diet and ecology, which remained similar to that of early *Homo*, at least at lower latitudes. Perhaps complex artifacts were a secondary consequence

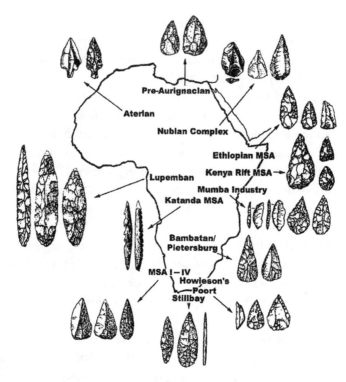

FIGURE 4.4 Regional variation in small bifaces and points in Africa, which becomes evident after 250,000 years ago and may reflect an intensification of social networking. (From Sally McBrearty and Alison S. Brooks, "The Revolution That Wasn't: A New Interpretation of the Origin of Modern Human Behavior," *Journal of Human Evolution* 39 [2000]: 498, fig. 5, with permission of Elsevier)

of enhanced computational faculties related primarily to social gaming (although later possibly playing a significant role in the Neanderthals' diet and ecology). They may have had some value as a form of display and/or contributed to interpersonal violence.

Regardless of the part they may or may not have played in the social life of *Homo heidelbergensis*, complex artifacts offer another potentially important parallel with the *culture* of modern humans. They represent an organized body of structure, based on the translation of neuronal information, that was shared among individuals and transmitted through many generations (that is, transcended individual brains in space and time). It probably was accompanied by an organized body of information in the form of vocalizations and/or gestures, also based on the translation of neuronal information and also shared among individuals and transmitted through many generations. Moreover, the two may have been integrated (for example, artifacts, artifact components, and technological operations categorized vocally) and collectively comprised a simpler version of modern human culture. This body of information and information-based structure may have influenced the organization of the brain during the extended maturation of *Homo heidelbergensis* children in the same way that exposure to syntactic language affects modern children during the critical period.[60]

The Neanderthals and Denisovans are likely crucial to understanding *Homo heidelbergensis* and the subsequent evolution of modern humans. To begin with, the Neanderthals have yielded a wealth of skeletal remains and occupation debris, much of it relatively recent in comparison with the archaeological evidence for *Homo heidelbergensis*. While the Neanderthals clearly reflect the effects of genetic isolation and drift, as well as adaptation to northern climates, over several hundred thousand years,[61] they offer important insights into aspects of cognition and behavior in their last common ancestor with modern humans. The archaeology of the Neanderthals reveals a striking contrast with *Homo sapiens* in terms of the complexity of the underlying computations. Neanderthal artifacts remain confined to two levels of hierarchical organization and never exhibit the highly recursive pattern of modern human artifacts. They offer a credible proxy for computational complexity in *Homo heidelbergensis*.[62]

Both Neanderthals and Denisovans are providing aDNA for genomic comparisons with the DNA of modern humans. Although the Neanderthal and Denisovan genomes contain substitutions specific to their own lineages, they probably lack most of or all the derived substitutions that underlie the unique patterns of brain development and cognitive faculties of *Homo sapiens*. In 2012, an international team of geneticists found substitutions relating to the prefrontal cortex in modern humans that were absent in both the Neanderthal and Denisovan genomes.[63] As with other recent research on the evolutionary genetics of modern humans, the focus was on epigenetics and not on the DNA that codes for protein sequences.[64]

THE RECENT AFRICAN ORIGIN OF *HOMO SAPIENS*

Current dating of human skeletal remains suggests that the modern human anatomical pattern was established in Africa between 300,000 and 150,000 years ago, and possibly somewhat earlier.[65] Many or most of the defining features of *Homo sapiens* can be identified in human fossils (or subfossils) from various parts of the continent (including North Africa) presently dated to this time period. Nevertheless, the anatomical variability within this group of fossils is significant—even extreme—and there are doubts among some paleoanthropologists (including Chris Stringer) that all of them should be classified as modern humans.[66] Many of these fossils are often labeled "near modern" human.[67] Living people probably are descended directly from a subset of the African population or subpopulation in this time range.[68]

It was the dating of African fossils assigned to anatomically modern humans to a time before they are present outside Africa that provided the original basis for the RAO model in the early 1980s (chapter 2). The genetic data offered an important body of supporting evidence for the model, beginning in the late 1980s, with respect to the overall pattern of decreasing diversity as a function of distance from Africa (also evident in the skeletal remains), and the estimated time to the most recent common ancestor for all maternal and paternal lineages, currently thought to have coalesced in Africa around 190,000 to 143,000 years ago (see table 2.1). The global dispersal of modern humans does not appear to have antedated about 75,000 years ago (based on genetic data) and may not have begun until 60,000 years ago.

The exception to this pattern of geographic distribution of dated fossils assigned to *Homo sapiens* is the Levant, where "near modern" human skeletal remains are present as early as 120,000 to 90,000 years ago (*Skhūl* and *Qafzeh* [Israel]).[69] Recently, fossils classified as modern human have been dated to between 116,000 and 106,000 years ago in southern China (*Zhiren Cave*) and possibly earlier at sites in Southeast Asia (chapter 5).[70] The fossils in the Levant have long been regarded as evidence of a short-term expansion out of Africa into an adjoining portion of Eurasia (African "doorstep"), followed by the reoccupation of the area by Neanderthals around 80,000 years ago. The new finds in East Asia may indicate that "near modern" humans spread more extensively in southern Eurasia than previously believed.[71] This raises another set of questions, however. Were these anatomically modern people still present in East Asia when the immediate ancestors of living humans migrated out of Africa (for example, mtDNA haplogroups M and N), or were they replaced by non-modern humans (for instance, Denisovans) as in the Levant? In theory, it should be possible to recognize an early, abortive expansion out of Africa in the genetic data—unless it left no traces in aDNA or the living human genome (box 4.1).[72]

The period between the appearance of modern human anatomy in Africa (ca. 300,000–150,000 years ago) and the beginning of the global dispersal out of Africa (ca. 75,000–60,000 years ago) remains an area of controversy and uncertainty in the study of the origin of modern humans.[73] Much of the focus falls on the relationship between the anatomy of modern humans and the cognitive faculties of living people that underlie syntactic language and complex artifacts. Did these two fundamental aspects of modern humanity evolve in concert, or did the cognitive faculties evolve at another time—either before or after the appearance of modern anatomy? And if the cognitive faculties of living humans were critical to their successful and rapid colonization of Eurasia and Australia (chapters 5 and 6), does the absence of a wider dispersal out of Africa before 75,000 years ago indicate that these faculties evolved much later than the anatomy?

Despite the spatial and temporal variability exhibited by the skeletal remains (and doubts that all can be assigned to the same taxon), the fossil data confirm the presence of modern human anatomy in Africa during this period. By contrast, the archaeological record—which provides the primary source of information on cognition (or what Darwin referred to as "mental powers")—currently offers an ambiguous picture. There are tantalizing signs of increased technological complexity (for example, stone-bladelet production at *Pinnacle Point* [South Africa] that dates to about 165,000 years ago).[74] There are barbed harpoons of bone dated to 90,000 years from *Katanda* (Democratic Republic of Congo).[75] Personal ornaments (such as perforated marine shells) have been recovered from a number of sites in sub-Saharan and North Africa that date to between 135,000 and 60,000 years ago, including sites occupied by anatomically modern humans in the Levant.[76] The most impressive items are pieces of rock or mineral pigment with engraved designs, which include an early specimen from the Levant.[77]

While these finds are characteristic of recent hunter-gatherers equipped with fully modern cognitive faculties, none of them exhibit the complex hierarchical structure of visual art, which may contain three or more organizational levels (analogous to the "phrases within phrases" structure of syntactic language).[78] And none of them necessarily reflect the

BOX 4.1
The "Near Modern" Human Dispersal Out of Africa

Anatomically modern humans dispersed out of Africa into Eurasia at least twice. In addition to the global migration that began 75,000 to 60,000 years ago, Northeast African *Homo sapiens* expanded into the adjoining region of the Levant roughly 120,000 to 90,000 years ago. The contrast between the two dispersals is stark. Whereas the later migration entailed the rapid settlement of southern and northern Eurasia—including the Arctic—and voyages to Australia/New Guinea, the early dispersal was limited to the Arabian Peninsula and Levant (although there is some evidence that it may have reached other areas in South Asia). And whereas the later dispersal produced the cultural and linguistic diversity of the current world population, the early migration was unsuccessful and seems to have left little trace in living people.

The early dispersal out of Africa is important for understanding the origin of modern humans, nevertheless. It is best represented by a set of *Homo sapiens* fossils found in Israel, the dating of which caused considerable confusion among paleoanthropologists in the 1980s, because they antedate some Neanderthal fossils in the Levant and elsewhere. In retrospect, it is apparent that the dating of these anatomically modern human fossils was another piece of the emerging picture of a "recent African origin" for all living people. Eventually, the early presence of "near modern" humans outside Africa was perceived as an added complication to this picture—but one with the potential to provide some critical information to the evolution of *Homo sapiens*.

The "near modern" fossils were recovered from the sites of Skhūl (Mount Carmel) and Qafzeh in central Israel and currently are believed to date to between 120,000 and 90,000 years ago (on the basis of electron spin resonance [ESR] and thermoluminescence [TL]).[1] The remains exhibit substantial variability but, as a group, are widely considered anatomically modern with some "archaic" features. The latter include a well-developed supraorbital torus and a prognathic face on the Skhūl V specimen, as well as some aspects of the teeth and mandibles.[2] Although some speculated that the archaic features were the result of hybridization between modern humans and local Neanderthals, the fossils lack any of the specialized traits of the classic Neanderthals, and the postcranial bones from Skhūl are strikingly modern in appearance.[3]

Archaeologists were initially surprised by the stone artifacts found with the human fossils from both sites because they were more typical of Neanderthals ("Levalloiso-Mousterian") than of modern humans. Eventually, a similar pattern was documented in Africa (that is, anatomically modern humans associated with Levallois cores and tools made on stone flakes). Recently, cores resembling those of the Nubian Complex were reported from southern Israel (Negev Highlands), suggesting a link with Northeast Africa at that time (ca. 120,000 years ago).[4] Also significant is the evidence for personal ornaments in the form of marine-shell beads from both Skhūl and Qafzeh, which reinforces the connection with North Africa (for example, *Grotte des Pigeons* [Morocco]) and parallels the wider evidence for personal ornaments in the African archaeological record before 75,000 years ago (see table 4.4).[5]

The Nubian Complex now is documented in the southern Arabian Peninsula (*Aybut Al Auwal*) at 106,000 years ago, apparently indicating that the expansion of "near modern" humans out of Africa extended to this region as well as to the Levant.[6] As described in chapter 5, recent discoveries of human skeletal remains (primarily isolated teeth) assigned to *Homo sapiens* in southernmost China—and dated to between 127,000 and 70,000 years ago—have raised the possibility that the same expansion reached as far as Southeast Asia.[7] It is unclear, however, if these fossils belong to the same or another "near modern" human population (some argue that they represent earlier representatives of the global dispersal).[8]

The controversy over the interpretation of the recent finds in southern China underscores the importance of establishing the spatial and temporal boundaries of the dispersal of "near modern" humans out of Africa that probably began around 120,000 years ago. The introduction of *Homo sapiens* fossils and/or their artifacts into Eurasia from Africa at that time has the potential to create confusion in the interpretation of the fossil/archaeological record—especially in southern Asia, where much uncertainty remains regarding the local non-modern population (chapter 5). The "near modern" dispersal provides a perspective on the anatomy and ecology of anatomically modern humans before the beginning of the global dispersal between 75,000 and 60,000 years ago. It may reveal critical differences between the people who spread out of Africa 120,000 years ago and our own immediate ancestors, who invaded more challenging habitats and climate zones and replaced the non-modern inhabitants of Eurasia, and thus shed light on what happened in Africa between 300,000 and 60,000 years ago.

1. Chris Stringer, *Lone Survivors: How We Came to Be the Only Humans on Earth* (New York: Times Books, 2012), 251; Richard G. Klein, *The Human Career: Human Biological and Cultural Origins*, 3rd ed. (Chicago: University of Chicago Press, 2009), 476–481.
2. Michael H. Day, *Guide to Fossil Man: A Handbook of Human Palaeontology*, 3rd ed. (Chicago: University of Chicago Press, 1977), 89–102; Bernard Vandermeersch, "The First *Homo sapiens* in the Near East," in *The Transition from Lower to Middle Palaeolithic and the Origin of Modern Man*, ed. Avraham Ronen, BAR International Series S151 (Oxford: British Archaeological Reports, 1982), 297–299.
3. Günter Bräuer, "The Evolution of Modern Humans: A Comparison of the African and Non-African Evidence," in *The Human Revolution: Behavioural and Biological Perspectives in the Origins of Modern Humans*, ed. Paul Mellars and Chris Stringer (Princeton, N.J.: Princeton University Press, 1989), 123–154; Klein, *Human Career*, 476–481. A multivariate statistical analysis of face shape and the relationship of the face to the cranial vault found that the Skhūl and Qafzeh fossils grouped with anatomically modern human samples and not with European and Near Eastern Neanderthals. See Chris Stringer, "The Origin of Early Modern Humans: A Comparison of the European and Non-European Evidence," in *Human Revolution*, ed. Mellars and Stringer, 232–244.
4. Mae Goder-Goldberger, Natalia Gubenko, and Erella Hovers, " 'Diffusion with Modifications': Nubian Assemblages in the Central Negev Highlands of Israel and Their Implications for Middle Paleolithic Inter-Regional Interactions," *Quaternary International* 408 (2016): 121–139.
5. Francisco d'Errico et al., "Archaeological Evidence for the Emergence of Language, Symbolism, and Music—An Alternative Multidisciplinary Perspective," *Journal of World Prehistory* 17 (2003): 1–70; Marian Vanhaeran et al., "Middle Paleolithic Shell Beads in Israel and Algeria," *Science* 312 (2006): 1785–1788.
6. Jeffrey I. Rose et al., "The Nubian Complex of Dhofar, Oman: An African Middle Stone Age Industry in Southern Arabia," *PLoS ONE* 6 (2011): e28239; Goder-Goldberger et al., " 'Diffusion with Modifications.' "
7. See, for example, Christopher J. Bae et al., "Modern Human Teeth from Late Pleistocene Luna Cave (Guangxi, China)," *Quaternary International* 354 (2016): 169–183; Wu Liu et al., "The Earliest Unequivocally Modern Humans in Southern China," *Nature* 526 (2015): 696–700; and Yanjun Cai et al., "The Age of Human Remains and Associated Fauna from Zhiren Cave in Guangxi, Southern China," *Quaternary International* 434 (2017): 84–91.
8. Robin Dennell and Michael D. Petraglia, "The Dispersal of *Homo sapiens* Across Southern Asia: How Early, How Often, How Complex?" *Quaternary Science Reviews* 47 (2012): 15–22; Jane Qiu, "The Forgotten Continent," *Nature* 535 (2016): 218–220.

computational complexity that underlies the production of tailored clothing or self-acting devices, such as snares and traps (automata). Direct or indirect evidence for such structural/functional and computational complexity is found in the archaeological record at roughly 75,000 to 60,000 years ago, when modern humans began their global spread out of Africa. Its absence at earlier dates may indicate that the cognitive faculties of living people did not fully evolve in *Homo sapiens* until shortly before or at the beginning of the global dispersal (or perhaps had not been fully applied to the making of artifacts until that time).

There are several possible explanations for what happened before modern humans migrated out of Africa between 75,000 and 60,000 years ago. One is that anatomically modern humans had evolved cognitive faculties fully commensurate with those of living people around 300,000 years ago (or perhaps even earlier), but the archaeological record has simply failed to yield unambiguous evidence for this before 75,000 years ago. This explanation is highly plausible, given the limitations and biases of the archaeological record. It should be noted that most of the finds mentioned earlier have turned up in the past few years (so more dramatic discoveries in the near future are likely), and definitive evidence for modern cognition currently does not antedate 45,000 years ago anywhere. Under this scenario, the timing of the wider dispersal out of Africa is unrelated to human cognitive faculties, but was constrained by climate, competition with Neanderthals and/or Denisovans, or another factor or combination of factors.[79]

An alternative explanation is that before 75,000 to 60,000 years ago, most or all subpopulations of *Homo sapiens* in Africa lacked fully modern human cognitive faculties, which either evolved or developed in one of these subpopulations between 300,000 and 75,000 years ago. This group (probably concentrated in southern Africa) subsequently spread throughout the continent and began to move out of Africa after 75,000 years ago. According to this model, the global dispersal (voyage to Australia/New Guinea, occupation of Siberia, and so on) was not possible without the full range of computational abilities—including the capacity for designing mechanical instruments and facilities—expressed in the later archaeological record of Eurasia (and Africa). The model is similar to one ("neural hypothesis") proposed in the 1990s by Richard G. Klein and has been described by Paul Mellars.[80]

A Two-Stage Process?

To begin with, it is by no means clear that a "discrete event" in human evolution took place between 300,000 and 150,000 years ago (the time at which the defining anatomical features of modern humans are identified in the African fossil record). Günter Bräuer, one of the original advocates of the RAO model, sees "continuous evolutionary change" in anatomy beginning more than 500,000 years ago in Africa.[81] A "mosaic" pattern of modern and non-modern traits is still evident in the fossils after 150,000 years ago. The estimated coalescence dates for maternal and paternal lineages are significantly younger than 300,000 years and represent either a separate evolutionary event or a genetic bottleneck (or simply low effective population size) in the African population.[82] In a major paper, Sally McBrearty and Alison Brooks argued that both the African fossil evidence and the archaeological data support a gradual evolution of modern anatomy and cognition, beginning no later than 300,000 to 250,000 years ago.[83]

Recent discoveries and analyses of the archaeological data suggest a pattern in the African record between 300,000 and 60,000 years ago that parallels somewhat the spatial and temporal variability found in the fossil anatomical data (especially evident during the later phases of this interval). Some technological innovations (for example, bladelet production) appear in the archaeological record, subsequently vanish, and then reappear much later.[84] In a broad-scale analysis of the archaeological record in southernmost Africa during the past 130,000 years, Alex Mackay, Brian Stewart, and Brian Chase found a pattern of recurring coalescence and fragmentation, correlated with climate. Intervals of cooler and wetter climate are associated with coalescence—indicated by wider sharing of materials, production techniques, and stone-artifact types, and interpreted as a sign of expanded and intensified social interaction. Personal ornaments also are associated with occupations during cooler and wetter periods, supporting the inference of increased networking (box 4.2).[85]

Perhaps the simplest explanation of the fossil and archaeological record pertaining to the evolution and dispersal of modern humans is that it was a *two-stage process*. During the first stage, the defining anatomical traits of *Homo sapiens* began to accumulate in the African population, reflecting a

BOX 4.2
A Hunter-Gatherer Social Network

Besides the evidence for composite artifacts, the most interesting items in the African archaeological record between 500,000 years ago and the global dispersal of modern humans pertain to social life. After 300,000 years ago, stone points and small bifaces exhibited a pattern of regional stylistic variation. And after the appearance of modern human anatomy, personal ornaments and items decorated with geometric designs began to materialize (in the Levant as well as in Africa). Both phenomena may reflect the intensification of social networking among individuals who were not close kin (siblings, parents, and children). Large-scale networks seem to have been a feature of modern human groups as they colonized relatively unproductive environments in Africa and beyond, and have been described among recent hunter-gatherers.[1]

One of the more thoroughly studied hunter-gatherer networks is that of the Ju/'hoansi (formerly known as the !Kung San) of the northern Kalahari Desert in southwestern Africa. The Ju/'hoansi were still practicing a traditional hunter-gatherer economy in the mid-twentieth century that entailed the collection of a variety of plant foods (especially the fruit and nuts of the mongongo tree) and a steady diet of meat (more than 20 percent) obtained from hunting and snaring. Their technology included ostrich-eggshell canteens, nets, fire drills, bows and arrows, and mechanical snares.[2] Despite the desert-margin setting (net primary productivity = 459), the Ju/'hoansi enjoyed a high population density (estimated at 10 to 16 persons per 100 square kilometers) in comparison with northern interior hunter-gatherers.[3]

Local groups (about 25 individuals) were associated with water holes, and each group identified a territory (n!ore) around a specific water hole. The territories ranged in size from about 300 to 600 square kilometers. During times of food stress (caused by drought), local groups arranged access the resources of another n!ore through their social networks.[4] The networks were based on marriage ties, trade, and fictive kinship. Marriage partners were drawn from other groups, and the average distance between the birthplaces of mates was 70 kilometers, ensuring a low level of endogamy.[5] The trading network (!hxaro) was based on the exchange of goods between established trading partners. Each adult had dozens of !hxaro partners (including spouses) and exchanged various items with his or her partners over a period of weeks or years. Most partners lived within 40 kilometers of each other, but sometimes partners resided as much as 200 kilometers away. The goods exchanged included ostrich-eggshell beads, blankets, pots, arrows, and clothing (but not food items), some of which would have visibility in the archaeological record.[6]

1. See, for example, Robert L. Kelly, *The Lifeways of Hunter-Gatherers: The Foraging Spectrum* (Cambridge: Cambridge University Press, 2013), 137–165.
2. Richard B. Lee, *The !Kung San: Men, Women, and Work in a Foraging Society* (Cambridge: Cambridge University Press, 1979).
3. Kelly, *Lifeways of Hunter-Gatherers*, 183, table 7-3.
4. Richard B. Lee, "!Kung Spatial Organization: An Ecological and Historical Perspective," *Human Ecology* 1 (1972): 125–147.
5. John Yellen and Henry Harpending, "Hunter-Gatherer Populations and Archaeological Inference," *World Archaeology* 4 (1972): 244–252.
6. Lee, *!Kung San*, 365–366; Polly Wiessner, "Risk, Reciprocity and Social Influences on !Kung San Economics," in *Politics and History in Band Societies*, ed. Eleanor Leacock and Richard B. Lee (Cambridge: Cambridge University Press, 1982), 61–84; Kelly, *Lifeways of Hunter-Gatherers*, 155–156.

gradual trend that extended back to 500,000 years ago. By 300,000 to 150,000 years ago, a sufficient number of these traits (defined on the basis of recent and living humans) are present in the fossil sample to warrant a somewhat arbitrary dividing line between modern humans and their immediate predecessors. As noted, some younger specimens retain non-modern traits, and at least a few of them may not technically belong to *Homo sapiens*.

The driving force behind the trend toward modern anatomy is assumed to have been continual selection pressure for enhanced social networking, as argued in chapter 3, although some traits may simply reflect random genetic drift. The growing investment of material and energy in manipu-lating reciprocal relationships is manifest both in larger brain volume ("social brain hypothesis"), which may reflect increased information-stor-age capacity and enhanced computational abilities, and in the production of personal ornaments, which represent the most striking change in the archaeological record before 75,000 years ago. But the most significant ana-tomical change before 150,000 years ago may lie in delayed maturation, which is associated with an extended learning period that is critical to the acquisition of language (specifically syntactic language). The fact that the pattern of delayed maturation and brain growth is documented in North Africa around 300,000 years ago suggests that this change took place dur-ing the first stage.[86] Also significant is the bulging parietal region and rounded occipital ("globular braincase") of the modern human cranium, reflecting early growth of the two areas of the brain—one of which is the cerebellum—associated with language.[87]

Assuming that the current lack of earlier evidence for structural and computational complexity in artifacts comparable to that in artifacts of liv-ing people is not due simply to the biases of the archaeological record, fully modern human cognitive faculties may have emerged *after* anatomi-cally modern humans spread throughout Africa (and at least briefly into part[s] of Eurasia). During this hypothesized second stage, a subpopulation of anatomically modern humans in Africa applied the complex computa-tions that underlie syntactic language to the making of artifacts, including instruments and facilities. Such an event may be documented in southern Africa, where there is evidence for the exploitation of birds about 77,000 years ago and indirect evidence for the use of snares between roughly 65,000 and 62,000 years ago.[88]

The subpopulation eventually spread throughout Africa and—equipped with a capacity for designing innovative and complex technologies, such as watercraft and small-mammal traps—rapidly colonized all of Eurasia (including cold habitats unoccupied by other forms of *Homo*) and Australia (inaccessible to other forms of *Homo* because of the water barrier) (figure 4.5).[89] The genetic data, as well as the archaeological evidence, suggest that this subpopulation originated in southern Africa and was ancestral to

CHRONOLOGY	EVENT	SUPPORTING EVIDENCE
75,000–60,000 years ago	Beginning of global dispersal of modern humans out of Africa and into southern and northern Eurasia, Australia/New Guinea, and Beringia	Archaeological sites in Australia/New Guinea by 50,000 years ago and in northern Eurasia by 45,000 years ago 
About 80,000–70,000 years ago	Major population expansion in Africa (probably originating in southern Africa)	Evidence of "archaic human" introgression in genetics of living Africans (pygmies) 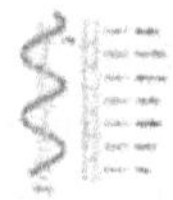
About 80,000 years ago	Neanderthals replace "near modern" humans in Levant during cold period (MIS 4)	Neanderthal skeletal remains in Levant dated to this period suggest replacement of "near modern" humans
120,000–90,000 years ago	Limited dispersal of "near modern" humans out of Africa and into Levant (and possibly southern Asia) during warm period (MIS 5)	"Near modern" human remains in Levant (and possibly southern China) dated to this period; Nubian Complex artifacts on Arabian Peninsula 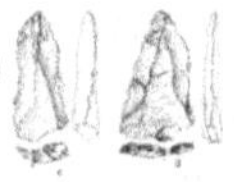
300,000–150,000 years ago	Initial appearance of anatomically modern humans in Africa (including North Africa) exhibiting mosaic pattern of modern and archaic traits (many described as "near modern")	Highly variable sample of anatomically modern human skeletal remains in southern, eastern, and northern Africa (see table 4.2 and figure 4.6) 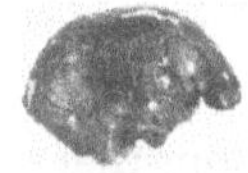

FIGURE 4.5 The evolution and dispersal of modern humans: a chronology of major events. (Modified from Paul Mellars, "Why Did Modern Human Populations Disperse from Africa ca. 60,000 Years Ago? A New Model," *Proceedings of the National Academy of Sciences* 103 [2006]: 9384, fig. 6)

all living humans (that is, mtDNA haplogroup L0). Based on the analysis of several living African populations, some geneticists concluded that there is evidence for an "archaic" genetic introgression from a non-modern African population (box 4.3), analogous to the introgression from Neanderthals in living non-Africans.[90]

Although Klein suggested that a genetic mutation may have been the trigger for a major shift in cognition at this time ("neural hypothesis"),[91] changes in non-genetic information conceivably underlie the "Cambrian explosion" of artifactual life that took place after 75,000 years ago. The slower pace of maturation and brain growth evident before 150,000 years ago probably was the (regulatory) genetic mutation most crucial to the evolution of the cognitive faculties of modern humans. It created the critical period for children and adolescents—apparently lacking among other forms of *Homo*—during which most synaptic connections in the brain are formed and reinforced. The lengthy critical learning period is essential to the acquisition of syntactic language, and thus to the ability to perform the complex computations that underlie syntactic language.

Technological complexity seems to be built on the more fundamental computational faculty associated with syntactic language, and this probably explains why the computational complexity of syntactic language is comparable among all living peoples, while the structural/functional and computational complexity of technology varies widely, depending on the historical trajectory of particular groups of people. The subpopulation of anatomically modern humans living in southern Africa before 75,000 years ago may simply have developed new and complex technologies at a faster pace than other subpopulations in Africa, much as various societies in later human prehistory and history outpaced their neighbors in terms of technological innovation. It could explain why the maternal and paternal lineages coalesced at a much earlier point in time (that is, without any hint of a major genetic event associated with the wider dispersal out of Africa).

Anatomy and Development

Fossils of modern humans from Africa antedating 60,000 years ago are listed in table 4.2, and their geographic locations are shown in figure 4.6. Most of the major fossils are included, along with several younger finds associated with important archaeological evidence for modern behavior.

BOX 4.3
Archaic Introgression in Africa and Modern Human Origins

Most attention to interbreeding between modern humans and other forms of *Homo* has been focused on the Neanderthals and Denisovans in Eurasia (see box 2.2). There also is evidence of interbreeding between modern humans and other forms of *Homo* in Africa, however, and it is relevant to larger issues of their origin and dispersal.

Unlike in northern Eurasia, there is no aDNA to confirm a *genetic introgression* from archaic humans in African *Homo sapiens*, due to poor conditions of preservation in tropical and subtropical climates (further compounded by a significantly greater time depth for the period of interest [300,000–60,000 years ago]). An entirely different methodology is applied to the problem of ancient admixture between modern and non-modern humans in Africa before—and, as it turns out, *after*—the beginning of the global dispersal. In 2000, computational biologist Jeffrey D. Wall proposed that traces of archaic introgression may be detected by identifying unusual patterns of variation in the genome of living humans.[1] Several years later, he and Vincent Plagnol reported such patterns in both European and West African samples.[2] Genetic admixture between ancestral Europeans and Neanderthals was confirmed with aDNA in 2010.[3]

A study published the following year reported signs of archaic introgression in large samples of 61 noncoding regions of autosomal DNA (that is, not mtDNA or Y-DNA) in three contemporary African populations, including two modern hunter-gatherer peoples (Biaka Pygmies and San). The conclusion was that roughly 2 percent of the genome of living Africans is derived from admixture with non-modern humans (comparable to the estimated percentage of Neanderthal genes in living non-Africans).[4] More recently, similar results were obtained with a model-based whole-genome analysis of samples from two West African Pygmy populations (Biaka and Baka).[5] The authors suggested that "frequent but low-level interbreeding between archaic and modern humans or their ancestors might have occurred in the past in Africa."[6]

Evidence of archaic introgression in Africa conceivably could shed light on the origin of modern humans and the question of whether the "near modern" humans (for example, those of Jebel Irhoud) were replaced by the immediate ancestors of living humans in a manner similar to that of the Neanderthals and Denisovans.[7] To date, the results are ambiguous, however. Despite the suggested recurring pattern of admixture between ancestral modern humans and others, the only specifically identified archaic introgression in Africa is estimated to have occurred roughly 30,000 years ago—long after the beginning of the global dispersal—and with a lineage that had diverged long before the spread of "near modern" humans in Africa. Presumably, genetic admixture between such closely related representatives of *Homo*, which generally are lumped into the same species (although some paleoanthropologists question the inclusion of all of them in *Homo sapiens*),[8] would be more difficult to detect than admixture between lineages that had diverged hundreds of thousands of years ago.

The evidence for archaic introgression in Africa may have broader implications for modern human origins. As discussed in box 2.2, modern humans may have acquired some genes from the Neanderthals and possibly the Denisovans that facilitated their adaptation to local conditions in Eurasia. Perhaps periodic gene flow among various human populations or "near modern" human subpopulations in Africa, some of which inhabited markedly different climate zones and habitats, was an important factor in the evolution of modern human anatomy and cognition both before and after 300,000 years ago.[9]

1. Jeffrey D. Wall, "Detecting Ancient Admixture in Humans Using Sequence Polymorphism Data," *Genetics* 154 (2000): 1271-1279. More specifically, traces of ancient admixture may be evident in statistically significant patterns of *linkage disequilibrium* (non-random association of alleles at different loci) that are consistent with genetic introgression from another human taxon.
2. Vincent Plagnol and Jeffrey D. Wall, "Possible Ancestral Structure in Human Populations," *PLoS Genetics* 2 (2006): e105.
3. Richard E. Green et al., "A Draft Sequence of the Neandertal Genome," *Science* 328 (2010): 710-722.
4. Michael F. Hammer et al., "Genetic Evidence for Archaic Admixture in Africa," *Proceedings of the National Academy of Sciences* 108 (2011): 15123-15128. Although only three populations were sampled, the authors identified a region of the genome likely derived from archaic introgression that is widespread in contemporary African populations.
5. PingHsun Hsieh et al., "Model-Based Analyses of Whole-Genome Data Reveal a Complex Evolutionary History Involving Archaic Introgression in Central African Pygmies," *Genome Research* 26 (2016): 291-300.
6. Ibid., 296.
7. Philipp Gunz et al., "Early Modern Human Diversity Suggests Subdivided Population Structure and a Complex Out-of-Africa Scenario," *Proceedings of the National Academy of Sciences* 106 (2009): 6094-6098; Hammer et al., "Genetic Evidence for Archaic Admixture in Africa," 15126.
8. As noted in the main text, a number of paleoanthropologists doubt that some of the "near modern" specimens exhibiting archaic features (for example, Omo-Kibish II and Laetoli 18) should be classified as *Homo sapiens*.
9. Chris Stringer, *Lone Survivors: How We Came to Be the Only Humans on Earth* (New York: Times Books, 2012), 250-262.

As described in chapter 2, the growing number of dates on these sites, which are based on a variety of methods, indicates beyond a reasonable doubt that *Homo sapiens* evolved in Africa, because modern human remains in Eurasia postdate 150,000 years ago (and most postdate 60,000 years ago). The pattern confirms conclusions based on the genetics of living people—that is, the recent origin in Africa of *Homo sapiens*.

Where measurable, cranial capacity for the specimens listed in table 4.2 reveals yet another increase in brain volume. From the *Homo heidelbergensis* mean of roughly 1,250 cc, cranial capacity rose to about 1,440 cc, reflecting an expansion of slightly less than 16 percent. When the corresponding decrease in overall body size is accounted for (from an estimated average of 71–65 kilograms to 67–63 kilograms; smaller for living people), the increase is even more significant. Lieberman found that the *Encephalization Quotient* (EQ), which provides a measure of brain size relative to overall body size, jumped from between 4.3 and 4.7 in *Homo heidelbergensis* to 5.3 in *Homo sapiens*.[92] And while a cranial capacity of roughly 1,400 cc

TABLE 4.2
African Fossils Assigned to *Homo sapiens*, Dating to More Than 60,000 Years

Site	Estimated Age (method)	Description	Cranial Capacity (cc)	Comments
Jebel Irhoud (Morocco)	300,000 years (TL, ESR, U-series, biostratigraphy)	Adult skulls; skullcap; adult and juvenile mandibles; vertebra; juvenile humeri; femora; isolated teeth	1,305 and 1,450	Elongated braincase (archaic); continuous supraorbital torus on Jebel Irhoud 1; analysis of teeth indicates modern growth rate
Guomde (Kenya)	270,000 years (ESR, U-series, stratigraphy)	Partial skull; femur		Cranium exhibits continuous supraorbital torus (archaic)
Florisbad (South Africa)	300,000– 150,000 years (ESR, biostratigraphy)	Skull fragments; isolated tooth		Thick cranial walls and moderately prognathic face, but other archaic features (for example, supraorbital torus) absent
Eliye Springs (Kenya)	More than 200,000 years (stratigraphy)	Cranium	Ca. 1,210	Low cranial capacity; archaic features include low cranial vault; occipital angle within range of *Homo erectus*
Omo-Kibish (Ethiopia)	195,000– 104,000 years (U-series, argon/argon)	Two partial skulls; skull fragments; postcranial bones	1,435 (Omo II)	Original find by Richard Leakey (1967); Omo II exhibits archaic features (for example, mid-sagittal keel and angulated occipital bone)
Herto (Ethiopia)	160,000 years (argon/argon)	Adult skull; partial juvenile skull; adult skull fragments	1,450	Stratigraphic provenience problematic; angulated occipital bone (archaic)
Singa (Sudan)	170,000– 150,000 years (U/Th, ESR, biostratigraphy)	Adult skull (complete)	1,550	Cranial volume high for lower latitudes; highly convex frontal combined with thick supraorbital torus, especially lateral segments

Site	Estimated Age (method)	Description	Cranial Capacity (cc)	Comments
Laetoli (Tanzania)	120,000 years (U-series, biostratigraphy)	Adult skull (complete)	1,367 (Laetoli 18)	Frontal bone exhibits archaic features (flat narrow squama and thick rounded lateral segment of supraorbital torus)
Klasies River Mouth (South Africa)	115,000–60,000 years (ESR, biostratigraphy)	Five mandible fragments; two maxilla fragments; other cranial fragments; vertebrae; postcranial bones		Zygomatic bone large and robust; teeth small; ulna exhibits low coronoid (archaic)
Border Cave (South Africa)	170,000–50,000 years (U-series, ESR, ostrich eggshell)	Adult skull; infant skeleton; mandible fragments; several postcranial bones	1,510	Cranial volume high for lower latitudes; no archaic features reported
Dar es-Soltan II (Morocco)	128,000–40,000 years (archaeological stratigraphy)	Partial adult skull; mandible fragment; partial child's skull; juvenile mandible		Associated with Aterian artifacts; Dar es-Soltan V exhibits supraorbital torus
Taramsa Hill (Egypt)	80,000–50,000 years (OSL)	Partial child's skeleton (8–10 years old)		Poor preservation; supraorbital and neurocranial anatomy similar to Qafzeh and Skhūl ("near modern")
Blombos Cave (South Africa)	Ca. 75,000 years (OSL)	Isolated teeth		Associated with ocher pieces engraved with geometric designs

Sources: Based on Erik Trinkaus. "Early Modern Humans," *Annual Review of Anthropology* 34 (2005): 207–230; Günter Bräuer, "The Evolution of Modern Humans: A Comparison of the African and Non-African Evidence," in *The Human Revolution: Behavioural and Biological Perspectives on the Origins of Modern Humans*, ed. Paul Mellars and Chris Stringer (Princeton, N.J.: Princeton University Press, 1989), 123–154, and "The Origin of Modern Anatomy: By Speciation or Intraspecific Evolution?" *Evolutionary Anthropology* 17 (2008): 24, table 1; Richard G. Klein, *The Human Career: Human Biological and Cultural Origins*, 3rd ed. (Chicago: University of Chicago Press, 2009), 466–469, table 6.2; Daniel E. Lieberman, *The Evolution of the Human Head* (Cambridge, Mass.: Harvard University Press, 2011), 534, table 13.1; Jean-Jacques Hublin et al., "New Fossils from Jebel Irhoud, Morocco and the Pan-African Origin of *Homo sapiens*," *Nature* 546 (2017): 289–292; Daniel Richter et al., "The Age of the Hominin Fossils from Jebel Irhoud, Morocco, and the Origins of the Middle Stone Age," *Nature* 546 (2017): 293–296.

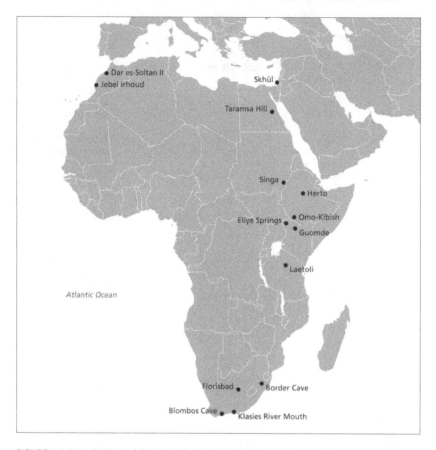

FIGURE 4.6 Map of Africa and the Levant, showing the location of fossils assigned to *Homo sapiens*, dated to between 300,000 and 75,000 years ago. (Base map adapted from en.wikipedia.org/wiki/File: BlankMap-World6.svg)

(with a wide range of variation) is given for modern humans, this represents an average for the global population. As noted, the current mean for tropical Africa (1,250–1,299 cc) is lower, due to thermoregulatory constraints (figure 4.7).

The increase in brain size that coincided with the appearance of the modern human anatomical pattern is potentially important. It conceivably was the trigger for the subsequent changes in cognitive function. The continued expansion of the brain after 500,000 years ago—presumably driven by the same factors that drove earlier increases in brain size—clearly pushed it temporarily beyond the optimum level for the equatorial zone.

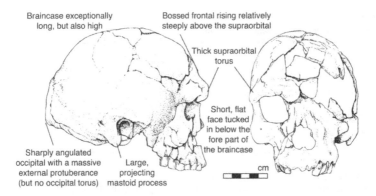

Braincase exceptionally long, but also high

Bossed frontal rising relatively steeply above the supraorbital

Thick supraorbital torus

Short, flat face tucked in below the fore part of the braincase

Sharply angulated occipital with a massive external protuberance (but no occipital torus)

Large, projecting mastoid process

cm

FIGURE 4.7 *Homo sapiens* cranium recovered from *Herto* (Ethiopia) and dated to around 160,000 years ago. The estimated cranial capacity is 1,450 cc, which is high for living humans in the same climate setting. (From Richard G. Klein, *The Human Career: Human Biological and Cultural Origins*, 3rd ed. [Chicago: University of Chicago Press, 2009], 476, fig. 6.26. Reprinted with the permission of the University of Chicago Press)

Added energy demands, as well as the warm-climate setting, may have been a factor in the later reduction in size (accompanied by overall reduction in body size). But the critical variable for the evolution of modern cognitive faculties may have been the constraints on brain size at birth imposed by the dimensions of the birth canal. Despite the change in pelvic anatomy, the birth canal probably did not widen, and the consequence was that an even higher percentage of brain development occurred after birth (with greater environmental input during early growth).[93]

Aside from cranial capacity, the defining skeletal features of *Homo sapiens* are listed in table 4.3 (see also figure 2.9). Many of them are related to the uniquely modern globular form of the braincase, which is reflected in the changing shape of the frontal, parietal, and occipital bones. As the frontal bone became more vertical, the parietal and occipital bones developed a more rounded form to accommodate the larger parietal and temporal lobes and the expanded cerebellum.[94] As noted, the expansion of the parietal and temporal lobes, as well as the cerebellum, are almost certainly tied to the evolution of the language faculty. More specifically, they are tied to the computations that underlie syntactic language and permit the generation of hierarchically structured sequences of digital information units. And because complex computation in the human brain apparently requires "material symbols," or virtual objects, it involves cerebellar function.

TABLE 4.3
Skeletal Anatomy of *Homo sapiens*: Defining Traits

1. *Frontal bone* is relatively vertical

2. *Cranial vault* is high, more or less parallel-sided, usually with some outward bulging in parietal region (reflecting "globular" braincase)

3. *Occipital* contour is rounded and lacking a prominent transverse torus (also reflecting "globular" braincase)

4. *Cranial* base is highly flexed (about 15° more flexed than earlier *Homo*)

5. *Brow ridge*, or *supraorbital torus*, development is greater in males than in females and rarely forms a continuous bar, but rather comprising two parts (bipartite)

6. *Face* is smaller, relatively flat, and "tucked in" below the anterior portion of the braincase

7. *Canine fossa*, or hollowing of the bone below each orbit, is present

8. *Mandible* usually exhibits an "inverted T-shaped" chin, or *mental eminence*

9. *Thorax*, or chest, is barrel-shaped (sharply contrasting with Neanderthals)

10. *Pelvis* lacks lateral flare and exhibits relatively vertical iliac blades

Sources: Based on Daniel E. Lieberman, "Speculations About the Selective Basis for Modern Human Craniofacial Form," *Evolutionary Anthropology* 17 (2008): 55–68; Ian Tattersall and Jeffrey H. Schwartz, "The Morphological Distinctiveness of *Homo sapiens* and Its Recognition in the Fossil Record: Clarifying the Problem," *Evolutionary Anthropology* 17 (2008): 51; Richard G. Klein, *The Human Career: Human Biological and Cultural Origins*, 3rd ed. (Chicago: University of Chicago Press, 2009), 622–626; Ian Tattersall, "Human Origins: Out of Africa," *Proceedings of the National Academy of Sciences* 106 (2009): 16018.

The other part of the brain often thought to be closely associated with modern human cognition is the *prefrontal cortex*. It performs executive functions, integrating signals from other parts of the neocortex, and—based on brain-imaging methods such as positron emission tomography (PET) scans—is implicated in the performance of novel tasks. It probably plays a role in the creative aspects of language, as well as working memory and social skills.[95] Although changes in the size or organization of the prefrontal cortex associated with modern human anatomy lack visibility in the fossil record, comparative genomic analysis reveals a series of substitutions unique to modern humans (that is, absent in the Neanderthal and Denisovan genomes) that are related to the development of the prefrontal cortex.[96]

Other diagnostic features apparently are not tied—either directly or indirectly—to the size and shape of the modern human brain. The smaller, less-projecting face is recognized in earlier human fossils (for example,

Lake Ndutu), as is the canine fossa (for example, Broken Hill [Kabwe]) (see figure 4.1).[97] By contrast, the chin is absent in some of the early skeletal remains assigned to *Homo sapiens*, including specimens from *Omo-Kibish* (Ethiopia) (which first set the RAO model in motion) and *Jebel Irhoud* (Morocco), and this feature is more pronounced in younger fossils. Although the chin sometimes is thought to have a structural role in the architecture of the lower jaw, a pattern of sexual dimorphism (among recent populations) in the expression of the chin supports the notion that it is a product of sexual selection.[98]

The delay in growth and maturation that distinguishes living humans so markedly from the living apes and early *Homo* is evident in a well-preserved juvenile mandible from Jebel Irhoud 3 that now is dated to about 300,000 years ago. The spacing of incremental growth lines on the teeth (perikymata) is comparable to that of a living child of the same estimated age (7.8 years) and indicates relatively slow growth (see box 2.4).[99] An emphasis on delayed biological development as a defining characteristic of modern humans is a recent phenomenon in paleoanthropology,[100] but it has roots in early-twentieth-century speculation about neotony in human evolution, including the "fetalization theory of anthropogeny" proposed by Louis Bolk (1866–1930) and recounted by Stephen Jay Gould in his book *Ontogeny and Phylogeny*.[101]

Perhaps the most significant discovery in this context is that the "globularization phase" of cranial development in modern humans is unique to *Homo sapiens*. The comparative analysis of the crania of modern human and Neanderthal infants reveals that the globular shape of the former, which emerges during the first year of postnatal life, is a result of a derived pattern of early brain growth. Especially important is the rapid growth of the cerebellum (240 percent increase in size) during the first year, which—along with growth of parts of the neocortex—generates the rounded form of the occipital and other parts of the endocranium. As in the case of the prefrontal cortex, there is supporting evidence for the derived pattern of growth from comparative genomics.[102]

In sum, comparative fossil anatomy, including developmental anatomy, suggests that the evolution of modern humans, which probably occurred no later than 150,000 years ago in Africa, coincided with significant changes in the growth and development of the brain. The changes may have been triggered by the continued expansion of cranial volume after 500,000 years ago. They produced—probably through epigenetic effects on growth and development—a uniquely modern pattern of delayed maturation and expansion of portions of the neocortex and cerebellum. The pattern is plausibly tied to the language faculty—or, more narrowly, to the computations that underlie language—including the acquisition of language, entailing the expansion of the traditionally identified language areas in the neocortex, as well as the cerebellum. It presumably is related to the critical period of language acquisition (roughly between five and 12 years of age), during which the child acquires syntactic language through exposure to spoken language by adults.

The heightened degree of cerebral plasticity that evolved in *Homo sapiens* carried some potential risks because many of the neurological disorders that seem to be as unique to modern humans as syntactic language (for example, schizophrenia) can be traced to childhood traumas that affected the developing brain.[103] To an extent unprecedented among the metazoa, the brain of modern humans is organized by a body of nongenetic information. The beneficial effects of the modern human pattern of brain growth and development—which must have dramatically enhanced the computational faculty in a complicated and unpredictable social setting—clearly outweighed the potential risks of developmental neuropathology. The risks probably were minimal in the orderly world of hunter-gatherer society and became apparent only in the complex societies of later epochs.

Computational Complexity in the Archaeological Record

Modern humans, as described in chapter 1, exhibit a special form of computation among the metazoa. While many animals, especially among the mammals, perform complex computations with neuronal information—received primarily through the visual system—living humans manipulate another form of (digital) information in the brain, creating complex, hierarchically organized arrangements or structures. The variety of such arrangements is potentially infinite. Because many of the units have referents (that is, are symbols for objects or processes), the computations may be *about* objects or processes outside the immediate spatial and temporal setting of the brain.

Students of language make a fundamental distinction between language as a communication system and "the computations underlying this system."[104] The latter, which sometimes is termed *internal language*, represents the most common manifestation of human computation. (Simple arithmetic calculations that do not require external computational aids may be included in internal language.) It is apparent, however, that the computational system is based on the same set of information units (words and numbers) employed for communication.

In one form or another (spoken or unspoken), these digital units of information are perceptible objects (or "material symbols") that provide the necessary *cognitive scaffolding* for computations that often require multiple hierarchical levels and embedded components containing multiple elements.[105] They are not units of neuronal information per se (that is, synaptic connections) but a *new type of information* generated from neuronal information that undoubtedly has evolutionary roots in the alarm calls and other vocalizations of nonhuman primates. They may have been employed primarily for communication in early *Homo*, but seem likely to have provided the basis for a computational system in *Homo heidelbergensis* (underlying the complexity of their artifacts). Modern humans soon created other new types of information to function as external computational aids (for example, notation), followed by written language, mathematics, and other information technologies.

The computational complexity of modern humans (without technological aids) is logically measured by syntactic language, although it also can be measured by simple arithmetic computation.[106] In general terms, the complexity of language may be characterized by its elaborate hierarchical structure—sounds into words, words into phrases, phrases into sentences, and sentences into narratives. Sentences often comprise three or more hierarchically organized levels with embedded phrases. In groundbreaking papers published in the 1950s, Noam Chomsky noted that no syntactic language can be fully described, because it generates a potentially infinite set of sentences from a finite set of elements ("discrete infinity"). The structure of language can be characterized only by a *grammar*, which specifies the rules by which the sentences are generated, and Chomsky proposed a general model for grammar that "transforms" a core set of simple sentences into a potentially unlimited variety of more complex sentences.[107]

Although Chomsky's views on language evolved in later years, the generative or "recursive" hierarchy remained a central concept.[108] It is a nondeterministic system that parallels an evolving genome (or the immune system) by producing random variations of digital information, subject to positive or negative selection.[109] In the unstable and competitive world of human social relationships, the computational power of internal language offered what Daniel Dennett labeled a *virtual machine* in the brain for predicting the behavior of others.[110] It also would have generated unpredictable behavior on the part of the individual performing the computations. Presumably, the long-term reproductive benefits to the latter outweighed the risks.

In any case, it is apparent that the computations that underlie syntactic language transformed existing systems of communication and technology, with far-reaching effects on the human population. Integrated with a sensorimotor system that probably had evolved much earlier (possibly in early *Homo*) for transmitting and receiving vocal signals, modern human computation produced an open-ended form of communication with a potentially infinite variety of hierarchically organized messages. Integrated with a sensorimotor system that evolved in early *Homo* for manipulating objects and features in the landscape (that is, visually coordinated hands), modern human computation produced a potentially infinite variety of hierarchically organized artifacts and features (for example, dwellings).

If the computations of internal language represent a "virtual machine" in the brain, computations of comparable complexity eventually produced a literal machine outside the brain. An automaton may be described as a material form of computation, or a "living" algorithm. It also may be described as a simple form of "artificial life" that is more functionally and structurally complex than a composite tool or weapon (since it typically includes at least three hierarchical levels of organization). Unlike those of language, however, the computations performed by an automaton are, for the most part, deterministic and predictable (although its design is the product of a nondeterministic process). A related category of technology comprises artifacts with moving parts that are not self-acting, such as a bow and arrow. These mechanical artifacts work only in concert with functions of the human body, but may be considered "partial machines" (box 4.4).

The computations that underlie spoken and unspoken language remain invisible in the fossil and archaeological record until the development of

BOX 4.4
Automata Theory and Hunter-Gatherer Technology

As with the computer, the theory of machines, or automata, has its roots in the early nineteenth century with Charles Babbage (1791–1871), who devised a system of "mechanical notation" to describe the multiple states of a machine.[1] A more widely used system of "kinetic notation" was developed in the late nineteenth century.[2] The modern theory of automata arose in concert with the programmable digital computer in the 1950s, along with a new form of graphic notation for machines ("machine state diagram").[3]

A machine contains the fundamental elements of a computation (input, defining functions or operations, and output), but performs its computation(s) independently of an organism. Unlike the evolutionary computations of an evolving lineage and an individual organism, a machine may compute with various types of information (all of which are products of human technological algorithms) or simply with materials (or with both information and materials).[4] And although there has been some debate on the subject, the computations of an automaton are generally deterministic in character (also unlike other evolutionary computations, which contain an essential randomizing function).

Certain categories of "untended facilities" produced by hunter-gatherers (for example, self-acting snares) meet the mathematical definition of a machine or, more specifically, a deterministic *finite-state machine*, where:[5]

Q is a finite set of states $(S_1, S_2 \ldots S_n)$,
X is the input,
Y is the output,
$\delta : Q \times X \rightarrow Q$, the next state function,
$\lambda : Q \times X \rightarrow Y$, the next output function.

The distinguishing features of a finite-state machine (as opposed to other forms of technology) are that it contains both multiple states $(S_1, S_2 \ldots S_n)$ and the means to transition from one state to another (δ, λ), including a state that will yield the output (Y).

Many hunter-gatherer groups have made and used self-acting devices that meet the definition of a simple finite-state machine.[6] An example is a rabbit snare designed by the *Tanaina* (southwestern Alaska) and illustrated—with a machine-state diagram—in box figure 4.4.[7] The snare may be classified as a "two-state machine" that is either un-sprung (S_1) or sprung (S_2) with its captured prey (Y). The input (X) is represented by the rabbit before it enters the snare. The operation of the machine is described by the combined "transition functions" (δ, λ). It performs a deterministic computation with materials (rather than information), lacks memory storage, and does not require a program; the defining functions of the computation are built into the structure of the automaton.

Automata tend to be structurally or organizationally complex (comprising multiple hierarchical levels) because they tend to be functionally complex (see box 2.5), which, in turn, reflects the complexity of the problems they are designed to solve (multiple subproblems). The significance of the machine in hunter-gatherer technology lies not in its underlying structural/functional complexity, but in the demonstrable fact that *automata represent a new level of evolutionary computation in living systems*. An

(continued)

(continued)

a

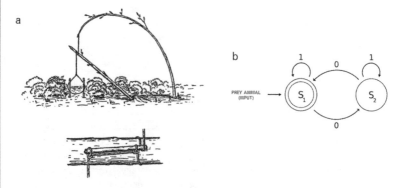

b

BOX FIGURE 4.4 A self-acting facility, or automaton, in the form of a rabbit snare made by the Tanaina (a), showing detail of the mechanical trigger pin, and (b) a diagram of a finite-state machine with no memory storage: state one (S_1) represents the un-sprung snare, and state two (S_2) represents the sprung snare with captured prey. ([a] From Cornelius Osgood, *The Ethnography of the Tanaina* [New Haven, Conn.: Yale University Press, 1937], 93, fig. 20. Courtesy of Yale University Publications in Anthropology; [b] modified from https://commons.wikimedia.org/wiki/File:DFAexample.svg)

automaton *is* an algorithm (produced in accordance with another algorithm computed at the organismal level with non-genetic information). Moreover, while the simple machines of hunter-gatherers lack memory storage and a program (that is, they compute with materials rather than information), they provide the model for more complex automata, including those that compute with forms of information created by humans (for example, punch cards and vacuum tubes).

1. Charles Babbage, "On a Method of Expressing by Signs the Action of Machinery," *Philosophical Transactions of the Royal Society* 116 (1826): 250–265.
2. Franz Reuleaux, *The Kinematics of Machinery: Outlines of a Theory of Machines*, trans. Alexander B. W. Kennedy (New York: Dover, 1963).
3. George H. Mealy, "A Method for Synthesizing Sequential Circuits," *Bell System Technical Journal*, September 1955, 1045–1079; Edward F. Moore, "Gedanken-Experiments on Sequential Machines," in *Automata Studies*, ed. C. E. Shannon and J. McCarthy (Princeton, N.J.: Princeton University Press, 1954), 129–153.
4. See, for example, Marvin L. Minsky, *Computation: Finite and Infinite Machines* (Englewood Cliffs, N.J.: Prentice Hall, 1967).
5. Philip C. Jackson Jr., *Introduction to Artificial Intelligence*, 2nd ed. (New York: Dover, 1985), 45; Harry R. Lewis and Christos H. Papadimitriou, *Elements of the Theory of Computation*, 2nd ed. (Upper Saddle River, N.J.: Prentice Hall, 1998); Elaine Rich, *Automata, Computability and Complexity: Theory and Applications* (Upper Saddle River, N.J.: Pearson Prentice Hall, 2008).
6. Wendell H. Oswalt, *An Anthropological Analysis of Food-Getting Technology* (New York: Wiley, 1976).
7. Cornelius Osgood, *The Ethnography of the Tanaina* (New Haven, Conn.: Yale University Press, 1937), 92–95.

writing, although some examples of notation (engraved on bone or antler) dating to the final millennia of the Ice Age exhibit a relatively complex hierarchical organization that may reflect computation on a level commensurate with that of syntactic language.[111] The most compelling argument for the presence of syntactic language in early modern humans (before 100,000 years ago), however, is that all languages seem to derive from a common ancestor; a recent analysis of phonemics specifically identified southern Africa as the most likely point of origin (which is consistent with genetic data on the geographic source of all living people).[112] It seems unlikely that syntactic language evolved independently in various populations following the dispersal of modern humans out of Africa between 75,000 and 60,000 years ago.

The complexity of the computations that underlie the artifacts made by modern humans is readily apparent before 40,000 years ago, however, and may be at least tentatively inferred from the archaeological record before 60,000 years ago. Examples of visual art, which represents an entirely novel form of analog information created by modern humans, dating to between 45,000 and 35,000 years ago (Europe) exhibit a hierarchical organization with three or more levels and the potential for infinite variation (see figure 2.13).[113] Comparable design complexity is implied by the use of automata and other types of mechanical artifacts, which also typically possess at least three hierarchical levels of organization, dating to between 65,000 and 40,000 years ago (southern Africa). Between the time modern human anatomy is present in the fossil record (ca. 300,000–150,000 years ago) and the beginning of the global dispersal (75,000–60,000 years ago), the archaeological evidence for computational complexity comparable to that of living people is problematic (table 4.4).

The artifacts listed in table 4.4 that are more than 75,000 years old do not necessarily reflect the same level of complexity as the computations that underlie language products. Many of them represent personal ornaments in the form of perforated marine shells, also common between 50,000 and 30,000 years ago (figure 4.8). Although they may have been strung together with other objects into complex arrangements, this cannot be confirmed at present. Several sites, most notably *Blombos Cave* and *Diepkloof Rock Shelter* (South Africa), have yielded objects with geometric designs (figure 4.9).[114] Like the ornaments, they appear to have been a form of display lacking both semantic information and technological

TABLE 4.4

Archaeological Record of *Homo sapiens* in Africa and the Levant, Before 40,000 Years Ago

Site	Estimated Age (method)	Description	Comments
Pinnacle Point (South Africa)[a]	164,000 years (OSL, U-series)	Use of pigment and bladelet technology; heat treatment	Earliest evidence of thermal treatment of raw material
Skhūl (Israel)[b]	135,000–100,000 years (ESR, U-series)	Marine-shell beads (*Nassarius gibbosulus*)	
Qafzeh (Israel)[c]	100,000–90,000 years	Marine-shell beads (*Nassarius gibbosulus*); engraved rock	
White Paintings Shelter and others (Botswana)[d]	94,000–66,000 years (OSL, TL)	Silcrete imported 295 kilometers	Wide foraging range and/or trade
Katanda (Democratic Republic of Congo)[e]	90,000–60,000 years (ESR, OSL)	Barbed and unbarbed points (bone); dagger-like object (bone)	Associated with freshwater-fish remains (catfish)
Grotte des Pigeons (Morocco)[f]	82,000 years (OSL, TL, U-series)	Marine shell beads (*Nassarius gibbosulus*)	Wear indicates that some beads were suspended and covered with red ocher
Blombos Cave (South Africa)[g]	78,000–75,000 years (TL, OSL)	Red-ocher fragments with engraved geometric design; bone awls and points; marine-shell beads (*Nassarius*)	Non-stone artifacts polished
Border Cave (South Africa)[h]	78,000–70,000 years (ESR)	Perforated shell (*Conus*) in grave (funerary object)	Associated with infant burial
Sibudu Cave (KwaZulu-Natal)[i]	Ca. 65,000–62,000 years (OSL)	Remains of blue duikers, bushpigs, monkeys, hares, and other taxa associated with snaring	Inferred use of snares (oldest automata?)
Diepkloof Rock Shelter (South Africa)[j]	60,000 years (OSL, TL, ^{14}C)	Ostrich-eggshell container fragments with engraved geometric designs	Four repetitive linear motifs, including hatched band

Site	Estimated Age (method)	Description	Comments
Border Cave (South Africa)[k]	44,000–42,000 years (^{14}C)	Bone arrow points; notched bones; digging sticks; bone awls; shell beads; ostrich-eggshell beads	Early mechanical technology (?) (bow and arrow); notched bones indicate notation (?)
Oued Djebbana (Algeria)[l]	More than 35,000 years (^{14}C)	Marine-shell bead (*Nassarius gibbosulus*)	Located about 200 kilometers from coast

[a]Curtis Marean et al., "Early Use of Marine Resources and Pigment in South Africa During the Middle Pleistocene," *Nature* 449 (2007): 905–908; Kyle S. Brown et al., "Fire as an Engineering Tool of Early Modern Humans," *Science* 325 (2009): 859–862.

[b]Marian Vanhaeren et al., "Middle Paleolithic Shell Beads in Israel and Algeria," *Science* 312 (2006): 1785–1788.

[c]Francesco d'Errico et al., "Archaeological Evidence for the Emergence of Language, Symbolism, and Music—An Alternative Multidisciplinary Perspective," *Journal of World Prehistory* 17 (2003): 1–70.

[d]David J. Nash et al., "Going the Distance: Mapping Mobility in the Kalahari Desert During the Middle Stone Age Through Multi-Site Geochemical Provenancing of Silcrete Artefacts," *Journal of Human Evolution* 96 (2016): 113–133.

[e]John E. Yellen et al., "A Middle Stone Age Worked Bone Industry from Katanda, Upper Semliki Valley, Zaire," *Science* 268 (1995): 553–556.

[f]Abdeljalil Bouzouggar et al., "82,000-Year-Old Shell Beads from North Africa and Implications for the Origins of Modern Human Behavior," *Proceedings of the National Academy of Sciences* 104 (2007): 9964–9969.

[g]Christopher S. Henshilwood et al. "An Early Bone Tool Industry from the Middle Stone Age at Blombos Cave, South Africa: Implications for the Origins of Modern Human Behaviour, Symbolism and Language," *Journal of Human Evolution* 41 (2001): 631–678.

[h]Francesco d'Errico and Lucinda Blackwell, "Earliest Evidence of Personal Ornaments Associated with Burial: The *Conus* Shells from Border Cave," *Journal of Human Evolution* 93 (2016): 91–108.

[i]Lyn Wadley, "Were Snares and Traps Used in the Middle Stone Age and Does It Matter? A Review and a Case Study from Sibudu, South Africa," *Journal of Human Evolution* 58 (2010): 179–192.

[j]Pierre-JeanTexier et al., "A Howiesons Poort Tradition of Engraving Ostrich Eggshell Containers Dated to 60,000 Years Ago at Diepkloof Rock Shelter, South Africa," *Proceedings of the National Academy of Sciences* 107 (2010): 6180–6185.

[k]Francesco d'Errico et al., "Early Evidence of San Material Culture Represented by Organic Artifacts from Border Cave, South Africa," *Proceedings of the National Academy of Sciences* 109 (2012): 13214–13219; Paola Villa et al., "Border Cave and the Beginning of the Later Stone Age in South Africa," *Proceedings of the National Academy of Sciences* 109 (2012): 13208–13213.

[l]Vanhaeren et al., "Middle Paleolithic Shell Beads in Israel and Algeria."

function—a common phenomenon in all living human societies (that is, social display in artifactual form). While artifacts for social display in living human societies, such as jewelry and decorative headgear, often exhibit a complex, hierarchical design, these examples are comparatively simple.

A similar pattern is apparent among the various technologies presented in table 4.4. For the most part, they are simple tools and weapons, although the points, which include some barbed examples from Katanda, may have been components of a more complex system (figure 4.10). This also applies

FIGURE 4.8 Personal ornaments in the form of perforated marine shells appear in the archaeological record at least 135,000 years ago. (From Marian Vanhaeren et al., "Middle Paleolithic Shell Beads in Israel and Algeria," *Science* 312 [2006]: 1786, fig. 1. Copyright AAAS 2006)

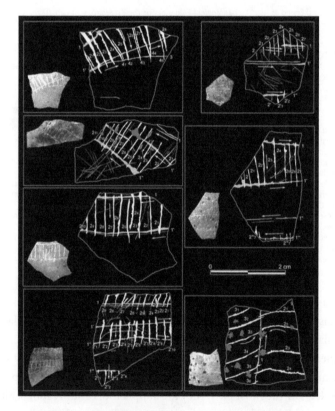

FIGURE 4.9 Fragments of engraved ostrich eggshell from Diepkloof Rock Shelter that date to around 60,000 years ago. (From Pierre-Jean Texier et al., "A Howiesons Poort Tradition of Engraving Ostrich Eggshell Containers Dated to 60,000 Years Ago at Diepkloof Rock Shelter, South Africa," *Proceedings of the National Academy of Sciences* 107 [2010]: 6183, fig. 3. Reprinted with the permission of the authors)

FIGURE 4.10 Worked bone, including barbed harpoons, from the site of Katanda that date to between roughly 90,000 and 60,000 years ago. (From John E. Yellen et al., "A Middle Stone Age Worked Bone Industry from Katanda, Upper Semliki Valley, Zaire," *Science* 268 [1995]: 555, fig. 1. Copyright AAAS 1995)

to the ostrich-eggshell containers from Diepkloof Rock Shelter, which may have been carried or suspended in a relatively complex apparatus. Small stone blades or bladelets appear for the first time in the archaeological record (for example, Pinnacle Point) and may be the only stone technology confined to *Homo sapiens*.[115] The most tantalizing evidence may be the indications of freshwater fishing—apparently associated with the barbed points at Katanda—because fishing often involves or requires the use of complicated and/or innovative implements and devices (such as harpoons, nets, gorges, and weirs/traps).[116]

The primary significance of the personal ornaments and decorated items in the archaeological record of *Homo sapiens* before 75,000 years ago is social. Like the emergence of regional stylistic variations among the points and small bifaces after 300,000 years ago (and possibly even the ancient hand axes), these artifacts probably reflect the investment of time and materials in managing reciprocal alliances with individuals outside the immediate family. Their rather sudden appearance following the evolution of modern human anatomy—specifically developmental anatomy—is consistent with the hypothesis that the anatomy reflects a major upgrade in computational complexity, driven by the pressures of social networking and game playing. Supporting evidence from southern Africa for expanded social networks during this period may be found in the analysis of archaeological data for coalescence and fragmentation and the long-distance transport of silcrete artifacts (see box 4.2).[117]

A "Major Transition" in Evolution?

By the time modern humans began their second expansion out of Africa between 75,000 and 60,000 years ago, they had evolved the diagnostic skeletal features of *Homo sapiens* and, most likely, the full suite of cognitive faculties found in all living people. The latter included both syntactic language and the computational faculty for manipulating objects and materials of comparable complexity. The presence of syntactic language is inferred from the phylogeny of all known languages, which may be traced to a common ancestor (that is, syntactic language dispersed globally with modern humans), while the structural and computational complexity of artifacts and features in the archaeological record was comparable to that of recent hunter-gatherers by 40,000 years ago at the latest (and, more likely, by 65,000 years ago).

The ability to both speak a syntactic language and design computationally complex instruments and devices, such as fish weirs and snares (automata), probably was critical to successful colonization of areas unoccupied by other forms of *Homo*. The widespread alliance networks that recent hunter-gatherers developed to cope with scarce and unpredictable resources in cold and arid environments are based on the complex reciprocal relationships that individuals negotiate with the computational power of syntactic language. Automata and other forms of mechanical technology are found among all recent hunter-gatherers in cold environments and appear to be essential for obtaining enough to eat in such habitats. The absence of other forms of *Homo* in the coldest and driest parts of northern Eurasia— including the anatomically modern humans who briefly dispersed into the Levant and possibly Southeast Asia between 120,000 and 90,000 years ago—presumably reflects their lack of syntactic language or functionally complex technology or both.

If the *Homo sapiens* jaw from Jebel Irhoud implies the presence of syntactic language—it shows that modern humans had evolved a delay in maturation that created the critical period for learning syntactic language—then language was present as early as 300,000 years ago.[118] The belated appearance of evidence for artifacts and features of comparable structural/functional and computational complexity (at the beginning of the second and wider dispersal 75,000–60,000 years ago) may indicate that the evolution of modern human cognitive faculties took place over an extended period of time (and in at least two stages).[119] It could explain why only the second dispersal was successful and ultimately global in scope. Alternatively, the apparent gap in the archaeological record may simply reflect multiple sources of bias.

Regardless of when all their anatomical features and cognitive faculties evolved, modern humans represent what John Maynard Smith and Eörs Szathmáry labeled a "major transition" in evolution, characterized by *fundamental changes in how information is stored, transmitted, and translated.*[120] Moreover, modern humans represent the first such transition in terms of non-genetic information. They also constitute a quantum jump to a new level of hierarchical complexity with emergent properties. The creation of automata produces a new level of evolutionary computation.

The importance of the evolutionary transition to anatomically and cognitively modern humans lies in their capacity for the translation of non-genetic information into structure, including other forms of information based on that structure. Modern humans translated neuronal information (which evolved 500 million years ago in the early metazoa) into vocal sounds real and imagined, which are themselves digital information units. These perceptible "structures" are "manipulated" in the "virtual machine" of the neocortex (with a major role for the cerebellum) to perform complex computations about their referents (for example, other people).[121] Modern humans also translate neuronal information into a unique form of information in analog form as visual art.

Modern humans also translate non-genetic information in the brain into complex, hierarchically organized structures in the form of artifacts used to acquire food, defend against predators, adapt to climatic conditions, and so on. These technologies may include moving components (mechanical technology) and even function independently of the human body (automata). An automaton represents a higher level of organizational complexity—and a new level of evolutionary computation—because the translation of non-genetic information into complex functioning structure ("artificial life") is analogous to the translation of genetic information into a functioning organism.

The computations that underlie both syntactic language and complex technology clearly receive input in the form of random recombination of neuronal information in the brain (and this also applies to musical sounds, cooking recipes, visual art, and other media). They are nondeterministic computations. Accordingly, modern humans can generate a potentially infinitive range of possible alternatives with respect to language products and artifacts. The consequences for human populations were profound as they redesigned themselves as organisms (with technological "traits" such as tailored clothing) and eventually began to redesign their environments.

The enlarged brain of modern humans bears some analogy to the eukaryote genome, with its massively expanded (over prokaryotes) information-storage capacity. Each brain stores an immense quantity of non-genetic information, most of which is received during its protracted postnatal growth and development. The non-genetic information contributes to the organization of the brain, especially during the critical period for language acquisition, in a manner analogous to the translation of genetic information into a developing organism. Modern humans have created artifacts that store various forms of information outside the brain (the earliest known possible example is digital notation on notched bones dating to 44,000 years ago from *Border Cave* [South Africa]).[122]

Because syntactic language doubles as a system of computation and communication, it is a major channel of information transmitted from one brain to another and allows collective computation between two or more brains. Collective computation clearly occurs with materials and objects as well, including technology and visual art (entailing the transmission of information in analog form), and it expands the quantity and variety of input. Technology is computed collectively over many generations by multiple brains, most of which do not communicate directly with one another; it represents an organizational level higher than the computation of the individual brain and may be described as a super-brain.[123]

The immense mass of information (in various forms) shared among members of a modern human social group was termed "culture" by pioneering ethnologists of the late nineteenth and early twentieth centuries (see box 1.3).[124] Following the classic definition, "culture" comprises structure based on translated information (that is, artifacts and features), as well as the information, which—like all forms of information, including DNA sequences—is symbolic. The development of a culture over many generations, and the individuals who compute with the mass of information and information-based structure comprising it, is analogous to an evolving lineage in biology. And the mass of structure based on collective computation with non-genetic information generated by modern humans after

75,000 years ago represents something analogous to the "Cambrian explosion" of metazoan life 500 million years ago.

Modern humans have deep evolutionary roots in the order Primates. Their visual system has its origin in the prosimians, and their specialized organs for translating non-genetic information into structure (and other forms of information) evolved from the grasping forelimbs and vocalizations of prosimians and catarrhines. Roughly 2 million years ago, a shift to pair-bonding and central-place foraging in drier habitats, where resources were less common and probably less predictable than in tropical environments, created selection pressure for a steady increase in brain volume (chapter 3).

Both the added information-storage capacity and the computational power of larger brains likely conferred long-term reproductive benefits in the competitive social setting of male–female pairs and a collective-foraging society. Anatomical evidence from fossils suggests that some form of language (presumably non-syntactic) evolved at this point,[125] and that it represented a new form of information (digital) based on firing synapses. Although its primary function may have been the transmission of information related to foraging, it conceivably had a collective-computation function, analogous to the "dance" language of the honeybee colony (and swarm), which shares information about resource locations (and potential nest sites).[126] This would have created an early and simple version of the super-brain, signs of which may be evident in the translation of shared information (analog) into structure in the form of a hand ax by 1.75 million years ago.[127] In any case, the transmitted units of information were "material symbols" in the form of sounds produced with the vocal tract and translated from neuronal information.

The modern human pattern began to emerge after 500,000 years ago, as brain volume approached—and subsequently exceeded—optimum size and alternative means of enhanced social gaming began to evolve. This may be the simplest explanation of syntactic language, which offers a means of performing complex computations with digital-information units borrowed from an ancient vocal-communication system. The key to evolving this bizarre feature lies in regulatory genes and developmental biology. By slowing maturation and brain growth (and thus increasing cerebral plasticity), anatomically modern humans evolved the means to translate a large quantity of non-genetic information into structure within the developing brain—constructing a virtual machine for complex, non-deterministic computation.

Despite the deep evolutionary roots of their specialized anatomical features and the earlier history of the transmission and translation of non-genetic information (which, following Maynard Smith and Szathmáry, may themselves be termed "major transitions" in evolution), the African origin and wider dispersal of *Homo sapiens* between 300,000 and 60,000 years ago represents a "major transition" in evolution. During this interval, modern humans reached a threshold with respect to nondeterministic computations with non-genetic information, analogous to the "major transition" from prokaryotes to eukaryotes and multicellular life (chapter 1). The appearance and spread of modern humans unleashed an entirely new phase of the evolution of living systems, which may be defined more broadly to include functioning structures based on information in the form of automata. In the near future, it is likely that some of these automata will perform nondeterministic computations (artificial intelligence) and reproduce themselves.[128]

MODERN HUMANS IN AFRICA

Living people probably are descended from a subset of the anatomically modern human population that inhabited Africa (and, at least briefly, part of Eurasia) between roughly 300,000 and 75,000 years ago. At some point before 75,000 to 60,000 years ago, this subpopulation spread from its place of origin—which appears to have been in southern Africa—throughout the continent. And after 75,000 to 60,000 years ago, a subset of those migrants dispersed out of Africa and into Eurasia.

The failure of the first, or "near modern," movement into Eurasia, where it is apparent that at least in one area—the Levant—anatomically modern humans were replaced by Neanderthals, combined with the lack of definitive evidence for artifacts of comparable structural and computational complexity to those of recent hunter-gatherers, has persuaded some paleoanthropologists that the earlier population of *Homo sapiens* did not possess the full suite of cognitive faculties found in living people.[129] The difference also could explain why the earlier population was effectively replaced by the direct ancestors of living people. Alternatively, as noted briefly in the preceding section, those in the second migration may simply have been the first to apply the same level of computational complexity found in syntactic language to making artifacts, including tools and weapons.

Until recently, reconstructing the spread of *Homo sapiens* within Africa was impossible. This was due to the paucity of fossil remains and archaeological sites and the fact that they provide few clues about the point of origin and dispersal routes of the founding lineages. During the past few years, however, large-scale analyses of the genetics of living populations have been undertaken in Africa. The results reveal a complex and confusing picture because living Africans are the most genetically diverse people in the world, and they exhibit a high degree of population substructure. Most African populations also show high levels of mixed ancestry, reflecting extensive migrations across the continent during recent millennia.[130] Nevertheless, it is possible to discern some broad patterns of paleogeography that indicate the likely place of origin of modern humans, and there is supporting evidence from linguistics.

Both the genetics of living Africans and the linguistic data suggest that *Homo sapiens* evolved in southern Africa. Anatomically modern humans also may have had an early presence in East Africa. The Bushmen, or *San*, of southern Africa are the most genetically diverse of living peoples, and they speak click languages, which are thought to be the oldest surviving languages on the planet.[131] Two groups in East Africa (*Hadza* and *Sandawe*), who also speak click languages, appear to have diverged from the southern African population at a very early date. And the Pygmies (*Mbuti*) in Central Africa have genetic ties to the original population, although they have adopted a new language.[132]

The warm-climate conditions of tropical and subtropical Africa have thus far precluded the recovery and analysis of aDNA from skeletal remains of early modern humans.[133] Accordingly, ground-truthing inferences drawn from the study of the genetics of living Africans must await the discovery and high-resolution dating of newly discovered fossils and archaeological sites that provide a record of the spreading population of modern humans between about 300,000 and 75,000 to 60,000 years ago. It may be more difficult to distinguish modern from non-modern humans in this setting than it is outside Africa after 60,000 years ago. The associated archaeological remains may be more helpful—they may yield evidence of novel technologies and other manifestations of modern cognitive faculties that are lacking in the sites of the people who were replaced.

The Genetics of Modern African Populations

A major study of living populations in Africa was published in 2009 by Sarah Tishkoff and her colleagues. They measured genome-wide nuclear genetic diversity with a sample of more than 2,400 individuals drawn from more than 110 geographically distinct populations (as well as a large sample of non-African populations, including African Americans, for comparative purposes). The sampled individuals were genotyped on the basis of 1,327 polymorphic markers, comprising "microsatellites" (short sequences of repetitive DNA), insertion/deletions, and single-nucleotide polymorphisms (SNPs).[134]

The results, which were combined with geographic/linguistic data and subjected to multivariate statistical analysis (principal components analysis), yielded six major clusters:

- Speakers of Nigerian-Kordofanian languages, distributed from far western through Central Africa to the speakers of Bantu languages in South Africa (and representing the most geographically widespread of the clusters)
- Bushmen and speakers of Khoisan languages in noncontiguous regions of Central and southern Africa, respectively
- Speakers of Afro-Asiatic languages in northern Africa, including Mali, as well as parts of East Africa (Ethiopia and northern Kenya)
- Speakers of Nilo-Saharan languages in southern Sudan, and speakers of Chadic and Nilo-Saharan languages in Nigeria, Cameroon, and central Chad
- Speakers of Nilo-Saharan and Cushitic languages in Sudan, Kenya, and Tanzania, and speakers of Bantu languages in Kenya, Tanzania, and Rwanda
- Hadza, a small group now living in Tanzania with ancient ties to the speakers of click languages in southern Africa[135]

The work of Tishkoff and her colleagues, as well as other studies of living Africans, have pointed to southern Africa (specifically, the current homeland of the Bushmen near the coastal border of Namibia and Angola) as the locus of highest genetic diversity and likely place of origin for modern humans.[136] The analysis of larger samples from the Bushmen and the

Hadza revealed that the former carry more than 700,000 unique SNPs, suggesting "high and ancient phylogenetic divergence."[137] The Bushmen also are characterized by high frequencies of the mitochondrial DNA (mtDNA) haplogroup L0 ("mitochondrial Eve"), which is estimated to have diverged from L1 (or the L1'5 branch) around 140,000 years ago or earlier.[138] Haplogroup L1'5 is ancestral to all other living human maternal lineages (see figure 2.12).

An early movement into East Africa apparently is reflected by the presence of the Hadza and Sandawe in Tanzania. Both speak click languages, although they are not closely related to the southern African click languages. As noted, the Hadza (who currently number about 1,000 people) represent a separate cluster in Africa, but the Sandawe include individuals belonging to mtDNA haplogroup L0 (L0d clade) and Y-chromosome DNA haplogroup A-M91 ("mitochondrial Adam").[139] And an early movement into eastern Central Africa may be indicated by the presence in the Congo of the Pygmies (Mbuti), who share a close genetic relationship with the click-language speakers in southern Africa (but who have acquired a Nigerian-Kordofanain language).[140]

The remaining groups of living people in Africa represent relatively recent population movements within the continent as well as some admixture with Eurasian populations (originally derived from Africa). The latter is most apparent among the North African groups whose paternal lineages are thought to have emerged as late as 15,000 years ago.[141] Proto-Chadic Afro-Asiatic–speaking people probably spread from the central Sahara region to the Lake Chad Basin about 7,000 years ago, while Cushitic speakers migrated from southern Ethiopia into Kenya and Tanzania after 4,000 years ago.[142] The most significant later population movement is the spread of the Bantu-speaking agriculturalists between 5,000 and 3,000 years ago (especially well represented by paternal, or Y-chromosome, lineage E1b1a). Originating in the highlands of Nigeria and Cameroon, the Bantu peoples spread both east and south, apparently intermarrying with local women in these parts of Africa.[143]

The Archaeological Evidence

Although the genetics of living North Africans reflect population movements that occurred after the Last Glacial Maximum (less than 20,000 years ago), the fossil record indicates that modern humans were present in North Africa by roughly 300,000 years ago (Jebel Irhoud). The pattern is further supported by *Homo sapiens* fossils in the Levant between 120,000 and 90,000 years ago (Skhūl and Qafzeh), which presumably reflects a limited extension of the colonization of North Africa into Eurasia.[144] There is archaeological evidence for settlement of the Arabian Peninsula during this period as well (see box 4.1).[145] Comparative analysis of the Neanderthal genome suggests that some gene flow between modern humans and Neanderthals took place in this setting (see box 2.2).[146]

The North African archaeological record may help explain why modern humans failed to disperse into Australia/New Guinea and the higher latitudes before 75,000 years ago. Although sites in North Africa and the Levant provide a significant part of the evidence for modern human cognitive faculties between 150,000 and 75,000 years ago (see table 4.4), it remains unclear if the people who made the shell beads and other artifacts found in these sites were fully modern cognitively or if, as suggested earlier, they were superseded by the direct ancestors of living humans.[147] The North African archaeological record, which contains several stone-tool industries that date to between 150,000 and 60,000 years ago, may eventually provide evidence of multiple migrations from the south.

New dates from *Dar es-Soltan I* (Morocco) suggest that the Aterian industry (associated with skeletal remains of modern humans at Dar es-Soltan II) is older than 110,000 years.[148] The Aterian, which is characterized by the production of stemmed points and is widespread in North Africa, was previously believed to be younger. In Northeast Africa, sites along the Upper Nile River yield assemblages with Levallois points (Nubian Complex) that are dated as early as 120,000 years ago. A younger industry containing Levallois blades (and the burial of a modern human child) at *Taramsa Hill* (Egypt) is dated to between 80,000 and 50,000 years ago.[149] The Taramsa industry exhibits similarities with the Initial Upper Paleolithic of the Levant, which is associated with the later global dispersal (chapter 6). Also recently dated to this time range (ca. 75,000 years ago) is an assemblage containing blades and tools on blades excavated many years ago from the lower levels of *Haua Fteah* (northern Libya) (and labeled "Pre-Aurignacian").[150]

Before the Global Dispersal: Technology and Economy

After 75,000 years ago (possibly as late as 60,000 years ago), modern humans dispersed out of Africa, across southern Asia, and to Australia/New Guinea within a few millennia. Within another few thousand years, they colonized higher latitudes in Eurasia, and their remains, dated to 45,000 years ago, are found as far as latitude 57° North in western Siberia.[151] The most important factor in the rapid colonization of this wide range of habitats and climate zones was a faculty for creating novel and sometimes complex technologies, which allowed modern humans to adapt to local conditions by designing their own "traits" (for example, insulated clothing).

Predictably, the earliest and richest source of evidence for new technologies is stone artifacts, which are the most likely to be preserved in large quantities. Excavations in a cave at Pinnacle Point on the South African coast reveal that modern humans were manufacturing small stone blades, or bladelets (less than 10 millimeters wide), by about 165,000 years ago.[152] The technology is well known from other southern African sites dating to about 70,000 years ago as the Howiesons Poort industry, which also contains geometric segments and backed pieces, some apparently used as tips or barbs on wooden spears.[153] The high-quality raw materials used to fashion these items often were imported from some distance (a pattern reported for industries of comparable age in East Africa).[154] Pinnacle Point also yields evidence for a sophisticated process of heat treatment of stone, which was systematically applied to the raw material to improve its flaking properties.[155]

Modern humans in Africa before 60,000 years ago increased their use of aquatic resources and may have developed new technologies (perhaps including watercraft) to exploit some of them.[156] This phenomenon could have played a role in the dispersal across southern Asia, which appears to have followed a coastal route. At Katanda, as noted earlier, freshwater-fish remains were found in association with barbed points (ca. 90,000 years old) (see figure 4.10), and marine-fish remains have been recovered from Blombos Cave (ca. 75,000 years old).[157] The remains of freshwater fish also are reported from several sites in Northeast Africa.

The collection of shellfish in southern coastal settings—including Pinnacle Point, Blombos Cave, and *Klasies River Mouth* (South Africa)—is better documented, although no special technologies would have been required and the same pattern is documented among non-modern people in Europe as early as 150,000 years ago.[158] Nevertheless, the analysis of shellfish remains in South African sites provides clues about population size and economic level, and has been injected into the debate over modern human cognitive faculties in this context. The size of the shellfish, which include limpets and Cape turban shells, is large and contrasts sharply with specimens of the same taxa in younger sites—reduced in size by intensive collection. The inference is that populations of modern humans remained small and stable before the global dispersal (that is, lacked any special cognitive faculties that may have transformed their economy and promoted population growth).[159]

Significantly, there is indirect evidence for the use of automata in the form of snares by 65,000 to 62,000 years ago (*Sibudu Cave* [KwaZulu-Natal]) and more direct evidence for the use of mechanical technology in the form of the bow and arrow by around 44,000 to 42,000 years ago (Border Cave) in southern Africa.[160] They support the notion that the complex computations underlying the technology of the people who first occupied places like western Siberia have their source—like the people themselves—in Africa (specifically, southern Africa). Another impressive innovation is the ostrich-eggshell water container or canteen, which is indicated by decorated eggshell fragments recovered from Diepkloof Rock Shelter and dated to about 60,000 years ago. Similar containers (with a hole punctured in the top of the egg) were made by later peoples in southern Africa, including the Bushmen.[161] They are a technological adaptation to arid environments and represent the earliest known container in human prehistory.

Marine-shell beads may be the most diagnostic artifact of modern humans before 75,000 years ago and may reflect the most important effect of increased computational complexity before the dispersal. The complexity of the underlying computation pertains not to the design of personal ornaments, which may have been relatively simple, but to the expanded social networks that they may represent. The beads, as well as other items such as exotic stone, may have been exchanged between individuals to symbolize a reciprocal social bond, as in the *!xharo* exchange system of the Ju/'hoansi (see box 4.2).[162] This, in turn, may reflect the effect of an enhanced faculty for establishing and managing a wider range of long-term social relationships with people outside the family and residential band.

Chapter Five

GLOBAL DISPERSAL

Southern Asia and Australia

In the case of the Andaman Islanders it is possible that they have been
entirely isolated in their island home, and have not been affected by contact
with other races.

—A. R. RADCLIFFE-BROWN (1922)

Across southern Asia and Australia, isolated groups of hunting-and-
gathering people survived until recent or modern times. They include the
Birhor of eastern India; the *Semang*, who inhabit lowland rain forest on
the Malay Peninsula; and the native people of the Andaman Islands. The
most geographically widespread are the Aboriginal people of Papua New
Guinea and Australia, while smaller groups, such as the *Aëta*, live on the
islands of the Philippines.

All these peoples carry genetic markers that tie them to the dispersal of
modern humans out of Africa between 75,000 and 60,000 years ago.[1] They
belong to mitochondrial DNA (mtDNA) and Y-chromosome DNA (Y-DNA)
haplogroups linked closely to the African lineages that are ancestral to all
non-African peoples throughout the planet. The genetics are reinforced by
language: most of the South Asian groups speak ancient Austro-Asiatic languages, while the Australians speak an altogether separate family of
languages. The overall pattern is reminiscent of that in sub-Saharan Africa
(chapter 4), where late-surviving hunter-gatherers, most of whom are speakers of the ancient click languages, reflect the initial geographic distribution
and spread of modern humans.

Many paleoanthropologists believe that the Eurasian lineages initially
dispersed out of Africa and directly into tropical Asia by way of the southern margin of the Arabian Peninsula. The timing of the dispersal conceivably was related to—if not controlled by—climate changes between 100,000

and 50,000 years ago that periodically rendered the Arabian Peninsula habitable after intervals of extreme aridity.[2] Some argue that the initial movement into southern Asia followed a *coastal route* and reflected adaptations to rich tropical coastal environments. If so, much of the fossil and archaeological record of the earliest modern humans in Asia may lie underwater—flooded by rising sea levels after 15,000 years ago.[3]

The idea that modern humans initially dispersed eastward out of Africa and into southern Asia ("southern dispersal hypothesis") emerged relatively early in debates about the recent African origin of *Homo sapiens* (chapter 2).[4] It was driven primarily by the dating of the oldest archaeological sites in Australia, which were widely assumed—correctly it appears—to be confined to modern humans because of the technical challenges of crossing more than 70 kilometers of open water to reach the island continent. The implications of the Australian archaeological record for watercraft and marine navigation encouraged the focus on coastal migration.[5] Both existing linguistic evidence and accumulating genetic data from southern Asia were cited in discussions of the hypothesis, but supporting archaeological and human fossil data from southern Asia were slow to materialize.

The evidence (derived from both ancient DNA [aDNA] and the genetics of living people) for Neanderthal introgression in all non-African populations suggests that the latter are descended from a group that was initially present in northern Eurasia.[6] Traces of an early movement out of Africa are lacking in the Arabian Peninsula, where older subclades of mtDNA haplogroup N (the oldest of all non-African lineages) are absent in the living population.[7] Arid climates 75,000 to 60,000 years ago may have rendered the region uninhabitable during the period when modern humans apparently moved into Eurasia.[8] At the same time, new evidence has been accumulating for an early (re)appearance of *Homo sapiens* in the Levant (ca. 55,000 years ago), as well as movement earlier than previously believed into other parts of northern Eurasia (chapter 6). In sum, the initial dispersal in southern Asia may have started somewhere in northwestern Eurasia, such as the Levant, rather than on the southern coast of the Arabian Peninsula (figure 5.1).[9]

In addition to early and later interbreeding with Eurasian Neanderthals, the modern humans migrating across Asia interbred with the Denisovan population (see box 2.2). Denisovan genes are carried in the living Melanesian population (along with the Neanderthal genes) and provide a significant

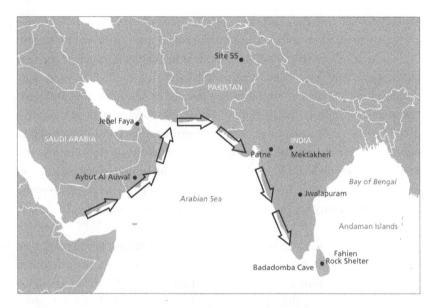

FIGURE 5.1 Map of the Arabian Peninsula and the Indian subcontinent, showing the location of fossils and archaeological sites assigned to *Homo sapiens*, dated to between 75,000 and 30,000 years ago, and the hypothesized coastal migration route out of Africa (*arrows*). (Base map adapted from en.wikipedia. org/wiki/File:BlankMap-World6.svg)

clue to the geographic distribution of the Denisovans at the time of the dispersal. Presumably, the Denisovans were present in Southeast Asia, where the introgression with modern humans took place.[10]

Regardless of the relationship between the dispersals in northern and southern Eurasia, the latter differed from the former in important ways. Modern humans moving into tropical Asia 60,000 years ago or earlier did not confront the formidable challenges posed by higher-latitude environments, which were more seasonal, generally cooler, and typically much poorer in resources than lower-latitude habitats. Many of the technological innovations essential to occupation of northern Eurasia were unnecessary in southern Asia and Australia. It may be significant that Native Australians lacked the bow and arrow, as well as traps and snares, while the Andaman Islanders had no technology for making fire.

If the initial movement of people across southern Asia was channeled along coastlines, modern humans may have largely avoided competing with the indigenous *Homo* population, which apparently never developed a maritime economy, until the former were already well established in southern Asia. Thus the ability to exploit the resources of the coastal zone—including some innovative and complex technologies—may have been critical to the southern dispersal. As described in chapter 4, much of the new technology may have been engineered in Africa by 75,000 years ago. Archaeological evidence supports its presence in southern Asia and Australia, at least indirectly. The remains of pelagic species, such as tuna, in a 40,000-year-old occupation on the northern coast of East Timor suggest deep-water fishing,[11] while traces of modern humans in New Guinea and the Philippines by 50,000 years ago indicates the sailing of watercraft over significant distances.[12] The absence of more direct evidence for fishing and boating technology is not surprising, because it is notoriously difficult to find in any prehistoric archaeological context.

Dating the dispersal in southern Asia and Australia is problematic and controversial.[13] The earliest dated archaeological sites in the conjoined landmass of Australia and New Guinea (termed *Sahul*) are about 50,000 years old, suggesting that the oldest modern humans in India must be a least a few thousand years older.[14] However, archaeological remains in India and Southeast Asia that may be unequivocally assigned to *Homo sapiens* are less than 50,000 years old. Artifact assemblages buried below a 74,000-year-old volcanic-ash layer in southern India exhibit similarities to contemporaneous assemblages in southern Africa,[15] but their attribution to modern humans is uncertain.

Estimates of a somewhat earlier movement from Africa into southern Asia by anatomically (and cognitively) modern humans (that is, 75,000–60,000 years ago) are based on genetic data. As noted in chapter 2, current assessments of the time to the most recent common ancestor (TMRCA) of the non-African lineages are greater than 70,000 years ago (see table 2.1).[16] This suggests that the dispersal out of Africa began during a relatively cool and dry period that correlates with Marine Isotope Stage 4 (MIS 4) in the global paleo-climate record. The divergence of "Eurasian" lineages such as mtDNA N and Y-DNA DE and CF could have taken place in Africa, however, and the arrival of modern humans in southern Asia before 60,000 (or 70,000) years ago must be confirmed with some unequivocal fossil or archaeological data.[17]

Human fossil remains assigned to *Homo sapiens* and dated to more than 60,000 years ago have been accumulating in southern China in recent years (table 5.1). Although they may eventually provide confirmation of modern

TABLE 5.1

South and Southeast Asian and Australian Fossils Assigned to *Homo sapiens*, Before 25,000 Years Ago

Site	Estimated Age (method)	Description	Comments
Bailong Cave (southern China)	160,000 years (dated stalagmites)	Isolated teeth	Previously reported radiometric dates much younger
Punung (Java, Indonesia)	126,000–81,000 years (OSL, U-series, biostratigraphy)	Isolated tooth (premolar)	Assignment to *Homo sapiens* problematic (based on size)
Luna Cave (southern China)	127,000–70,000 years (U-series, biostratigraphy)	Isolated teeth (2 molars)	Metric comparison suggests *Homo sapiens*, but overlap with Neanderthals
Fuyan Cave (southern China)	120,000–80,000 years (thorium dating of speleothem)	Isolated teeth ($n = 47$)	Morphologically and metrically "unequivocal assignment to *Homo sapiens*"
Zhiren Cave (southern China)	116,000–106,000 years (U-series, OSL, paleomagnetism)	Mandible	Mandible exhibits some archaic features, but is assigned to *Homo sapiens*
Liujiang (Guangxi, southern China)	68,000–61,000 years (U-series)	Cranium; postcranial bones	Problematic association of human remains (intrusive?) and dated cave breccia
Callao Cave (Luzon, Philippines)	67,000 years (U-series, ESR)	Third metatarsal	Indicates marine travel and watercraft before 60,000 years ago
Tam Pà Ling (northern Laos)	63,000–46,000 years (radiocarbon, TL, OSL, U-series)	Cranial fragments (frontal, parietal, temporal, occipital, maxilla)	Suggested date of about 50,000 calibrated radiocarbon years ago[a]
Tabon Cave (Palawan Island, southern Philippines)	58,000–24,000 years (U-series)	Mandible fragment (younger); tibia fragment (older)	Tibia fragment dated to between 58,000 and 37,000 years ago
Laibin (Guangxi, southern China)	More than 44,000–39,000 years (U-series on carbonate)	Cranial fragments (maxilla, occipital); isolated teeth	Maximum age of 112,000 years ago
Niah Great Cave (Sarawak, Borneo, Indonesia)	Ca. 46,000–34,000 years (radiocarbon [ABOx], U-series)	Partial skull; femur; proximal tibia; talus	Associated with stone artifacts; broad-based rain-forest diet; freshwater fish (evidence of traps and snares?)

(continued)

(continued)

Site	Estimated Age (method)	Description	Comments
Lake Mungo (Willandra Lakes, Australia)	Ca. 43,000–41,000 years (radiocarbon, OSL)	Three partial skeletons	
Fahien Rock Shelter (Sri Lanka)	Ca. 38,000 years (radiocarbon)	Two infants; one child; one sub-adult; one adult female ($n = 7$)	
Badadomba Cave (Sri Lanka)	28,000 years	Cranial fragments; mandible; isolated teeth; postcranial fragments	Associated with microlithic artifacts
Moh Khiew Cave (Thailand)	26,000 years	Partial skull	

[a]Fabrice Demeter et al., "Anatomically Modern Human in Southeast Asia (Laos) by 46 ka," *Proceedings of the National Academy of Sciences* 109 (2012): 14375–14380.

Sources: Graeme Barker et al., "The 'Human Revolution' in Lowland Tropical Southeast Asia: The Antiquity and Behavior of Anatomically Modern Humans at Niah Cave (Sarawak, Borneo)," *Journal of Human Evolution* 52 (2007): 243–261; Richard G. Klein, *The Human Career: Human Biological and Cultural Origins*, 3rd ed. (Chicago: University of Chicago Press, 2009), 620–622; Demeter et al., "Anatomically Modern Human in Southeast Asia (Laos) by 46 ka"; Jim Allen and James F. O'Connell, "Both Half Right: Updating the Evidence for Dating First Human Arrivals in Sahul," *Australian Archaeology* 79 (2014): 86–108; Christopher J. Bae et al., "Modern Human Teeth from Late Pleistocene Luna Cave (Guangxi, China)," *Quaternary International* 354 (2014): 169–183; Sheela Athreya, "Modern Human Emergence in South Asia: A Review of the Fossil and Genetic Evidence," in *Emergence and Diversity of Modern Human Behavior in Paleolithic Asia*, ed. Yousuke Kaifu et al. (College Station: Texas A&M University Press, 2015), 61–79; Wu Liu et al., "The Earliest Unequivocally Modern Humans in Southern China," *Nature* 526 (2015): 696–700; Yanjun Cai et al., "The Age of Human Remains and Associated Fauna from Zhiren Cave in Guangxi, Southern China," *Quaternary International* 434 (2017): 84–91.

human dispersal in southern Asia between 75,000 and 60,000 years ago, they currently seem more likely to document a Southeast Asian presence of the "near modern" humans who expanded into the Levant between 120,000 and 90,000 years ago, before being replaced by Neanderthals (chapter 4). A cranium from *Liujiang* tentatively dated to 68,000 to 61,000 years ago may represent the later spread of cognitively modern people, but the association between the dates and the fossil is problematic.[18] More recently, modern human remains have been recovered from *Luna* and *Zhiren Caves* and dated to between about 127,000 and 70,000 years ago.[19] They have been attributed to both fully modern humans and Denisovans, but the simplest explanation is that they represent the early (and ultimately unsuccessful) colonization event documented in the Levant by an archaic form of anatomically modern humans.[20]

CLIMATE CHANGE AND THE ARABIAN PENINSULA

Climate changes between 100,000 and 50,000 years ago and their effects on the Arabian Peninsula may have been an important variable in the timing and route of modern humans spreading across southern Asia (box 5.1). At various times during this interval, shifting weather patterns brought increased moisture to the otherwise arid region, creating systems of rivers and lakes and enriching plant and animal life. During the wet phases, human populations occupied the Arabian Peninsula, and direct movements between Africa and south Asia became possible.[21] During the arid phases, humans may have been forced to abandon the region, leaving an empty niche for others to occupy with the return of more favorable conditions. The ability to exploit coastal marine resources (beyond the collecting of shellfish along the shore) may have conferred advantages on modern humans that other forms of *Homo* did not enjoy.[22]

Although today the interior of the Arabian Peninsula is largely desert, the southern coastal zone receives some rainfall. The pattern is especially pronounced in the southwestern portion of the peninsula—in the Yemen Highlands—but also is found in the northeastern end, near the Gulf of Oman. During warmer periods of the Pleistocene, even the interior was much wetter. Satellite-based imagery reveals extensive paleo-river channels, alluvial fans, and former lake basins across the landscape. At times of lowered sea level, the shallow Persian Gulf itself became a large river basin.[23] These intervals of wetter climate probably reflected the northward shift of the Inter-Tropical Convergence Zone, which would have brought summer monsoon rains from the Indian Ocean.[24]

The genetics of the present-day inhabitants of the Arabian Peninsula provide little information about the dispersal of modern humans out of Africa between 75,000 and 60,000 years ago. Although African and South Asian contributions to the genome are evident, they appear to represent more recent population history.[25] The current population primarily reflects the spread of people from western Eurasia after the Last Glacial Maximum

BOX 5.1
Climate Change and the Dispersal of Modern Humans

Climate change may have had a major role in both the early dispersal of "near modern" humans out of Africa and into the Levant and Arabian Peninsula about 120,000 to 90,000 years ago, and the global dispersal of *Homo sapiens* that began 75,000 to 60,000 years ago. The effects of climate change on dispersal are likely to have been twofold: improved conditions for plants and animals (especially increased moisture) may have generated human population growth and pressure for expansion, while also rendering neighboring areas more accommodating to the expanding population.[1]

The early dispersal of "near modern" humans coincided with the Last Interglacial climatic optimum and the warm periods that followed (or Marine Isotope Stage 5 [MIS 5]) and may have terminated in the Levant with a cold phase (MIS 4) and the return of the Neanderthals. The movement of *Homo sapiens* from Northeast Africa into the southern Arabian Peninsula during the later phases of MIS 5 (recently documented with archaeological data) may be related to a northward shift of the Inter-Tropical Convergence Zone in the Indian Ocean. The changes in climate are reconstructed from the measurement of stable isotopes, especially oxygen isotopes, in marine-sediment cores. When substantial quantities of ocean water are locked up in the form of terrestrial ice sheets by cold temperatures, sea water is enriched with the heavy isotope ^{18}O (measured against the lighter isotope ^{16}O).[2]

The global dispersal may have begun during a cold period (MIS 4), but the occupation of northern Eurasia dates to a warmer interval (MIS 3). This interval was characterized by a series of alternating brief warm and cool phases between roughly 60,000 and 30,000 years ago (followed by the Last Glacial Maximum [MIS 2]) that are identified in a finely calibrated stable-isotope record retrieved from the Greenland ice sheet, in which the ratio of oxygen isotopes is reversed and cold periods correspond to increased levels of ^{16}O (the lighter isotope).[3] The warm intervals are labeled "Greenland Interstadials" (GI), and the cold intervals are termed "Greenland Stadials" (GS). The GI/GS periods may be correlated with terrestrial sediments that contain dated skeletal remains and artifacts of modern humans in sites across northern Eurasia—often with supporting paleo-climate data (for example, plant pollen/spores)—to assess their relationship to the spatial and temporal distribution of human populations (box figure 5.1).[4]

As discussed in chapter 6, the initial spread of modern humans in Europe and northern Asia is correlated with GI 12, which was a period of significant warmth between about 47,500 and 44,500 years ago. The primary impact of GI 12 climates on modern humans may have been population growth fueled by the rise in biological productivity. A later phase of expansion took place during the succeeding but briefer GI 11 (ca. 43,500–42,500 years ago), while there is evidence of a settlement hiatus in some places during the cold period (GS 9 [Heinrich Event 4]) that followed the Campanian Ignimbrite volcanic eruption around 40,000 years ago. And there is a correlation between (1) archaeological and genetic evidence for population expansion at both ends of northern Eurasia, and (2) the protracted warm interval (GI 8) that followed GS 9 roughly 38,000 years ago.

Finally, it should be noted that the extreme cold of the Last Glacial Maximum of MIS 2 (ca. 28,000–16,000 years ago) probably played a major role in the dispersal of modern humans in the Western Hemisphere by isolating the parent lineages of most living Native Americans in one of several human geographic refugias—possibly in

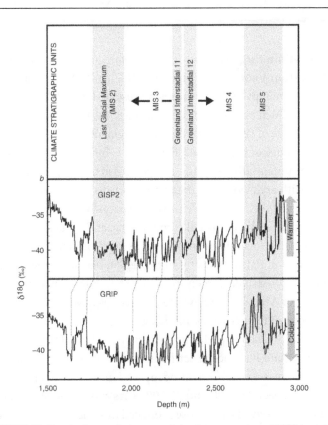

BOX FIGURE 5.1 Greenland oxygen-isotope record, showing a comparison of GISP2 (*center*) and GRIP (*left*) ice cores. Climate-stratigraphic units are shown on the right. (Adapted from P. M. Grootes et al., "Comparison of Oxygen Isotope Records from the GISP2 and GRIP Greenland Ice Cores," *Nature* 366 [1993]: 552, fig. 1)

Beringia—before they spread throughout North and South America (chapter 7).[5] The movement out of Beringia 15,000 to 14,000 years ago into mid-latitude North America was controlled by the effect of warming climates on the ice sheets that covered most of northern North America during both MIS 3 and MIS 2.[6]

1. J. R. Stewart and C. B. Stringer, "Human Evolution Out of Africa: The Role of Refugia and Climate Change," *Science* 335 (2012): 1317–1321.
2. J. John Lowe and Michael J. C. Walker, *Reconstructing Quaternary Environments*, 3rd ed. (Abingdon: Routledge, 2014).
3. Ibid.
4. Bernhard Weninger and Olaf Jöris, "A [14]C Age Calibration Curve for the Last 60 ka: The Greenland-Hulu U/Th Timescale and Its Impact on Understanding the Middle to Upper Paleolithic Transition in Western Eurasia," *Journal of Human Evolution* 55 (2008): 772–781.
5. See, for example, Andrew Kitchen, Michael M. Miyamoto, and Connie J. Mulligan, "A Three-Stage Colonization Model for the Peopling of the Americas," *PLoS ONE* 3 (2008): e1596.
6. Chris R. Stokes, Lev Tarasov, and Arthur S. Dyke, "Dynamics of the North American Ice Sheet Complex During Its Inception and Build-up to the Last Glacial Maximum," *Quaternary Science Reviews* 50 (2012): 86–104.

(LGM), or within the past 20,000 years. The majority of maternal lineages are derived from a northern source (mtDNA haplogroups N and R), while the most common paternal lineage is represented by Y-DNA haplogroup J (58 percent), which emerged in the Near East less than 20,000 years ago.

The challenge of reconstructing the movements of *Homo sapiens* along the southern edge of the Arabian Peninsula is further compounded by the lack of any potentially relevant human fossils. The problem has been addressed with archaeological data and efforts to tie these data to (1) stone artifacts outside the Arabian Peninsula, especially in Africa, on the basis of technological and typological comparisons, and (2) past climate change and its effects on local environments. The pace of archaeological survey and excavation of Paleolithic sites in the region has increased considerably in recent years—driven largely by interest in the dispersal of modern humans out of Africa—and new site discoveries have been tied to the satellite mapping of former rivers and lakes.

Several major discoveries in the past few years have dramatically altered the picture of late Pleistocene settlement on the Arabian Peninsula. Collectively, they reveal that humans occupied southern Arabia during major wet phases between 130,000 and 40,000 years ago. Some of the stone artifacts are clearly derived from those produced in Africa, while others exhibit at least general similarities to contemporaneous industries in Africa. At least some of the artifacts most probably were made by anatomically modern humans.

In 2011, Jeffrey Rose and his colleagues reported the discovery of the Nubian Complex—originally from Northeast Africa and described briefly in chapter 4—in a buried and dated context at *Aybut Al Auwal* in the Dhofar region of Oman.[26] The site is on the north-facing slope of the high ridge that parallels the coast in this part of Oman. Artifacts were recovered from stream deposits along a former tributary channel of a larger drainage. The location of the site probably is tied to the presence of both local groundwater springs and exposed chert beds. The artifacts are typical of the later Nubian Complex and date to about 106,000 years ago—placing them in a wet phase that corresponds to Marine Isotope Stage 5c (MIS 5c) (figure 5.2).

A rock shelter near the Gulf of Oman (*Jebel Faya*) yielded artifacts dating to an earlier wet interval that corresponds to the Last Interglacial (MIS 5e), as well as a more recent period (early part of MIS 3).[27] The earlier

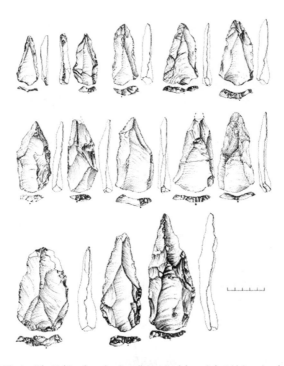

FIGURE 5.2 Artifacts of the Nubian Complex (Levallois points) from Aybut Al Auwal and other sites in the Dhofar region of Oman. (From Jeffrey I. Rose et al., "The Nubian Complex of Dhofar, Oman: An African Middle Stone Age Industry in Southern Arabia," *PLoS ONE* 6 [2011]: e28239, fig. 10. Reproduced from *PLoS ONE* under the terms of the Creative Commons Attribution [CCBY] license)

occupation dates to about 125,000 years ago, and stone artifacts include flakes and blades stuck from prepared cores and bifacial points similar to those of contemporaneous sites in Northeast Africa. The later artifacts are dated to about 40,000 years ago and include some generic Upper Paleolithic forms, such as end scrapers and burins.

Despite these and other recent discoveries, debate continues over the Arabian Peninsula as a route for modern human dispersal into southern Asia.[28] The finds from Aybut Al Auwal and Jebel Faya demonstrate a human presence on or near the coast of the Arabian Sea—at least during wetter periods—between 130,000 and 40,000 years ago. The artifacts of the later Nubian Complex most likely were produced by *Homo sapiens* from Northeast Africa, and they are correlated with the spread of African hamadryas baboons into Arabia.[29] They appear to coincide with the movement of "near modern" humans into the Levant between 120,000 and 90,000 years ago

and presumably are related to the same early dispersal out of Africa.[30] Similar artifacts in dated contexts have yet to be identified in India and Southeast Asia or farther east, suggesting that the people who made these
artifacts may not have expanded farther. The later artifacts postdate the
early phases of the global dispersal that began 75,000 to 60,000 years ago
and provide no evidence that the second dispersal followed this route (and
some genetic data suggest that the northern route is more likely).

THE INDIAN SUBCONTINENT AND SOUTHEAST ASIA

Unlike the Arabian Peninsula, South and Southeast Asia contain substantial genetic evidence for a modern human presence before 50,000 years ago.
Despite recent population intrusions from other parts of the world, southern Asia is home to a number of groups that are genetically linked to early
movements out of Africa. Some of these groups were mentioned at the
beginning of the chapter—the surviving and recent hunter-gatherers of
India and Southeast Asia—while others include people who adopted agri-
culture at some point in their past. Among the former, one group (Andaman Islanders) maintained a coastal-foraging economy and may provide
some insight into how modern humans adapted to coastal environments
more than 50,000 years ago.

The South Asian population exhibits a high degree of genetic diversity—
second only to that of Africa—which is consistent with its hypothesized
early role in the global dispersal of modern humans.[31] Mitochondrial DNA
haplogroups N and M—which represent the earliest maternal lineages to
have left Africa—are common (with an estimated TMRCA of 93,200–
61,400 years).[32] Several paternal lineages that are widespread among modern tribal groups (for example, Y-DNA haplogroups H and F) also are
closely tied to African lineages (recently defined South Asian subclade H0
is thought to have diverged from Y-DNA H about 51,000 years ago).[33] As
does that in Africa, the living population reflects the impact of more recent
invasions from the north—specifically, from western and Central Asia.[34]
The genetic input from northern Eurasia is reflected especially in the frequencies of certain Y-DNA haplogroups, indicating that the invasions were
primarily carried out by males.[35] In India, the frequency of western and
Central Asian lineages decreases along a north–south gradient.[36]

In southern Asia, an important factor in preserving genetic (and linguistic) traces of an early dispersal from Africa is the traditional endogamy among the Indian caste populations.[37] Adherence to strict marriage rules has limited gene flow to the various tribal groups scattered across eastern India and farther east, and they exhibit particularly high frequencies of mtDNA haplogroup M, which represents an early South Asian lineage (estimated coalescence date of 50,000 years).[38] And within this lineage, sublineage M2, with unusually high nucleotide diversity (indicating an early origin), is common.[39] Especially significant, the paternal lineages that are widespread among the caste populations are correspondingly rare in the tribal groups. Among the latter, Y-DNA haplogroup O-M95, which has a South Asian origin, is extremely common.[40] Many of the tribal groups speak Austro-Asiatic languages, which represent the oldest established language family in the region.

A whole-genome analysis published in early 2016 found four major ancestral components in the mainland South Asian population, plus a fifth group corresponding to inhabitants of islands off the southeastern coast of India.[41] More than 350 individuals from various mainland and island populations were sampled, and more than 800,000 single-nucleotide polymorphisms (SNPs) were incorporated into the analysis. The four major ancestral groups on the Indian mainland are North Indian, South Indian, Tibeto-Burman, and Austro-Asiatic. The fifth group comprises people from the Andaman and Nicobar Islands. While the North Indian and Tibeto-Burman groups are associated with later movements of people from other parts of Asia, ancestral South Indians and speakers of Austro-Asiatic languages are linked to early movements out of Africa, as are the Andaman and Nicobar Islanders.[42]

Some groups in Southeast Asia were labeled "Negrito" many years ago on the basis of shared physical characteristics (for example, short stature, very dark skin, and tightly curled hair) that were thought to indicate both an early presence in Southeast Asia and a link with some African people (especially the *Mbuti* [Pygmies]). Analysis of their craniodental characters reveals a more complex picture, reflecting the impact of relatively recent population movements in Southeast Asia.[43] Among the Negrito groups are the Semang (Orang-Asli) of Malaysia, Aëta of the Philippines, and Andaman Islanders. Traditionally, these societies pursued a foraging economy in the tropical forests of Southeast Asia, with the exception of the Andaman Islanders.

Archaeological investigations have failed to produce evidence of early settlement on the Andaman Islands, and it appears that the ancestors of the native inhabitants may have occupied the islands within the past few thousand years.[44] It is conceivable that at some point in their past, they practiced a forest economy similar to that of other Negrito groups. But both genetic and linguistic data suggest that the Andaman Islanders have been isolated from other populations for a long time. Ancient mtDNA extracted from the remains of nineteenth- and early-twentieth-century islanders represented only haplogroup M, including sublineage M2.[45] Both genetic and phenotypic diversity is high, indicating a lengthy history on the islands. The languages of the Andaman Islanders both are highly differentiated and compose a separate family with no obvious links to other known languages.[46]

Even if the Andamanese seem unlikely to represent a survival of the original foraging economy pursued by coastal dwellers in southern Asia more than 50,000 years ago, because of the enormous time depth involved, they provide insights into how modern humans could have lived in this setting—and why they may have expanded eastward at a rapid pace. The diet was based on the hunting and gathering of both marine and terrestrial foods (box 5.2). Despite the absence of some technologies that are otherwise widespread among nonagricultural peoples of both the tropics and the higher latitudes, the economy supported a relatively high population density (estimated at roughly 2.25 persons per square mile during the mid-nineteenth century).[47] This was almost certainly a consequence of an unusually rich habitat—especially one that incorporated off-shore marine resources (including a large marine mammal).

Unlike many other hunter-gatherers, the Andaman Islanders were semi-sedentary, occupying village sites for extended intervals. While the food-gathering technology included the bow and arrow, much of the equipment found among other foraging peoples for harvesting fish, birds, and other small vertebrates (for example, fishhooks, traps, and weirs) was unknown. Most striking, the Andaman Islanders lacked the technology for making fire—a rare omission among recent and living hunter-gatherers.[48] They did, however, produce a simple and effective watercraft in the form of a wooden dugout canoe fitted with an outrigger.

BOX 5.2

The Andaman Islanders and the Dispersal of Modern Humans

Among the hunter-gatherers of southern Asia who carry genetic and linguistic markers that tie them to the dispersal of modern humans out of Africa more than 50,000 years ago, the native people of the Andaman Islands may be unique. Unlike other groups in southern Asia that represent remnants of the dispersal, the Andaman Islanders rely on marine resources. They have been isolated for an extended period of time on islands, now part of India, located in the eastern Bay of Bengal. The Andaman Islanders provide insights into how modern humans may have lived on the South Asian coast 50,000 years ago.

It is unclear when the Andaman Islanders first occupied their current home—it may have been long after the dispersal out of Africa. They appear to have been isolated genetically from other groups for many millennia, however, and have maintained a hostile attitude toward intruders to the present day. The British established a penal colony on the islands in 1789 that was abandoned a few years later but reestablished in 1858. Eventually, the British developed amicable relations with many of the Andaman Islanders, although the introduction of various diseases wiped out much of the local population in the late nineteenth century. Information about their way of life was recorded by E. H. Man, a British government official between 1869 and 1880 and by the anthropologist A. R. Radcliffe-Brown from 1906 to 1908.[1]

While they have adopted some new technologies in recent millennia, the Andaman Islanders have maintained a hunter-gatherer economy based heavily on marine resources, including the dugong (large sea mammal); turtles; a wide variety of fish; and crabs, crayfish, and mollusks. They also hunt wild pigs and civet cats, along with smaller terrestrial vertebrates. A critical piece of technology is the dugout canoe, which is carved out of a tree trunk and fitted with an outrigger float to prevent capsizing (box figure 5.2). Hunting equipment includes the bow and arrow and barbed harpoons. In recent times, weapons have been tipped with points that are now made of iron but formerly were fashioned from shell or bone. The Andaman Islanders do not build traps or snares (they do not exploit locally abundant birds), but they traditionally used nets for catching some fish and turtles. They do not make fire, which is extremely rare among recent hunter-gatherer peoples. They carefully maintain their hearths and

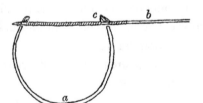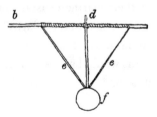

BOX FIGURE 5.2 Transverse section of the dugout canoe and outrigger made by the Andaman Islanders: (a) hull; (b) boom; (c) cane binding over boom; (d) stick attaching boom to float; (e) stays of cane; (f) float. (From A. R. Radcliffe-Brown, *The Andaman Islanders* [Cambridge: Cambridge University Press, 1922], 487, fig. 44. Reprinted with the permission of Cambridge University Press)

(continued)

(continued)

apply heat treatment to stone for improved working quality. They produce a variety of containers, including clay cooking pots that may be of relatively recent origin (although they are similar to late Upper Paleolithic clay pots of southern China). In a comparative review of their food-getting technology, Wendell Oswalt found it to be relatively complex for hunter-gatherers of the tropical zone.[2]

The Andaman Islanders decorate various items with simple geometric designs but do not make representational art.[3] They fashion ornaments from various materials, including marine shells, and they practice body painting with red ocher and clay. Dancing is an important part of their social life. They maintain a strikingly communal society, sometimes inhabiting a communal structure instead of separate family dwellings. Food is shared, and children often are adopted by neighbors.

1. A. R. Radcliffe-Brown, *The Andaman Islanders* (Cambridge: Cambridge University Press, 1933).
2. Wendell H. Oswalt, *An Anthropological Analysis of Food-Getting Technology* (New York: Wiley, 1976).
3. M. Streenathan, V. R. Rao, and R. G. Bednarik, "Palaeolithic Cognitive Inheritance in Aesthetic Behavior of the Jarawas of the Andaman Islands," *Anthropos* 103 (2008): 367–392.

If modern humans dispersing out of Africa more than 60,000 years ago developed a coastal economy similar to that of the Andaman Islanders—and, with the possible exception of the bow and arrow, they probably possessed the requisite technology to support it—they would have enjoyed a potential for rapid population growth in comparison with nonagricultural societies. Both the relatively high population density and the semi-sedentary lifestyle would have generated a high rate of increase. At the same time, their marine-coastal adaptation would have channeled the population expansion into a relatively narrow corridor, promoting an unusually rapid process of colonization. Competition for resources with other humans would have been minimized by the apparent absence of a coastal adaptation in pre-modern *Homo* in southern Asia.

Modern Humans in the Fossil Record

Dated fossils and archaeological sites related to the dispersal of modern humans into South and Southeast Asia have been slow to materialize. But in the past few years, a number of major discoveries have been reported. They triggered a debate between advocates and critics of an early-dispersal model, which postulates a movement of modern humans into southern Asia before the massive Toba volcanic eruption (ca. 74,000 years ago) on the island of Sumatra.[49] The ash was deposited widely across southern Asia and provides a useful chrono-stratigraphic marker for archaeological remains. Although predictably, some of the debate over the early-dispersal model concerns the dating of human fossils, much of the argument revolves around the interpretation of the archaeological record—especially the issue of local continuity in the South Asian industries over time.

A recent discovery that suggests a slightly earlier dispersal is a human foot bone (third metatarsal) recovered from *Callao Cave* on northern Luzon (Philippines) that has been directly dated by U-series and ESR to about 67,000 years ago.[50] The bone exhibits some unusual anatomical features and is "provisionally attributed" to *Homo sapiens*. Its most striking characteristic is its small size, which is comparable to metatarsals of both living Negritos and *Homo floresiensis*.[51] No artifacts were found associated with the Callao specimen, further contributing to its mysterious character. Perhaps the most significant aspect of this find is the indication of a relatively early sea crossing, which may preclude attribution to non-modern humans (figure 5.3).[52]

As noted earlier, fossil remains assigned to *Homo sapiens* and dated to more than 60,000 years ago have been turning up in southern China (see table 5.1). Most notably, they include the Liujiang cranium and associated postcranial bones, which were recovered more than 50 years ago from a cave near Liuzhou City.[53] The skull is unquestionably that of a modern human, but its stratigraphic position in the cave deposits (later dated by U-series) has never been clear. Isolated teeth assigned to *Homo sapiens* from *Bailong Cave* were dated several years ago to around 160,000 years ago, but the new estimate contradicts a suite of much younger dates (less than 30,000 years) based on radiocarbon and uranium-thorium.[54]

More reliably dated finds have been reported recently from several other caves in the region, including a mandible and two molars from Zhiren Cave dated to between 116,000 and 106,000 years ago (biostratigraphy, OSL, U-series, and paleomagnetism).[55] The mandible exhibits some non-modern features, and its taxonomic status is problematic.[56] Two isolated human teeth were recovered from sediments in Luna Cave (Bubing Basin, Guangxi) dated to between 127,000 and 70,000 years ago (biostratigraphy and U-series), along with a dozen stone artifacts.[57] One of the teeth (upper

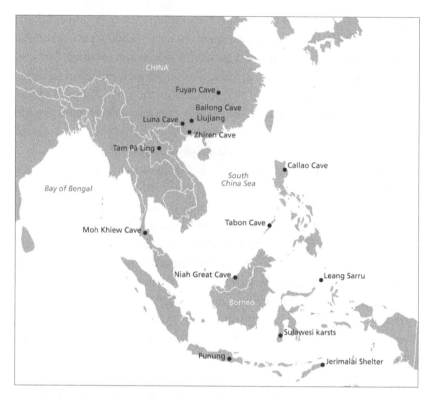

FIGURE 5.3 Map of Southeast Asia, showing the location of fossils and archaeological sites assigned to *Homo sapiens*, dated to between 160,000 and 26,000 years ago. (Base map adapted from en.wikipedia.org/wiki/File:BlankMap-World6.svg)

second molar) is assigned to *Homo sapiens* on the basis of morphology and metrics, while the other tooth (lower second molar) is tentatively assigned to modern humans. However, the metric comparisons of the teeth exhibit significant overlap with non-modern teeth, and long-standing problems with the attribution of isolated teeth to modern and non-modern humans ensure debate over their interpretation. At *Fuyan Cave*, 47 isolated teeth (described as "unequivocally modern human") were recovered from deposits dated to between 120,000 and 80,000 years ago (U/Th).

One way or another, the dated fossils accumulating in southern China are likely to alter the picture of the origin and dispersal of modern humans. On the one hand, at least some of these remains may confirm a movement out of Africa by the immediate ancestors of living Eurasians as early as

75,000 years ago (currently based on estimates of TMRCA with the genetic data). On the other hand, the presence of archaic features on other specimens suggests that they may represent a Southeast Asian extension of the *Homo sapiens* population (sometimes described as "near modern") that moved into the Levant between 120,000 and 90,000 years ago. This possibility raises several new questions: Were these people replaced by the local non-modern population (presumably Denisovans) in the same way the "near modern" population in the Levant is thought to have been replaced by Neanderthals? Did they interbreed with the Denisovans, and, if so, are there traces of an introgression in the ancient or modern genomes of either taxon?

Well-dated human fossils document the presence of *Homo sapiens* in Southeast Asia by 50,000 years ago (see table 5.1). The most famous are from the *Niah Great Cave* (Sarawak, Borneo), which is part of a large complex of caves and smaller caverns situated near the coast.[58] In 1958, fragments of a partial cranium were excavated from immediately below deposits dating to about 40,000 radiocarbon years ago. Although some paleoanthropologists later questioned the stratigraphic position of the remains, recent investigations determined that they were not intrusive from younger levels.[59] Some postcranial bones, possibly from the same individual, also were recovered, and stone artifacts (quartzite flakes and flake and core tools) were found in the overlying horizon. A description of the cranium—now thought to represent a young adult female—was published in 1960. It is similar in many respects to crania of Aboriginal Australians, as well as those of the Andaman Islanders and Negritos.[60]

A major find was reported from northern Laos in 2012. Fragments of a human cranium and maxilla were excavated from deposits in the cave of *Tam Pa Ling* dated to between 51,000 and 46,000 years ago (radiocarbon and luminescence). U-series dating of one of the cranial fragments suggested a maximum age of about 63,000 years. Unlike the mandible from Zhiren Cave, the remains from Tam Pa Ling are characterized as "fully modern" and appear to represent the earliest well-dated fossil traces of the global dispersal (not an earlier "near modern" dispersal) in Southeast Asia.[61]

In the early 1960s, excavation at *Tabon Cave*, on the southwestern coast of Palawan (Philippines), yielded cranial and postcranial remains of several individuals. Additional archaeological materials and human remains were recovered between 2000 and 2007. New U-series dates on several of the human bones range from 47,000 to 16,500 years, while dates on charcoal from the occupation layers are up to 30,500 radiocarbon years ago.[62] The artifacts include abundant flakes and some retouched pieces (primarily scrapers). The human remains exhibit significant variation in body size, recalling the find from Callao Cave on Luzon. As in the case of the Niah specimen, researchers consistently have noted similarities between the cranial fragments and the anatomy of Aboriginal Australians (including fossil Australians).[63]

In South Asia, modern human remains dated to more than 30,000 years ago have been recovered from two sites in southern Sri Lanka. At *Fahienlena*, no fewer than seven individuals are represented in rock-shelter deposits. Charcoal associated with the oldest specimens dates to around 38,000 calibrated years ago. The skeletal remains have been characterized as "robust," with features found among the later Sri Lankan population.[64] At the small rock shelter *Batadomba-lena*, which was discovered and originally investigated in the late 1930s, the later excavation of Layer 7c produced cranial fragments, a mandible, and some postcranial bones, associated with microliths and subsequently dated to about 36,000 calibrated years ago.[65] The mandible is thought to be that of an adult female but is described as "unusually massive" (and compared with Middle Pleistocene *Homo*).[66]

Modern Humans in the Archaeological Record

Archaeologists currently are debating the status of artifacts in tropical Asia that may or may not represent *Homo sapiens*. These include the assemblages recovered from below the ash from the Toba supervolcanic eruption in southern India, as well as slightly younger materials above the ash. The debate inhabits a larger problem: the recognition of modern human cognition and behavior in the archaeological record or, more simply, the archaeological definition of modern humans in a global context.

The problem is acute in southern Asia and Australia, where definitions of modern humans developed in Europe and Africa do not seem to apply to the local archaeological record (although, as will be discussed, some recent discoveries have reduced the contrast).[67] Some of the differences with Europe (and northern Asia) probably reflect the tropical setting— heavier use of wood as a raw material, which is rarely preserved anywhere, versus regular use of bone, antler, and ivory, often preserved in sites at higher latitudes. Another factor is uncertainty about who made the artifacts that underlie the archaeological remains of modern humans.

Artifacts excavated from below the Toba ash at *Jwalapuram Locality 3*, in the Jurreru Valley (India), comprise both flakes and blades struck from prepared cores and a few retouched items, including atypical scrapers, retouched blades, and one burin. A piece of red ocher, exhibiting striations from use, was found with the stone artifacts. The assemblage is dated to 77,000 years ago by OSL, which supports its stratigraphic position beneath the approximately 74,000-year-old tephra. Artifacts recovered from above the ash (and dated to the same age as the latter) at Locality 3 and other Jwalapuram sites are similar to the assemblage below the tephra (table 5.2).[68]

Michael Petraglia and others note similarities between the Jwalapuram assemblages and the Middle Stone Age of Africa, specifically the Howiesons Poort industry of South Africa (chapter 4), and argue that they probably represent modern humans in southern Asia by 75,000 years ago.[69] A number of paleoanthropologists, including Paul Mellars, dispute this suggestion, observing that similar artifacts may be found in a broad range of industries produced by modern and non-modern forms of *Homo* across western Eurasia.[70] The Jwalapuram assemblages conceivably are related to the early dispersal of "near modern" humans, which may have extended into southern Asia (see box 4.1).

Until recently, *Site 55* at Riwat (northern Pakistan) represented the earliest credible traces of modern humans on the Indian subcontinent. The site, which contains both stone artifacts and associated features, is buried in loess and is more typical of open-air localities in northern Eurasia.[71] The artifacts and features were excavated from a level dated by TL to more than 45,000 years. The features include a linear arrangement of stones (labeled a "wall-footing") and a pit containing a stone slab. Most of the artifacts were made from quartzite cobbles; both flakes and blades are present, and few are retouched. Although the small size of some of the blades (which, according to the investigators, "could be classed as microliths") and the features may be diagnostic of modern humans, the dating of Site 55 is problematic.

In 2013, Sheila Mishra, Naveen Chauhan, and Ashkok Singhvi reported radiocarbon and OSL dates of more than 45,000 years ago on a microblade industry at the open-air site of *Mehtakheri* on the Narmada River (central

TABLE 5.2
Archaeological Record of *Homo sapiens* in South and Southeast Asia, Before 30,000 Years Ago

Site	Estimated Age (method)	Description	Comments
Jwalapuram Locality 3 (southern India)[a]	77,000 ± 6000 years (OSL, tephrochronology)	Stone blades with platform preparation; retouched blades; tanged point; burin; red-ocher fragment	Buried below the Youngest Tuba Tuff (YTT) volcanic ash (74,000 years ago)
Jwalapuram Localities 3, 17, and 21 (southern India)[b]	74,000 ± 7000 years (OSL, tephrochronology)	Some blades and bladelets; faceted platform; end scrapers, side scrapers; burins	Artifacts overlie YTT volcanic ash: similar to assemblage below ash
Site 55 (Riwat, northern Pakistan)[c]	Ca. 45,000 years (TL on loess)	Small blade assemblage associated with pit and other features	TL dating should be considered a rough estimate
Mehtakheri (Narmada River, central India)[d]	45,000 years (OSL, radiocarbon)	Microlithic industry with microblade cores, microblades, and backed blades	Earliest dated microlithic artifacts in Asia
Jerimalai Shelter (East Timor)[e]	42,000 years (radiocarbon)	Cores; flakes; scrapers (chert); worked and stained shells; large quantity of fish remains (tuna and others); shellfish	Drilling and grinding technology; exploitation of deep-water fish (tuna) and inshore species
Niah Great Cave (Sarawak, Borneo)[f]	Ca. 46,000–34,000 years (radiocarbon, U-series)	Cores, flakes, and flake tools (quartzite); variety of terrestrial and aquatic faunal remains	Use of traps/snares (?); freshwater fishing
Sulawesi karsts (Indonesia)[g]	40,000–35,000 years (U-series)	Rock art includes hand stencil and painting of pig-deer	Oldest figurative art in South and Southeast Asia
Jwalapuram Locality 9 (southern India)[h]	35,000 years (radiocarbon?)	Microlithic industry with backed segments; stone beads	Similar to African Middle Stone Age assemblages
Leang Sarru (Talaud Islands)[i]	Ca. 35,000 years (radiocarbon)	Cores, flakes, and scrapers (chert); shellfish	Exploitation of marine resources
Batadomba-lena (south-central Sri Lanka)[j]	Ca. 31,000–27,000 years (radiocarbon)	Microlithic industry associated with multiple human burials; shell beads	Continuous sequence of microlithic assemblages into Holocene

Site	Estimated Age (method)	Description	Comments
Patne (northern India)[k]	Ca. 30,000 years (radiocarbon)	Microlithic industry associated with ostrich-eggshell fragment	Ostrich eggshell engraved with geometric designs

[a]Michael Petraglia et al., "Middle Paleolithic Assemblages from the Indian Subcontinent Before and After the Toba Super-Eruption," *Science* 317 (2007): 114–116; Michael Haslam et al., "The 74,000 BP Toba Super-eruption and Southern Indian Hominins: Archaeology, Lithic Technology, and Environments at Jwalapuram Locality 3," *Journal of Archaeological Science* 37 (2010): 3370–3384.

[b]Petraglia et al., "Middle Paleolithic Assemblages from the Indian Subcontinent."

[c]Robin W. Dennell et al., "A 45,000-Year-Old Open-air Paleolithic Site at Riwat, Northern Pakistan," *Journal of Field Archaeology* 19 (1992): 17–33.

[d]Sheila Mishra, Naveen Chauhan, and Ashkok K. Singhvi, "Continuity of Microblade Technology in the Indian Subcontinent Since 45 ka: Implications for the Dispersal of Modern Humans," *PLoS ONE* 8 (2013): e69280.

[e]Sue O'Connor, "Crossing the Wallace Line: The Maritime Skills of the Earliest Colonists in the Wallacean Archipelago," in *The Emergence and Diversity of Modern Human Behavior in Paleolithic Asia*, ed. Yousuke Kaifu et al. (College Station: Texas A&M University Press, 2015), 214–224; Michelle C. Langley, Sue O-Connor, and Elena Piotto, "42,000-Year-Old Worked and Pigment-Stained *Nautilus* Shell from Jerimalai (Timor-Leste): Evidence for an Early Coastal Adaptation in ISEA," *Journal of Human Evolution* 97 (2016): 1–16.

[f]Tom Harrisson and Lord Medway, "A First Classification of Prehistoric Bone and Tooth Artifacts Based on Material from Niah Great Cave," *Asian Perspectives* 6 (1962): 219–229; Kenneth A. R. Kennedy, "The Deep Skull of Niah: An Assessment of Twenty Years of Speculation Concerning Its Evolutionary Significance," *Asian Perspectives* 20 (1977): 32–50; Graeme Barker et al., "The 'Human Revolution' in Lowland Tropical Southeast Asia: The Antiquity and Behavior of Anatomically Modern Humans at Niah Cave (Sarawak, Borneo)," *Journal of Human Evolution* 52 (2007): 243–261; Tim Reynolds and Graeme Barker, "Reconstructing Late Pleistocene Climates, Landscapes, and Human Activities in Northern Borneo from Excavations in the Niah Caves," in *Emergence and Diversity of Modern Human Behavior in Paleolithic Asia*, ed. Kaifu et al., 140–157.

[g]M. A. Aubert et al., "Pleistocene Cave Art from Sulawesi, Indonesia," *Nature* 514 (2014): 223–227.

[h]Chris Clarkson et al., "The Oldest and Longest Enduring Microlithic Sequence in India: 35 000 Years of Modern Human Occupation and Change at the Jwalapuram Locality 9 Rockshelter," *Antiquity* 83 (2009): 326–348.

[i]Rintaro Ono, Santoso Soegondho, and Minoru Yoneda, "Changing Marine Exploitation during Late Pleistocene in Northern Wallacea: Shell Remains from Leang Sarru Rockshelter in Talaud Islands," *Asian Perspectives* 48 (2009): 318–341.

[j]Kenneth A. R. Kennedy et al., "Upper Pleistocene Fossil Hominids from Sri Lanka," *American Journal of Physical Anthropology* 72 (1987): 441–461; Sheela Athreya, "Modern Human Emergence in South Asia: A Review of the Fossil and Genetic Evidence," in *Emergence and Diversity of Modern Human Behavior in Paleolithic Asia*, ed. Kaifu et al., 61–79.

[k]S. A. Sali, *The Upper Palaeolithic and Mesolithic Cultures of Maharashtra* (Pune: Deccan College Post Graduate and Research Institute, 1989).

India).[72] The assemblage contains microblade cores, microblades, and two backed blades—all made on chert and chalcedony—along with quartzite flakes (figure 5.4). Although the assemblage has been compared with the Howiesons Poort industry of South Africa (and a similar industry in East Africa), the characteristic "backed segment" pieces of the African assemblages are absent at Mehtakheri. The source of the microblade technology is unclear—it may represent a local development—but it almost certainly is an archaeological proxy for modern humans (and possibly the only chipped-stone artifacts that are diagnostic of *Homo sapiens*).

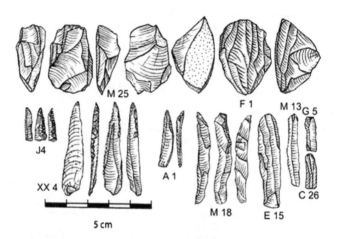

FIGURE 5.4 Microblade industry from Mehtakheri, including cores (M 25 and F 1), backed blades, and blades. (From Sheila Mishra, Naveen Chauhan, and Ashkok K. Singhvi, "Continuity of Microblade Technology in the Indian Subcontinent Since 45 ka: Implications for the Dispersal of Modern Humans," *PLoS ONE* 8 [2013]: e69280, fig. 10. Reproduced from *PLoS ONE* under the terms of the Creative Commons Attribution [CCBY] license)

Younger microlithic assemblages at Batadomba-lena, which are associated with modern human skeletal remains, and at Jwalapuram Locality 9 and *Patne* (India) contain backed-segment pieces similar to those found in African sites.[73] These assemblages date to about 35,000 years ago and later. The backed-segment pieces include geometric forms such as crescents, triangles, and trapezoids. They may have been used as stone tips and lateral barbs on arrows. Equally interesting are the associated personal ornaments, also similar to those found in African sites (chapter 4), which include perforated shell beads from Batadomba-lena, drilled stone beads from Jwalapuram Locality 9, and—most striking of all—ostrich-eggshell beads from Patne. Because of the obvious parallels between the South and East African artifacts, many perceive a direct historical connection between the two, although the current dating of the South Asian sites suggests a significant temporal gap.

There is a sharp contrast between the archaeological record associated with the dispersal of modern humans in Southeast Asia and that for the Indian subcontinent. No microlithic industry is known in Southeast Asia during this time period (that is, before 30,000 years ago).[74] The stone artifacts, some of which were found in layers containing human remains, are confined to core tools, flake tools, and flakes—often of quartzite. In some places, the lack of more sophisticated lithic technology (for example, microblade cores) may reflect the local absence of high-quality stone.[75]

Non-stone artifacts are rare,[76] but a bone artifact (dated to about 35,500 years ago) thought to represent the broken base of a hafted projectile point, was recovered from a site (*Matja Kuru 2*) on the island of Timor (figure 5.5).[77] Another site on Timor (*Jerimalai Shelter*) produced worked *Nautilus*-shell artifacts that exhibit traces of drilling and grinding, as well as pigment staining, dated to between 42,000 and 38,000 years ago. The scarcity of bone artifacts, as noted earlier, is likely to reflect a heavier emphasis on wood technology (and a corresponding bias against preservation).[78]

The discoveries on Timor have begun to change a long-standing perception that the early Southeast Asian archaeological record is lacking elements widely considered to be diagnostic of modern humans in Africa and northern Eurasia. Even more dramatic was the recent dating of rock art in Indonesia to more than 35,000 years ago.[79] Paintings of a female babirusa (or pig-deer) and another animal (possibly a pig) were dated with U-series on coralloid speleothems that later formed over the pigment (and provide

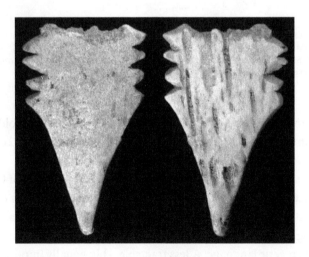

FIGURE 5.5 A bone artifact—thought to represent the butt end, not the point, of a hafted weapon—recovered from Matja Kuru 2 and dated to between roughly 36,500 and 34,500 calibrated years before the present. (Reproduced from Sue O'Connor, Gail Robertson, and K. P. Aplin, "Are Osseous Artefacts a Window to Perishable Material Culture? Implications of an Unusually Complex Bone Tool from the Late Pleistocene of Timor," *Journal of Human Evolution* 67 [2014]: 108–109, fig. 2, with permission of Elsevier)

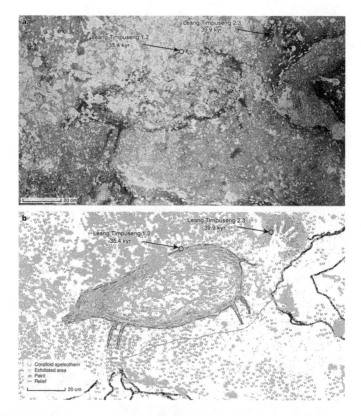

FIGURE 5.6 Rock art from a cave on Sulawesi (Indonesia) (*a*), depicting an animal (thought to represent a female babirusa) and a hand stencil, and (*b*) a tracing of the original. (From M. Aubert et al., "Pleistocene Cave Art from Sulawesi, Indonesia," *Nature* 514 [2014]: 224, fig. 2*a* and *b*. Reproduced by permission from Macmillan Publishers Ltd., *Nature*, copyright 2014)

a minimum age estimate for the painting) on cave walls. An older minimum estimate (more than about 40,000 years ago) was obtained on a hand stencil near the painting of the babirusa (figure 5.6).

As discussed in chapter 4, the complex structure of much of the visual art produced by modern humans (with multiple, hierarchically organized components and embedded levels) is the best available manifestation in the archaeological record before 30,000 years ago of the cognitive faculties that underlie syntactic language. A less tractable—but equally important—sign of these cognitive faculties is the complex and innovative technology associated with the dispersal of modern humans. If the visual art is at least indirectly related to organizational strategies that facilitated the dispersal (for example, widely distributed alliance networks), technological innovation probably was critical to the successful colonization of many new habitats. The archaeological visibility of these innovations is extremely low, however. In northern Eurasia, as well as in the lower latitudes, they often are inferred or deduced from other aspects of the archaeological record. Their presence is sometimes impossible to confirm, and the details of their design remain unknown.

As already noted, in South and Southeast Asia (and Australia/New Guinea), the technologies most important to the dispersal of modern humans may have been developed in Africa (and their prior existence may have been a key factor in the timing and route of the initial spread of *Homo sapiens* across tropical Asia). They probably include watercraft and instruments and/or devices for harvesting fish, which could have been essential for successful adaptation to coastal habitats.[80] As the ethnographic example of the Andaman Islanders shows, the bow and arrow may have been a major component of fishing technology (see box 5.2). The islanders also illustrate how occupants of the tropical coast could have prospered without an innovation critical to colder parts of northern Eurasia—fire-making technology.

The development of watercraft before 50,000 years ago was first inferred from the evidence of modern human settlement in Australia, which would have required a sea crossing of more than 70 kilometers from Southeast Asia, despite lowered ocean levels between 60,000 and 30,000 years ago (box 5.3). In recent decades, it has become apparent that people—presumably modern humans—made many voyages across substantial bodies of water in Southeast Asia before 30,000 years ago. The pattern suggests the widespread use of boats during the early phases of *Homo sapiens* settlement in the region and is logically tied to a coastal adaptation with an emphasis on marine resources.

To date, no remains of boats or possible boat parts have been recovered from any early prehistoric contexts, let alone sites in the tropical zone occupied more than 30,000 years ago. There are, therefore, no clues about the design or materials used for the craft that conveyed people across the waters of the Southwest Pacific and probably along the coasts of Southeast Asia. In theory, they may have been simple rafts composed of bamboo or wooden logs bound together. However, the voyages over many kilometers of open sea suggest a more navigable craft, such as a reed boat (ancient Egypt) or

BOX 5.3

Sea-Level History and the Global Dispersal of Modern Humans

At present, sea level is rising slowly as a result of the melting of glaciers and ice sheets around the world due to increased temperatures. Sea level has fluctuated significantly in the past when the volume of water locked up on land in the form of ice changed in response to changes in climate. During the period when modern humans dispersed out of Africa (ca. 75,000–60,000 years ago) and subsequently spread into the higher latitudes and Beringia (probably 50,000–35,000 years ago), the climate was cooler than that of the present, and sea level was correspondingly lower. As a result, continental shelves around the world often were exposed, expanding coastlines; and in some places, such as between New Guinea and northern Australia and between Chukotka and Alaska, large areas of shallow seafloor emerged, creating bridges that joined land that today is separated by water. These changes in global geography played an important role in the global dispersal of modern humans.

The past volume of ice on land can be estimated by measuring changes in the ratio of oxygen isotopes in deep-sea sediment cores (periods of higher ice volume correspond to increased enrichment of the heavy isotope ^{18}O), as well as annual increments in the ice itself, reflected in cores drilled into the Greenland and Antarctic ice sheets.[1] Past sea levels can be determined by dating ancient beaches or terraces formed at the time. A sea-level curve for the period between 140,000 and 20,000 years ago is shown in box figure 5.3, based in part on dated coral reefs off the Huon Peninsula (Papua New Guinea).[2] As the figure indicates, global sea level was at least 45 meters below that of the present between 75,000 and 35,000 years ago,

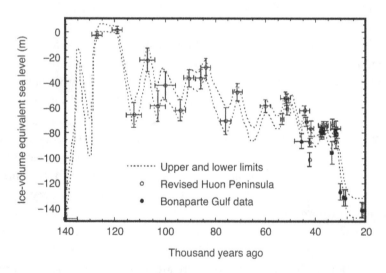

BOX FIGURE 5.3 Ice-volume equivalent sea level for the period between 140,000 years ago and the peak cold of the Last Glacial Maximum, as inferred from the age–elevation relationship of raised coral reefs on the Huon Peninsula (Papua New Guinea) and sediments from the tectonically stable Bonaparte Gulf (northwestern Australia). (Reproduced from Kurt Lambeck, Yusuke Yokoyama, and Tony Purcell, "Into and Out of the Last Glacial Maximum: Sea-Level Change During Oxygen-Isotope Stages 3 and 2," *Quaternary Science Reviews* 21 [2002]: 355, fig. 6, with permission of Elsevier)

exposing coastal areas across southern Asia that are now underwater and creating land connections between New Guinea and Australia and between Chukotka and Alaska.

1. J. John Lowe and Michael J. C. Walker, *Reconstructing Quaternary Environments*, 3rd ed. (Abingdon: Routledge, 2014).
2. Kurt Lambeck, Yusuke Yokoyama, and Tony Purcell, "Into and Out of the Last Glacial Maximum: Sea-Level Change During Oxygen-Isotope Stages 3 and 2," *Quaternary Science Reviews* 21 (2002): 343–360.

dugout canoe powered and guided with handheld wooden paddles. An example of a dugout canoe is provided by the Andaman Islanders, who carved the hull out of a tree trunk (with a stone adze) and constructed an outrigger to stabilize the boat in the water. Within the overall design of the boat, the outrigger assembly requires a third hierarchical level of organization (see box figure 5.2).[81]

The development of fishing technology is inferred from the presence of fish remains in archaeological sites. As described in chapter 4, there is evidence for freshwater fishing in a few African sites. In Southeast Asia, some sites are found in areas where the coast is deeply shelved and the modern shoreline is close to its position of 50,000 to 30,000 years ago. At Jerimalai Shelter (East Timor), Sue O'Connor excavated "significant quantities" of fish bone from test-unit levels dated to between 42,000 and 38,000 years ago. The most common taxa were tuna, which is a pelagic fish (that is, does not inhabit shallow water), but a wide range of other deep-water and inshore species were represented (pelagic species accounted for almost 50 percent of the total).[82] The method(s) and technology employed to catch the pelagic species (remains of which are dominated by smaller immature individuals) are unknown, but conceivably involved spear fishing close to the shore or the use of nets. It may be noted that the Andaman Islanders did not use fish hooks, but produced a variety of nets.[83]

In addition to the inshore and pelagic fish, the faunal remains from Jerimalai Shelter indicate a broad use of marine resources, including sea turtles, shellfish, crabs, and urchins, by 40,000 years ago. Occupations dating to around 35,000 years ago in *Leang Sarru*, a rock shelter on one of the Talaud Islands (southeast of the Philippines), yielded a variety of marine shellfish, but no fish or other vertebrates.[84] The exploitation of marine shellfish also is documented in the African sites (chapter 4), and even among non-modern humans in western Eurasia, and does not require complex or specialized technology, but probably played an important role in coastal economies.

A number of the sites that date to between 45,000 and 35,000 years ago on the Indian subcontinent, as well as some in Southeast Asia, indicate that—regardless of initial dispersal route—within a few thousand years at the most, some modern human groups had penetrated the interior regions of South and Southeast Asia. In many areas, this would have required the development of a tropical-forest economy, and some recent and modern foraging peoples, such as the Semang in Malaysia, may provide insights into how these groups adapted to local conditions between 50,000 and 40,000 years ago. The analysis of faunal remains from the Niah Great Cave (many kilometers from the coast at the time) reveals evidence for the consumption of a diverse array of rain-forest taxa, including bearded pig (most common), leaf monkey, macaque, orangutan, Malay bear, viverrids (for example, civet cats), monitor lizard, and extinct giant pangolin. The distribution of skeletal parts and age groups in the pig remains suggested the possible use of traps or snares, while freshwater-fish remains may indicate some fishing technology.[85]

Interactions with Other Forms of *Homo*

As modern humans spread across southern Asia after 75,000 years ago, they interbred with non-modern humans, and the analysis of ancient modern and non-modern genomes—combined with the genetics of living people—is shedding new light not only on the interbreeding among human taxa but also on the dispersal process. As noted in box 2.2, at least five episodes of genic exchange among modern humans, Neanderthal, and Denisovans have been identified, and at least two of them are pertinent to the movement of modern humans out of Africa and into southern Eurasia and Australia/New Guinea.[86]

All non-Africans carry some non-modern human DNA, and all Eurasians carry an estimated 2.5 percent of Neanderthal DNA, reflecting an interbreeding event that must have taken place before the widespread dispersal of *Homo sapiens* across northern and southern Eurasia.[87] Because this event involved gene flow from Neanderthals to modern humans, it is assumed to have occurred within the geographic range of the former (northern Eurasia) and likely occurred in the Levant. If so, it may indicate that modern humans did not disperse into southern Asia by way of the Arabian Peninsula, but through the Levant.[88] As they spread into southern Asia, however, modern humans also interbred with another non-modern form of *Homo* apparently concentrated in Southeast Asia at the time of their arrival.

Until recently, there was little discussion or debate about interactions and gene flow between dispersing modern humans and other forms of *Homo* in South and Southeast Asia. In Southeast Asia, this may have been largely a consequence of the widespread perception that a substantial evolutionary gulf separated modern humans from their predecessors in this part of the world. The view was based primarily on a comparison of the Lower and Middle Paleolithic record of Africa and western Eurasia (that is, finely shaped bifaces and prepared-core techniques) with the simple pebble choppers and flakes of East and South Asia. In 1948, Hallam Movius (1907–1987) labeled the latter "a region of cultural retardation."[89] The dating in 1996 of *Homo erectus* remains from Java to less than 55,000 years ago seemed to be consistent with this assessment.[90] The discovery a few years later of a small-bodied early *Homo* on the island of Flores (east of Java) who lived as recently as 13,000 years ago provided an even more startling example of the late survival of pre-modern humans in Southeast Asia.[91]

The discovery in 2010 of a previously unrecognized human taxon on the basis of aDNA extracted from a bone recovered from a cave in southwestern Siberia (Altai Mountains) quickly altered the picture.[92] The reconstructed Denisovan genome provided a means of identifying contact between the newly recognized non-modern *Homo* and both modern humans and Neanderthals. Despite the North Asian context of the discovery, a comparative analysis with living people revealed that interbreeding between Denisovans and modern humans most likely occurred in Southeast Asia. The highest genetic contribution of the Densiovans (up to 4.8 percent) is found among the living native peoples of Papua New Guinea and Australia. A smaller Denisovan contribution is found among some neighboring groups, including the Fijians, a Philippine Negrito group (*Mamanwa*), people on Flores Island, and others, but no Denisovan genes have shown up on the Asian mainland (figure 5.7).[93]

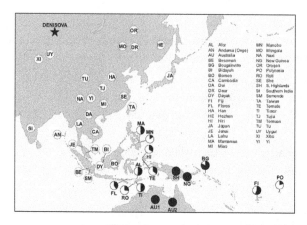

FIGURE 5.7 Denisovan genetic material in East Asian, Australian, and southwestern Pacific populations. The location of Denisova Cave in the Altai region of Siberia is shown at the upper left. (From David Reich et al., "Denisova Admixture and the First Modern Human Dispersals into Southeast Asia and Oceania," *American Journal of Human Genetics* 89 [2011]: 519, fig. 1, with permission of Elsevier)

The geographic pattern of Denisovan admixture in living populations provides some information about the dispersal of *Homo sapiens* in southern Asia. It sets the Australian and Papua New Guinea groups apart from those on the Asian mainland and in much of Southeast Asia, suggesting that the former arrived before any of the latter—including the speakers of Austro-Asiatic languages and the Andaman Islanders.[94] The small percentage of Denisovan genes found among some groups that live or have lived in proximity to New Guinea and Australia probably reflects gene flow from the people whose ancestors interbred with the local non-modern population.

The discovery of the Denisovan genome and its application to the problem of the dispersal of modern humans in southern Asia and Australia/New Guinea illustrates how fundamentally paleoanthropology has changed since 1962, when Morris Goodman published his paper on molecular phylogenetics (chapter 1).[95] Although more aDNA analyses of their genome have been undertaken, the Denisovans remain almost unknown in terms of skeletal anatomy.[96] Their presence in Southeast Asia is based entirely on the genetics of living people, while there is confusion about the relationship of the human fossil record in this part of the world to the origin and dispersal of modern humans.

As described earlier, at least some of the fossils recovered from southern China and dated to more than 60,000 years ago conceivably represent a Southeast Asian extension of the "near modern" population that occupied the Levant between 120,000 and 90,000 years ago. Others may represent Denisovans, but—since their anatomy is virtually unknown—confirming this hypothesis may require the recovery of some aDNA from the remains.[97]

Adding further confusion is the *Narmada* skullcap from north-central India, which now is tentatively dated to between 100,000 and 50,000 years ago.[98] The Narmada specimen exhibits a combination of modern and non-modern features and has been placed provisionally in the taxon *Homo heidelbergensis*,[99] but it, too, may represent a Denisovan. Although this find is not associated directly with artifacts, India contains a rich Middle Paleolithic record similar to that of the regions occupied by Neanderthals and Denisovans (for example, prepared-core techniques, some blade production, and small bifaces).[100]

COLONIZING SAHUL: NEW GUINEA AND AUSTRALIA

At the time that modern humans first settled Australia, global sea level was somewhere between 50 and 75 meters below that of the present day.[101] In addition to expanded coastlines around the world, wide areas of shallow seafloor emerged, connecting lands that now are separated by water. The shallow "Sahul Shelf" between New Guinea and northern Australia, and Bass Strait, which divides southeastern Australia from the island of Tasmania, were exposed, creating an enlarged continental mass termed Sahul.

Although the native peoples of New Guinea include the more recently arrived speakers of Austronesian languages (who immigrated several thousand years ago), both the indigenous Melanesians (Papuan) of New Guinea and all the Aboriginal peoples of Australia most likely are direct descendants of the original colonizers of Sahul. Significantly, all of them carry genetic markers that reflect the encounter with the pre-modern human population of Southeast Asia. Aside from the delayed dispersal of Native peoples in the Americas (chapter 7), they represent the largest remnant of the initial spread of modern humans out of Africa. And while the notion that Aboriginal Australians are "Paleolithic survivals" has been rejected— as the archaeologist Peter Hiscock notes, their material record exhibits much spatial and temporal variation—they lack some technologies that also may have been absent among the original colonizers.[102]

Analyses of mtDNA and Y-DNA among the living native populations of New Guinea and Australia reveal patterns similar to those found among the Austro-Asiatic peoples of southern Asia—links to the oldest Eurasian lineages and high genetic diversity. Maternal lineages include all three of the major haplogroups that diverged from African lineage L3 (mtDNA N, M, and R), as well as several other early non-African lineages (mtDNA O, P, and S).[103] Paternal lineages include C-M130, K-M526, and M-M186.[104] The genetic diversity is matched by high linguistic diversity, especially in New Guinea, where no fewer than 800 Papuan languages are found (in addition to a large number of Austronesian languages). In fact, the differences between the Papuan and the Australian maternal and paternal lineages have encouraged the suggestion that they represent separate colonization events.[105] However, the two populations clustered together in a whole-genome analysis, conducted in 2011, of a hair sample from an early-twentieth-century Australian Aborigine male.[106] This analysis also indicated that the genetically closest groups to both are the Munda (Austro-Asiatic) speakers in South Asia and the Aëta (Negrito) in the Philippines.

The earliest sites have yet to yield human skeletal remains, but several partial skeletons were recovered from burials dated to roughly 40,000 years ago at *Lake Mungo* in the Willandra Lakes region of southeastern Australia. Although damaged and burned, two crania (WLH1 and WLH3) were described as gracile with high vaults, thin walls, and flat faces, but without a well-developed torus (brow ridge).[107] Ancient DNA was reportedly extracted from WLH3 in 2001, but now appears to have been a result of contamination from living people.[108] A less securely dated cranium from the same region (WLH50) that may be much younger (ca. 25,000 years ago?) is relatively robust-looking with thick walls, pronounced postorbital constriction, and a prominent torus.[109] The contrast between the gracile and robust specimens may be due to sexual dimorphism.

The Chronology of Colonization

The dating of the colonization of Sahul has been a major topic of research and debate in Australian archaeology. Much argument has surrounded the accuracy of various chronometric methods, especially luminescence dating, as well as the postdepositional movement of artifacts in buried contexts. The effort to obtain reliable dates on the earliest sites in Australia drove the development of new pretreatment techniques for old radiocarbon samples.[110] Most archaeologists seem to have rejected past claims of early occupation between 100,000 and 80,000 years ago or more, and there is some consensus that sites in New Guinea and Australia can be firmly dated to no more than 50,000 years ago. The early sites nevertheless reveal a pattern of established settlement across a vast and varied landscape, hinting at an earlier initial arrival.[111] Estimates based on the coalescence of mtDNA and Y-DNA lineages have consistently fallen around 50,000 years,[112] but the authors of the 2011 whole-genome study calculated an earlier "split from the ancestral Eurasian population" of 75,000 to 62,000 years ago, while the mean coalescence date for mtDNA haplogroup P recently was recalculated (with aDNA) at around 56,800 years ago (see table 2.1).[113]

One factor that simplifies the dating of the arrival of modern humans in Sahul is the presumed absence of any pre-modern humans on the continent. As a result, even the most ephemeral traces of occupation, such as several stone flakes—if in a datable context—provide a reliable proxy for *Homo sapiens*. As shown in table 5.3, which lists the early archaeological sites of New Guinea and Australia, including Tasmania, the oldest known human skeletal remains are about 43,000 years old (Lake Mungo), while archaeological data indicate settlement at least a few thousand years earlier.[114]

Among the earliest firmly dated sites are open-air occupations of the *Ivane Valley* in the New Guinea Highlands, roughly 2,000 meters above sea level. The sites document a well-established upland interior-forest economy almost 50,000 years ago, reflecting the exploitation of yams (revealed by traces of starch grains on the artifacts) and *Pandanus* nuts. The tools include waisted stone axes, which may have been used to open up patches of the forest floor to sunlight, promoting the growth of useful plants.[115] The Ivane Valley sites reinforce the suspicion that the initial settlement of Sahul, presumably on the northwestern coast, took place before 50,000 years ago (figure 5.8).[116]

The Technology of Colonization

As noted earlier, the colonization of Sahul required a sea voyage of more than 70 kilometers, but both the route and the means remain unknown. It has been argued that the journey could have been made with a bamboo

TABLE 5.3

Archaeological Record of *Homo sapiens* in Sahul, Before 30,000 Years Ago

Site	Estimated Age (method)	Description	Comments
Huon Peninsula (northeastern New Guinea)[a]	61,000–44,500 years (TL, U-series)	Waisted axes; core; flakes	Open-air occupation on coral terraces; artifacts buried in dated tephras
Nauwalabila I (northern Australia)[b]	60,300–53,400 years (OSL)	Flakes (quartzite and chert); steep-edged scraper; possible grinding stone	Radiocarbon dates from upper levels agree with TL/OSL ages
Malakunanja II (northern Australia)[c]	Ca. 60,000–50,000 years (OSL, TL)	Small pit (more than 45,000 years ago); artifacts but no fauna	Controversy over accuracy of TL dates
Parnkupirti (northwestern interior Australia)[d]	Ca. 50,000–45,000 years (OSL)	Large core (silcrete)	Open-air site buried in stream sediment; associated with lake
Lake Mungo (southeastern Australia)[e]	50,000–46,000 years (radiocarbon, OSL)	Flakes (silcrete) from oldest levels; younger burials	Human occupation related to aridity
Ivane Valley (New Guinea Highlands, Papua New Guinea)[f]	49,000–43,000 years (radiocarbon)	Waisted axes (?) (schist and metabasalt); cores; flaking debris; bifacially flaked tools	Occupations at high elevation (2,000 meters above sea level); starch grains on artifacts
Devil's Lair (southwestern coast Australia)[g]	Ca. 48,000 years (radiocarbon, OSL, ESR, U-series)	Isolated artifacts: flakes (calcrete and quartz) and hammerstone (?) (limestone) (lowest levels); hearths (upper levels)	Isolated artifacts, including retouched flake, from lowest level indicates occupation more than 50,000 years ago
Nawarla Gabarnmang (north-central Australia)[h]	47,000 years (radiocarbon)	Large painted rock shelter: ground-edge ax, quartzite, and sandstone artifacts	Oldest securely dated site in Australia
Puritjarra Rock Shelter (west-central interior Australia)[i]	45,000–40,000 years (radiocarbon, TL, OSL)	Lowest occupation level: isolated flakes (silcrete), block (sandstone), and red-ocher fragment	Early period of occupation characterized by short visits
Riwi Cave (northwestern Australia)[j]	41,300 years (radiocarbon)	*Dentalium* shell beads; hearth	Associated with small fauna and plant remains
Carpenter's Gap 1 (northwestern Australia)[k]	44,000 years (radiocarbon [ABOx])	Small flakes (mostly quartz); flaking debris; bifacial point tip; small bifacial point (quartz)	Ground-edge-ax production

Site	Estimated Age (method)	Description	Comments
Parmerpar Meethaner (interior Tasmania)[l]	Ca. 40,000 years (radiocarbon)	Large flakes (quartzite); thumbnail scrapers; burned bone	Timing of occupation probably related to lowered sea level
Cuddie Springs (southeastern Australia)[m]	Ca. 40,000 years (radiocarbon, OSL)	Flaked stone artifacts; ground stone artifacts (upper level)	Associated with extinct megafauna (*Diprotodon, Macropus giganteus*)
Mandu Mandu Creek Rock Shelter (western Australia)[n]	37,000 years (radiocarbon?)	Red-ocher fragments; 22 perforated cone shells	Shell beads may have been strung together as a necklace

[a]James F. O'Connell and Jim Allen, "Dating the Colonization of Sahul (Pleistocene Australia–New Guinea): A Review of Recent Research," *Journal of Archaeological Science* 31 (2004): 835–853; Jim Allen and James F. O'Connell, "Both Half Right: Updating the Evidence for Dating First Human Arrivals in Sahul," *Australian Archaeology* 79 (2014): 86–108.

[b]Richard G. Roberts et al., "The Human Colonisation of Australia: Optical Dates of 53,000 and 60,000 Years Bracket Human Arrival at Deaf Adder Gorge, Northern Territory," *Quaternary Geochronology* 13 (1994): 575–583; O'Connell and Allen, "Dating the Colonization of Sahul (Pleistocene Australia–New Guinea)."

[c]Richard G. Roberts, Rhys Jones, and M. A. Smith, "Thermoluminescence Dating of a 50,000 Year Old Human Occupation Site in Northern Australia," *Nature* 345 (1990): 94–97; Allen and O'Connell, "Both Half Right."

[d]Peter Veth et al., "Excavations at Parnkupirti, Lake Gregory, Great Sandy Desert: OSL Ages for Occupation Before the Last Glacial Maximum," *Australian Archaeology* 69 (2009): 1–10.

[e]J. M. Bowler et al., "New Ages for Human Occupation and Climatic Change at Lake Mungo, Australia," *Nature* 421 (2003): 837–840; O'Connell and Allen, "Dating the Colonization of Sahul (Pleistocene Australia–New Guinea)."

[f]Glenn R. Summerhayes et al., "Human Adaptation and Plant Use in Highland New Guinea 49,000 to 44,000 Years Ago," *Science* 330 (2010): 78–81.

[g]Chris S. M. Turney et al., "Early Occupation at Devil's Lair, Southwestern Australia 50,000 Years Ago," *Quaternary Research* 55 (2001): 3–13; O'Connell and Allen, "Dating the Colonization of Sahul (Pleistocene Australia–New Guinea)"; Peter Hiscock, "Occupying New Lands: Global Migrations and Cultural Diversification with Particular Reference to Australia," in *Paleoamerican Odyssey*, ed. Kelly E. Graf, Caroline V. Ketron, and Michael R. Waters (College Station: Texas A&M University Press, 2013), 3–11.

[h]Bruno David et al., "Nawarla Gabarnmang, a 45,180 ± 910 cal BP Site in Jawoyn Country, Southwest Arnhem Land Plateau," *Australian Archaeology* 73 (2011): 73–77; Allen and O'Connell, "Both Half Right."

[i]M. A. Smith, "Characterizing Late Pleistocene and Holocene Stone Artefact Assemblages from Puritjarra Rock Shelter: A Long Sequence from the Australian Desert," *Records of the Australian Museum* 58 (2006): 371–410.

[j]Jane Balme, "Excavations Revealing 40,000 Years of Occupation at Mimbi Caves, South Central Kimberley, Western Australia," *Australian Archaeology* 51 (2000): 1–5; L. K. Fifield et al., "Radiocarbon Dating of the Human Occupation of Australia Prior to 40 ka BP—Successes and Pitfalls," *Radiocarbon* 43 (2001): 1139–1145.

[k]Sue O'Connor, "Carpenter's Gap Rockshelter 1: 40,000 Years of Aboriginal Occupation in the Napier Ranges, Kimberley WA," *Australian Archaeology* 40 (1995): 58–59; Peter Hiscock et al., "World's Earliest Ground-Edge Axe Production Coincides with Human Colonisation of Australia," *Australian Archaeology* 82 (2016): 2–11.

[l]Richard Cosgrove, "Late Pleistocene Behavioural Variation and Time Trends: The Case from Tasmania," *Archaeology of Oceania* 30 (1995): 83–104.

[m]Judith Field and John Dodson, "Late Pleistocene Megafauna and Archaeology from Cuddie Springs, Southeastern Australia," *Proceedings of the Prehistoric Society* 65 (1999): 275–301; Richard G. Roberts et al., "New Ages for the Last Australian Megafauna: Continent-Wide Extinction About 46,000 Years Ago," *Science* 292 (2001): 1888–1892; Judith H. Field, "Trampling Through the Pleistocene: Does Taphonomy Matter?" *Australian Archaeology* 63 (2006): 9–20.

[n]Kate Morse, "Shell Beads from Mandu Mandu Creek Rock-Shelter, Cape Range Peninsula, Western Australia, Dated Before 30,000 B.P.," *Antiquity* 67 (1993): 877–883.

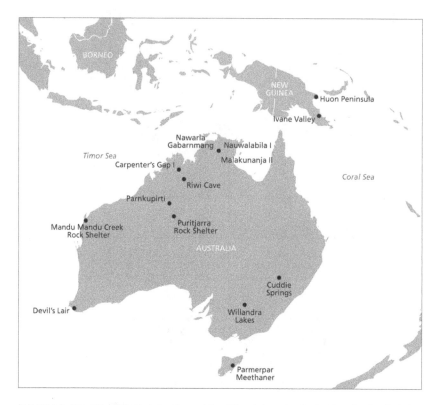

FIGURE 5.8 Map of Australia (including Tasmania) and New Guinea, showing fossils and archaeological sites assigned to *Homo sapiens*, dated to between roughly 60,000 and 37,000 years ago. (Base map adapted from en.wikipedia.org/wiki/File:BlankMap-World6.svg)

raft, propelled by wind and current, although more navigable watercraft (for example, outrigger canoe with paddles) seem likely, given the evidence for a maritime economy in Southeast Asia before 40,000 years ago.[117] In the 1970s, anthropologist Joseph Birdsell (1908–1994) worked out several possible "island-hopping" routes, including a northern route by way of Sulawesi and a southern route through East Timor. Only the northern route would have provided continuous visibility of land.[118]

Regardless of the sophistication of the watercraft and fishing equipment that may have arrived with the early coastal settlers, the technology of the people who colonized the interior of Sahul was simple. And despite the regional and temporal variability mentioned earlier, it remained relatively simple—especially in comparison with that of modern human foragers of the higher latitudes. As shown in table 5.3, archaeological sites older than

30,000 years have yielded little beyond stone cores, flakes, and occasional scrapers, along with some heavier stone items (hammerstones and blocks). Waisted stone axes are known from some of the northern sites, while "thumbnail" scrapers are common in Tasmania.[119] Evidence for ground-edge ax production in deposits dated to between 48,875 and 43,941 years ago at *Carpenter's Gap 1* (northwestern Australia) was reported in 2016.[120] Items related to art and/or ornamentation include fragments of red ocher and perforated shells.

Although the simplicity of the material record in Australia provoked a debate about the archaeological definition of *Homo sapiens*, the adaptations of recent foraging peoples offer insights into the early archaeology of Sahul. Recent foragers in the rich northern coastal zone and the Western Desert provide examples of adaptations to the types of environments occupied between 50,000 and 30,000 years ago that entailed a relatively simple material culture with low archaeological visibility.

The *Tiwi* were a foraging people who inhabited a tropical-forest environment on islands off the northern coast of Australia. Plant and animal resources were abundant and predictable throughout the year. Wendell Oswalt, who developed a simple but effective method of measuring technological complexity (see box 2.5), counted a total of only 11 artifacts related to food-getting—almost all of them comprising a single component (for example, digging stick). Their most complex subsistence technology was the hafted stone ax (assembled from four parts). With the exception of the ax head, none of the Tiwi artifacts would be visible in the archaeological record.[121]

By contrast, peoples of the Western Desert, such as the *Aranda* (*Arunta*), developed a more complex food-getting technology that included—according to Oswalt's calculations—a total of 16 items. Some comprised multiple parts, most notably the throwing board and spear, which is assembled from nine components.[122] The Aranda and other peoples of the Western Desert nevertheless devised a relatively low-tech adaptation to an environment that presented many of the same challenges to dispersing modern humans as did the higher latitudes.

In a habitat where resources were widely scattered and often unpredictable, recent foragers of the Western Desert ranged over a wide area, moving frequently and often for long distances—even hundreds of kilometers.[123] As Richard Gould noted, their technology was light and portable, and included rather ingenious multipurpose items, such as the throwing board, which also was designed to start fires, work wood (with an embedded stone flake), and mix pigments.[124] Only the stone flake and, possibly, traces of red ocher would be visible in the archaeological record. Heavier stone items, such as grinding slabs, were simply left at regular campsites and reused whenever the sites were reoccupied. Other tools or materials were expedient—acquired, used, and discarded on the spot, such as sharp pieces of stone or a twist of grass.[125]

Both the peoples of the northern coast and those of the Western Desert based their subsistence economy primarily on organization and information. Most of the complex technologies critical to the dispersal in northern Eurasia (chapter 6) were not needed, especially by the Tiwi and other recent foragers in tropical-coastal regions. The archaeological record probably reflects a similar phenomenon as modern humans dispersed—relatively swiftly, it appears—across a wide variety of habitats and climate zones in Sahul between 50,000 and 30,000 years ago. In addition to mountain forests in New Guinea, the habitats included subalpine grassland in Tasmania, where the technology of later peoples was famously simple, despite the high latitude (below 40° South) and low mean annual temperature.[126] The organizational adaptations probably included social networks that extended far beyond the immediate family, and are represented archaeologically by the ornaments and red ocher.

The Extinction of Megafauna

A major issue of the colonization of Sahul is the possible role played by modern humans in the extinction of the Australian megafauna. The same question has been debated in regard to North and South America, where megafaunal extinctions also followed the arrival of *Homo sapiens* (chapter 7). By 40,000 years ago, no fewer than 20 genera of giant marsupials, monotremes, birds, and reptiles that had inhabited the continent before humans arrived were extinct. Among them are the massive *Diprotodon*, *Sthenurus*, a giant kangaroo (*Macropus* sp.), and a flightless bird more than 2 meters tall (*Genyornis*). Some megafauna, such as kangaroo and wallaby, evolved into a smaller species, analogous to postglacial bison in North America. A number of researchers argue that modern humans drove these taxa to extinction by over-hunting and/or destroying habitat by fire.[127]

In contrast to North America, Australia lacks sites where megafauna were killed and butchered by human hunters, although it should be noted that such (open-air) sites would be three times older than those in North America and may have low visibility in the archaeological record.[128] In fact, only *Cuddie Springs* (southeastern Australia) contains artifacts associated with the remains of megafauna (see table 5.3). The site represents the floor of an ephemeral lake, where the remains of various megafauna—including *Diprotodon*, *Sthenurus*, *Genyornis*, and giant kangaroo—were buried with artifacts in two separate occupation layers.[129] Critics of the hypothesis that humans played a significant role in the extinctions note that—in addition to the lack of kill sites—Cuddie Springs indicates several millennia of coexistence between humans and megafauna. Moreover, deeper levels of the site reveal that some megafauna already were locally extinct before humans arrived.[130]

Recently, however, burned fragments of *Genyornis* eggshell dated to about 54,000 to 43,000 years ago have been recovered from sites in western and southeastern Australia. They apparently represent traces of human food debris, and their age corresponds to the time period during which the giant bird became extinct.[131] It is difficult to escape the conclusion that the colonization of Sahul by modern humans had an impact on both plants and animals, and probably contributed to the extinction of at least some megafauna. The analysis of a lengthy paleo-environmental record from a former lake/swamp in northeastern Australia revealed a dramatic increase in charcoal—indicating increased landscape burning—and change in vegetation unrelated to climate shifts at about 45,000 years ago.[132] Regardless of the role of hunting, humans probably had an effect on megafauna habitat through their use of fire.

The significance of the extinction of megafauna for the colonization of Sahul is that—as in North America—it may reflect a rapid settlement process based on the exploitation of a narrow ecological niche distributed across the continent. Unaccustomed to human predators, the megafauna presumably made easy prey. Adaptation to local environments—such as the Western Desert, New Guinea Highlands, and subalpine grassland in Tasmania—may have occurred only after modern humans spread throughout Sahul and drove the remaining megafauna to extinction through a combination of habitat destruction and hunting.[133]

Chapter Six

GLOBAL DISPERSAL

Northern Eurasia

Man has spread widely over the face of the earth, and must have been exposed,
during his incessant migrations, to the most diversified conditions . . . must
have passed through many climates and changed their habits many times.

—CHARLES DARWIN (1871)

In 2008, the fragment of a human upper-leg bone (femur) was discovered on the bank of the Irtysh River in western Siberia. The bone apparently had eroded out of stream deposits of sand and gravel dating to between 50,000 and 30,000 years ago near the town of *Ust'-Ishim*, although the exact provenience remains unknown. It eventually was subject to various analyses, including ancient DNA (aDNA), and the findings were published in 2014.[1] The results contributed some information important to reconstructing the global dispersal of modern humans.

Radiocarbon-dated to about 45,000 years ago, the femur from Ust'-Ishim provides a snapshot of some of the earliest modern humans to occupy northern Eurasia (figure 6.1). Not surprisingly, the individual belonged to two of the oldest known lineages outside Africa: mitochondrial DNA (mtDNA) haplogroup R (maternal) and Y-chromosome DNA (Y-DNA) haplogroup K(xLT) (paternal). According to recent estimates, both lineages date to more than 50,000 years ago and are not far removed from their African parent groups.[2] In fact, the application of a molecular clock model to the mtDNA sequence obtained from the bone suggests an even earlier age (ca. 49,000 years) than does the radiocarbon date.[3]

Given the Ust'-Ishim bone's close connection with Africa, its geographic setting is striking. It was found at latitude 57° North in western Siberia—in an area that experiences a January mean temperature of –15°C—and provides evidence of how quickly modern humans from the

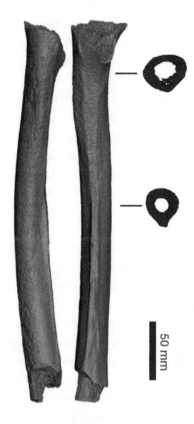

FIGURE 6.1 The femur from Ust'-Ishim: (*left*) lateral view; (*center*) posterior view; (*upper right*) cross section at 80 percent level; (*lower right*) cross section at mid-shaft. (From Qiaomei Fu et al., "Genome Sequence of a 45,000-Year-Old Modern Human from Western Siberia," *Nature* 514 [2014]: 446, fig. 1c–*f*. Reproduced by permission from Macmillan Publishers Ltd., *Nature*, copyright 2014)

equatorial zone adapted to environments characterized by low winter temperatures, extreme seasonal fluctuations, and generally low biotic productivity.

Stable-isotope analysis of the bone offers a clue to how the new arrivals dealt with scarce resources; the nitrogen-isotope values indicate the heavy consumption of freshwater aquatic foods, suggesting an expansion of the diet to include fish and/or waterfowl (resources that were not a dietary staple among the earlier occupants of Eurasia).[4] This, in turn, suggests some new and potentially complex technology, such as fishhooks, gorges, nets, and/or throwing darts. As discussed in chapters 4 and 5, at least some of this technology may have been developed before modern humans invaded the higher latitudes.

Significantly, the bone dates to an interval of unusually warm climate (Greenland Interstadial 12 [GI 12]) that began about 48,000 years ago and lasted for at least a few thousand years. There is evidence from other sources that modern humans first occupied parts of northern Eurasia during this interval, and it indicates that—despite their ability to devise technological solutions to low temperatures and scarce resources—their initial expansion into the higher latitudes may have been controlled by climate. By contrast, climate seems to have been a relatively unimportant factor in the dispersal across southern Asia (although it probably was a factor in the colonization of arid regions, such as the Arabian Peninsula and the Australian desert).

Perhaps the most important piece of information extracted from the Ust'-Ishim bone is the absence of a close genetic relationship with the Andaman Islanders. This suggests that the earliest known modern humans in northern Asia were not derived from a population that had first occupied southern Asia, but from people who followed a more northern route (north of the Himalayan Plateau) into Siberia. A multivariate statistical analysis of the genome revealed no close ties to any non-African population, indicating that the individual antedated the divergence of European and Asian lineages. It also revealed some genetic input from the Neanderthals (2.3 ± 0.3 percent), which is estimated to have occurred between 60,000 and 50,000 years ago, but no traces of admixture with the elusive Denisovans.[5] The absence of Denisovan genes is additional (negative) evidence for a connection to southern Asia (chapter 5).

The Ust'-Ishim bone supports the notion that modern humans colonized northern Eurasia in one or more migratory waves separate from the early occupation of southern Asia and Australia. And while the dating of the bone documents the settlement of northern Eurasia by 45,000 years ago, other lines of evidence point to even earlier movements into Europe and northern Asia—roughly 50,000 years ago—apparently postdating the southern dispersal by only a few thousand years. The modern humans who first colonized northern Eurasia have been described as a "meta-population" that preceded the divergence of Europeans, Asians, and groups ancestral to most Native Americans.[6] It is possible that the lineages identified from the Ust'-Ishim genome played a major role in the initial dispersal (for example, mtDNA haplogroup R may be a northern analogue to haplogroup M in southern Asia), but confirming this will require more samples of aDNA from human remains of comparable age.

Reconstructing the dispersal of modern humans in northern Eurasia differs in several respects from reconstructing that in southern Asia. The most obvious difference, illustrated by the Ust'-Ishim bone, is that aDNA has been successfully extracted from human remains that are tens of thousands of years old. While a comparable database on the genetics of living people is available for southern Asia and Australia, it is only in the higher latitudes that climate conditions have permitted the preservation and recovery of DNA from ancient bones and teeth.[7] They provide a means of ground-truthing models of the migration of modern human populations (with respect to time and place) generated from the genetics of living people. The process of collecting aDNA samples from human remains tied to the global dispersal has only just begun, but has accelerated in the past several years and now is increasingly yielding whole-genome analyses (for example, Ust'-Ishim), with an exponential increase in the volume of information. A list of modern human remains from sites in northern Eurasia that antedate 30,000 years is presented in table 6.1, which includes information on the results of aDNA analyses.

Another difference lies in the archaeological record, which is more diagnostic of modern humans in northern Eurasia than it is in South and Southeast Asia. (It may be recalled that the early archaeological record of Australia and New Guinea is identified as that of modern humans largely because no other forms of *Homo* are assumed to have been present in Sahul [chapter 5].) The stone technology produced by *Homo sapiens* in northern Eurasia is more easily distinguished from that of other human species than are the artifacts found in sites in southern Asia. Although modern humans in northern Eurasia continued to make many of the same types of stone artifacts made by their predecessors, they also manufactured large quantities of blades and points from prepared cores. These artifacts were widespread across northern Eurasia between 50,000 and 40,000 years ago, and they appear to be a proxy for modern humans—possibly representing the meta-population suggested by the genetics of living people.

The archaeological record of modern human dispersal in northern Eurasia also is richer in non-stone artifacts and faunal remains. More favorable conditions for the preservation of bone at higher latitudes (especially compared with the tropical-forest environments of South and Southeast Asia) presumably account for some of the difference in artifact inventories, but the extensive use of material substitutes for wood in northern habitats where hardwood trees were scarce or absent undoubtedly contributes to the pattern. Many sites occupied by modern humans in northern Eurasia,

TABLE 6.1

North Eurasian Fossils and Ancient DNA Assigned to *Homo sapiens*, Before 30,000 Years Ago

Site, Layer	Estimated Age (method)	Description	Genetics
Manot Cave (Israel)[a]	54,700 ± 5500 years (U/Th)	Human partial cranium (calvaria)	No aDNA reported
Ust'-Ishim (Irtysh River, western Siberia, Russia)[b]	46,880–43,210 years (radiocarbon)	Femur shaft	mtDNA haplogroup R; Y-DNA haplogroup K
Grotta del Cavallo (southern Italy)[c]	45,000–43,000 years (radiocarbon)	Isolated teeth (deciduous molars)	No aDNA reported
Kent's Cavern (England)[d]	44,200–41,500 years (radiocarbon)	Maxilla (KC4)	mtDNA could not be extracted
Huanglong Cave (central China)[e]	44,000–34,000 years (ESR, older U-series)	Isolated teeth ($n = 7$)	No aDNA reported
Ksâr ʿAkil, Layer XXV (Lebanon)[f]	42,400–41,700 years (radiocarbon)	Partial human maxilla ("Ethelruda," or Ksâr ʿAkil 2)	No aDNA reported
Üçağızlı Cave (south-central Turkey)[g]	Ca. 44,000–39,000 years (radiocarbon)	Isolated teeth ($n = $ ca. 10)	No aDNA reported
Tianyuan Cave, Layer III (northern China)[h]	42,000–39,000 years (radiocarbon)	Mandible; axis; scapulae; humeri; radii; phalanges; femora; tibiae; tarsals	mtDNA haplogroup B
Peştera cu Oase (Romania)[i]	Ca. 42,000–37,800 years (radiocarbon)	Mandible (Oase 1); cranium (Oase 2)	mtDNA haplogroup N (Oase 1)
Grotta di Fumane (northern Italy)[j]	41,100–38,500 years (radiocarbon)	Isolated tooth (upper incisor) (Fumane 2)	mtDNA haplogroup R (Fumane 2)
Kostenki 14 and 17 (European Russia)[k]	More than 40,000 years (radiocarbon, tephrochronology)	Isolated tooth (K 14-IVb); isolated tooth (K 17-II)	No aDNA reported
Kostenki 14, Layer III (European Russia)[l]	38,700–36,200 years (radiocarbon, tephrochronology)	Complete skeleton in burial pit	mtDNA haplogroup U2; Y-DNA haplogroup C-M130
Kostenki 1, Layer III (European Russia)[m]	38,000–33,000 years (radiocarbon)	Tibia fragments ($n = 2$); pelvic fragment; isolated tooth	No aDNA reported (but stable-isotope analysis)

Site, Layer	Estimated Age (method)	Description	Genetics
Buran-Kaya III (Crimea)[n]	36,900–35,500 years (radiocarbon)	Cranial fragments; isolated teeth (more than 160 fragments)	No aDNA reported
Sungir' (northern Russia)[o]	35,200–34,100 years (radiocarbon)	Complete skeletons in burial pit	mtDNA haplogroup I (?) (modern contamination?)
Goyet Cave (Belgium)[p]	Ca. 35,000 years (radiocarbon)	Fragments of human bone (n = ca. 80) (excavated in 1860s)	mtDNA haplogroup M
Goat's Hole (southern England)[q]	Ca. 34,000–33,250 years (radiocarbon)	"Red Lady" of Paviland skeleton (male)	mtDNA haplogroup H
Mladeč (Czech Republic)[r]	Ca. 34,000–28,000 years (radiocarbon)	Isolated teeth (canine and molars); ulna	mtDNA analysis (haplogroup?)
Paglicci Cave (southern Italy)[s]	Ca. 33,000–31,250 years (radiocarbon)	Crania; maxilla; mandible; tibia (Paglicci 23)	mtDNA haplogroup U (Paglicci 133)
Dolní Věstonice (Czech Republic)[t]	31,155 years (radiocarbon)	Complete skeletons in multiple burials	mtDNA haplogroup U (U8)

[a]Israel Hershkovitz et al., "Levantine Cranium from Manot Cave (Israel) Foreshadows the First European Modern Humans," Nature 520 (2015): 216–219.

[b]Qiaomei Fu et al., "Genome Sequence of a 45,000-Year-Old Modern Human from Western Siberia," Nature 514 (2014): 445–449.

[c]Stefano Benazzi et al., "Early Dispersal of Modern Humans in Europe and Implications for Neanderthal Behaviour," Nature 479 (2011): 525–528.

[d]Tom Higham et al. "The Earliest Evidence for Anatomically Modern Humans in Northwestern Europe," Nature 479 (2011): 521–524.

[e]Wu Liu et al., "Huanglong Cave: A Late Pleistocene Human Fossil Site in Hubei Province, China," Quaternary International 211 (2010): 29–41; Guanjun Shen et al., "Mass Spectrometric U-Series Dating of Huanglong Cave in Hubei Province, Central China: Evidence for Early Presence of Modern Humans in Eastern Asia," Journal of Human Evolution 65 (2013): 162–167.

[f]Katerina Douka et al., "Chronology of Ksar Akil (Lebanon) and Implications for the Colonization of Europe by Anatomically Modern Humans," PLoS ONE 8 (2013): 372931.

[g]Steven L. Kuhn et al., "The Early Upper Paleolithic Occupations at Üçağızlı Cave (Hatay, Turkey)," Journal of Human Evolution 56 (2009): 87–113.

[h]Hong Shang et al., "An Early Modern Human from Tianyuan Cave, Zhoukoudian, China," Proceedings of the National Academy of Sciences 104 (2007): 6573–6578; Hong Shang and Erik Trinkaus, The Early Modern Human from Tianyuan Cave, China (College Station: Texas A&M University Press, 2010); Qiaomei Fu et al., "DNA Analysis of an Early Modern Human from Tianyuan Cave, China," Proceedings of the National Academy of Sciences 110 (2013): 2223–2227.

[i]Erik Trinkaus et al., "An Early Modern Human from the Peştera cu Oase, Romania," Proceedings of the National Academy of Sciences 100 (2003): 11231–11236; Qiaomei Fu et al., "An Early Modern Human from Romania with a Recent Neanderthal Ancestor," Nature 524 (2105): 216–219.

[j]S. Benazzi et al., "The Makers of the Protoaurignacian and Implications for Neandertal Extinction," Science 348 (2015): 793–796.

[k]M. V. Anikovich et al., "Early Upper Paleolithic in Eastern Europe and Implications for the Dispersal of Modern Humans," Science 315 (2007): 223–226; John F. Hoffecker et al., "From the Bay of Naples to the River Don: The Campanian Ignimbrite Eruption and the Middle to Upper Paleolithic Transition in Eastern Europe," Journal of Human Evolution 55 (2008): 858–870.

(continued)

(continued)

[l]Johannes Krause et al., "A Complete mtDNA Genome of an Early Modern Human from Kostenki, Russia," *Current Biology* 20 (2010): 231–236; Andaine Seguin-Orlando et al., "Genomic Structure in Europeans Dating Back at Least 36,200 Years," *Science* 346 (2014): 1113–1118.

[m]M. M. Gerasimova, S. N. Astakhov, and A. A. Velichko, *Paleoliticheskii Chelovek, Ego Material'naya Kul'tura i Prirodnaya Sreda Obitaniya* (St. Petersburg: Nestor-Istoriya, 2007); John F. Hoffecker et al., "Kostenki 1 and the Early Upper Paleolithic of Eastern Europe," *Journal of Archaeological Science: Reports* 5 (2016): 307–326.

[n]Sandrine Prat et al., "The Oldest Anatomically Modern Humans from Far Southeast Europe: Direct Dating, Culture, and Behavior," *PLoS ONE* 6 (2011): e20834.

[o]A. B. Poltoraus, E. E. Kulikov, and I. A. Lebedeva, "Molekulyarnyi Analiz DNK iz Ostatkov Trekh Individuumov so Stoyanki Sungir' (Predvaritel'nye Itogi)," in *Homo Sungirensis: Verkhnepaleoliticheskii Chelovek: Ekologicheskie i Evolyutsionnye Aspekty Issledovaniya* (Moscow: Nauchnyi Mir, 2000), 351–358; Igor V. Ovchinnikov and William Goodwin, "Ancient Human DNA from Sungir?" *Journal of Human Evolution* 44 (2003): 389–392; Anat Marom et al., "Single Amino Acid Radiocarbon Dating of Upper Paleolithic Modern Humans," *Proceedings of the National Academy of Sciences* 109 (2012): 6878–6881.

[p]Cosimo Posth et al., "Pleistocene Mitochondrial Genomes Suggest a Single Major Dispersal of Non-Africans and a Late Glacial Population Turnover in Europe," *Current Biology* 26 (2016): 827–833.

[q]R. M. Jacobi and T. F. G. Higham, "The 'Red Lady' Ages Gracefully: New Ultrafiltration AMS Determinations from Paviland," *Journal of Human Evolution* 55 (2008): 898–907.

[r]David Serre et al., "No Evidence of Neandertal mtDNA Contribution to Early Modern Humans," *PLoS Biology* 2 (2004): 0313–0317; Eva M. Wild et al., "Direct Dating of Early Upper Palaeolithic Human Remains from Mladeč," *Nature* 435 (2005): 332–335.

[s]David Caramelli et al., "A 28,000 Years Old Cro-Magnon mtDNA Sequence Differs from All Potentially Contaminating Modern Sequences," *PLoS ONE* 3 (2008): e2700; Qiaomei Fu et al., "A Revised Timescale for Human Evolution Based on Ancient Mitochondrial Genomes," *Current Biology* 23 (2013): 553–559; Posth et al., "Pleistocene Mitochondrial Genomes Genomes Suggest a Single Major Dispersal of Non-Africans."

[t]Fu et al., "Revised Timescale for Human Evolution"; Posth et al., "Pleistocene Mitochondrial Genomes Genomes Suggest a Single Major Dispersal of Non-Africans."

including many open-air sites, have yielded artifacts of bone, antler, and ivory. They often represent more complex technologies (for example, eyed needles = insulated clothing) and thus provide a fuller picture of material culture. At the same time, the often abundant samples of faunal remains in north Eurasian sites offer a fuller picture of diet and economy.

Organizational and Technological Adaptation to Northern Environments

The dispersal of *Homo sapiens* in northern Eurasia required a suite of adaptations, both organizational and technological, to the colder, more seasonal, and less-productive habitats of the higher latitudes. Modern humans invaded places that no human predecessors had inhabited (at least on a year-round basis) because of their inability to adapt to colder and drier climates and/or the scarcity of accessible resources.[8] The archaeological record indicates, moreover, that the process of colonization was rapid. The Ust'-Ishim bone documents the presence of modern humans in the subarctic region of western Siberia within a few thousand years of their arrival in the Levant, while an archaeological site in northern Russia (*Mamonto-vaya kurya*) puts them on the Arctic Circle at roughly the same time.[9] Although these places seem to have been occupied during a warm interval (GI 12) climates probably were cooler and drier than those of the present day. And the problems of coping with low temperatures were exacerbated for the modern humans who occupied these regions because they retained the anatomical adaptations to tropical climates of their African ancestors, which would have rendered them especially vulnerable to frostbite and hypothermia.[10]

Owing to the effects of geography and atmospheric circulation, northern Eurasia exhibits both a latitudinal and a longitudinal climate gradient. The westernmost portion of the continental landmass receives a steady supply of warm and moist air derived from oceanic circulation in the North Atlantic. As a result, western Europe enjoys mild winter temperatures and the highest biological productivity in northern Eurasia. But as westerlies carry the air into the eastern half of Europe, across the Ural Mountains, and into Siberia, it becomes increasingly dry—widening the seasonal contrast in temperatures and reducing primary productivity. While the January mean for London is 4.3°C, it is –8°C at the same latitude (51° North) and elevation at longitude 39° East (Voronezh on the Don River in Russia). The decline in plant productivity is roughly 50 percent (and is even greater in the southern steppe zone). Beyond the Urals, winter temperatures continue to fall—the January mean is –16°C in Omsk, at latitude 55° North/ longitude 73° East—and plant productivity declines another 50 percent.[11]

Not surprisingly, evidence for the settlement of eastern Europe and Siberia before the arrival of modern humans remains limited. Skeletal remains of Neanderthals are present along the southern fringe of eastern Europe (Crimea and northern Caucasus), and their archaeological traces are found in the southernmost parts of the East European Plain, but earlier reports of Neanderthal sites on the central plain or in northern Russia are problematic.[12] These sites now appear equally—if not more—likely to have been occupied by modern humans. In Siberia, both Neanderthals and Denisovans were present in the Altai (a mountainous region in the southwestern corner of Siberia), but neither can be firmly documented farther north or east (for example, Lake Baikal area).[13]

As in the equatorial zone, the ethnographic record of recent foraging peoples offers insights into how modern humans adapted to the environments of northern Eurasia. Information on the ecology and technology of various groups in the higher, middle, and lower latitudes is presented in table 6.2. In general, foraging peoples of the temperate zone and farther north live at lower population densities (fewer persons per square kilometer) than do hunter-gatherers of the tropics and subtropics, although the inhabitants of tropical-desert environments also live at low densities. Foraging peoples of the higher latitudes tend to move frequently and often travel considerable distances over the course of the year, reflecting the wide distribution of their food resources (also evident among desert foragers). The percentage of meat and fish in their diet increases with latitude, ultimately reaching 100 percent in the Arctic. These sources of protein and fat represent a more efficient means of acquiring energy and nutrients, and they satisfy the higher caloric needs of people in cold places.[14]

And as a comparison between foraging peoples in colder environments and those in warmer environments illustrates, the former depend more than the latter on technology. In his analysis of food-getting technology ("subsistants") among various recent hunter-gatherers, Wendell Oswalt found a relationship between latitude and technological complexity (figure 6.2a).[15] Not only do people in colder places tend to have more instruments and facilities, but individual tools, weapons, and devices tend to be composed of more parts (and to exhibit a more complex hierarchical organization). Oswalt noted that the most complex subsistant in his sample was the *toggle-headed harpoon* with drag float, which is assembled from as many as 26 parts (by the *Iglulik*).[16] In another study, he observed that even more complex technology may be found under the category of clothing and housing.[17]

Conversely, the correlation between technological complexity and biological productivity is weak (see figure 6.2b). Foraging peoples who live in low-latitude deserts and desert margins, where plant and animal productivity is often comparable to that of high-latitude environments, lack the complex food-getting instruments and facilities of northern hunter-gatherers. As noted in chapter 5, foragers in places like the Western Desert of Australia (net primary productivity [NPP] = 90) rely more on organization and information than on technology.

Why the heavy emphasis on complex technology in northern environments? The answer seems to lie in the added caloric demands of a cold

TABLE 6.2
Recent Foraging Peoples: Primary Productivity, Diet, Population, and Food-getting Technology

Ethnic Group	Habitat	Primary Productivity (g/C/m²/yr)	Meat and Fish in Diet (%)	Fish in Diet (%)	Persons (per square kilometer)	Subsistants[a]	Technounits[a]	Automata
Iglulik (northern Canada)	Arctic tundra	40	100	50	0.005	42	225	Yes
Copper Inuit (northwestern Canada)	Arctic tundra	50	100	60	0.012	27	122	Yes
Northern Paiute (Great Basin)	Temperate desert	85	50	30	0.011	39	97	Yes
Aranda/Arunta (Australia)	Desert	90	40	0	0.03	16	42	No
Tanana (interior Alaska)	Subarctic forest	100	90	20	0.006	25	105	Yes
Deg Hit'an (northern interior Alaska)	Subarctic forest	110	90	50	0.025–0.04	55	296	Yes
Klamath (northwestern United States)	Temperate upland	145	70	50	0.25	43	151	Yes
Ju/'hoansi (northern Kalahari, southern Africa)	Desert margin	205	20	0	0.10–0.16	20[b]	(missing data)	Yes[b]
Tanaina (southern Alaska)	Northern forest and coast	210	90	50	0.04–0.06	40	224	Yes
Northern Tlingit (southeastern Alaska)	Temperate rain forest	285	90	60	0.1–0.4	28	121	Yes
Twana (western Washington)	Temperate rain forest	325	90	60	0.17–0.33	48	237	Yes

(continued)

(continued)

Ethnic Group	Habitat	Primary Productivity (g/C/m²/yr)	Meat and Fish in Diet (%)	Fish in Diet (%)	Persons (per square kilometer)	Subsistants[a]	Technounits[a]	Automata
Chenchu (eastern India)	Tropical savanna	670	15	5		20	55	No
Hadza (East Africa)	Savanna-woodland	680	35	0	0.15–0.24	24[c]	64[c]	No[c]
Ingura (northern Australia)	Tropical woodland	790	70	60	0.115	13	32	No
Tiwi (northern Australia)	Tropical forest	1,100	50	20		11	14	No
Andamanese (Andaman Islands)	Tropical coast	1,750	60	40	0.86	11	51	No

[a]Except where noted otherwise, based on Wendell H. Oswalt, An Anthropological Analysis of Food-Getting Technology (New York: Wiley, 1976).

[b]Based on Richard B. Lee, The !Kung San: Men, Women, and Work in a Foraging Society (Cambridge: Cambridge University Press,1979), 119–157.

[c]Based on Frank W. Marlowe, The Hadza: Hunter-Gatherers of Tanzania (Berkeley: University of California Press, 2010), 77–78, table 4.4, but with modifications in accordance with the approach in Wendell H. Oswalt, Habitat and Technology: The Evolution of Hunting (New York: Holt, Rinehart and Winston, 1973), and Anthropological Analysis of Food-Getting Technology; for example, arrow poison is reclassified as a technounit rather than a subsistant.

Sources: Based on data in Oswalt, Anthropological Analysis of Food-Getting Technology; Lee, !Kung San; Marlowe, Hadza; Robert L. Kelly, The Lifeways of Hunter-Gatherers: The Foraging Spectrum (Cambridge: Cambridge University Press, 2013).

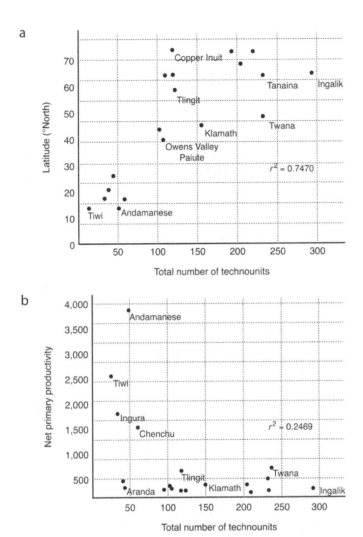

FIGURE 6.2 Technological complexity: (*a*) the relationship between latitude and the complexity of food-getting technology among recent hunter-gatherers, as measured in terms of total number of an artifact's components, or technounits; (*b*) the relationship between net primary productivity and the complexity of food-getting technology among recent hunter-gatherers, as measured in terms of technounits. ([*a*] Redrawn from Robin Torrence, "Time Budgeting and Hunter-Gatherer Technology," in *Hunter-Gatherer Economy in Prehistory: A European Perspective*, ed. Geoff Bailey [Cambridge: Cambridge University Press, 1983], 19, fig. 3.2)

TABLE 6.3
Average Daily Intake (Per Person) of Peoples Living at Different Latitudes

	Kikuyu (equator)		British (temperate)		Inuit (Arctic)	
	Grams	KiloJoules	Grams	KiloJoules	Grams	KiloJoules
Fat (38kJ/g)	22	836	110	4,180	162	6,156
Carbohydrate (17kJ/g)	390	6,630	400	6,800	59	1,003
Protein (17kJ/g)	100	1,700	100	1,700	377	6,409
Total		9,166		12,680		13,569

Note: The energy needs of the Inuit, living in the Arctic, are 48 percent higher than those of the Kikuyu, who inhabit the equatorial zone. 1 kJ = 0.2 calorie

Source: Based on G. A. Harrison et al., *Human Biology: An Introduction to Human Evolution, Variation, Growth, and Adaptability*, 3rd ed. (Oxford: Oxford University Press, 1988), 482, table 22.3.

climate, which force northern hunter-gatherers, such as the *Tanana* in interior Alaska (NPP = 100), to extract a significantly higher amount of energy from an equally poor habitat. As shown in table 6.3, people who live in the Arctic require roughly one-third more calories than those who inhabit the equatorial zone. Furthermore, there are few digestible plants in high-latitude environments (for example, boreal forest and arctic tundra), so most of the diet must be obtained from animals—and hunting or snaring game is generally more challenging than collecting plants.

Modern human foragers solved the formidable problem of finding enough to eat in such environments by expanding their dietary breadth to include small mammals, birds, and fish, and by creating the complex technology needed to harvest these food resources in sufficient quantities. The food-getting technology of recent hunter-gatherers in cold environments includes mechanical instruments and facilities, and among them— invariably—are self-acting devices or machines (see table 6.2). The automata of hunter-gatherers comprise snares and traps for catching small mammals (especially hare) and some birds (such as ptarmigan). These devices contain all the basic elements of a "finite-state machine" (more specifically, a "two-state machine") in automata theory (see box 4.4). They eliminate the potentially onerous time and energy costs of searching for, and killing or capturing, small and elusive prey (which are traded for the time and materials invested in constructing, deploying, and maintaining the machines).

Snares and traps may be deployed in large numbers across the landscape and used repeatedly to accumulate significant quantities of food (and materials, such as insulated hide in prime condition).[18]

In addition to snares and traps, modern human foragers in northern environments often developed complex technologies to harvest fish in substantial numbers. These include weirs and other fish-trapping devices designed to operate in a stream channel. Although they typically lack moving parts and are not technically automata, they "perform" a sequence of functions and solve a complex problem (which is decomposed into a set of sub-problems) (see figure 2.8). They eliminate the potentially high time and energy costs of obtaining fish with nets, hooks, or spears (see table 6.2).

Along similar lines, recent hunter-gatherers designed a variety of mechanical artifacts—such as spear-throwers, throwing darts, and fire-making drills—that include moving parts (that is, components that alter their relationship to other components when an artifact is used). As discussed in chapter 4, they may be considered "partial machines" that function like the moving parts of an organism. The significance of spear-throwers and other mechanical artifacts in the context of hunter-gatherers' adaptation to northern environments is that they increase the efficiency and effectiveness (success rate) of obtaining animal foods and thus contribute to the forager's ability to balance the ecological equation in a resource-poor setting. Likewise, fire-making equipment potentially saves significant time and fuel by avoiding the need to continually maintain a hearth.[19]

At least some of these mechanical artifacts—and, most probably, some of the technology for fishing—were developed by modern humans before they colonized the colder and drier regions of northern Eurasia, including those never previously occupied by humans. As noted in chapter 4, automata in the form of snares may have been present in southern Africa as early as about 65,000 to 62,000 years ago, and there is evidence for the use of the bow and arrow (also in southern Africa) before 40,000 years ago.[20] A variety of mechanical devices are found among recent hunter-gatherers of the equatorial and warmer temperate zones, especially in areas of dry climate where plant and animal productivity is low,[21] and in a few cases, they include automata (for example, Ju/'hoansi snare).[22]

As already noted, there is a significant trend toward more complex technology in the higher latitudes, where automata are universal (see table 6.2). These technologies apparently were necessary for the long-term survival of modern humans in the colder and drier parts of Eurasia, and their pre-sumed absence among other forms of *Homo* may explain why they were unable to occupy these environments. Because their design entails a quan-tum jump in *computational complexity* over the most complex forms known to date among other forms of *Homo* (see boxes 2.5 and 4.4), autom-ata and other mechanical technologies imply significant differences in cog-nitive faculties between non-modern and modern humans.

Recognizing the presence of automata and other complex technologies in the archaeological record is difficult because of their low visibility, which is particularly pronounced in sites that antedate the Last Glacial Maximum (LGM) (that is, more than 30,000 years ago). Although possible compo-nents of traps were identified in a site on the central East European Plain that dates to the final millennia of the Pleistocene,[23] evidence for this type of technology in earlier sites is indirect. It includes a large concentration of hare bones in an open-air site in eastern Europe that dates to more than 40,000 years ago, and stable-isotope data from human skeletal remains that indicate a heavy consumption of freshwater foods were reported from the 45,000-year-old Ust'-Ishim bone and younger remains (ca. 35,000–30,000 years ago).[24] The freshwater foods could include fish, waterfowl, or both. Evidence for mechanical projectile weaponry (spear-thrower and/or bow and arrow) is based primarily on morphometric analyses of stone points from the Levant (about 50,000 years ago) and Japan (37,000–32,000 years ago).[25]

The major technological adaptations that probably were necessary for successful dispersal in the colder regions of northern Eurasia as well as the evidence for these technologies in the archaeological record are presented in table 6.4. In addition to machines and mechanical technologies related to subsistence,[26] they include complex artifacts and features related to clothing and shelter. Insulated *clothing* may be the most complex technol-ogy developed by modern humans during the early phases of their disper-sal in northern Eurasia. If comparable to the multilayered winter clothing of recent foraging peoples of the Arctic, it could have comprised as many as 100 parts (organized in multiple hierarchical levels of subcomponents).[27] Applying a "thermal model" to the problem of a modern human thresh-old for insulated clothing, Ian Gilligan noted that—because there are a number of variables—there is "no single temperature point."[28] However, a short-term safety limit would be roughly –1°C, which clearly applies to winter-temperature levels in eastern Europe and Siberia, even under full

TABLE 6.4
Modern Human Dispersal in Northern Eurasia: Major Technological Adaptations

Technology	Purpose and Need	Evidence	Dating	Comments
Fire-making device	Thermoregulation; food preparation	Traces of rotary drill on stone (East European Plain and southwestern Siberia)	Ca. 42,000 years ago	Mechanical technology
Small-mammal traps and snares	Winter food supply; material for insulated clothing	Concentrations of small-mammal bones (East European Plain and Northeast Asia)	Ca. 42,000 years ago	Automata; three hierarchical levels
Spear-thrower; throwing darts	Improved efficiency and success in hunting	Stone points (Levant and Japan)	More than 40,000 years ago (?)	Mechanical technology
Storage pits	Storage of food for periods of reduced availability (cold months)?	Stone-lined pits in open-air sites (southern Siberia)	More than 40,000 years ago	Implies digging implements
Insulated and tailored clothing	Thermoregulation	Eyed needles (northern Caucasus)	Ca. 40,000 years ago	Potentially mechanical technology (drawstring); three or more hierarchical levels
Artificial shelters	Thermoregulation	Post-mold pattern (East European Plain)	32,000 years ago	May contain hearth
Domesticated dogs (?)	Improved efficiency and success in hunting	Canid morphology; genetics	Ca. 33,000 years ago (?)	May not antedate Last Glacial Maximum
Bow and arrow	Improved efficiency and success in hunting	Stone points (Levant and Japan)	More than 40,000 years ago (?)	Mechanical technology
Artificial memory systems	Computational aid	Bone and antler engraved with dots	Ca. 35,000 years ago	Later examples exhibit hierarchical organization
Cold storage	Preservation of food and fuel during warmer months	Traces of pits and remains of contents	30,000 years ago (?)	Application of low temperature in controlled setting

interglacial conditions. The most critical piece of archaeological evidence for insulated clothing is the eyed sewing needle, which has been recovered from early contexts (more than 40,000 years ago) in both eastern Europe and Siberia (but not in western Europe until the beginning of the LGM).[29]

Artificial *shelters with interior hearths* are another important technological adaptation to cold climates, especially in regions where natural shelters are scarce or absent. But while foraging people in higher latitudes eventually developed complex winter houses, the modern humans who colonized northern Eurasia either got by without shelters or constructed small, lightweight structures that have low archaeological visibility. This likely reflects a highly mobile settlement system with no long-term occupations. The earliest reliable evidence for artificial shelters dates to about 32,000 years ago, from the south-central East European Plain.[30] Although there is no archaeological evidence for fire-making technology in the context of modern human dispersal, there is reliable evidence of hand-operated drills before 40,000 years ago (on the central East European Plain and in southwestern Siberia).[31]

The occupation of environments characterized by pronounced—even extreme—seasonal fluctuations in resource availability (in time and space) probably required the *storage of food* for later consumption. Pits, which may have been used for the cold storage of perishable foods during the warmer months, are reported from a few sites in southern Siberia that may antedate 40,000 years. At one site, the sides of the pit were lined with stone.[32] Pits are rare in northern Eurasian sites older than 30,000 years, however, and much of the storage technology employed in these sites may have been primarily simple and aboveground, with little or no archaeological visibility.

The ecology of recent hunter-gatherers in northern environments also reflects important *organizational adaptations* to cold and dry habitats. They are similar to the organizational strategies of foragers in warm and dry regions, such as the northern Kalahari Desert, entailing widespread alliance networks maintained by marriage and exchange of gifts or raw materials (see box 4.2). They may include periodic aggregations of people to exploit seasonally available concentrations of resources (for example, salmon run). Unlike some of the complex technologies developed by hunter-gatherers in northern environments, the organizational adaptations have deep roots in the lower latitudes, including Africa. As described in chapter 4, there is evidence of expanded networks in the African Middle Stone Age as early as 130,000 years ago.[33]

In the archaeological record of northern Eurasia, the transport of objects and materials over hundreds of kilometers suggests long-distance movements of people or trade networks or—more probably—both. Sites in eastern Europe and southern Siberia contain materials imported from distances of up to or greater than 500 kilometers before 40,000 years ago.[34] The pattern of small, ephemeral occupations in this time range also is consistent with a highly mobile lifestyle. The profusion of personal ornaments (typically perforated shell or animal teeth) provides a material manifestation of the investment in maintaining relationships outside the immediate family. Sites on the East European Plain also yield evidence for organized mass-kills of large mammals. Small herds of horses and reindeer were driven into cul-de-sacs or, possibly, compounds before 40,000 years ago on the Don River in Russia and, especially horses, at a number of sites after 40,000 years ago.[35] Herd kills require advanced planning, extensive coordination, and the participation of multiple family groups.

CENTER OF DISPERSAL: THE LEVANT
AND CENTRAL ASIA

Between the eastern shore of the Mediterranean Sea and the Pamirs lies a region that became a center of dispersal for modern humans in northern Eurasia. The area embraces the Levant; the modern nation-states of Iraq, Iran, and Afghanistan; and the adjoining portion of Central Asia (Turkmenistan, Uzbekistan, and Tajikistan). At the junction of Africa, Europe, and Asia, it has been a major crossroads of human history.

Perhaps no other major region on Earth has seen as many population movements since the arrival of modern humans, which began roughly 120,000 years ago with an early dispersal of "near modern" humans out of northeastern Africa (chapter 4). The Levant subsequently was occupied by Neanderthals, who were almost certainly present when—60,000 years ago or earlier—modern humans again invaded the region. The primary evidence for the later encounter between the two populations lies in the genetics of living non-Africans, all of whom carry a small percentage of Neanderthal genes. The analysis of the aDNA from the femur found near Ust'-Ishim suggests that the modern humans who subsequently dispersed across northern Eurasia interbred with local Neanderthals during this interval—at the time of their migration into Eurasia (see box 2.2).[36]

The genetic impact of later events is apparent in a whole-genome analysis of Levantine peoples published in 2013.[37] The living populations reflect events that occurred after 25,000 years ago, beginning with a major divergence of Levantine and Arabian Peninsula/East African groups during the LGM, followed by another major split of Levantine and European groups near the end of the Pleistocene (after 16,000 years ago). Developments during the Neolithic were followed by a succession of states and empires, migrations, and invasions. Today, the Levant and Central Asia (and areas between the two) are characterized by a diverse mix of maternal and paternal lineages. The presence of mtDNA haplogroups N and R may reflect the movement of modern humans into the region after 60,000 years ago, but most of the lineages probably represent more recent events. As elsewhere, contrasts between the mtDNA and the Y-DNA data indicate different histories for the paternal lineages.[38]

Human Skeletal Remains

New fossil discoveries, new dates, and reassessments of older finds recently have altered the picture of the dispersal of modern humans in the Levant and Central Asia (see table 6.1). The most important new discovery is a modern human cranium recovered from *Manot Cave* (Israel) and published in January 2015.[39] The cranium consists of the uppermost portion of the frontal bone, nearly complete parietals, and the occipital bone. It was found in a side chamber of the cave and was not directly associated with any of the archaeological layers. A thin calcite patina that formed on the surface of the bone was dated by uranium-thorium, yielding a mean (minimum) estimate of about 55,000 years (figure 6.3).

The Manot Cave cranium is "relatively small and gracile in appearance," with thin walls and an estimated brain volume of roughly 1,100 cubic centimeters. It represents an adult (female?) modern human with some archaic features that presumably reflect its age. The parietal bosses are pronounced, and the side walls are parallel and vertically oriented, as in the crania of modern humans. The superior nuchal line extending across the occipital bone is well developed, unlike that of the Neanderthals. Also significant is the suprainiac fossa, which is unlike that in most Neanderthals and differs

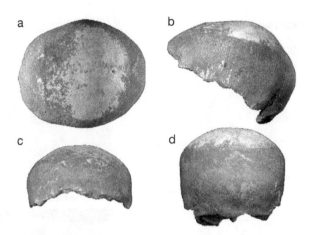

FIGURE 6.3 The cranium from Manot Cave: (*a*) superior view; (*b*) lateral view (note the occipital angle); (*c*) frontal view; (*d*) posterior view. (From Israel Hershkovitz et al., "Levantine Cranium from Manot Cave [Israel] Foreshadows the First European Modern Humans," *Nature* 520 [2015]: 217, fig. 2. Reproduced by permission from Macmillan Publishers Ltd., *Nature*, copyright 2015)

from the pattern in the earlier modern human remains of the Levant (ca. 100,000 years ago). Archaic features include the convexity of the occipital plane and a pronounced fossa between the nuchal lines.[40]

Manot Cave contains a lengthy sequence of occupation levels, including five horizons (Units 2–6) that appear to be too young to be associated with the cranium. These horizons are assigned to the *Levantine Aurignacian* industry. Below these levels lie two layers (Units 7 and 8) containing a small number of stone blades struck from prepared Levallois cores and a few retouched pieces, such as end scrapers made on flakes. They are assigned to the *Initial Upper Paleolithic* (IUP). Unit 7 yielded a radiocarbon date that calibrates to more than 50,000 years ago. The lower levels also contained some artifacts typical of the *Middle Paleolithic*, which is associated with both the earlier modern humans and the Neanderthals in the Levant. The archaeologists who excavated Manot Cave concluded that the cranium is associated with either the IUP or the Middle Paleolithic artifacts.[41]

Aside from the recent Manot Cave find, the most important modern human remains in the Levant are from *Ksâr ʿAkil* (Lebanon).[42] Ksâr ʿAkil is a rock shelter located near the Mediterranean coast about 10 kilometers north of Beirut. It contains a deep sequence of archaeological deposits subdivided into no fewer than 37 levels that represent the most complete Paleolithic record in the Levant (figure 6.4). Early excavations in 1937 and

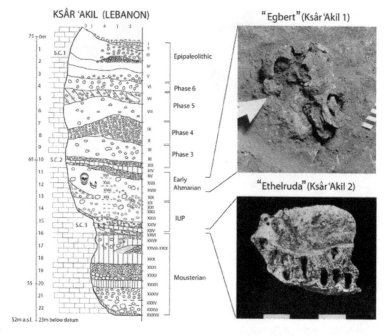

FIGURE 6.4 The stratigraphy of Ksâr ʿAkil. (From Katerina Douka et al., "Chronology of Ksar Akil [Lebanon] and Implications for the Colonization of Europe by Anatomically Modern Humans," *PLoS ONE* 8 [2013]: e72931, fig. 1. Reproduced from *PLoS ONE* under the terms of the Creative Commons Attribution [CCBY] license)

1938 produced a modern human skull and postcranial remains (commonly referred to as "Egbert") in Layer XVII/XVIII.[43] Egbert is associated with the *Early Ahmarian* industry, which is characterized by the production of numerous small blades and bladelets and is believed to have developed directly from the IUP. The remains recently have been dated to between 41,050 and 38,300 years ago, while new dates on shell from the archaeological layer are somewhat older (43,200–42,900 years ago).[44] Despite Egbert's age and significance, it has never been described in detail, in part due to poor preservation (and now is reported lost).

Several meters below Egbert, in Layer XXV, a partial maxilla was recovered in 1948 (referred to as "Ethelruda"). The find initially was believed to be Neanderthal, but was reassessed in the late 1990s as equally likely to be modern human.[45] New dates on the bone are 42,850 to 41,550 years ago, while Layer XXV is thought to be at least about 46,000 years old.[46] It is associated with the basal level of the IUP at Ksâr ʿAkil, which yielded a diagnostic IUP stone artifact (*Emireh point*). Although thought to be lost, like Egbert, Ethelruda recently was relocated and is now under renewed study.

Human remains also are known for this general time period from Central Asia, but their assignment to *Homo sapiens* is problematic. The most important finds are from *Obi-Rakhmat* (Uzbekistan), a rock shelter located in the western piedmont zone of the Tien Shan Mountains at an elevation of 1,250 meters.[47] Originally investigated in the 1960s, the site has been subject to renewed study since 1998 by an international team under the direction of Anatoly Derevianko. Human skeletal remains, comprising six permanent upper teeth and 150 cranial fragments, were recovered from the lower portion of the deposits (Stratum 16) and appear to date—based on a variety of methods—to at least 50,000 years ago and possibly between 70,000 and 90,000 years ago.[48] They are associated with artifacts similar to those of the IUP in the Levant (figure 6.5).[49]

The morphology of the Obi-Rakhmat remains, which are thought to represent a single individual (juvenile of roughly nine years), reflects a

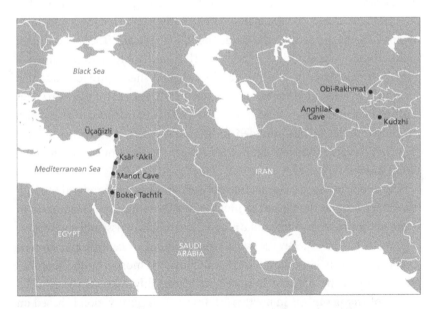

FIGURE 6.5 Map of the Levant and Central Asia, showing fossils and archaeological sites assigned to *Homo sapiens*, dated to between roughly 55,000 and 36,000 years ago. (Base map adapted from en.wikipedia.org/wiki/ File:BlankMap-World6.svg)

mixture of features found among modern and non-modern humans. The teeth are unusually large and exhibit some typical Neanderthal traits (for example, shovel-shaped incisor and large hypocone on the first molar). The cranial fragments, however, are more characteristic of *Homo sapiens*. Especially significant is the length of the left parietal bone, which falls outside the range for juvenile and adult Neanderthals (as well as that for other non-modern forms of *Homo*).[50] In view of the uncertain dating and mixed morphology of the remains, their association with the IUP is potentially important. The point is underscored by the fact that a human metatarsal from *Anghilak Cave* (southeastern Uzbekistan), dated to less than 40,000 years ago, is associated with typical Middle Paleolithic artifacts (for example, side scrapers).[51]

Modern Humans and the Initial Upper Paleolithic

Modern humans probably made the artifacts assigned to the IUP in the Levant. For some years, this was thought likely because Egbert, which clearly represents *Homo sapiens*, was found associated with the Early Ahmarian industry, which appears to have developed directly from the IUP. It was assumed that the makers of the Early Ahmarian tools were local descendants of the people who made the IUP artifacts.[52] The reassessment of Ethelruda, associated directly with the IUP, as possibly modern human— as well as new dates on the bone and associated shell—provided support for a link between modern humans and the IUP. The recent discovery and dating of the cranium from Manot Cave, while not associated directly with diagnostic artifacts, has provided additional support, because it confirms that *Homo sapiens* was present in the Levant as early as or earlier than the oldest known IUP sites.

The implications of a connection between modern humans and the IUP extend far beyond the Levant to much of northern Eurasia. In addition to those in Central Asia, sites containing IUP artifacts or assemblages of similar artifacts (that is, Levallois blades) are found in central and eastern Europe, southern Siberia, and northern China.[53] If most of or all these sites were occupied by modern humans, they probably represent the initial spread of *Homo sapiens* in northern Eurasia. This observation is based on the fact that the industries that precede the IUP (or similar assemblages) are associated directly or indirectly with non-modern forms of *Homo*.

Moreover, estimates of coalescence times for modern human lineages suggest that the initial spread of modern humans in northern Eurasia did not antedate the appearance of the IUP and similar industries (unless the lineages that initially spread across the northern half of the continent left no genetic trace among later peoples).[54]

Not one of the IUP sites (or sites with similar artifacts) outside the Levant, however, contains human skeletal remains that may be firmly attributed to *Homo sapiens*, and an alternative explanation for the spread of the IUP is that local non-modern groups (Neanderthals and, possibly, Denisovans) adopted the characteristic Levallois blade technology and often copious production of typical Upper Paleolithic tool types, such as end scrapers and burins. According to this scenario, it is the industries that followed the Levallois blade assemblages that represent modern humans. Outside the Levant, they would correspond to industries that are closely related to the Early Ahmarian (for example, *Proto-Aurignacian* in southern Europe) or artifact assemblages that are associated with modern human remains or include forms that are unknown in sites occupied by non-modern humans. The chief problem with this hypothesis is that both the genetics of living people (that is, estimated coalescence times for lineages) and the dated occurrences of aDNA suggest that the post-IUP sites are too young to represent the initial spread of *Homo sapiens* in northern Eurasia, which took place before 45,000 years ago.

"Initial Upper Paleolithic" is a term applied to an archaeological entity originally defined by Dorothy A. E. Garrod (1892–1968) on the basis of her excavations at Ksâr ʿAkil in the years before the Second World War.[55] Garrod proposed the label "Emiran" for the assemblages of stone artifacts in Layers XXV to XXI containing Levallois blades, end scrapers, and burins. The signature type in these assemblages was a large triangular Emireh point, and the industry also was recognized at other sites in the Levant.[56] In the early 1980s, Anthony E. Marks described a similar industry in the uppermost level at *Boker Tachtit* in the Negev highlands (southern Israel) that he labeled "Initial Upper Paleolithic."[57] In the 1990s, the industry also was recognized in the northern Levant at *Üçağızlı Cave* (southern Turkey), and the IUP label was applied to all these assemblages, as well as to others outside the Levant (table 6.5).[58]

Unlike the Early Ahmarian and subsequent industries, the IUP was produced with "hard-hammer" percussion (stone blanks were struck off cores

TABLE 6.5

Archaeological Record of *Homo sapiens* in the Levant and Central Asia, Before 30,000 Years Ago

Site, Layer	Estimated Age (method)	Description	Comments
Manot (western Galilee, Israel)[a]	54,700 ± 5500 years (U/Th)	Human partial cranium (calvaria) associated with Middle Paleolithic or Initial Upper Paleolithic artifacts	Oldest dated *Homo sapiens* in northern Eurasia
Obi-Rakhmat (northeastern Uzbekistan)[b]	48,800–36,000 years (radiocarbon, U-series, OSL)	Human remains associated with Levallois blade industry (IUP)	Human remains exhibit non-modern and modern traits; transition to IUP begins more than 50,000 years ago (?)
Boker Tachtit (Negev Desert, Israel)[c]	Ca. 47,000–46,000 years (radiocarbon)	Levallois point and blade industry (lower level); Early Ahmarian (upper level)	No human remains
Üçağızlı Cave (south-central Turkey)[d]	Ca. 44,000–39,000 years (radiocarbon)	IUP: Levallois blade industry (lower level); Early Ahmarian (upper level)	Associated shell ornaments
Ksâr ʿAkil, Layer XXV (Lebanon)[e]	42,400–41,700 years (radiocarbon)	Partial human maxilla ("Ethelruda," or Ksâr ʿAkil 2) associated with Emiran point	*Homo sapiens* remains associated with Levallois point and blade industry (IUP)
Khudzhi (western Tajikistan)[f]	Ca. 42,000–36,000 years (radiocarbon)		

Note: All radiocarbon dates are given in calibrated radiocarbon years before present.

[a]Israel Hershkovitz et al., "Levantine Cranium from Manot Cave (Israel) Foreshadows the First European Modern Humans," *Nature* 520 (2015): 216–219.

[b]Andrei I. Krivoshapkinm and P. Jeffrey Brantingham, "The Lithic Industry of Obi-Rakhmat Grotto, Uzbekistan," in *Actes du XIV Congres UISPP, 2-8 Septembre 2001*, BAR International Series 1240 (Oxford: Archaeopress, 2004), 203–214; Michelle Glantz et al., "New Hominin Remains from Uzbekistan," *Journal of Human Evolution* 55 (2008): 223–237; Andrei I. Krivoshapkin, Yaroslav V. Kuzmin, and A. J. Timothy Jull, "Chronology of the Obi-Rakhmat Grotto (Uzbekistan): First Results on the Dating and Problems of the Paleolithic Key Site in Central Asia," *Radiocarbon* 52 (2010): 549–554.

[c]Anthony E. Marks, "The Middle to Upper Paleolithic Transition in the Levant," in *Advances in World Archaeology*, ed. Fred Wendorf and Anglea E. Close (New York: Academic Press, 1983), 2:51–98; Anthony E. Marks and C. Reid Ferring, "The Early Upper Paleolithic of the Levant," in *The Early Upper Paleolithic: Evidence from Europe and the Near East*, ed. J. F. Hoffecker and C. A. Wolf, BAR International Series 437 (Oxford: British Archaeological Reports, 1988), 43–72.

[d]Steven L. Kuhn et al., "The Early Upper Paleolithic Occupations at Üçağızlı Cave (Hatay, Turkey)," *Journal of Human Evolution* 56 (2009): 87–113.

[e]Katerina Douka et al., "Chronology of Ksar Akil (Lebanon) and Implications for the Colonization of Europe by Anatomically Modern Humans," *PLoS ONE* 8 (2013): 372931.

[f]V. A. Ranov and S. A. Laukhin, "Stoyanka na Puti Migratsii Srednepaleoliticheskogo Cheloveka iz Levanta v Sibir," *Priroda* 9 (2000): 52–60.

with another piece of stone). And the iconic Emireh points were something of a holdover from earlier times—similar to the triangular points of the preceding Middle Paleolithic. Large blades, many of which were retouched into tools such as end scrapers and burins, gave the IUP its progressive look, even though they were struck off typical Middle Paleolithic cores with a hard hammer (figure 6.6).[59]

From the outset, the IUP was perceived as an industry that had developed gradually from the local Middle Paleolithic and, in turn, evolved into

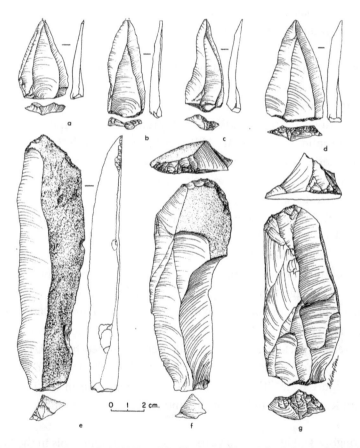

FIGURE 6.6 Stone artifacts of the Initial Upper Paleolithic from Level 4 at Boker Tachtit: (*a–d*) Levallois points; (*e–g*) end scrapers on blades. (From Anthony E. Marks and C. Reid Ferring, "The Early Upper Paleolithic of the Levant," in *The Early Upper Paleolithic: Evidence from Europe and the Near East*, ed. J. F. Hoffecker and C. A. Wolf, BAR International Series 437 [Oxford: British Archaeological Reports, 1988], 55, fig. 4. Reprinted with the permission of the authors)

the Early Ahmarian. Marks argued that the slow emergence of the IUP in the Levant was illustrated best in the temporal sequence of assemblages at Boker Tachtit.[60] The implication is that modern humans made not only the IUP, but also the preceding Middle Paleolithic (as suggested by the dating of the Manot Cave cranium).[61] If so, the relatively quick succession of changes in stone technology—over a period of a few millennia—would seem to reflect the unique cognitive faculties of *Homo sapiens* and a pattern not observable in the archaeological record of non-modern humans. At present, the whole process (from the end of the Middle Paleolithic to the beginning of the Early Ahmarian) seems to have taken place within a span of less than 5,000 years. Moreover, the IUP in the Levant exhibits some local variations that further underscore the pattern of innovation and creativity in toolmaking.[62]

There is, however, little evidence of the new technologies listed in table 6.5 in the Levant (or Central Asia) during the IUP. A bone awl was recovered from Layer XXIII at Ksâr ʿAkil, but it represents an isolated example of a non-stone tool in the IUP.[63] This may reflect the comparatively low latitude and mild climate: Ksâr ʿAkil is roughly 34° North, in an area that today experiences a comfortable January mean of 12°C. Although IUP sites in colder and drier places also lack evidence of major innovations, many date to an unusually warm interval that began about 48,000 years ago (GI 12). A contributing factor may be the availability and use of hardwoods (as opposed to antler and bone) for toolmaking, because wood has low visibility in the archaeological record.

By contrast, personal ornaments are well represented in the IUP of the Levant and suggest the same pattern of social networking evident in the archaeological record of modern humans in Africa before 50,000 years ago (and in the record of the dispersal in southern Asia) (chapters 4 and 5). Hundreds of perforated marine shells were recovered from IUP levels in Ksâr ʿAkil and Üçağızlı Cave, and an incised bird bone was found at the latter (figure 6.7).[64] The abundance of ornamentation in the IUP of the Levant leaves little doubt that this industry was the product of modern humans.

Roughly 43,000 to 42,000 years ago, makers of stone tools in the Levant shifted to a soft-hammer technique and began to produce the bladelets of the Early Ahmarian.[65] The bladelets often were retouched into narrow points (*El-Wad points*). The change in lithic technology may be connected

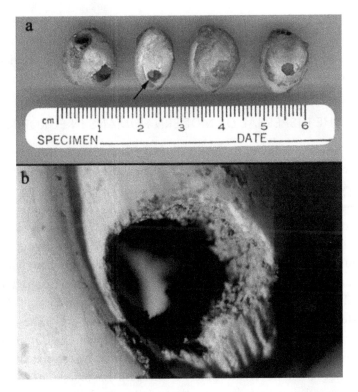

FIGURE 6.7 Personal ornaments from IUP levels at Üçağızlı Cave in the form of perforated marine shells (a), and (b) close-up of artificial perforation. (From Steven L. Kuhn et al., "Ornaments of the Earliest Upper Paleolithic: New Insights from the Levant," *Proceedings of the National Academy of Sciences* 98 [2001]: 7645, fig. 2. Copyright 2001 National Academy of Sciences)

with innovations in projectile weaponry (for example, spear-thrower and bow and arrow?), but this remains to be demonstrated. In any case, the Early Ahmarian soon spread northward into Europe, where it is associated with at least some of the major technological innovations discussed earlier, including insulated clothing and devices for catching small mammals.

Unlike the IUP, which spread eastward to Central Asia (for example, Obi-Rakhmat) and beyond into southern Siberia and northern China, the Early Ahmarian does not appear east of the Levant. Nor are there industries that may conceivably be related to the Early Ahmarian in Siberia or East Asia. There is a broadly similar industry in the region that lies between the Levant and Central Asia (*Baradostian*), but it appears to be younger and may be connected to the eastward expansion of a European industry

(*Aurignacian*) after 40,000 years ago. The development and spread of the Early Ahmarian may mark a historic split between the populations of western and eastern Eurasia, which apparently took place after the initial dispersal of the meta-population of modern humans (after 45,000 years ago).

THE DISPERSAL OF MODERN HUMANS IN EUROPE

A fundamental irony of paleoanthropology is that the birthplace of the field contains an anomalous record of the human past. Both evolutionary biology and Paleolithic archaeology have their deepest roots in western Europe, and much of the classificatory framework for human fossils and their artifacts was developed in France and Britain.[66] As described earlier, however, the mild winter temperatures and high biotic productivity of western Europe fade rapidly across the longitudinal climate gradient of northern Eurasia. The human fossil and archaeological records in eastern Europe and Siberia differ in many respects from those in western Europe, despite repeated efforts to fit them into the interpretive framework of the latter.

The earliest credible evidence for modern humans in Europe is an archaeological proxy for *Homo sapiens* (supported by estimated coalescence times for maternal and paternal lineages). The oldest dates on modern human fossils in Europe are several thousand years younger. The archaeological proxy is the IUP, which exhibits close parallels to the IUP of the Levant and Central Asia. It has yet to yield associated skeletal material that may be firmly assigned to *Homo sapiens* (or Neanderthals). The European IUP assemblages date to slightly less than 50,000 years ago, which is broadly coterminous with the estimated ages of the oldest mtDNA and Y-DNA lineages in Europe. The oldest dates on modern human fossils are roughly 45,000 years ago (see table 6.1).

As in many other parts of the world, the genetics of the living population of Europe reflect movements of people that took place after the initial arrival of modern humans. Traces of the dispersal recently have emerged from the analysis of aDNA recovered from dated skeletal remains, revealing the presence of the three maternal lineages tied most closely to African mtDNA haplogroup L3 (mtDNA M, N, and R) (see table 6.1).[67] None of these lineages (all of which yield estimated coalescence dates of about 50,000 years) is widely represented in Europe today. Mitochondrial DNA haplogroup N recently was identified in the mandible from *Peştera cu Oase* (Romania), which has been dated to roughly 40,000 years ago, while mtDNA R was found in an isolated tooth of comparable age from *Grotta di Fumane* (northern Italy).[68] As noted earlier, the femur from Ust'-Ishim is derived from mtDNA R, which appears to represent a maternal lineage that was broadly distributed across northern Eurasia by 45,000 years ago.

More common in the living population is mtDNA haplogroup U (including subclades U5 and U8), which is a subgroup of mtDNA N and also has an estimated coalescence date of about 50,000 years.[69] The subclade U2 has been identified from aDNA extracted from a 37,000-year-old skeleton from the central East European Plain.[70] Among paternal lineages, Y-DNA haplogroup I may be the oldest in Europe (and today is largely restricted to Europe),[71] but the lineage associated with the 37,000-year-old skeleton in eastern Europe is C-M130, which is one of the oldest and most widespread in southern Eurasia (chapter 5).[72]

Along with Siberia, Europe yields evidence for early technological adaptations related to the dispersal of modern humans into the higher latitudes (that is, technologies for coping with cold climates and reduced biological productivity). Significantly, this evidence surfaces in eastern Europe and after the GI 12 warm interval, which ended about 45,000 years ago. The innovations include eyed needles at about 40,000 years ago (northern Caucasus) and artificial shelters before 30,000 years ago (East European Plain). Evidence for the expansion of the diet, likely related to novel technology, includes the harvesting of small mammals before 40,000 years ago and the consumption of aquatic foods by about 35,000 years ago (both East European Plain). Evidence for long-distance transport of raw materials (over hundreds of kilometers) also is found on the East European Plain before 40,000 years ago.

The archaeological record of modern humans in Europe suggests some strong correlations between population movements and climate. While climate probably played an important role in the spread of modern humans in Siberia, the European record is better dated and more thoroughly tied to the Greenland ice-core stratigraphy (which is the primary reference for climate in the Northern Hemisphere during this period). The most striking correlation is between the appearance of the IUP in Europe and the beginning of GI 12 at roughly 48,000 years ago. There also seems to be a correlation between the Proto-Aurignacian industry in southwestern Europe and the warm interval that began about 43,500 years ago (GI 11), and there is evidence for major population movements following the cold period between 40,000 and 38,000 years ago (Heinrich Event 4 [HE4]).

The Initial Upper Paleolithic

The core area for the IUP in Europe lies in Moravia and the Balkans. A group of major open-air sites and caves in these regions contain assemblages of stone artifacts that are very similar to those assigned to the IUP in the Levant and Central Asia. In the Balkans, specifically in Bulgaria, they include the caves of *Bacho Kiro* (Layer 11) and *Temnata* (Layer VI), and in Moravia (Czech Republic), they include the open-air sites of *Brno-Bohunice* and *Stránská skála*. Farther north, in southern Poland, is *Dzierzyslaw I*.[73] The industry is known locally as the *Bohunician* (table 6.6 and figure 6.8).

The stone tools often were produced from Levallois blade cores with hard-hammer percussion and include many forms considered typical of Upper Paleolithic industries associated with modern humans (that is, end scrapers and simple burins), as well as Levallois points and side scrapers.[74] Some assemblages also contain bifacial leaf-shaped points (more common in the northern sites). As with the IUP in the Levant, non-stone artifacts are extremely rare.[75]

The presence of bifacial leaf-shaped points in the Bohunician may be highly significant because similar points are found in another group of sites in Hungary (known locally as the *Szeletian*) that appear to be of comparable age. Many are caves in the Bükk Mountains that appear to have been occupied briefly—perhaps as short-term hunting camps.[76] In addition to the points, the sites often contain typical Middle Paleolithic forms (for example, side scrapers). Rather than constituting a separate industry (and one often attributed to the Neanderthals), the Szeletian may simply represent a recurring set of artifacts made by the same people who made the Bohunician assemblages (a functional subset of the IUP).[77]

The IUP sites in central Europe date as early as the beginning of GI 12 (about 48,000 years ago), and some are associated with a buried soil that apparently formed during this warm interval.[78] Other assemblages significantly postdate GI 12 and are associated with a younger soil that could have

Table 6.6

Archaeological Record of *Homo sapiens* in Western and Central Europe, Before 35,000
Years Ago

Site, Layer	Estimated Age (method)	Description	Comments
Brno-Bohunice, lower paleosol (Czech Republic)[a]	48,200 ± 1900 years (TL) 41,350–38,250 years (^{14}C)	Initial Upper Paleolithic: Levallois blades and points; end scrapers; bifaces	Type-site for Bohunician industry of central Europe (IUP)
Stránská skála, upper paleosol (Czech Republic)[b]	Ca. 45,600– 40,500 years (^{14}C)	IUP: Levallois blades and points; end scrapers; side scrapers	Bohunician assemblage also recovered from lower paleosol
Grotta del Cavallo (southern Italy)[c]	45,000–43,400 years (^{14}C)	Shell beads	Uluzzian industry
Kent's Cavern (England)[d]	44,200–41,500 years (^{14}C)	Isolated artifacts (stone blades) buried below human maxilla (dated)	Aurignacian blades and IUP blades (?)
Goat's Hole (southern England)[e]	44,000–40,000 years (?) (artifacts antedate "Red Lady" burial)	Leaf-shaped points; nosed scrapers; busked burins	IUP? (leaf-shaped points); younger Aurignacian assemblage
Geißenklösterle, AH III (southern Germany)[f]	44,100–41,860 years (^{14}C)	Carinated and nosed scrapers; burins; bone, antler, and ivory artifacts	Assemblage now assigned to Early Aurignacian industry
Bacho Kiro, Layer 11 (Bulgaria)[g]	More than 43,000 years (^{14}C)	Levallois blades and points; retouched blades; end scrapers associated with human mandible	Assemblage now reassigned to Bohunician (IUP)
Willendorf II, AH 3 (Austria)[h]	Ca. 43,500 years (^{14}C)	Carinated scraper; nosed scraper; retouched blade; bladelets	Early Aurigancian
Temnata Cave (Bulgaria)[i]	Ca. 45,000– 40,000 years (^{14}C, tephrochronology)	Bladelets	IUP (lower level); Proto-Aurignacian industry
Castelcivita Cave (southern Italy)[j]	Ca. 45,000– 40,000 years (^{14}C, tephrochronology)	Bladelets; bone artifacts	Uluzzian (lower level); Proto-Aurignacian (upper level)
Grotta di Fumane (northern Italy)[k]	41,100–38,500 years (^{14}C)	Bladelets	Proto-Aurignacian
Riparo Bombrini (northern Italy)[l]	40,710–35,640 years (^{14}C)	Bladelets	Proto-Aurignacian

(continued)

(continued)

Note: All radiocarbon dates are given in calibrated radiocarbon years before present.

[a]D. Richter, G. Tostevin, and P. Škrdla, "Bohunician Technology and Thermoluminescence Dating of the Type Locality of Brno-Bohunice (Czech Republic)," *Journal of Human Evolution* 55 (2008): 871–885.

[b]Jiří A. Svoboda and Ofer Bar-Yosef, eds., *Stránská skála: Origins of the Upper Paleolithic in the Brno Basin, Moravia, Czech Republic* (Cambridge, Mass.: Peabody Museum of Archaeology and Ethnology, 2003).

[c]Stefano Benazzi et al., "Early Dispersal of Modern Humans in Europe and Implications for Neanderthal Behaviour," *Nature* 479 (2011): 525–528.

[d]Tom Higham et al. "The Earliest Evidence for Anatomically Modern Humans in Northwestern Europe," *Nature* 479 (2011): 521–524.

[e]R. M. Jacobi and T. F. G. Higham, "The 'Red Lady' Ages Gracefully: New Ultrafiltration AMS Determinations from Paviland," *Journal of Human Evolution* 55 (2008): 898–907.

[f]Thomas Higham et al., "Testing Models for the Beginnings of the Aurignacian and the Advent of Figurative Art and Music: The Radiocarbon Chronology of Geißenklösterle," *Journal of Human Evolution* 62 (2012): 664–676.

[g]Janusz K. Kozlowski et al., "Upper Paleolithic Assemblages," in *Excavation in the Bacho Kiro Cave (Bulgaria): Final Report*, ed. Janusz K. Kozlowski (Warsaw: Państwowe Wydawnictwo Naukowe, 1982), 119–162; Tsenka Tsanova and Jean-Guillaume Bordes, "Contribution au débat sur l'origine de l'Aurignacien: Principaux résultats d'une étude technologique de l'industrie lithique de la couche 11 de Bacho Kiro," in *The Humanized Mineral World: Towards Social and Symbolic Evaluation of Prehistoric Technologies in Southeastern Europe*, ed. Tsoni Tsonev and Emmanuela Montagnari Kokelj, ERAUL 103 (Liege: University of Liege, 2003), 41–50.

[h]Philip R. Nigst et al., "Early Modern Human Settlement of Europe North of the Alps Occurred 43,500 Years Ago in a Cold Steppe-Type Environment," *Proceedings of the National Academy of Sciences* 111 (2014): 14394–14399.

[i]Janusz K. Kozlowski, "The Middle and Early Upper Paleolithic Around the Black Sea," in *Neandertals and Modern Humans in Western Asia*, ed. Takeru Akazawa, Kenichi Aoki, and Ofer Bar-Yosef (New York: Plenum, 1998), 461–482; Francesco G. Fedele, Biagio Giaccio, and Irka Hajdas, "Timescales and Cultural Process at 40,000 BP in the Light of the Campanian Ignimbrite Eruption, Western Eurasia," *Journal of Human Evolution* 55 (2008): 834–857.

[j]P. Gambassini, ed., *Il Paleolitico di Castelcivita: Culture e ambiente* (Naples: Electa, 1997); Fedele, Giaccio, and Hajdas, "Timescales and Cultural Process at 40,000 BP."

[k]S. Benazzi et al., "The Makers of the Protoaurignacian and Implications for Neandertal Extinction," *Science* 348 (2015): 793–796.

[l]Ibid.

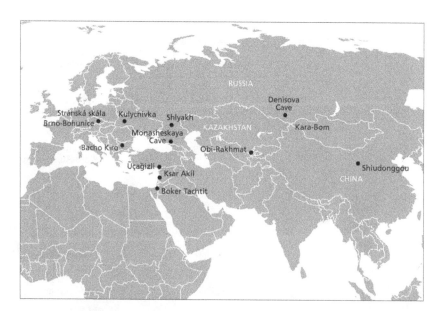

FIGURE 6.8 Map of archaeological sites (ca. 50,000–ca. 35,000 years ago) containing Levallois blades and points, many of which may be assigned to the IUP. (Base map adapted from en.wikipedia.org/wiki/File:BlankMap-World6.svg)

formed during GI 11. The IUP may have endured for more than 5,000 years.[79]

To date, the IUP has failed to yield any diagnostic skeletal remains of modern humans. The most substantial specimen is a fragment of the left side of a mandible containing a deciduous tooth (first molar) recovered from Layer 11 at Bacho Kiro.[80] The dimensions of the tooth are large for modern humans, but the taxonomic status of the bone is problematic. However, a Szeletian site in Slovakia (*Dzeravá skála*) yielded an isolated tooth (lower first or second molar) that is probably modern human.[81] It appears increasingly likely—primarily on the basis of the close similarities of the stone artifacts with those of the IUP of the Levant—that the IUP sites in central Europe represent the initial movement of modern humans into the region.[82]

Eastern Europe contains a number of sites with similar artifacts—and most of them seem to date to the same time range—but their link to the IUP and modern humans is more tenuous. A Bohunician assemblage is recognized at a site in western Ukraine (*Kulychivka*), but its age is unclear; it may be relatively young (less than 40,000 years old).[83] An older assemblage is known from an open-air site in southern Russia (*Shlyakh*) (see box figure 2.8). However, while Levallois blades and points resembling those of the IUP are present, very few typical tools (for example, burins and end scrapers) were found at Sklyakh, and the artifacts are regarded as Middle Paleolithic (that is, more commonly associated with Neanderthals) (table 6.7).[84]

Other sites in eastern Europe that date or may date to between 50,000 and 40,000 years ago contain similar assemblages. They are found scattered across the East European Plain, as well as along the southern margin of the plain in Crimea and the Caucasus Mountains.[85] As with sites in central Europe, they have yet to produce diagnostic human skeletal remains. Some fossils, including several isolated teeth, were recovered from *Monasheskaya Cave*, in the foothills of the northern Caucasus Mountains. They exhibit a mixture of modern and non-modern traits.[86]

The most surprising feature of the European IUP, including the potentially related sites in eastern Europe, is the lack of evidence for any of the innovative technologies associated with hunter-gatherers in higher latitudes (see table 6.4). This applies especially to the sites in southern Poland and on the East European Plain, where—despite the warmer climates of GI

TABLE 6.7
Archaeological Record of *Homo sapiens* in Eastern Europe, Before 30,000 Years Ago

Site, Layer	Estimated Age (method)	Description	Comments
Shlyakh, Layers 8 and 9 (southern Russia)[a]	Ca. 45,000–35,000 years (radiocarbon, paleomagnetism)	Levallois blades and points	Similar to Emiran industry; proxy for *Homo sapiens* (?)
Biryuch'ya Balka 2, Layers 3–6 (southern Russia)[b]	Ca. 45,000–35,000 years (radiocarbon, paleomagnetism)	Levallois blades (lowest level); bifacial points (upper level)	Similar to Emiran (?) (lowest level); typical Early Upper Paleolithic, with bifacial points (upper level)
Mamontovaya kurya (northern Urals)[c]	Ca. 44,000 years (radiocarbon)	Isolated flakes; scraper; biface (slate); incised bone (mammoth)	Located on Arctic Circle; short-term occupation
Mezmaiskaya Cave, Layer 1C (northwestern Caucasus)[d]	Ca. 38,000–37,000 years (radiocarbon)	Bladelets; retouched blades; ornaments; eyed needle	Similar to Early Ahmarian industry; oldest eyed needle in Europe
Kostenki 14, Layer IVb (central Russia)[e]	Ca. 44,000–42,000 years (radiocarbon, OSL, tephrochronology)	Bladelets; bladelet cores; bifaces; burins; antler mattocks; bone point; perforated fossil-shell ornaments	Isolated tooth (*H. sapiens*?); oldest figurative art (?); long-distance transport of raw materials (more than 500 kilometers)
Kostenki 17, Layer II (central Russia)[f]	Ca. 44,000–42,000 years (radiocarbon, tephrochronology, paleomagnetism)	Bladelets; burins; drilled fossil ornaments	Isolated tooth (*H. sapiens*?); traces of rotary drill
Kostenki 12, Layer III (central Russia)[g]	Ca. 44,000–42,000 years (radiocarbon, OSL, paleomagnetism)	Bifacial points; side scrapers; end scrapers (local chert and quartzite)	Kill-butchery site (horse and reindeer); short-term occupation
Buran-Kaya III (Crimea)[h]	Ca. 39,000–35,000 years (radiocarbon)	Bladelets	Associated *Homo sapiens* remains; oldest Early Gravettian date (ca. 39,000 years ago)
Sungir' (northern Russia)[i]	35,000 years (radiocarbon)	Bifacial points; side scrapers; end scrapers; retouched blades	Multiple burials (*H. sapiens*) with grave goods; cold-climate anatomy; large-mammal butchery (reindeer)
Kostenki 1, Layer III (central Russia)[j]	Ca. 35,000 years (radiocarbon, OSL)	Carinated scraper; retouched blades; backed bladelets	Diagnostic Aurignacian industry artifacts

Site, Layer	Estimated Age (method)	Description	Comments
Kostenki 15 (central Russia)[k]	Ca. 35,000 years (radiocarbon)	Side scrapers; end scrapers; decorated bone; eyed needle	Kill-butchery site (horse) associated with long-term occupation and burial (*H. sapiens*)
Mira, Layers I and II (south-central Ukraine)[l]	32,000 years (radiocarbon, OSL)	Bladelets (lower level)	Similar to Early Gravettian (lower level); artificial shelter (upper level)
Syuren' 1 (Crimea)[m]	Ca. 35,000 years (radiocarbon)	Carinated scrapers; bladelets	Diagnostic Aurignacian artifacts
Molodova 5, Layer 10 (western Ukraine)[n]	Ca. 35,000 years (radiocarbon)	Burins; retouched blades; carinated scraper; bacled bladelets	Diagnostic Aurignacian artifacts

Note: All dates are given in calibrated radiocarbon years before present.

[a] P. E. Nehoroshev, *Tekhnologicheskii metod izucheniya pervobytnogo rasshchepleniya kamnya srednego paleolita* (St. Peterburg: Evropeiskii Dom, 1999); John E. Hoffecker et al., "Geoarchaeological and Bioarchaeological Studies at Mira, an Early Upper Paleolithic Site in the Lower Dnepr Valley, Ukraine," *Geoarchaeology: An International Journal* 29 (2014): 61–77.

[b] A. E. Matyukhin, "Mnogosloinye Paleoliticheskie Pamyatniki v Ust'e Severskogo Dontsa," in *Rannyaya Pora Verkhnego Paleolita Evrazii: Obshchee i Lokal'noe*, ed. M. V. Anikovich (St. Petersburg: Russian Academy of Sciences, 2006), 157–182; Marcel Otte, A. E. Matyukhin, and Damien Flas, "La chronologie de Biryuchya Balka (Région de Rostov, Russie)," in *Rannyaya Pora Verkhnego Paleolita Evrazii*, ed. Anikovich, 183–192.

[c] Pavel Pavlov, John Inge Svendsen, and Svein Indrelid, "Human Presence in the European Arctic Nearly 40,000 Years Ago," *Nature* 413 (2001): 64–67.

[d] L. V. Golovanova, V. B. Doronichev, and N. E. Cleghorn, "The Emergence of Bone-Working and Ornamental Art in the Caucasian Upper Paleolithic," *Antiquity* 84 (2010): 299–320.

[e] M. V. Anikovich et al., "Early Upper Paleolithic in Eastern Europe and Implications for the Dispersal of Modern Humans," *Science* 315 (2007): 223–226.

[f] P. I. Boriskovskii, *Ocherki po Paleolitu Basseina Dona*, Materialy i Issledovaniya po Arkheologii SSSR 121 (Moscow: Nauka, 1963).

[g] Anikovich et al., "Early Upper Paleolithic in Eastern Europe"; John F. Hoffecker et al., "Evidence for Kill-Butchery Events of Early Upper Paleolithic Age at Kostenki, Russia," *Journal of Archaeological Science* 37 (2010): 1073–1089.

[h] Sandrine Prat et al., "The Oldest Anatomically Modern Humans from Far Southeast Europe: Direct Dating, Culture, and Behavior," *PLoS ONE* 6 (2011): e20834.

[i] O. N. Bader, *Sungir' Verkhnepaleoliticheskaya Stoyanka* (Moscow: Nauka, 1978); O. N. Bader and N. O. Bader, "Verkhnepaleoliticheskoe Poselenie Sungir,'" in *Homo Sungirensis: Verkhnepaleoliticheskii Chelovek: Ekologicheskie i Evolyutsionnye Aspekty Issledovaniya* (Moscow: Nauchnyi Mir, 2000), 21–29.

[j] A. N. Rogachev, *Mnogosloinye Stoyanki Kostenkovsko-Borshevskogo Raiona na Donu i Problema Razvitiya Kul'tury v Epokhy Verkhnego Paleolita na Russkoi Ravnine*. Materialy i Issledovaniya po Arkheologii SSSR 59 (Moscow: Nauka, 1957); John F. Hoffecker et al., "Kostenki 1 and the Early Upper Paleolithic of Eastern Europe," *Journal of Archaeological Science: Reports* 5 (2016): 307–326.

[k] Rogachev, *Mnogosloinye Stoyanki Kostenkovsko-Borshevskogo Raiona na Donu*; Hoffecker et al., "Evidence for Kill-Butchery Events of Early Upper Paleolithic Age at Kostenki, Russia."

[l] V. N. Stepanchuk, "The Archaic to True Upper Paleolithic Interface: The Case of Mira in the Middle Dnieper Area," *Eurasian Prehistory* 3 (2005): 23–41; Hoffecker et al., "Geoarchaeological and Bioarchaeological Studies at Mira."

[m] Yuri E. Demidenko and Pierre Noiret, "Radiocarbon Dates for the Siuren I Sequence," in *Siuren I Rock-Shelter: From Late Middle Paleolithic and Early Upper Paleolithic to Epi-Paleolithic in Crimea*, ed. Yuri E. Demidenko, Marcel Otte, and Pierre Noiret, ERAUL 129 (Liege: University of Liege, 2012), 49–53.

[n] A. P. Chernysh, "Mnogosloinaya Paleoliticheskaya Stoyanka Molodova I," in *Molodova I: Unikal'noe Must'erskoe Poselenie na Srednem Dnestre*, ed. G. I. Goretskii and I. K. Ivanova (Moscow: Nauka, 1982), 6–102

12—winter temperatures must have fallen well below 0°C. Moreover, as noted earlier, the European IUP appears to have lasted through at least one cold interval (Greenland Stadial 12 [GS 12]). If modern humans, recently arrived from the lower latitudes, produced the IUP and similar assemblages in central and eastern Europe, how did they cope with subfreezing temperatures and the reduced density of resources?

Future research and discoveries may show that the current lack of evidence for innovative technologies in the European IUP is a function of low archaeological visibility and sampling error. Some of the major open-air sites in Europe are thought to represent stone-quarry/workshop locations (for example, Stránská skála and Kulychivka), where people collected raw materials from natural outcrops or secondary concentrations of suitable rock.[87] Such sites seem unlikely to yield traces of technology related to protecting the occupants from cold temperatures and harvesting small vertebrates. Significantly, a bone perforator was recovered from the cave deposits at Bacho Kiro, which probably represents a habitation area (and where preservation conditions are better than at most open-air settings).[88] The dating of the IUP to the beginning of GI 12, combined with the absence of evidence for technological innovations, suggests nevertheless that climate (and its effects on biota) played an important role in the initial movement of modern humans into Europe.

Another factor may have been the geographic distribution of Neanderthals. One of the most startling implications of assigning the assemblages of Levallois blades and points in eastern Europe—generally assumed to have been made by Neanderthals because technically they are Middle Paleolithic assemblages—to modern humans is that it leaves little evidence for a Neanderthal presence on the East European Plain by 60,000 years ago. Accordingly, much of eastern Europe (and at least some portions of central Europe) may have been unoccupied by any humans at the beginning of GI 12. The implications are significant because the Neanderthals would have competed with modern humans for the same large-mammal prey (that is, ecological-niche overlap). The combination of a warming climate and an "empty niche" may underlie the spread of the IUP into Europe after 50,000 years ago, while the continued presence of the Neanderthals in western Europe at this time may explain why the IUP did not spread farther west.

A Second Wave?

Another wave of modern humans seems to have arrived in Europe by the beginning of the GI 11 warm period (ca. 43,500 years ago). Their sites, which are found in both western and eastern Europe, have yielded small stone bladelets and points, bone tools, and personal ornaments; one of those in eastern Europe contained what may be a piece of figurative art. Unlike those of the IUP, the artifacts are firmly associated with skeletal remains of *Homo sapiens*, and some aDNA provides a link with a specific maternal lineage. And, unlike the IUP, there is good evidence for major technological innovations related to climate and productivity.

In western Europe, the most important concentration of sites is found in Italy. Caves and open-air sites contain artifact assemblages commonly assigned to the Proto-Aurignacian industry (figure 6.9). Many of them are buried beneath a volcanic tephra deposited roughly 40,000 years ago by a massive volcanic eruption (*Campanian Ignimbrite* [CI]) that destroyed half of the Bay of Naples and spread an ash plume across much of southeastern

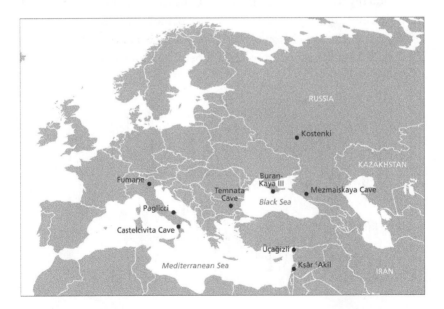

FIGURE 6.9 Map of western Eurasia, showing archaeological sites assigned to *Homo sapiens* that have yielded Proto-Aurignacian and similar artifacts, dated to between roughly 45,000 and 40,000 years ago. (Base map adapted from en.wikipedia.org/wiki/ File:Blank Map-World6.svg)

Europe.[89] The CI tephra, which has been dated by a variety of methods in various places, provides a reliable and widespread chrono-stratigraphic marker for the "second wave" of modern humans in southern and eastern Europe. Supporting radiocarbon dates suggest that the artifact assemblages date to about 43,000 to 41,000 years ago. The subsequent disappearance of the Proto-Aurignacian probably is related to the CI eruption.

Assemblages of Proto-Aurignacian artifacts typically are dominated by small stone bladelets produced by the soft-hammer technique and retouched along their margins. In some assemblages, bladelets account for more than 80 percent of retouched items. Artifacts of bone and antler (for example, awls and points) are common at many sites, especially in northern Italy. Also common are personal ornaments fashioned from marine shells.[90] At the Grotta di Fumane in northern Italy, an isolated tooth (upper deciduous incisor) associated with Proto-Aurignacian artifacts is assigned to modern humans on morphological grounds (relative enamel thickness) and yielded mtDNA from haplogroup R (figure 6.10).[91]

The artifacts of the Proto-Aurignacian are strikingly similar to those of the Early Ahmarian, but while the latter appears to have developed locally and gradually from the IUP in the Levant, the former represents an abrupt transition and probably reflects a movement of modern humans into southern Europe.[92] The association of the Proto-Aurignacian with mtDNA haplogroup R (also identified in western Siberia at about 45,000 years ago),

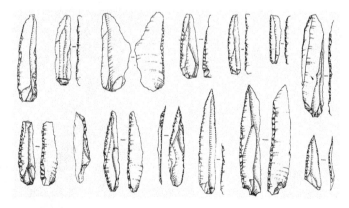

FIGURE 6.10 Proto-Aurignacian artifacts (retouched bladelets) from the Grotta di Fumane. (From Giorgio Bartolomei et al., "La Grotte de Fumane: Un site aurignacien au pied des Alpes," *Preistoria Alpina* 28 [1992]: 171, fig. 26.

though, suggests a link with an earlier and wider dispersal (that of the hypothesized meta-population across northern Eurasia). If and when Y-DNA data become available for the people who made the Proto-Aurignacian (and the Early Ahmarian), they may reveal that the Early Ahmarian represents a combination of invading paternal lineages and previously established local maternal lineages. The pattern was repeated often in the later population history of Eurasia and Africa.

Another group of sites in Italy (assigned to the *Uluzzian* industry), dated to GI 11 and later, also is associated with skeletal remains of modern humans.[93] The artifacts include not only a large number of typical Middle Paleolithic flake tools (that is, artifacts commonly associated with the Neanderthals and other non-modern humans), but also occasional blades and bladelets, as well as other items diagnostic of industries produced by modern humans.[94] Among the latter are bone awls and marine-shell pendants. The Uluzzian was long believed to represent a "transitional" industry created by local Neanderthals—perhaps influenced by the nearby presence of modern humans. However, the identification of two modern human teeth from an Uluzzian context in southernmost Italy (*Grotta del Cavallo*) settled the issue in 2011.[95] The Uluzzian may be analogous to the earlier described Szeletian industry, containing a high percentage of expedient flake tools made on poor-quality stone.

At the beginning of the twentieth century, the legendary French archaeologist Henri Breuil (1877–1961) established the classic Aurignacian industry of southwestern Europe as the initial phase of the Upper Paleolithic (and tied to the appearance of modern humans).[96] Especially characteristic are the thick "carinated scraper" (from which small blades or microblades were struck) and the split-base bone point (actually made on antler). Both forms show up occasionally in the Proto-Aurignacian of Italy, and new dates on major sites in Austria and southern Germany indicate that the Aurignacian also dates to GI 11.[97] These sites lie outside the area affected by the CI volcanic eruption; significantly perhaps, the artifacts of the classic Aurignacian not only are found in the millennia following 40,000 years ago, but subsequently spread into the areas devastated by the CI ash fall.

The most important traces of what appears to have been a second wave of modern humans spreading into Europe lie in the eastern half of the continent. In the Caucasus Mountains and especially on the central East European Plain, *Homo sapiens* adapted to winter temperatures and resource abundance that must have been considerably lower than those in western Europe at this time (GI 11 and its aftermath). Assemblages containing blade-lets similar to those of the Early Ahmarian are found on both the southern and northern slopes of the Caucasus Mountains and suggest the direct movement of modern humans from the Levant and into eastern Europe (rather than through the Balkans). A fragment of an eyed needle was recov-ered from one of these assemblages at *Mezmaiskaya Cave* (northwestern Caucasus). It represents the oldest evidence known to date for insulated clothing in western Eurasia.[98]

On the western bank of the Don River in central Russia, open-air sites concentrated around the villages of Kostenki and Borshchevo contain a number of occupation levels antedating 40,000 years ago. The area lies within the ash-fall zone of the CI volcanic eruption, and the stratigraphic context is identical to that of the Proto-Aurignacian—the occupations are buried beneath the 40,000-year-old tephra horizon.[99] The sites are found along spring-fed drainages that empty into the main valley, and the analy-sis of associated animal remains indicates that many were locations at or near which horses, reindeer, and occasionally mammoths were killed and butchered.[100] Presumably, the springs attracted the animals (perhaps espe-cially in winter), which, in turn, drew people to the area on a regular basis. The concentrations of bone are invariably associated with artifacts—such as stone points, sharp flakes, scrapers, and heavier implements—typically found at sites in other parts of the world where large mammals were killed and butchered.[101] Most were made on local stone of poor quality and appear to have been expedient tools, quickly discarded after use.

The bone concentrations and associated artifacts represent places where groups of animals were driven into some form of trap and killed en masse. At the site of *Kostenki 12*, reindeer and horses apparently were herded from the main valley and into the narrow space at the mouth of a large ravine system. This would have required some careful planning and coordination among a group of people; it also may have required some technological aids in the form of a fence or barrier (wooden posts or, possibly, brush piles) to funnel the animals into the trap (figure 6.11).[102]

Evidence for other organizational adaptations that may have been criti-cal to life on the immense East European Plain, which stretches north from the Black Sea to the Arctic Ocean, are found at Kostenki sites that proba-bly represent living areas. At *Kostenki 14*, in the upper reaches of the same

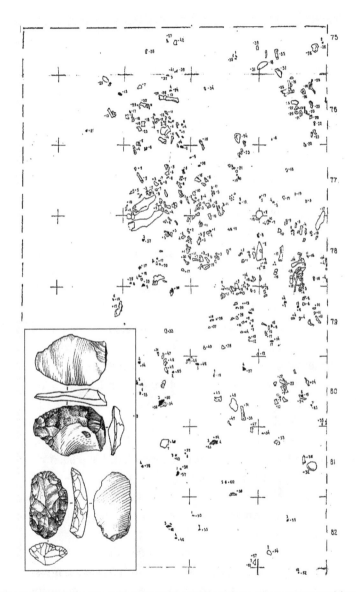

FIGURE 6.11 Concentration of several hundred reindeer and horse bones (also some mammoth bones) and associated stone artifacts (*inset*) on Layer III at the site of Kostenki 12, dated to more than 40,000 years ago. (Adapted from M. V. Anikovich, V. V. Popov, and N. I. Platonova, *Paleolit Kostenkovsko-Borshchevskogo Raiona v Kontekste Verkhnego Paleolita Evropy* [St. Petersburg: Russian Academy of Sciences, 2008], 83, fig. 36)

ravine system, artifacts and associated animal remains more typical of extended habitation include perforated shells imported more than 500 kilometers from the south.[103] This likely reflects both long-distance movements (that is, expanded foraging range in an environment where resources are thin) and trade with other groups, which suggests widespread alliance networks. Such networks could have helped buffer the inevitable effects of local fluctuations in resources (for example, reindeer-herd collapse). The occupants of Kostenki also imported high-quality stone from a source located roughly 150 kilometers to the southwest.

At Kostenki 14, imported stone was used to produce small bladelets, which offer the strongest link to the Early Ahmarian and related industries of the Levant, but the non-stone artifacts reveal more about adaptation to the East European Plain. They include bone awls and points, as well as what appear to be digging tools ("mattocks") fashioned from fragments of antler. And Kostenki 14 yielded what is thus far unique to modern human settlement of northern Eurasia before 40,000 years ago: a possible example of representational art in the form of a fragment of a human figurine carved from mammoth ivory.[104]

The habitation areas at Kostenki also reveal evidence for a significant expansion of diet beyond large mammals—achieved through the design and implementation of new technologies. Years ago, the lowest layer at Kostenki 14 yielded a concentration of more than 1,500 hare bones. Most had been fractured, and some exhibited traces of tool cuts.[105] They probably reflect large-scale harvesting (and likely for fur as well as food) because as individual prey, small mammals provide little return for the time and energy expended in hunting them. Hares might have been caught with nets, but this also consumes time and energy (and usually requires multiple participants). A more efficient method entails the designing and building of self-acting traps and snares, which can be deployed in large numbers across the landscape (see box 4.4).[106]

The expansion of the diet to include other small vertebrates, such as birds and fish, may have taken place at this time as well, but supporting evidence is weak. Some bird remains were recovered from the lowest levels of Kostenki 14, but they do not necessarily represent human food debris.[107] High stable-nitrogen-isotope (^{15}N) values measured in a human bone from another site on the opposite side of the ravine mouth (*Kostenki 1*) suggest the heavy consumption of freshwater foods (fish and/or waterbirds), but the bone was recovered from a layer above the CI tephra and thus is younger than 40,000 years old.[108] As noted at the beginning of the chapter, however, the 45,000-year-old human femur from Ust'-Ishim in western Siberia yielded a high ^{15}N value.

Evidence for technological adaptation to a cold climate on the East European Plain before 40,000 years ago also is weak. To date, eyed needles and traces of artificial structures have been found only in younger deposits at Kostenki and other sites on the southern and central plain. There is evidence, however, that hints at fire-making technology (which all recent hunter-gatherers of the Northern Hemisphere developed) in the form of traces of a rotary drill. With the aid of magnification, such traces are visible on stone objects from the lowest occupation level at *Kostenki 17* that were perforated with a hand-operated drill. A rotary drill is the most common device for generating fire (and, if used with a bow and/or socket, may be considered mechanical technology).

Another hint of adaptation to cold climate is the site of Mamontovaya kurya (northeastern Russia), located near the Arctic Circle and apparently contemporaneous to the Kostenki sites. The site contains a few simple artifacts (as well as a fragment of mammoth tusk marked with a series of fine incisions) and probably represents a brief warm-season occupation.[109] Nevertheless, it is the earliest known human adventure in the Arctic—and one that took place at a time when climates were somewhat cooler than in the present day—and further underscores the significance of the dispersal of modern humans in Europe during this period (43,500–40,000 years ago).

"Volcanic Winter" and Its Aftermath

The Campanian Ignimbrite volcanic eruption, roughly 40,000 years ago, had both short-term and long-term effects on Europe and beyond. As the ash plume spread across millions of square kilometers in southern and eastern Europe, it must have had a devastating impact on biota, including people. In places on the banks of the Don River (central Russia) where it was not subsequently deflated or washed away, the tephra layer measures up to 20 centimeters in thickness—almost 2,000 kilometers from its source.[110] The immediate effect on plant and animal life can be imagined.

The long-term effect on climate in the Northern Hemisphere was equally significant. The immense sulfur-rich ash plume inaugurated one of most intense periods of cold in the past 100,000 years—designated the Heinrich Event 4—which lasted for roughly 2,000 years.[111] Trees became scarce in some of the mildest parts of southwestern Europe, where there is evidence for the copious use of bone fuel in the hearths. On the East European Plain, the hiatus in settlement appears to have extended through this entire period. And when traces of human occupation reappear at Kostenki after about 38,000 years ago, they include stone artifacts typical of the Aurignacian industry, suggesting that post-HE4 settlement of the central plain was derived from western Europe. (This inference is supported by other sites of similar age with typical Aurignacian forms on the southwestern plain.) A contemporaneous human burial from Kostenki 14 is assigned to mtDNA haplogroup U and Y-DNA haplogroup C-M130.[112]

Because of its comparatively high biotic productivity, western Europe is a logical source for a post-HE4 population expansion following the return of warmer and wetter climates about 38,000 years ago (GI 8). On the basis of artifact density, associated faunal debris, and occupation area, Paul Mellars and Jennifer French estimated a 10-fold increase in population density during Aurignacian times in southwestern France.[113] The migration of this population out of western Europe may correspond to the spread of mtDNA haplogroup U, and the sudden appearance of Aurignacian artifacts in the Levant during this interval suggests that it included some back-migration into the original source area of the European population.[114]

NORTHERN ASIA AND THE ARCTIC

The Himalayan Plateau played a major role in the dispersal of modern humans across Eurasia. This uniquely elevated feature—sometimes described as the "roof of the world"—sits in the approximate center of the continent, where it effectively divided migrating populations into northern and southern route. As modern humans expanded eastward and out of the Levant, they were diverted northward into Siberia and Mongolia.[115]

The initial movement into northern Asia seems to have been part of the broader dispersal out of the Levant and Central Asia and across northern Eurasia that began roughly 50,000 years ago (or shortly thereafter, with the beginning of GI 12). This event presumably produced the north Eurasian meta-population comprising groups with close links to African parent lineages. Significantly, mtDNA haplogroup R has been identified among the earliest dated modern human remains in both Europe and Siberia (see table 6.1). Archaeologically, these groups are represented by the Levallois blade technology of the IUP, although the Asian sites also contain artifacts more typical of the later industries of modern humans (for example, bone points).[116] They also have yielded some evidence for technological adaptations to cold climates and reduced biological productivity that has yet to surface in the European IUP sites.

Within a few millennia, however, the Asian sites exhibit clear signs of independent development and divergence from the meta-population. By 45,000 years ago, if not earlier, the sites in southern Siberia reveal local changes in stone technology and tools that bear little resemblance to the Early Ahmarian and related industries of western Eurasia. In addition, estimated divergence dates for the major Asian lineages (that is, the breakup of the meta-population) are of comparable age. By 40,000 years ago, mtDNA haplogroup B was present in northern China, where it has been identified with aDNA extracted from human remains in *Tianyuan Cave*.[117]

Northern Asia contains the driest and coldest terrestrial environments known outside Antarctica. By the time eastward-flowing air from the North Atlantic reaches Siberia, it retains relatively little moisture. Climates in the interior of Northeast Asia are extremely continental, and some areas experience a January mean temperature of –30°C or lower. Plant productivity is generally low, and in the widespread boreal forest zone—where plants have evolved formidable defenses against herbivory—animal productivity is especially low.

With the exception of the Altai region of southwestern Siberia and northern China, *Homo sapiens* probably was the first human species to occupy this part of the world (figure 6.12). Both organizational and technological adaptations to local climates and biotas undoubtedly were necessary from the outset (and perhaps preceded such adaptations in western Eurasia by a few millennia). And perhaps it is significant that the earliest known year-round occupation of the Arctic began in northern Asia.

The Initial Upper Paleolithic

As in Europe, the oldest known evidence for modern humans in northern Asia is an archaeological proxy in the form of assemblages of Levallois blades that date to more than 45,000 years ago. They are similar to—and

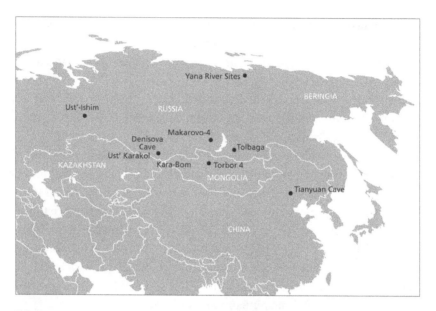

FIGURE 6.12 Map of Northeast Asia, showing fossils and archaeological sites assigned to *Homo sapiens*, dated to between roughly 55,000 and 35,000 years ago. (Base map adapted from en.wikipedia.org/wiki/File:Blank Map-World6.svg)

presumably derived from—assemblages of IUP artifacts in Central Asia and, ultimately, in the Levant. The earliest known human remains are slightly younger and are represented by the 45,000-year-old femur from Ust'-Ishim in western Siberia. The major archaeological sites are listed in table 6.8.

Key sites are found in the Altai Mountains, including the famous *Denisova Cave* (700 meters above sea level) and the open-air locality of *Kara-Bom* (1,100 meters). Both sites contain Levallois blades and points buried in layers that date to more than 45,000 years ago, along with some typical Upper Paleolithic stone-tool types such as end scrapers, burins, and gravers.[118] The emphasis on these forms in modern human sites reflects a major shift in technology. The microwear polish often seen on the end scrapers indicates their habitual use in the intense preparation of animal hides, an important early step in the production of clothing. The burins and gravers were used for working hard materials such as bone, antler, and ivory. These materials—rarely used by non-modern people for making artifacts—were shaped into various implements and devices, often comprising multiple components. To a much higher degree than their predecessors, modern humans were using simple stone tools to fashion more complex non-stone artifacts.

TABLE 6.8

Archaeological Record of *Homo sapiens* in Northern Asia, Before 30,000 Years Ago

Site, Layer	Estimated Age (method)	Description	Comments
Denisova Cave, Layer 11 (foothills of northern Altai Mountains)[a]	Ca. 50,000–35,000 years (radiocarbon, luminescence)	Levallois blades; microblades; points; side scrapers; end scrapers; burins; needles; drilled ornaments	Oldest eyed needle (?) associated with Denisovan bone (mixed in cryo-turbated sediments?)
Kara-Bom, Layers 2d–2a (foothills of northern Altai Mountains)[b]	Ca. 50,000–35,000 years (radiocarbon)	Early Upper Paleolithic: blade cores; retouched blades; points on blades; end scrapers; burins; bifaces; side scrapers	Pre-EUP assemblages as proxy for *Homo sapiens* (?)
Tolbor 4, Layers 6 and 5 (Mongolia)[c]	More than 41,050 years (uncalibrated radiocarbon)	Initial Upper Paleolithic: Levallois blades; carinated end scrapers on blades; side scrapers	Earliest IUP assemblage in Mongolia as proxy for *Homo sapiens* (?)
Makarovo-4 (Lake Baikal, south-central Siberia)[d]	More than 45,000 years (radiocarbon)	Blades (predominant); retouched blades; end scrapers; side scrapers; cobble choppers	Very early Upper Paleolithic assemblage
Ust' Karakol, Layers 11–8 (foothills of northern Altai Mountains)[e]	44,600–33,400 years (radiocarbon)	Blade cores; side scrapers; points on blades; bifaces; burins; retouched blades	Very early UP assemblage
Tolbaga (Lake Baikal, south-central Siberia)[f]	43,000–29,500 years (radiocarbon)	Retouched blades; points on blades; burins; side scrapers; needle fragments; animal carving (?)	Representational art (?) heavily disturbed by slope action; storage pits (?); dwelling (?)
Varvarina Gora (Lake Baikal, south-central Siberia)[g]	41,600–31,100 years (radiocarbon)	Blade cores; end scrapers; side scrapers; gravers; burins; retouched blades; two bone points	Storage pits; traces of large dwelling (?)
Shuidonggou, Localities 1 and 2 (north-central China)[h]	Ca. 41,000–34,000 years (radiocarbon)	Levallois blades; end scrapers; ostrich-eggshell beads; needle fragment	IUP in north-central China, but possibly younger than Siberian sites
Malaya Syia (Krasnoyarsk region, south-central Siberia)[i]	40,800–38,600 years (radiocarbon)	Blade cores; cobble choppers; end scrapers; retouched blades; antler points	Traces of dwelling (?); cold-climate fauna; pollen-spore data also indicate cold climate

(continued)

(continued)

Site, Layer	Estimated Age (method)	Description	Comments
Yana River sites (Lower Yana River, northeastern Asia/western Beringia)[j]	32,000 years (radiocarbon)	Blade cores; side scrapers; backed points; end scrapers; burins; needles; awls; beads; ornaments; ivory vessels	Colonization of Arctic and Beringia; harvesting of small mammals (traps/snares?)

Note: All radiocarbon dates are given in calibrated radiocarbon years before present.

[a]A. P. Derevianko, M. V. Shunkov, and S. V. Markin, *Dinamika Paleoliticheskikh Industrii v Afrike i Evrazii v Pozdnem Pleistotsene i Problema Formirovaniya* Homo sapiens (Novosibirsk: Institute of Archaeology and Ethnography, SB RAS Press, 2014); Evgeny P. Rybin, "Middle and Upper Paleolithic Interactions and the Emergence of 'Modern Behavior' in Southern Siberia and Mongolia," in *Emergence and Diversity of Modern Human Behavior in Paleolithic Asia*, ed. Yousuke Kaifu et al. (College Station: Texas A&M University Press, 2015), 470–489.

[b]Ted Goebel, Anatoli P. Derevianko, and Valerii T. Petrin, "Dating the Middle-to-Upper-Paleolithic Transition at Kara-Bom," *Current Anthropology* 34 (1993): 452–458; Ted Goebel, "The Early Upper Paleolithic of Siberia," in *The Early Upper Paleolithic Beyond Western Europe*, ed. P. Jeffrey Brantingham, Steven L. Kuhn, and Kristopher W. Kerry (Berkeley: University of California Press, 2004), 162–195; Rybin, "Middle and Upper Paleolithic Interactions and the Emergence of 'Modern Behavior' in Southern Siberia and Mongolia."

[c]Jacques Jaubert, "The Paleolithic Peopling of Mongolia: An Updated Assessment," in *Emergence and Diversity of Modern Human Behavior in Paleolithic Asia*, ed. Kaifu et al., 453–469.

[d]Goebel, "Early Upper Paleolithic of Siberia"; Kelly E. Graf, "Siberian Odyssey," in *Paleoamerican Odyssey*, ed. Kelly E. Graf, Caroline V. Ketron, and Michael R. Waters (College Station: Texas A&M University Press, 2013), 65–80; Rybin, "Middle and Upper Paleolithic Interactions and the Emergence of 'Modern Behavior' in Southern Siberia and Mongolia."

[e]Goebel, Derevianko, and Petrin, "Dating the Middle-to-Upper-Paleolithic Transition at Kara-Bom"; Goebel, "Early Upper Paleolithic of Siberia"; Graf, "Siberian Odyssey."

[f]Goebel, "Early Upper Paleolithic of Siberia"; Graf, "Siberian Odyssey"; Ian Buvit et al., "Last Glacial Maximum Human Abandonment of the Transbaikal," *PaleoAmerica* 1 (2015): 374–376.

[g]Goebel, "Early Upper Paleolithic of Siberia"; Graf, "Siberian Odyssey"; Buvit et al., "Last Glacial Maximum Human Abandonment of the Transbaikal."

[h]Feng Li et al., "The Development of Upper Palaeolithic China: New Results from the Shuidonggou Site," *Antiquity* 87 (2013): 368–383.

[i]Goebel, "Early Upper Paleolithic of Siberia"; Graf, "Siberian Odyssey."

[j]V. V. Pitulko et al., "The Yana RHS Site: Humans in the Arctic Before the Last Glacial Maximum," *Science* 303 (2004): 52–56; Vladimir Pitulko et al., "Human Habitation in Arctic Western Beringia Prior to the LGM," in *Paleoamerican Odyssey*, ed. Graf, Ketron, and Waters, 13–44.

Associated with the IUP assemblage in Layer 11 at Denisova Cave are a variety of non-stone items, including both eyed needles and awls made of bone and a remarkable array of personal ornaments. The ornaments include drilled pendants made from the teeth of foxes and other mammals, pipe-shaped beads fashioned from small tubular bones decorated with incised lines, and a ring produced from ornamented mammoth tusk.[119] The eyed needle may be the oldest known in the world. There is, however, some uncertainty regarding the dating and association of these items because Layer 11, which yielded dates ranging from about 50,000 to 30,000 years ago, contains materials from multiple occupations, including those of non-modern humans; the Denisovan genome was obtained on a bone from Layer 11.[120] The sediments and their contents probably have been mixed by repeated freezing and thawing within the confines of the cave.

Regardless of the dating of the needle from Denisova Cave, there can be little doubt that the modern human inhabitants of Siberia had developed insulated clothing and other technological adaptations to extreme winter conditions by 45,000 years ago. As noted at the beginning of the chapter, the Ust'-Ishim femur was recovered from an area where the current January mean temperature is lower than –15°C. As in eastern Europe, evidence for a rotary drill in the making of ornaments indicates the use of the core component of fire-making technology. Traces of an artificial shelter (delineated with rocks) are reported from the open-air site of *Kamenka A*, located near Lake Baikal.[121] Non-stone artifacts have turned up in IUP assemblages where the potential for mixture appears lower, including bone points at *Ushlep 6*, in the northern Altai region.[122]

Although the expansion of the diet to include small mammals, birds, and fish has yet to be documented in associated faunal remains, stable-isotope analysis of the Ust'-Ishim bone suggested the heavy consumption of freshwater foods.[123] This almost certainly indicates technological adaptations to reduced biological productivity, although the specifics are unknown (for example, nets, gorges, or weirs?). Organizational adaptations to scarce resources, such as expanded foraging ranges and widespread social networks, are suggested by the long-distance movement of materials and the considerable investment in ornamentation. A bracelet from Layer 11 at Denisova Cave was made from stone brought from a source at least 250 kilometers from the site. Even more impressive—and evoking images of sub-Saharan Africa—is the presence of ostrich-eggshell beads in the same layer, presumably transported from Mongolia (the nearest possible source of ostrich eggs) over as much as about 500 kilometers.[124] While the association of these items with the IUP remains problematic, they probably date to well before 30,000 years ago.

The IUP may have extended into northern China, although it currently dates to a somewhat younger time range there. At the open-air sites of *Shuidonggou*, on a tributary of the Yellow River (and the edge of the Ordos Desert) in north-central China at latitude 38° North, Levallois blades (some of which were struck with soft hammers) are found with end scrapers and few burins and other stone tools. Other items include ostrich-eggshell beads and a bone-needle fragment. New dates on the assemblages of Levallois blades fall in the range of 41,000 to 34,000 years ago.[125]

Adaptation to the Arctic

The analysis of the genetics of living peoples suggests a relatively early split of west and east Eurasian lineages (more than 40,000 years ago), and this appears to be confirmed by aDNA. If the 45,000-year-old Ust'-Ishim individual represents a lineage that was part of the original meta-population derived from western Eurasia, the skeleton from Tianyuan Cave confirms the divergence of one of the major Asian maternal lineages (mtDNA haplogroup B) by 40,000 years ago.[126]

A divergence from the west Eurasian pattern also is evident in the archaeology of sites in northern Asian that are dated to after 45,000 years ago. Although the stone artifacts are unlike the bladelet-dominated assemblages of the Early Ahmarian and related industries, they are more typical of the Upper Paleolithic in northern Eurasia in general. At *Makarovo-4*, in the Lake Baikal region, *pointed blades* with bifacially thinned and tanged bases appeared at about this time. They are particularly characteristic of the post-IUP industry in southern Siberia. End scrapers, including "micro–end scrapers," and burins are common.[127] Bone awls and needles are reported from *Tolbaga* and *Varvarina Gora* in the same region.[128] Both of these sites yield evidence for storage (food?) in the form of stone-lined pits. The storage of foods harvested during the warmer months is a common strategy among recent hunter-gatherers who live in northern environments for coping with seasonal fluctuations in resource availability.[129]

If the later phases of the warm GI 12 (ca. 48,000–45,000 years ago) and succeeding GI 11 (ca. 43,500–42,500 years ago) represent periods of modern human expansion in Europe and northern Asia, there is little evidence that the cold HE4 (ca. 40,000–38,000 years ago) had the same severe impact on the Asian population that it did on west Eurasian groups. But GI 8, the warm period that followed HE4, seems to have triggered another expansion at both ends of the continent. In East Asia, the most striking development after 38,000 years ago is the seemingly rapid colonization of the Japanese archipelago, where an astonishing number of sites—roughly 500 open-air localities—are dated to the Early Upper Paleolithic (EUP), or before 30,000 years ago. Typical stone tools include geometric forms ("trapezoids") and pointed blades. A unique feature of the EUP in Japan is the widespread use of deep pits for trap-hunting.[130]

Greenland Interstadial 8 also may have encouraged the first year-round occupation of the Arctic. Although the oldest known sites date to no earlier than about 32,000 years ago,[131] problems of archaeological visibility and sampling render it unlikely that they represent the earliest colonization of latitudes above 66° North, and it may be noted that modern humans recently have been documented as far as 57° North before the end of GI 12 (although not necessarily on a year-round basis).[132] By 32,000 years ago, the Northern Hemisphere had entered another cold period (HE3), and it seems likely that—given their pattern of climate-mediated dispersal in northern Eurasia—modern humans had occupied the Arctic during an earlier warm interval.

Arguably the most important archaeological discovery of the twenty-first century, the open-air sites found near the mouth of the Yana River in northeastern Siberia not only document the presence of modern humans in the Arctic before 30,000 years ago but provide a relatively detailed picture of life in this setting.[133] The sites are found at roughly 71° North (several hundred kilometers above the Arctic Circle), where the January mean temperature is about –38°C. Today, the area lies within the zone of wet tundra, with woody shrubs and some trees (larch). Six sites now are identified on the second terrace level on the northern side of the river, near the confluence of a large tributary. Three of them contain buried occupation debris radiocarbon-dated to roughly 32,000 years ago. The artifacts are in alluvial silt with traces of grass rootlets and mammal bones, and the silt exhibits substantial frost disturbance.[134]

Although the effects of frost action have obscured most of the occupation floor patterns (for example, traces of artificial shelters), former hearths—comprising red ash overlying "charred fatty dirt with tiny grains of charcoal"—are present. Thousands of stone tools—including both simple expedient forms such as scrapers on flakes and partially flaked bifaces, as well as a number of finely crafted implements on small flakes and blades—were recovered from areas around the hearths. The most impressive finds are the non-stone artifacts, which include eyed needles (sometimes decorated), possible needle cases, awls, pendants, mammoth-bone beads (possibly sewn onto clothing), ivory rods, and decorated ivory vessels.[135] If properly identified, the needle cases would be the oldest known in the world.

When the discovery on the Yana River was initially reported a decade ago, it was assumed to represent a short-term camp occupied by people who trekked north each summer to exploit seasonally available resources (such as migratory waterfowl). The same pattern seemed to be evident in the European Arctic (that is, seasonal forays into the tundra by people whose economy was based primarily on resources in the temperate zone).[136] But the accumulated evidence from the Yana River sites suggests a year-round presence at high latitudes, and a northern-interior economy comparable to that of the Yukaghir and Athapaskans. The large-mammal remains are primarily of bison and reindeer, and the latter include skulls with shed and unshed antlers, indicating multiple seasons of death. The small-mammal remains are dominated by hare, and many are represented by complete skeletons, suggesting harvesting (with snares or traps) for pelts rather than food.[137] The profusion of household items, including vessels and sewing needles, argues for family groups—not a seasonal hunting party—and extended habitation (figure 6.13).

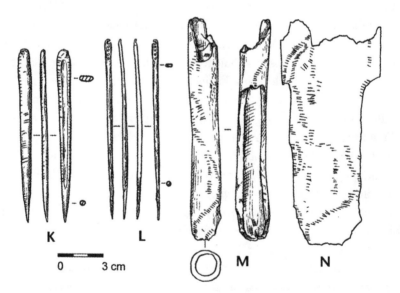

FIGURE 6.13 Non-stone artifacts from Yana River, including (K) a polished awl with incisions, (L) a large eyed needle with dot marks near the eye, and (M) an ornamented needle case made from bone with a pattern of small incisions, which is shown unrolled (N). (From Vladimir Pitulko et al., "Human Habitation in Arctic Western Beringia Prior to the LGM," in *Paleoamerican Odyssey*, ed. Kelly E. Graf, Caroline V. Ketron, and Michael R. Waters [College Station: Texas A&M University Press, 2013], 32, fig. 2.12. Reproduced with permission from Texas A&M University Press)

The Yana River sites also document the early presence of modern humans in the western half of the subcontinent of *Beringia*, which was created by the exposure of the shallow platform between Chukotka and Alaska during periods of lowered sea level. It is unclear if their settlement extended across Beringia to what is now part of North America, although no barriers to such colonization are known. But access to the rest of the Western Hemisphere was completely blocked by what was at the time the largest mass of glacial ice on the planet.

The presence of people in Beringia before the beginning of the LGM may be critical to explaining the genetics of Native American populations, most of which appear to be the descendants of groups from Northeast Asia that were isolated from their parent lineages during the LGM.[138] It seems increasingly likely that these groups were isolated in Beringia itself, until climates began to warm after 17,000 years ago. As the glacial ice masses retreated between 15,000 and 14,000 years ago, the ancestors of Native Americans resumed the global dispersal of modern humans.

Chapter Seven

GLOBAL DISPERSAL

Beringia and the Americas

[T]hey crossed not by sailing on the sea but by walking on land.

—JOSÉ DE ACOSTA (1590)

Twenty-eight thousand years ago, the dispersal of modern humans in Eurasia came to an end with the greatest environmental disaster in human history. The Last Glacial Maximum (LGM) was a period of intense and sustained cold in the Northern Hemisphere that led to the expansion of massive ice sheets and widespread effects on plant and animal life. It also affected environments of the middle and lower latitudes.

Global ice volume during the LGM is estimated to have been roughly 50 million cubic kilometers greater than that of the present day.[1] Northwestern Europe was largely covered by ice, while the immense coalesced ice mass over northern North America also experienced substantial growth.[2] At the peak of the LGM (roughly 21,000 years ago), sea level fell to 134 meters below where it is now.[3] Mean global temperature is estimated to have been 5.1°C lower and precipitation roughly 10 percent lower than they are today.[4]

The impact on plant productivity (or net primary productivity [NPP]) and the secondary effects on animal biomass were significant. The key variable was moisture (cold air has less capacity to hold moisture than does warm air), but reduced carbon dioxide in the atmosphere also lowered plant production. Computer modeling of LGM vegetation suggests a major expansion of deserts (36 percent greater than those of the present interglacial) and contraction of grasslands (40 percent less than those today). Tropical forest was an estimated 20 percent less extensive and extra-tropical forest an estimated 45 percent less than those in the recent past. Only shrubs expanded relative to their postglacial distribution. *Global NPP is estimated to have been slightly more than 50 percent of what it was during the postglacial epoch* (before the industrial era).[5]

The impact on the global human population is most clearly illustrated by the genetics of living peoples, with some supporting ancient DNA (aDNA). As already mentioned in earlier chapters, the living people of Africa, Europe, and Asia carry genetic evidence of population bottlenecks related to the effects of LGM climate on environments. Areas in the temperate zone and lower latitudes that now are arid—such as North Africa, the Arabian Peninsula, and upland zones in South Asia—apparently were abandoned.[6] Much of western and eastern Europe was evacuated, as populations retreated south into multiple isolated refugia, and parts of northern Asia also seem to have been abandoned.[7]

The evidence of population bottlenecks in the genetic data is supported by patterns in the archaeological (and human fossil) record. At stratified sites in some of the aforementioned areas, a significant hiatus in occupation may be observed in deposits dating to the LGM.[8] On a regional basis, a gap or depression in the radiocarbon chronology corresponding to the LGM (or to the coldest phase of the LGM) may be recognized, while credible evidence for occupations of LGM age are confined to places where climates were warmer and/or wetter at the time (that is, the refugia). The pattern is well documented in Australia (figure 7.1).[9]

Why were human populations forced to abandon so much of their habitat during the LGM? The simplest explanation is that plant and animal productivity fell below the minimum threshold level for sustaining a human population. If groups in regions that later were vacated already were foraging across extensive areas—tens of thousands of square kilometers—each year (chapter 6), the sharp decline in NPP probably pushed them over the edge of sustainability. In low-latitude deserts, the scarcity of water alone may have been sufficient to preclude human settlement. In the higher latitudes, other factors may have come into play. Despite the development of insulated winter clothing, the extreme low temperatures may have been too much for people who retained the warm-climate anatomy of their recent ancestors.[10] The absence of wood in some places also may have been an obstacle to year-round settlement; experimental research reveals that a modest quantity of wood is necessary to render bone practical as a fuel.[11]

FIGURE 7.1 Major areas abandoned by human populations during the peak cold and aridity of the Last Glacial Maximum (ca. 21,000 years ago). (Base map adapted from en.wikipedia.org/wiki/ File:Blank Map-World6.svg)

Native Americans are descended from a group of lineages that were genetically isolated from Asian parent populations at the time of the LGM (see table 2.1).[12] At least some of these lineages apparently were confined to a geographic refugium during the LGM (or during most of the LGM). With one exception, they did not reunite with other Asian groups after the LGM, but dispersed directly into North and South America, indicating that they were isolated in either Beringia or another location in Northeast Asia that provided access to the Western Hemisphere—but not to other human populations—after the climate warmed (ca. 16,000 years ago).[13]

Despite its high latitude (mostly above 60° North), Beringia contained a plausible LGM refugium for the ancestral Native American population.[14] Several lines of evidence indicate that the south-central lowland, most of which was inundated by the rising Bering Sea after the LGM, supported a shrub-tundra environment with some trees. An analysis of the insect fauna in sediment cores extracted from the former land surface reveals surprisingly mild temperatures between 28,000 and 18,000 years ago.[15] The reason for this almost certainly lies in the effect of the North Pacific circulation, which brought relatively warm and moist air to southern Beringia.[16] Other areas above 60° North either were covered with ice or were exposed to extreme cold and aridity during this period. There is little archaeological evidence for the occupation of Beringia during the LGM, but this may be due primarily to limited visibility, especially if sites were concentrated in the central lowland, which now is under water.[17]

After climates began to ameliorate 17,000 to 16,000 years ago, groups in Africa, Eurasia, and Australia expanded out of their refugia and reoccupied the areas abandoned during the LGM. As NPP and animal biomass rose, human populations experienced rapid growth, which is reflected in their genetics (increased diversity). Archaeological evidence, including some skeletal remains, documents humans' return to northern Europe, central Siberia, and the Arctic. In the middle and lower latitudes, sites reappeared in North Africa and the interior of Australia. In places, the archaeological record contains traces of large settlements comprising multiple dwellings and new technologies (for example, ceramic vessels in East Asia).[18]

In Beringia, shrub-tundra vegetation expanded out of the south-central lowland and into what is now Chukotka (and farther west) and central Alaska, reflecting increased moisture in these areas. If the ancestral Native American population was isolated in Beringia during the LGM, it presumably would have followed the spreading shrub tundra east- and westward. The genetic and linguistic data suggest movements in both directions, and archaeological sites in Northeast Asia and Alaska yield artifacts that date to around 14,000 to 13,000 years ago and lack any obvious antecedents outside Beringia.[19] These assemblages may represent the descendants of the Beringian population during the LGM and the ancestors of the people who initially colonized North and South America.

Dispersal in the Americas was triggered by the retreat of the massive glaciers that had covered most of northern North America, opening up a coastal and—somewhat later—an interior route southward after 15,000 years ago. Genetic data indicate a dispersal date of about 15,000 years ago or slightly less,[20] and it appears increasingly likely that the migration began on the Northwest Pacific coast.[21] Although no archaeological traces of people older than 13,300 years are known on the Northwest coast (or from the interior "ice-free corridor" that eventually opened up in western Canada), isolated sites in mid-latitude North America antedate 14,000 years.[22] An increasing number of aDNA samples extracted from skeletal remains older than 10,000 years provide a means of ground-truthing the predictions of the genetic analyses in North and Central America (table 7.1).

TABLE 7.1
Human Remains and Ancient DNA from Early Sites in North America

Site	Estimated Age (method)	Description	Genetics
Paisley 5 Mile Point Caves (Oregon, U.S.)[a]	14,340 years (radiocarbon)	Five human coprolites	mtDNA haplogroup A; mtDNA haplogroup B
Arlington Springs (Santa Rosa Island, California, U.S.)[b]	Ca. 13,000 years (radiocarbon)	Burial of complete skeleton (adult male)	mtDNA haplogroup B (probably contamination)
Hoyo Negro (Yucatán Peninsula, Mexico)[c]	Ca. 12,910–11,750 years (U/Th, radiocarbon)	Nearly complete skeleton, including skull with complete dentition	mtDNA D1
Anzick (Montana, U.S.)[d]	12,707–12,556 years (radiocarbon)	Cranium; four bones (infant)	mtDNA D4H3a; Y-DNA Q-L54 (xM3)
Arch Lake (Texas, U.S.)[e]	11,640–11,260 years (radiocarbon)	Burial of complete skeleton (adult female)	aDNA could not be extracted
Upward Sun River (central Alaska, U.S.)[f]	Ca. 11,500 years (radiocarbon)	Two burials of infants	mtDNA B2; mtDNA C1b
Wizard's Beach (Nevada, U.S.)[g]	Ca. 10,560–10,250 years (radiocarbon)	Skull with partial dentition; partial skeleton	mtDNA C1
On Your Knees Cave (southeastern Alaska, U.S.)[h]	Ca. 10,500–10,250 years (radiocarbon)	Mandible; other fragments (carnivore damage)	mtDNA D43a; Y-DNA Q1a3a1a
Hourglass Cave (Colorado, U.S.)[i]	Ca. 10,000 years (radiocarbon)	Partial skeleton, including calvarium, isolated teeth, femora, humerus, ribs, and others	mtDNA haplogroup B

Site	Estimated Age (method)	Description	Genetics
Windover (Florida, U.S.)[j]	Ca. 10,000 years (radiocarbon)	Multiple skeletons (cemetery)	mtDNA haplogroup X (probably contaminated)
Kennewick (Washington, U.S.)[k]	8,690–8,400 years (radiocarbon)	Partial skeleton	mtDNA haplogroup X2a; Y-DNA haplogroup Q-M3

Note: All radiocarbon dates are given in calibrated radiocarbon years before present.

[a]M. Thomas P. Gilbert et al., "DNA from Pre-Clovis Human Coprolites in Oregon, North America," Science 320 (2008): 786–789; Dennis L. Jenkins et al., "Geochronology, Archaeological Context, and DNA at the Paisley Caves," in Paleoamerican Odyssey, ed. Kelly E. Graf, Caroline V. Ketron, and Michael R. Waters (College Station: Texas A&M University Press, 2013), 485–510.

[b]Jon M. Erlandson, Madonna L. Moss, and Matthew Des Lauriers, "Life on the Edge: Early Maritime Cultures of the Pacific Coast of North America," Quaternary Science Reviews 27 (2008): 2232–2245.

[c]James C. Chatters et al., "Late Pleistocene Human Skeleton and mtDNA Link Paleoamericans and Modern Native Americans," Science 344 (2014): 750–754.

[d]Morten Rasmussen et al., "The Genome of a Late Pleistocene Human from a Clovis Burial Site in Western Montana," Nature 506 (2014): 225–229.

[e]Douglas W. Owsley et al., Arch Lake Woman: Physical Anthropology and Geoarchaeology (College Station: Texas A&M University Press, 2010).

[f]Ben A. Potter et al., "New Insights into Eastern Beringian Mortuary Behavior: A Terminal Pleistocene Double Infant Burial at Upward Sun River," Proceedings of the National Academy of Sciences 111 (2014): 17060–17065; Justin C. Tackney et al., "Two Contemporaneous Mitogenomes from Terminal Pleistocene Burials in Eastern Beringia," Proceedings of the National Academy of Sciences 112 (2015): 13833–13838.

[g]Frederika A. Kaestle and David Glenn Smith, "Ancient Mitochondrial DNA Evidence for Prehistoric Population Movements: The Numic Expansion," American Journal of Physical Anthropology 115 (2001): 1–12.

[h]Brian M. Kemp et al., "Genetic Analysis of Early Holocene Skeletal Remains from Alaska and Its Implications for the Settlement of the Americas," American Journal of Physical Anthropology 132 (2007): 605–621.

[i]Anne C. Stone and Mark Stoneking, "Genetic Analyses of an 8000-Year-Old Native American Skeleton," Ancient Biomolecules 1 (1996): 83–87.

[j]David Glenn Smith et al., "Mitochondrial DNA Haplogroups of Paleoamericans in North America," in Paleoamerican Origins: Beyond Clovis, ed. Robson Bonnichsen et al. (College Station: Texas A&M University Press, 2005), 243–254.

[k]Morten Rasmussen et al., "The Ancestry and Affiliations of Kennewick Man," Nature 523 (2015): 455–458.

Both the genetics of living Native Americans and the archaeological record reflect an unexpectedly quick dispersal throughout the Western Hemisphere. People reached the southern tip of South America in less than 2,000 years—a cave in Patagonia contains artifacts dated to about 13,000 years ago.[23] Unlike the modern humans who spread across southern Asia, Native Americans did not have to compete with other forms of Homo. And in contrast to the people who colonized northern Eurasia, they did not have to cope with increasingly cold temperatures and resource-poor habitats. In some respects, the colonization of the Americas reversed the pattern of the earlier dispersal into northern Eurasia, as foraging groups from Beringia spread southward into a land of warmth and abundance.

Native American Genetics and the Beringian
Standstill Hypothesis

Forty-five years ago, the evolutionary geneticist Richard Lewontin observed that Native Americans (along with Native Australians and Oceanians) exhibit significantly lower genetic diversity than do Africans and Eurasians (chapter 1). He suggested that the phenomenon may be related to their population history.[24] Since the early 1990s, a growing body of research has filled out the picture. Native Americans are a subset of the Eurasian genome, just as the latter is a subset of the African genome, reflecting the same pattern of reduced diversity as a function of distance from its parent genome.[25]

Native Americans nevertheless exhibit a substantial amount of genetic diversity, which is matched by a high degree of linguistic diversity,[26] and this seeming paradox has driven much of the debate during the past several decades about their origin. The evidence for a severe reduction in genetic diversity ("genetic bottleneck") among Native Americans relative to Eurasians is lacking: an analysis conducted in 2007 of single-nucleotide polymorphisms (SNPs) among 24 Native American populations found only a 6 to 7 percent reduction of heterozygosity relative to the global average.[27] Significantly, a recent analysis of mitochondrial DNA (mtDNA) variation with a comparatively large aDNA sample ($n = 92$ ancient mitogenomes) revealed much greater genetic diversity than previously estimated for the pre-contact American population on the basis of living Native Americans.[28] The catastrophic impact of European colonization has masked the true degree of diversity among the original Native American population.

In 1986, Joseph H. Greenberg (1915–2001), Christy Turner II (1933–2013), and Stephen Zegura published a controversial paper on the origin of Native Americans based on a synthesis of the linguistic, dental, and—then rather limited—genetic data. They identified three major groups within the Native American population—Aleut-Eskimo (or Aleut-Inuit), Na-Dené (speakers of *Athapaskan-Eyak*, *Tlingit*, and *Haida*), and "Amerind" (all other Native Americans)—and explained the differentiation as the result of three separate migrations (beginning at the end of the Pleistocene, or about 13,000 years ago) from Northeast Asia to the Americas.[29] Linguists were especially critical of the "three-wave model" because it failed, in their view, to acknowledge the degree of linguistic diversity among Native Americans.[30] Some of them argued that more than 13,000 years would have been needed to account for the existing variation among North and South American languages.

In several papers written both before and after the publication of the "three-wave model," Emőke J. E. Szathmary concluded that genetic and anatomical differences among Native Americans had been overstated.[31] She suggested that biological differentiation among various groups had developed *after* their arrival in the Western Hemisphere and speculated that the coalescence of the North American ice sheets during the LGM had isolated people in Beringia from people in mid-latitude North America and farther south. Szathmary's suggestion that people had inhabited the Americas before the LGM never found much support in the archaeological record, but her conclusion that all Native Americans, including the Aleut-Inuit and Na-Dené, represented a unified whole (relative to other populations) has been bolstered by some—although not all—recent genetic and dental analyses.

In the 1990s, studies of mtDNA began to yield a comprehensive portrait of the maternal lineages among living Native Americans, revealing links with and estimated divergence times from their source populations in Asia. The four major lineages—mtDNA haplogroups A, B, C, and D—are subsets of haplogroups widely represented today in Central and Northeast Asia.[32] A fifth lineage, mtDNA haplogroup X, which is much less common in living Native American populations, is rare but present in Eurasia (including the Altai region of southwestern Siberia).[33] In 2015, mtDNA X was identified in the skeletal remains from *Kennewick* (Washington), which are dated to between 8,690 and 8,400 years ago.[34]

Likewise, the major paternal lineages—Y-chromosome DNA (Y-DNA) haplogroups Q, C, and P—are subsets of Asian haplogroups, and it may be noted that Q and C also appear to have ties with ancient lineages in the Altai region.[35] In general, the Western Hemisphere is characterized by less extensive population movements after the initial dispersal than are Africa and Eurasia, but paternal lineages exhibit recent admixture from foreign sources, especially in some regions (for example, Greenland).[36]

Some of the estimated divergence times of Native American mtDNA lineages from their Asian source populations reported in the 1990s were surprisingly old—in the range of 30,000 years ago or more, which exceeded estimates based on the archaeological data (that is, dated sites in North and South America) by an order of magnitude.[37] In 1997, geneticists Sandro Bonatto and Francisco Salzano, who estimated a divergence between Asian and American maternal lineages at 43,000 to 30,000 years ago, suggested that some Native Americans had become isolated in Beringia during the LGM:

> The "three-wave" and most other colonization models regard Beringia just as a "corridor" or a "bridge" for a migration to America, centering the origin of the different migration(s) somewhere in Asia. Our results support a different model that, on the contrary, puts Beringia in a central role, where the population that originated the Native Americans settled and diversified before the further colonization of the rest of the American continent. It could be called an "out of Beringia" model.[38]

Although there was no archaeological support in 1997 for a pre-LGM occupation of Beringia, later discoveries documented year-round settlement near the mouth of the Yana River in northwestern Beringia by 32,000 years ago. Following Szathmary's earlier suggestion, Bonatto and Salzano proposed that the Beringians had been isolated from other Native Americans by the coalescence of the North American ice sheets. While compelling archaeological evidence for Native Americans in mid-latitude North America before the LGM remains elusive, the idea of Beringia as a place "where the population that originated the Native Americans settled and diversified" acquired increased credibility after the discovery of the sites near the Yana River (chapter 6).

A decade after the publication of Bonatto and Salzano's paper, an international team of geneticists proposed a somewhat different model, Beringian Standstill Hypothesis (or Beringian Incubation Model), based on the analysis of a large sample of complete mtDNAs from Asia and the Americas. Erika Tamm and her colleagues argued that the ancestral Native American population had been isolated in Beringia from its Asian parent lineages during the LGM, only later dispersing in the Western Hemisphere after the retreating glaciers opened access to mid-latitude North America.[39] The Beringian Standstill Hypothesis was grounded on the observation that the maternal lineages accumulated a large number of mtDNA mutations *after* their divergence from the Asian parent lineages (more than 25,000 years ago) but *before* the population dispersed throughout North and South America and local groups began to accumulate their own set of mutations (less than 15,000 years ago). The authors noted that some movement of people from Asia into Beringia took place after the LGM and that one subclade (C1a) found today in Asia apparently represents a group that migrated back into Northeast Asia after 14,000 years ago (figure 7.2).[40]

The following year, Andrew Kitchen, Michael Miyamoto, and Connie Mulligan reconstructed the history of Native American populations on the basis of statistical analyses of both mitochondrial- and nuclear-DNA sequences. They identified an early phase of slow population growth, roughly 40,000 years ago, associated with the initial expansion into Northeast Asia and Beringia. Between about 36,000 and 16,000 years ago, the "standstill" population was stable and isolated (as indicated by a significant accumulation of new mutations). The authors calculated an effective population size of approximately 1,000 to 5,400 individuals and, although unable to specify a geographic region, suggested that "it may have occupied the large region from Siberia to Alaska, most of which is currently underwater."[41] After 15,000 years ago, the population began a phase of rapid growth associated with the equally rapid dispersal throughout the Americas.

The Beringian Standstill Hypothesis reconciles mtDNA analyses with the archaeological data (pre-LGM divergence of Native Americans from their Asian source populations, but post-LGM dispersal in mid-latitude

FIGURE 7.2 The Beringian Standstill Hypothesis, which postulates (*1*) an early occupation of Beringia, followed by (*2*) a genetically isolated refugium population during the Last Glacial Maximum and (*3–4*) post-LGM expansions into both Northeast Asia and North America. (From Erika Tamm et al., "Beringian Standstill and Spread of Native American Founders," *PLoS ONE* 9 [2007]: e829, fig. 2. Reproduced from *PLoS ONE* under the terms of the Creative Commons Attribution [CCBY] license)

North America and South America). It explains the otherwise puzzling degree of genetic and linguistic diversity of the people who dispersed rapidly throughout the Americas after 15,000 years ago (they were not derived from a small subset of the Northeast Asian population 15,000 years ago, but from a relatively large and diverse group that had been isolated from its Asian source for thousands of years). And it accounts for the absence of archaeological evidence for a direct connection between Northeast Asia and the Americas at the time that the dispersal.[42]

Beringian Standstill suggests a parallel between the peopling of the Americas and the reoccupation of areas in Africa, Eurasia, and Australia abandoned during the arid LGM. It adds Beringia (or some portion of Beringia) to the list of refugia occupied by various groups between roughly 28,000 and 16,000 years ago. Although Tamm and her colleagues never addressed the ecology of the Standstill Hypothesis, paleobotanists and paleozoologists had been discussing the existence of a wet shrub-tundra refugium in central Beringia for decades. Computer modeling of NPP in central Beringia during the LGM indicates that it probably was sufficient to support the "standstill" population estimated by Kitchen, Miyamoto, and Mulligan. [43]

A whole-genome survey of Native Americans published in 2012 analyzed more than 370,000 SNPs from a sample of 55 local populations and found that the Inuit and speakers of Na-Dené languages could be separated from the remaining indigenous American groups.[44] The results, the authors concluded, supported the "three-wave model" proposed by Greenberg, Turner, and Zegura three decades ago. Other analyses of autosomal DNA have yielded results more consistent with Szathmary's idea of a unified Native American population.[45] More important in this context may be a reanalysis in 2016 of teeth (a major component of the three-wave model) of Native Americans, Asians, and Native Australians by G. Richard Scott and his colleagues that yielded a pattern similar to that of the mtDNA data reported by Tamm and her colleagues: an inferred early separation of Native Americans from Asians and relatively uniform measures of distance among the former.[46]

The Beringian Standstill Hypothesis remains to be confirmed by data from Beringia (that is, evidence of the ancestral Native American population occupying one or more Beringian refugia during the LGM), but has provided new avenues of research for all the disciplines—genetics, physical anthropology, archaeology, and linguistics—engaged in this work.[47] Research with mtDNA has identified as many as 16 Native American founding maternal lineages, underscoring the size and diversity of the population that dispersed out of Beringia 15,000 years ago.[48] Analyses of Y-DNA also found evidence for diversification of paternal sublineages in Beringia before the dispersal.[49] And a linguistics study conducted in 2014 concluded that Na-Dené speakers migrated out of Beringia and back into northern Asia (Altai region) after the LGM, thus supporting one element of the model.[50] A reanalysis of materials excavated decades ago from several small caves in the southern Yukon (northeastern Beringia), published in January 2017, concluded that they contained evidence of human occupation during the LGM, although it remains to be shown that the occupants were part of the "standstill" population.[51]

THE LOST WORLD OF BERINGIA

Throughout the long and bitter academic battle over the settlement of the Americas, one fundamental assumption has remained unchanged until now. Archaeologists, linguists, and others have assumed that the Native American population is derived from a known Old World group that migrated across the Bering Land Bridge or by another route. The idea that Native Americans may have originated in central Beringia itself—from a place that no longer exists and from a population that has yet to be documented by a single bone or artifact—is a new one. The hypothesized Beringian population would, of course, have been derived from a Northeast Asian group, but would have developed independently for thousands of years, presumably acquiring a unique local character.

The emergence of an out-of-Beringia model for the origin of Native Americans reflects the considerable impact of genetics research on paleoanthropology, especially on the study of modern humans—a major theme of this book. It parallels the use of ancient and modern DNA data to reconstruct the Denisovans and their interactions with modern humans in Asia (chapter 2) and the migration of people affiliated with mtDNA haplogroup M into Europe before 40,000 years ago (chapter 6). The model integrates the settlement of the Americas with the out-of-Africa dispersal of modern humans, because it ties the Native American population to the occupation of Beringia and the Arctic more than 30,000 years ago.[52]

Beringia: A Working Definition

The term *Beringia* was proposed in 1937 by the Swedish botanist Eric Hultén (1894–1981). From the outset, it was linked to the idea that the flooded lowland between Northeast Asia and Alaska had been a refugium for shrub-tundra vegetation. Hultén had observed that the same mesic tundra plants were present on both sides of the Bering Strait, as well as on the Aleutians. Because it was unlikely that the same species had evolved independently in isolation in three places, he postulated a center of dispersal on the plain that now lies beneath the Bering Sea, which he named Beringia. During cold periods, when sea level was lower, tundra plants retreated to the central lowland; when climates warmed and sea level rose, they dispersed out of their refugium to Chukotka, Alaska, and the Aleutians.[53]

By 1937, both the existence of a land connection between Northeast Asia and Alaska and its relationship to falling sea levels and expanding glaciers during the Pleistocene had been established.[54] In fact, a land bridge between Asia and North America had been suggested as early as 1590 by the Spanish missionary and naturalist José de Acosta (1539–1600).[55] More than a century later (1728), Vitus Bering sailed into the strait that bears his name. By the late 1890s, the shallow depth of the seafloor in and around the Bering Strait had convinced geologist G. M. Dawson that Chukotka and Alaska had been joined by dry land in the recent past, and the thesis was bolstered by discoveries of mammoth remains on islands in the Bering Sea and in the Aleutian chain. Dawson also emphasized the lack of evidence for past glaciation in the region. And although early speculation on underlying causes was focused on tectonics, in the 1930s, Quaternary scientists tied increases in global ice volume to corresponding reductions in sea level, with clear implications for the Bering Strait.[56]

In the 1960s, David M. Hopkins (1921–2001) expanded the geographic definition of Beringia to include land that adjoins the submerged central plain in both Chukotka and western Alaska.[57] The 1960s also saw the first efforts to reconstruct the LGM vegetation of Beringia with paleobotanical data (Hultén's work had been based entirely on the distribution of modern plants). Analysis of pollen cores from various locations in western and northern Alaska (as well as Saint Lawrence Island) revealed herbaceous tundra "like that of modern Barrow."[58] This conclusion was disputed by paleontologist R. Dale Guthrie, who noted the environmental implications of large grazing mammals such as steppe bison and horse, remains of which had been recovered from last glacial deposits in central Alaska. Guthrie argued that more steppic conditions must have prevailed in Beringia during the LGM ("steppe tundra").[59]

By the early 1980s, Hopkins and others had further expanded the geographic definition of Beringia to take in the lower Mackenzie River in the Northwest Territories (Canada), aligning its eastern boundary with the approximate margin of the LGM ice.[60] Several archaeologists proposed extending the western boundary as far as the Lena River or the Verkhoyansk Mountains in eastern Siberia.[61] This broader definiton of Beringia—expanded far beyond the exposed central lowland recognized by Hultén—now has been widely adopted.[62] Also in the 1980s, core samples of sediments were extracted from the seafloor between Northeast Asia and Alaska, yielding pollen, insects, and plant remains of LGM age. They confirmed Hultén's hypothesis of a mesic tundra refugium in the center of Beringia.[63] Guthrie suggested that it probably had created an ecological barrier for mammals adapted to the drier steppes of western and eastern Beringia and could explain the otherwise puzzling absence of some taxa, such as Siberian woolly rhinoceros, in eastern Beringia or that of American camel in western Beringia. The paleo-entomologist Scott Elias noted the same phenomenon in fossil beetle taxa of LGM age (for example, the absence of steppe weevils in Alaska) (figure 7.3).[64]

The sediment cores from the former surface of the Bering Land Bridge also yielded radiocarbon dates that were younger than expected, indicating that it was still possible to walk from Asia to North America less than 13,000 years ago. This brought the chronology of the land bridge into line with a curve for global sea-level change published at the end of the 1980s.[65] More recent work on global sea level confirms that the land bridge was present before 45,000 years ago (see box 5.3). At its fullest extent (that is, when sea level had fallen to more than 130 meters below that of the present day), it became what Hopkins described as a "monotonously flat" plain stretching for more than 1,500 kilometers from the ice-bound shore of the Arctic Ocean to the rim of the North Pacific continental shelf.[66] The waters of the Bering Strait were never an obstacle to the spread of modern humans into North America.

The geohydrology of the exposed plain—poorly drained lowlands and meandering streams—probably helped maintain a shrub-tundra

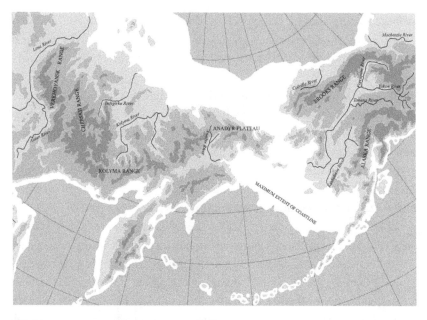

FIGURE 7.3 Map of Beringia, which commonly is defined now as extending from the Verkhoyansk Range in Northeast Asia to the Mackenzie River in northwestern Canada. (From John F. Hoffecker and Scott A. Elias, *Human Ecology of Beringia* [New York: Columbia University Press, 2007], fig. 1.1)

environment in central Beringia, although the influx of relatively warm and moist air from the North Pacific presumably was the key factor.[67] Support for Hultén's refugium hypothesis has continued to accumulate in the past few years. There is evidence for the survival of some arboreal taxa in central Beringia throughout the LGM, including birch, alder, poplar, and spruce. The analysis of a sediment core recently drilled into the southern edge of the land bridge also found ample evidence of sphagnum moss—another indicator of wetter climates in the region.[68] The presence of wood as a potential fuel source may have been an important factor in creating a refugium for humans in Beringia after 28,000 years ago (that is, the beginning of the LGM).

Initial Colonization

One consequence of the expanded geographic definition of Beringia is that it encompasses the Yana River in Northeast Asia and thus includes Beringia in the area of colonization by modern humans in northern Eurasia between 50,000 and 30,000 years ago. As described in chapter 6, the sites near the mouth of the Yana River document a high-tech northern-interior economy in the Arctic more than 30,000 years ago. These sites were occupied when climates probably were cooler than they are now, since they date to a pronounced cold period known as Heinrich Event 3 (ca. 32,500–29,500 years ago) (see box 5.1). The larch trees that grow in the area today likely were absent, but some woody shrubs may have been available as a fuel supplement to fresh bone (accounting for tiny fragments of charcoal in the hearths).[69]

As noted in chapter 6, both the evidence for a human presence at high latitudes as early as 45,000 years ago (for example, *Ust'-Ishim*) and the record of recurring warm intervals between 45,000 and 32,000 years ago suggest that older, as yet undiscovered, sites may exist in the Arctic, including some within the larger map of Beringia. In 2016, traces of human occupation (inferred from damage to a mammoth carcass) were reported from the central Siberian Arctic (latitude 72° North), along with similar evidence for a human presence on the Yana River (*Bunge-Toll*), both sites dated to more than 45,000 years ago. [70] This, in turn, raises the question of whether modern humans may have spread farther into Beringia before the beginning of the LGM. Environments in central Beringia probably were similar to those in the lower Yana River area (that is, northwestern Beringia), offering the same array of large and small mammals and at least some wood for fuel and material (for the construction of artifacts and facilities). Farther east—in what is now central Alaska—climates likely were somewhat drier than in western and central Beringia, but large grazing animals such as bison and horse were present.

There is, in fact, some evidence for a human presence in east-central Beringia before the LGM (and before the occupation of the Yana River sites), but it is problematic. In 2002, a fragment of worked mammoth tusk was found in gravel deposits along the Inmachuk River near the village of Deering in northwestern Alaska. The tusk had been chopped and grooved in several places in order to remove pieces of ivory. Although it yielded a date of about 40,000 calibrated radiocarbon years, it may have been picked up—still fresh from a permafrost context—and worked by more recent inhabitants of the area.[71]

If there is no apparent reason why modern humans could not have occupied Alaska before the LGM, the absence of compelling evidence for people in mid-latitude North America and farther south before and during the LGM is not difficult to explain. Computer modeling of the glaciers that covered much of North America between 65,000 and 17,000 years ago suggests that coastal and interior routes from Beringia to mid-latitude North America were completely blocked by ice throughout this interval.[72]

An LGM Refugium in Beringia?

The global-sea-level record indicates that between 32,000 and 28,000 years ago, sea level fell sharply—from a depth of roughly 75 meters to more than 120 meters below that of the present day (see figure 5.9). The bathymetry of the Bering Sea reveals that the effect of the LGM on sea level exposed a significantly larger area of the Bering Land Bridge. Much of the expansion occurred in the southeastern margin of the land bridge (around Bristol Bay in southwestern Alaska) at comparatively low latitude (roughly 56–58° North). Computer modeling of Beringian environments during the LGM suggests a relatively high level of plant productivity in this region (due to the effects of moisture from the North Pacific, as noted earlier) and the formation of numerous wetlands on the "monotonously flat" plain (figure 7.4), while models of LGM ocean temperature suggest minimal reduction for the Bering Sea and Gulf of Alaska.[73]

In the context of the global retreat of human populations to various refugia in Africa, Eurasia, and Australia during the LGM, the expansion of habitat in south-central Beringia is significant. Beringia may have been the only place on Earth to have produced a substantial area of new terrestrial habitat—in this case, comparatively rich shrub tundra with some arboreal vegetation—for hunter-gatherers during the LGM. The creation of wetlands probably increased the seasonal concentration of migratory waterfowl on the southeastern land bridge.[74] Paleobiological data indicate that the shrub-tundra zone in the south-central land bridge probably extended into low-lying adjoining regions, such as the Anadyr Valley in Chukotka and the Yukon Delta in southwestern Alaska.[75] There also is evidence for shrub-tundra vegetation in lowland northwestern Beringia during the LGM.[76] Furthermore, hunter-gatherer groups that wintered in the milder areas of the south-central lowland (where wood was available for fuel) could have hunted large mammals (mammoth, horse, bison, and

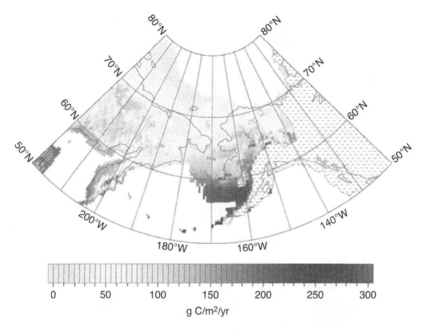

FIGURE 7.4 Net primary productivity (measured as g/C/m²/yr, which is approximately 45 percent of total organic matter) for Beringia during the LGM (ca. 21,000 years ago), as predicted by the BIOME4 model. According to the model, the most productive areas would have been on the southeastern margin of the Bering Land Bridge, much of which was not exposed until the LGM. (Model output produced by Nancy H. Bigelow and Amy Hendricks, University of Alaska, Fairbanks, and reproduced from John F. Hoffecker et al., "Beringia and the Global Dispersal of Modern Humans," *Evolutionary Anthropology* 25 [2016]: 73, fig. 7)

others) in the more steppic zones of western and eastern Beringia during warmer months.

As already noted, the Beringian Standstill Hypothesis—the presence of the ancestral Native American population in Beringia between about 28,000 and 16,000 years ago—remains to be confirmed (or refuted), although there is some archaeological evidence for human occupation during the LGM from the eastern edge of Beringia. At *Bluefish Caves* (northern Yukon, Canada) from 1977 to 1987, Jacques Cinq-Mars excavated about 100 stone artifacts and more than 35,000 bones of mammals (primarily caribou, horse, bison, and mammoth) and other vertebrates, including birds and fish from loess deposits of LGM age.[77] Some of the artifacts, including a microblade core and a microblade, were recovered from the lower portion of the loess (more than 1 meter below the surface). Most of the newly reported radio-carbon dates on the bones fall between about 24,000 and 17,500 years ago.

While the precise temporal relationship between the bones and the artifacts is unclear (both are likely to have been disturbed by frost action since their deposition), their co-occurrence throughout the loess unit suggests that they are roughly contemporaneous. Moreover, new taphonomic analyses of the large-mammal bones reveal that a few of them exhibit stone-tool damage.[78]

To confirm the "standstill" model, however, it must be shown that any inhabitants of Beringia during the LGM represent the ancestral Native American population—not another group that occupied the region during the LGM. The most useful data would be aDNA from dated human remains. The finds from Bluefish Caves indicate that people probably were living in Beringia during the LGM, but the microblade technology is not characteristic of either the pre-LGM inhabitants of Beringia (Yana River sites) or the Native American groups that dispersed throughout North and South America after 15,000 years ago. In sum, it is unclear how the occupation of Bluefish Caves fits into the larger context of the dispersal of modern humans into Beringia and the Americas.

More generally, the plausibility of Beringian Standstill may be addressed with some data on the ecology of hunter-gatherers in high-latitude environments. Given estimates of the population size thought to have been isolated in Beringia, how credible is Beringia as a resource base during the LGM? Was there enough to eat between 28,000 and 16,000 years ago? On the basis of the mtDNA data, Mulligan and Kitchen estimated an effective population (N_e) of 1,000 individuals, which translates to a total population of about 5,000. They found no indications of a genetic bottleneck in the "standstill" population, but concluded that it either was stable or was growing at a modest rate (box 7.1).[79]

Net primary productivity for Beringia during the LGM has been estimated with the *BIOME4* computer model (see figure 7.4). Model output suggests that the south-central portion of the land bridge would have supported a relatively high NPP at about 21,000 years ago (ca. 250–300 $g/C/m^2/yr$). Recent hunter-gatherers, such as the *Deg Hit'an*, in northern-interior environments with comparable NPP, are estimated to have enjoyed a population density of at least 2.5 persons per 100 square kilometers (see table 6.2). At this level, the estimated "standstill" population of 10,000 would have required a territory of 250,000 square kilometers, which apparently would have been available in south-central Beringia.

BOX 7.1
Estimating "Effective Population Size" with Genetic Techniques

Genetic analyses of living humans not only allow the reconstruction of maternal- and paternal-lineage history among all peoples on Earth, but also can be used to estimate the size of specific populations at various points in the past, such as the population thought to have been isolated in Beringia during the LGM. Genetic techniques have been developed to estimate what Sewall Wright (1889–1988) defined as *effective population size*, or N_e, which is the "number of breeding individuals in an idealized population that would show the same amount of dispersion of allele frequencies under random genetic drift or the same amount of inbreeding as the population under consideration."[1] Effective population size sometimes is estimated separately for each sex (for example, N_{ef} = female effective population size). Total population size in large mammals is estimated at five times N_e ($N = 5 \times N_e$).[2]

A technique for estimating N_e in a recent population is based on the temporal change of allele frequencies. Increased change in allele frequencies due to genetic drift is expected as a population becomes smaller and generation times are extended. Another approach to estimating recent population size is based on the measurement of "heterozygote excess" (due to increased heterozygosity in parents as a function of sampling error in a small population) in offspring.[3] A third technique uses "linkage disequilibrium," or the statistical association between alleles at different loci. In theory, N_e should decline as the measured correlation between allele frequencies increases. This technique may be applied to more ancient populations when tightly linked loci are sampled.[4]

Mutation rate also may be applied to estimates of effective population size in ancient populations. Because genetic diversity (θ) is a function of mutation rate (u) and effective population size, the equation

$$\theta = 4N_e u$$

can be solved for N_e if the mutation rate is known.[5] Using mtDNA data (applicable to only the maternal lineages), Andrew Kitchen, Michael Miyamoto, and Connie Mulligan estimated N_{ef} of about 2,700 for the Beringian Standstill population, with assumed mutation rates (per site/per year) of 1.7×10^{-8} for the coding portion, and 4.7×10^{-7} for the non-coding "hypervariable" segments, of mtDNA.[6]

1. Sewall Wright, "Evolution in Mendelian Populations," *Genetics* 16 (1931): 97–159.
2. Alan R. Templeton, "Human Races: A Genetic and Evolutionary Perspective," *American Anthropologist* 100 (1998): 632–650.
3. J. Wang, "Estimation of Effective Population Sizes from Data on Genetic Markers," *Philosophical Transactions of the Royal Society B: Biological Sciences* 360 (2005): 1395–1409.
4. Philip W. Hedrick, *Genetics of Populations*, 4th ed. (Sudbury, Mass.: Jones and Bartlett, 2011), 235–237.
5. Wang, "Estimation of Effective Population Sizes from Data on Genetic Markers," 1404–1405.
6. Andrew Kitchen, Michael M. Miyamoto, and Connie J. Mulligan, "A Three-Stage Colonization Model for the Peopling of the Americas," *PLoS ONE* 3 (2008): e1596.

After the LGM: Out of Beringia?

After 17,000 to 16,000 years ago, as climates became warmer and wetter and glaciers began to recede, people started to reclaim the large areas abandoned during the LGM. Occupations reappear in dry interior regions of Australia at this time, as well as in North Africa.[80] The hiatus in northern European sequences ends, and many new sites appear in northwestern Europe and on the central East European Plain—including, on the plain, the famous houses of mammoth bone, which date to the post-LGM and sometimes are found in groups resembling the villages of the postglacial epoch. The expansion out of southern refugia also is reflected in the genetics of living Europeans.[81]

In Siberia, the post-LGM period is characterized by a large increase in the number of sites relative to earlier periods. Many occupations in the Upper Yenisei Valley and other parts of southern Siberia date to this interval. The mass-production of microblades from small wedge-shaped cores—which represents a highly economic use of good-quality stone—became especially common.[82] The microblades were inserted into grooves carved into the sides of bone or antler points to create a composite weapon with razor-sharp edges.

The occupation of the Lena Basin above latitude 55° North (presumably abandoned during the LGM) is dated to about 16,800 to 15,100 years ago at *Dyuktai Cave*, on the Aldan River. The lowest levels contain a diagnostic form of wedge-shaped microblade core (Yubetsu technique) and microblades. There are hints of a broad-based northern-interior economy at Dyuktai Cave. In addition to those of large mammals such as horse and bison, the faunal remains include hare, birds, and fish. The presence of charcoal indicates that some wood was available.[83]

A similar assemblage has been found in the lowest layer of *Swan Point* (central Alaska), an open-air site on the Tanana River dated to about 14,000 years ago. It is dominated by wedge-shaped microblade cores (Yubetsu technique) and burins. The latter include transverse and dihedral forms, and their presence reflects a heavy emphasis on the working of bone, antler, and ivory (including carving slots for microblades) in an environment where hardwood was scarce. The hearths are filled with bone ash containing small fragments of charcoal from woody shrubs (willow and birch).[84] The pattern indicates primary dependence on fresh bone for fuel, but with a wood supplement to help extend the life of the hearth (and possibly to help start the fire, given the high ignition temperature of bone).[85] The Dyuktai/Yubetsu microblade technology is an unusually credible archaeological proxy for a specific group of people, and appears to document the spread of these people from the Lena Basin into Beringia after 15,000 years ago.

While the artifacts from Bluefish Caves suggest that microblade makers were present in Beringia during the LGM, between 15,000 and 13,000 years ago the archaeological record is dominated by another industry (or industries) that lacks any obvious source outside Beringia. It is a plausible archaeological proxy for the mysterious "standstill" population in the post-LGM landscape, which is the most likely source of groups that initially settled the rest of the Americas. They were first recognized in the form of assemblages containing stemmed points in Kamchatka (southwestern Beringia) by Nikolai Dikov, who speculated on possible links to the early sites in western North America ("Western Stemmed Tradition").[86] In the 1980s, a contemporaneous and seemingly related industry was identified in central Alaska (Nenana Complex).[87] Its most characteristic artifact type is a teardrop-shaped point (often referred to as a "Chindadn point"). The major sites are listed in table 7.2.

The earliest occurrence of the Nenana Complex was reported from the open-air site of *Berelekh* (northeastern Siberia [northwestern Beringia]), near the mouth of the Indigirka River (figure 7.5). This site comprises a large bed of mammoth bones and a human occupation that apparently postdates the bones by at least a few decades. Berelekh, initially investigated several decades ago and now largely destroyed, was reexamined in 2009. Artifacts were recovered in buried context, and new dates were obtained on a variety of materials, including worked bone and ivory (ca. 14,900–13,500 years ago). Among the stone artifacts are a typical Chindadn point (teardrop-shaped with incomplete bifacial flaking) and the fragment of a stemmed point. The faunal remains are dominated by hare, but ptarmigan and waterfowl also are represented; mammoth bones and tusks apparently were scavenged for materials by the occupants of the site (figure 7.6).[88]

The stemmed-point fragment may indicate a tie to the lowest occupation levels at *Ushki* (central Kamchatka, Russia), where stemmed points are common in the stone-artifact assemblages. Like many of the Chindadn

TABLE 7.2

Archaeological Record of *Homo sapiens* in Beringia, Before 13,000 Years Ago

Site, Layer	Estimated Age (method)	Description	Comments
Inmachuk River (northeastern Seward Peninsula, Alaska, U.S.)[a]	Ca. 41,000 years (radiocarbon)	Worked mammoth-tusk fragment	Tusk could have been worked in recent times
Yana River (northeastern Siberia, Russia)[b]	Ca. 32,000 years (radiocarbon)	Side scrapers; plano-convex tools; pointed tools; needles; ivory vessels; pendants; beads	Oldest reliable evidence of occupation in Beringia
Bluefish Caves (northern Yukon, Canada)[c]	Ca. 24,000 years (radiocarbon)	Microblade core, microblade, and other artifacts associated with cut-marked mammal bones	Only evidence of occupation in Beringia during the LGM (?)
Berelekh (northeastern Siberia, Russia)[d]	Ca. 14,900–13,500 years (radiocarbon)	Nenana Complex: Chindadn point; stemmed-point fragment; bifaces	Oldest known Nenana Complex assemblage; hare = 90% of fauna
Swan Point, lowest layer (Tanana Basin, Alaska, U.S.)[e]	14,150–13,870 years (radiocarbon)	Dyuktai Culture: microblade cores; microblades; burins	Similar to Dyuktai Cave (Siberia)
Little John, lowest layer (southwestern Yukon, Canada)[f]	14,050–13,720 years (radiocarbon)	Nenana Complex: Chindadn points	Easternmost occurrence of Nenana Complex; Wiki Peak obsidian
Dry Creek, Layer I (south-central Alaska, U.S.)[g]	Ca. 13,500–13,300 years (radiocarbon)	Nenana Complex: Chindadn points; end scrapers	Large-mammal hunting in foothills of northern Alaska Range (sheep, elk)
Mead, lowest layers (Tanana Basin, Alaska, U.S.)[h]	13,440–13,200 years (radiocarbon)	Triangular point base; bifaces; side scrapers; burins	Relationship to other sites unclear; Wiki Peak obsidian
Walker Road (south-central Alaska, U.S.)[i]	13,300–12,800 years (radiocarbon)	Nenana Complex: Chindadn points; end scrapers	Traces of small artificial shelter; Wiki Peak obsidian
Ushki 1 and 5, Layer VII (central Kamchatka, Russia)[j]	Ca. 13,200–12,700 years (radiocarbon)	Stemmed points: bifaces	Related to Berelekh assemblage (?); traces of artificial shelters

Site, Layer	Estimated Age (method)	Description	Comments
Moose Creek, Layer I (south-central Alaska, U.S.)[k]	Ca. 13,700–12,880 years (radiocarbon)	Nenana Complex: Chindadn point	

Note: All dates are given in calibrated radiocarbon years before present.

[a]Carol Gelvin-Reymiller et al., "Technical Aspects of a Worked Proboscidean Tusk from Inmachuk River, Seward Peninsula, Alaska," *Journal of Archaeological Science* 33 (2006): 1088–1094.

[b]V. V. Pitulko et al., "The Yana RHS Site: Humans in the Arctic Before the Last Glacial Maximum," *Science* 303 (2004): 52–56; Vladimir Pitulko et al., "Human Habitation in Arctic Western Beringia Prior to the LGM," in *Paleoamerican Odyssey,* ed. Kelly E. Graf, Caroline V. Ketron, and Michael R. Waters (College Station: Texas A&M University Press, 2013), 13–44.

[c]Jacques Cinq-Mars, "La place des grottes du Poission-Bleu dans la prehistoire béringienne," *Revista de Arqueología Americana* 1 (1990): 9–32; Lauriane Bourgeon, "Bluefish Cave II (Yukon Territory, Canada): Taphonomic Study of a Bone Assemblage," *PaleoAmerica* 1 (2015): 105–108; Lauriane Bourgeon, Ariane Burke, and Thomas Higham, "Earliest Human Presence in North America Dated to the Last Glacial Maximum: New Radiocarbon Dates from Bluefish Caves, Canada," *PLoS ONE* 12 (2017): e0169486.

[d]Vladimir V. Pitulko, Aleksandr E. Basilyan, and Elena Y. Pavlova, "The Berelekh 'Graveyard': New Chronological and Stratigraphical Data from the 2009 Field Season," *Geoarchaeology: An International Journal* 29 (2014): 277–299.

[e]Charles E. Holmes, "Tanana River Valley Archaeology Circa 14,000 to 9000 B.P.," *Arctic Anthropology* 38 (2001): 154–170; Ben A. Potter, Charles E. Holmes, and David R. Yesner, "Technology and Economy Among the Earliest Prehistoric Foragers in Interior Eastern Beringia," in *Paleoamerican Odyssey,* ed. Graf, Ketron, and Waters, 81–103.

[f]Norman Alexander Easton et al., "Chindadn in Canada? Emergent Evidence of the Pleistocene Transition in Southeast Beringia as Revealed by the Little John Site, Yukon," in *From the Yenisei to the Yukon: Interpreting Lithic Assemblage Variability in Late Pleistocene/Early Holocene Beringia,* ed. Ted Goebel and Ian Buvit (College Station: Texas A&M University Press, 2011), 289–307.

[g]William R. Powers and John F. Hoffecker, "Late Pleistocene Settlement in the Nenana Valley, Central Alaska," *American Antiquity* 54 (1989): 263–287; Kelly E. Graf et al., "Dry Creek Revisited: New Excavations, Radiocarbon Dates, and Site Formation Inform on the Peopling of Eastern Beringia," *American Antiquity* 80 (2015): 671–694.

[h]Potter, Holmes, and Yesner, "Technology and Economy Among the Earliest Prehistoric Foragers in Interior Eastern Beringia."

[i]Ted Goebel et al., "Walker Road," in *American Beginnings: The Prehistory and Palaeoecology of Beringia,* ed. Frederick Hadleigh West (Chicago: University of Chicago Press, 1996), 356–362.

[j]N. N. Dikov, *Arkheologicheskie Pamyatniki Kamchatki, Chukotki i Verkhnei Kolymy* (Moscow: Nauka, 1977); Ted Goebel, Michael R. Waters, and Margarita Dikova, "The Archaeology of Ushki Lake, Kamchatka, and the Pleistocene Peopling of the Americas," *Science* 301 (2003): 501–505.

[k]Georges A. Pearson, "Early Occupation and Cultural Sequence at Moose Creek: A Late Pleistocene Site in Central Alaska," *Arctic* 52 (1999): 332–345.

points, they are not fully flaked on both sides. The occupations date to roughly 13,000 years ago and provide an unusually rich picture of life in Beringia during this period. Dikov uncovered traces of artificial structures and a burial; the deceased had been interred with numerous stone points and hundreds of beads. Unfortunately, the human remains were almost completely weathered away (generally, little bone was preserved at Ushki). Hints of a broad-based subsistence economy were found in avian gastroliths (gizzard stones), and recent investigations yielded evidence of salmon.[89] The material culture of recent hunter-gatherers in similar settings

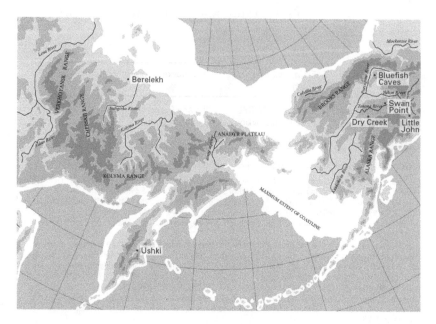

FIGURE 7.5 Archaeological sites in Beringia, dated to between roughly 15,000 and 12,000 years ago. (Adapted from John F. Hoffecker and Scott A. Elias, *Human Ecology of Beringia* [New York: Columbia University Press, 2007], fig. 1.1)

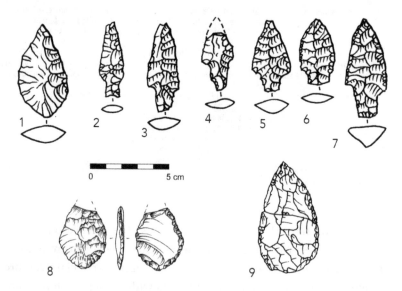

FIGURE 7.6 Bifacial points belonging to industries that may represent an archaeological proxy for the post-LGM "standstill population" in Beringia: (*1–7*) Ushki; (*8*) Berelekh; (*9*) Walker Road, including (*2–7*) stemmed points and (*8–9*) Chindadn points. (From N. N. Dikov, *Drevnie Kul'tury Severo-Vostochnoi Azii* [Moscow: Nauka, 1979], 35, fig. 3; Vladimir V. Pitulko, Aleksandr E. Basilyan, and Elena Y. Pavlova, "The Berelekh 'Graveyard': New Chronological and Stratigraphical Data from the 2009 Field Season," *Geoarchaeology: An International Journal* 29 [2014]: 294, fig. 11 [reproduced by permission of John Wiley & Sons]; William R. Powers and John F. Hoffecker, "Late Pleistocene Settlement in the Nenana Valley, Central Alaska," *American Antiquity* 54 [1989]: 279, fig. 8 [reproduced by permission of the Society for American Archaeology])

suggests that most of their undoubtedly complex technology is archaeologically invisible (that is, was made out materials that have not preserved).

The largest concentration of Nenana Complex sites in eastern Beringia is found in the northern foothills of the Alaska Range, along terraces of the Nenana River. Three sites in the valley contain Nenana Complex assemblages buried in loess that is dated to about 13,500 to 12,800 years ago. A few badly weathered remains of sheep and elk were identified by Guthrie in the lowest level at *Dry Creek*. These sites probably were occupied in late summer or fall to hunt large mammals that were attracted to the foothills at this time of the year. Traces of a small artificial shelter were found at *Walker Road*, which yielded classic Chindadn points.[90]

The Beringian archaeological record suggests that when the Siberian microblade makers moved into Beringia after 15,000 years ago, they encountered an established local population. As already noted, the archaeological remains of this group (that is, assemblages containing Chindadn and stemmed points) are a credible, if unconfirmed, proxy for the "standstill" population hypothesized to have inhabited a shrub-tundra refugium in south-central Beringia throughout the LGM. If so, this population seems to have expanded out of the central lowland—which was beginning to shrink as sea levels rose—in both directions. In some respects, the archaeological record parallels the dispersal of tundra plants out of the central refugium, as climates warmed following the LGM, envisioned by Hultén. This is illustrated by the younger Nenana Complex sites in the foothills of the Alaska Range, which were not occupied until shrub tundra spread to higher elevations.[91]

Despite the high latitude (Berelekh is at 70°30′ North), the Beringians were already living in a postglacial world. If they relied on the exploitation of more steppic habitat outside the central lowlands during the LGM, the large-mammal resources in these areas were shrinking in response to the expansion of shrub tundra. Their primary large-mammal prey were postglacial taxa (for example, elk, sheep, and caribou) supplemented by small mammals, birds, and fish, all of which are found in northern-interior settings of the Holocene. Migratory waterfowl may have been especially important, as millions of ponds and wetlands emerged from the thawing landscape.[92] The subsistence technology probably was comparable to that of recent northern-interior foragers, and the Ushki sites indicate extended habitation (at least in some places).

According to the genetic models, the Beringian "standstill" population expanded rapidly after 15,000 years ago. Most of the people migrated eastward into mid-latitude North America (mtDNA haplogroups A, B, C, D, and X), but at least one maternal lineage (C1a) moved westward into Northeast Asia.[93] In addition to rapid population growth (ultimately an estimated 16-fold increase), two other factors may have encouraged the flow of Beringians into recently deglaciated coastal and interior areas of northwestern North America between 14,000 and 13,000 years ago.[94] One was the inundation of the central lowland (Bering Land Bridge), which significantly reduced their resource base. The other—perhaps even more distressing— factor was the influx of Siberians (the microblade makers) from the Lena Basin. The Siberians may have represented mtDNA subclade D2, thought to have moved from Northeast Asia into Beringia and northern North America at the end of the Pleistocene.[95]

A brief but pronounced cold interval (Younger Dryas) took place between 12,900 and 11,300 years ago, and it is signaled by a dramatic change in the archaeological record of Beringia. Most of the Nenana Complex artifacts are replaced (usually after a brief hiatus) by assemblages containing wedge-shaped microblade cores, microblades, and burins.[96] The change appears to represent not a shift in technology, but a complete replacement of the Beringians by a different group of people with clear ties to Northeast Asia (the original source of such artifact assemblages). These assemblages also appear on the Northwest coast and in northwestern Canada, and they probably were made by the ancestors of the people who later occupied central and northeastern Canada (after the retreat of the glaciers in the mid-Holocene) and coastal Greenland (aDNA from the latter was assigned to mtDNA subclade D2a1) (figure 7.7).[97]

In northern Alaska and the Yukon, sites of later Younger Dryas age contain lanceolate projectile points similar to those found on the North American Plains for this time period. They apparently represent a back-migration of people originally derived from Beringia who followed Plains bison herds as they spread northward with the expansion of steppic habitat under colder and drier conditions.[98] Some of the bison hunters seem to have filtered down into central and southwestern Alaska, but their artifacts are absent from Northeast Asia. By the end of the Younger Dryas, the rising Pacific Ocean had flooded the Bering Strait and Beringia had ceased to exist.

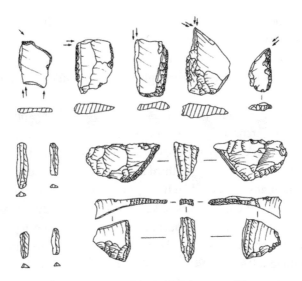

FIGURE 7.7 Artifacts from Dry Creek representing the microblade industry that dominates the archaeological record of Northeast Asia and Alaska after the inundation of central Beringia. (From William R. Powers and John F. Hoffecker, "Late Pleistocene Settlement in the Nenana Valley, Central Alaska," *American Antiquity* 54 [1989]: 274, fig. 5. Reproduced by permission of the Society for American Archaeology)

Two infant burials from a site in central Alaska (*Upward Sun River*) have yielded ancient mtDNA that dates to the final centuries of the Younger Dryas. Subclades of two of the major maternal lineages were identified (mtDNA B2 and C1b), neither of which is well represented in living peoples in the region (but is common among Native Americans in other places).[99] Although the presence of these subclades in central Alaska more than 11,000 years ago has been attributed to the genetic diversity of the post-LGM Beringian population,[100] an alternative explanation is that they represent the bison hunters from the North American Plains who migrated north during the Younger Dryas.[101]

THE DISPERSAL IN NORTH AND SOUTH AMERICA

Although still disputed by a number of archaeologists, most living Native Americans are descended from a population that moved southward from Beringia between 15,000 and 14,000 years ago. The timing of the dispersal is based on estimated coalescence dates for the major lineages.[102] It coincides at least broadly with independent estimates of the retreat of the

Laurentide and Cordilleran ice sheets, which eventually opened up both coastal and interior routes to mid-latitude North America.[103] And it is supported by radiocarbon dates of between 15,000 and 14,000 years ago on archaeological sites scattered throughout mid-latitude North and South America.

The late arrival of modern humans in the Western Hemisphere provides a much younger archaeological record of the initial dispersal in the Americas than Eurasia or Australia. The earliest known sites in mid-latitude North and South America are only one-third as old as the estimated time of arrival of modern humans in places like India and the Levant, and they reflect the initial stages of settlement. Sites older than 13,000 years are largely, if not wholly, confined to ephemeral occupations. Most of them represent locations near a water source where a large mammal was killed and butchered. They contain few, if any, artifacts—typically simple flake tools and/or heavy stones used to dismember and deflesh the carcass.[104] Comparable traces of the initial dispersal in Eurasia and Australia are thus far unknown, probably due to extreme low visibility in the archaeological record.

In a classic paper published in 1988, archaeologists Robert Kelly and Lawrence Todd proposed that the seemingly peculiar pattern of the early North American sites was best explained as that of a "colonizing" population in a landscape devoid of other humans.[105] They emphasized (1) dependence on migratory large mammals (versus broad-based subsistence tailored to specific local habitats), (2) short-term occupations (and limited use of natural shelters), (3) highly transportable technology (analogous to that of foragers in the Australian desert [chapter 5]), and (4) absence of long-term food storage. In the years since the publication of the paper, much additional empirical support for the model has emerged from new research and discoveries.

Both the genetics of living Native Americans and the archaeological record indicate a very rapid dispersal across North and South America. Within a few centuries, people were present in all major regions of North America, including Mexico and Florida. By 13,000 years ago, they had reached the southern tip of South America, as well as the eastern lowlands of the continent.[106] If anything, the current estimate of less than 2,000 years, based on the genetics of living Native Americans, may be too conservative. The dates from the archaeological sites imply 1,500 years at the most.

The ephemeral character of the early occupations—especially in North America—suggests that the population spread thinly across large areas during the initial phase of settlement. As Kelly and Todd suggested, the emphasis on wide-ranging large mammals (at a time when regional variations in North American plant and animal life were less pronounced than they are today) probably encouraged high mobility. The development of local dietary patterns, extended habitations, and increased population density took place at a later time. Highly distinctive adaptations to the Desert Southwest, Pacific coast, Central High Plains, Eastern Woodlands, and other regions evolved within a few millennia.[107] The process of local adaptation was accelerated by the extinction of many of the large mammals, as climates changed during the final millennia of the Pleistocene.

In contrast to the people who invaded northern Eurasia before the LGM, the Beringians who moved south about 14,500 years ago did not have to adjust to increasingly severe winter climates and lower biological productivity. Instead, they found themselves in places where the winters were increasingly mild and the game animals plentiful (and probably easy to obtain, because they were not adapted to human predators). They appear to have abandoned their high-tech northern-interior way of life immediately—there are no hints of the extended occupations or broad-based diet (with implications for complex technologies) typical of the Beringian shrub tundra. This, too, must have contributed to the relative speed of the dispersal.

The Coastal and Interior Corridors

The generally high visibility of early sites in the Western Hemisphere is not found, at least to date, in the areas that provided access from eastern Beringia to mid-latitude North America (that is, south of the receding but still massive glaciers that covered most of Canada). Neither the Northwest coast nor the interior "ice-free corridor" along the eastern margin of the Canadian Rockies has yielded any dated traces of human occupation older than 13,300 years ago. The presence of reliably dated sites older than 14,000 years in mid-latitude regions of North and South America indicates that people were present in at least one of these corridors 1,000 years or more before.[108]

The dearth of older sites on the Northwest coast is almost certainly due to the inundation by the rising Pacific Ocean of much of the unglaciated land occupied by humans between 15,000 and 14,000 years ago. The same problem confronts researchers looking for evidence of any early coastal migration in southern Asia (chapter 5). Although some portions of the former coastline have been elevated as a result of isostatic rebound,[109] much of it is underwater and currently either difficult or impossible to survey for archaeological remains. There have, however, been efforts to prospect near-shore settings, and they have yielded some results.[110]

The oldest firmly dated traces of humans on the Northwest coast were found in two caves located on Haida Gwaii (Queen Charlotte Islands, British Columbia). *K1 Cave* (Moresby Island) yielded stone-point fragments and evidence of bear hunting dated to roughly 12,900 to 12,400 years ago. *Gaadu Din 1* (Huxley Island)) contained a complete bifacial spear point and other artifacts that date to about 12,800 to 11,300 years ago.[111] The artifacts do not exhibit any obvious parallels to the Beringian industries described earlier (figure 7.8).

Slightly younger sites on the Northwest coast provide evidence of adaptation to a habitat that may have created the basis for a rapid population spread down the Pacific coast. At *Kilgii Gwaay*, which lies in the intertidal zone between Moresby and Kunghit Islands, Daryl Fedje and his colleagues excavated an occupation dated to about 10,700 years ago. The saturated deposits yielded organic artifacts and debris, including shell, plant seeds, leaves, bone tools, animal hair, and wooden stakes (*in situ*). The faunal remains comprise marine mammals (sea otter, harbor seal, and others), black bear, river otter, and a variety of fish (for example, rockfish, halibut, and lingcod).[112] Farther north, at *On Your Knees Cave* (southeastern Alaska), human remains of comparable age exhibited isotopic values indicating a diet high in marine foods.[113]

The oldest traces of occupation by humans along the eastern margin of the Canadian Rockies are found in the southernmost portion of the "ice-free corridor" at *Wally's Beach* (southwestern Alberta), recently dated to about 13,300 years ago. The site is typical of occupations in North and South America before 13,000 years ago. It represents a place where individual large mammals were killed and butchered. The remains of seven horses (*Equus conversidens*) and one camel (*Camelops hestemus*), exhibiting signs of butchery, were associated with 29 stone artifacts. The artifacts, which include heavy chopping tools and modified flakes, are not diagnostic of any specific complex or tradition.[114]

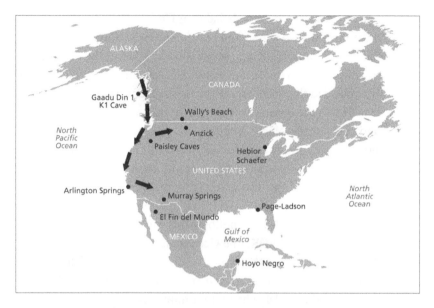

FIGURE 7.8 Map of North America, showing human remains and archaeological sites, dated to more than 12,000 years ago. (Base map adapted from en.wikipedia.org/wiki/File:Blank Map-World6.svg)

Wally's Beach probably has little to do with the initial movement of people from Beringia to mid-latitude North America. It is more likely to contain traces of one or more groups that had been moving around on the northern Plains for generations. Roughly 400 fluted points have been collected from surface locations within the "ice-free corridor," and at least of few of them are typical—in terms of size and shape—of the Clovis Complex, which now dates as early as Wally's Beach. The typical Clovis points are likely to represent people who drifted north many centuries after the initial settlement of the continent. Most of the remaining fluted points are typical of later periods.[115] Significantly, no Chindadn points are reported from the region in either surface or buried contexts.

The inference that both the coastal and interior corridors played a role in the early settlement of the Western Hemisphere is based on genetic analyses. In 2009, Ugo Perego and his colleagues sequenced the mitogenomes of two rare haplogroups among living Native Americans. Although mtDNA haplogroups D4h3 and X2a appear to have diverged from their parent maternal lineages in Beringia no later than 14,000 years ago, their geographic distributions differ.[116] Haplogroup D4h3 is found on the Pacific coast of North America (with subclades in Chile), while X2a is concentrated in the Great Lakes region and North American Plains. In 2009, the geographic pattern among living people was reflected in the distribution of aDNA, although the sample size was small (see table 7.1). The authors of the study concluded that members of the haplogroups had followed different routes—D4h3, the coastal corridor; X2a, the interior corridor—from Beringia to mid-latitude North America.

By 2016, however, new discoveries had altered the picture. In 2014, aDNA from a burial at the Clovis site of *Anzick* (Montana), dated to between 12,700 and 12,600 years ago, was assigned to mtDNA haplogroup D4h3a, suggesting a link between the Pacific coast and the people who made the widespread Clovis Complex.[117] The following year, haplogroup X2a was identified in aDNA from the Kennewick remains, dated to between 8,690 and 8,400 years ago, placing the earliest known occurrence of this lineage in the Pacific Northwest.[118] And in 2016, new research on the "ice-free corridor" established that a viable habitat for humans along the entire length of the corridor (ca. 1,500 kilometers) was not present until about 12,600 years ago (that is, during the Younger Dryas).[119] It now appears likely that mid-latitude North America (and South America) was initially settled by people who migrated down the Pacific coast from Beringia.

Small Sites and Large Mammals: North America Before 13,000 Years Ago

According to the current chronology—based on calibrated radiocarbon dates—North America was widely settled between about 14,500 and 13,000 years ago. Sites are found in the interior Northwest (Washington and Oregon), northern Great Plains (Montana), Great Lakes (southern Wisconsin and Ohio), mid-Atlantic (Pennsylvania), Southeast (northern Florida), southern High Plains (Texas), and northern Mexico (Sonora). Most of the earlier sites represent kill-butchery locations that contain the remains of one or more large mammals and a few simple artifacts (some sites yield only the mammal remains) or a few artifacts in a natural shelter (table 7.3).

The scarcity and occasional absence of artifacts associated with the large-mammal remains underscores the importance of taphonomic analyses of the bones. By itself, the presence of the artifacts—even if typical of a kill-butchery site—does not prove that the mammal remains represent an

TABLE 7.3

Archaeological Record of *Homo sapiens* in North America, Before 13,000 Years Ago

Site	Estimated Age (method)	Description	Comments
Hebior (southern Wisconsin, U.S.)[a]	Ca. 14,775 years (radiocarbon)	Remains of one mammoth associated with two bifaces	Kill-butchery site without diagnostic artifacts
Page-Ladson (northern Florida, U.S.)[b]	Ca. 14,550 years (radiocarbon)	Mastodon remains associated with biface fragment and flakes	Kill-butchery site without diagnostic artifacts
Schaefer (southern Wisconsin, U.S.)[c]	Ca. 14,480 years (radiocarbon)	Remains of one mammoth associated with two stone flakes	Kill-butchery site without diagnostic artifacts
Lindsay (Montana, U.S.)[d]	Ca. 14,300 years (radiocarbon)	Mammoth remains exhibit evidence of butchery	Kill-butchery site without artifacts (based on taphonomy)
Paisley Five Mile Point Caves (Oregon, U.S.)[e]	Ca. 14,340 years (radiocarbon)	Isolated stone artifacts associated with five human coprolites	mtDNA haplogroups A and B identified from coprolites
Manis (Washington, U.S.)[f]	Ca. 13,815 years (radiocarbon)	Remains of one mastodon with bone-point tip embedded in rib	Kill-butchery site without artifacts
Firelands (Ohio, U.S.)[g]	13,485–13,575 years (radiocarbon)	Femur of ground sloth exhibits traces of tool cuts	Kill-butchery site without artifacts
Aubrey (northeastern Texas, U.S.)[h]	Ca. 13,400 years (radiocarbon)	Large artifact assemblage; bison remains	Oldest dated Clovis points; habitation area and kill site
El Fin del Mundo (Sonora, Mexico)[i]	Ca. 13,390 years (radiocarbon)	Remains of two gomphotheres associated with seven Clovis points, scraper, and flakes	Early dated Clovis points; evidence of hunting of gomphotheres
Wally's Beach (southwestern Alberta, Canada)[j]	Ca. 13,300 years (radiocarbon)	Horse and camel remains associated with core tools and flakes	Multiple kill-butchery localities (horse and camel)
Friedkin (central Texas, U.S.)[k]	Ca. 15,500–13,200 years (OSL)	More than 16,000 artifacts (bifaces, flake tools, debitage)	No diagnostic artifacts; earlier OSL dates too old (?)

(continued)

(continued)

Site	Estimated Age (method)	Description	Comments
Lubbock Lake (western Texas, U.S.)[l]	Ca. 13,100–12,100 years (radiocarbon)	Remains of mammoth exhibit traces of tools	Kill-butchery locations
Lange Ferguson (South Dakota, U.S.)[m]	Ca. 13,100–12,900 years (radiocarbon)	Juvenile and adult mammoth associated with Clovis points	Kill-butchery site; diagnostic artifacts (Clovis)

Note: All radiocarbon dates are given in calibrated radiocarbon years before present.

[a]David F. Overstreet, "Late-Glacial Ice-Marginal Adaptation in Southeastern Wisconsin," in *Paleoamerican Origins: Beyond Clovis*, ed. Robson Bonnichsen et al. (College Station: Texas A&M University Press, 2005), 183–195; Eileen Johnson, "The Taphonomy of Mammoth Localities in Southeastern Wisconsin (USA)," *Quaternary International* 142-143 (2006): 58-78; Michael R. Waters and Thomas Wier Stafford Jr., "The First Americans: A Review of the Evidence for the Late-Pleistocene Peopling of the Americas," in *Paleoamerican Odyssey*, ed. Kelly E. Graf, Caroline V. Ketron, and Michael R. Waters (College Station: Texas A&M University Press, 2013), 541–560.

[b]James D. Dunbar, "Paleoindian Archaeology," in *First Floridians and Last Mastodons: The Page-Ladson Site in the Aucilla River*, ed. S. David Webb (Dordrecht: Springer, 2006), 403–435; Jessi J. Halligan et al., "Pre-Clovis Occupation 14,550 Years Ago at the Page-Ladson Site, Florida, and the Peopling of the Americas," *Science Advances* 2 (2016): e1600375.

[c]Overstreet, "Late-Glacial Ice-Marginal Adaptation in Southeastern Wisconsin"; Johnson, "Taphonomy of Mammoth Localities in Southeastern Wisconsin"; Waters and Stafford, "First Americans."

[d]Waters and Stafford, "First Americans."

[e]Dennis L. Jenkins et al., "Clovis Age Western Stemmed Projectile Points and Human Coprolites at the Paisley Caves," *Science* 337 (2012): 223-228.

[f]Michael R. Waters et al., "Pre-Clovis Mastodon Hunting 13,800 Years Ago at the Manis Site, Washington," *Science* 334 (2011): 351-353.

[g]Michael R. Waters et al., "Late Pleistocene Horse and Camel Hunting at the Southern Margin of the Ice-Free Corridor: Reassessing the Age of Wally's Beach, Canada," *Proceedings of the National Academy of Sciences* 112 (2015): 4263-4267; Brian G. Redmond et al., "New Evidence for Late Pleistocene Human Exploitation of Jefferson's Ground Sloth (*Megalonyx jeffersonii*) from Northern Ohio, USA," *World Archaeology* 44 (2012): 75-101

[h]C. Reid Ferring, *The Archaeology and Paleoecology of the Aubrey Clovis Site (41DN479) Denton County, Texas* (Denton: Department of Geography, University of North Texas, 2001).

[i]Guadalupe Sanchez et al., "Human (Clovis)–Gomphothere (*Cuvieronius* sp.) Association ~13,390 Calibrated yBP in Sonora, Mexico," *Proceedings of the National Academy of Sciences* 111 (2014): 10972-10977.

[j]Waters et al., "Late Pleistocene Horse and Camel Hunting at the Southern Margin of the Ice-Free Corridor."

[k]Michael R. Waters et al., "The Buttermilk Creek Complex and the Origins of Clovis at the Debra L. Friedkin Site, Texas," *Science* 331 (2011): 1599-1603

[l]Eileen Johnson, "Cultural Activities and Interactions," in *Lubbock Lake: Late Quaternary Studies on the Southern High Plains*, ed. Eileen Johnson (College Station: Texas A&M University Press, 1987), 120-158; Vance T. Holliday, *Paleoindian Geoarchaeology of the Southern High Plains* (Austin: University of Texas Press, 1997).

[m]L. Adrien Hannus, "Flaked Mammoth Bone from the Lange/Ferguson Site, White River Badlands Area, South Dakota," in *Bone Modification*, ed. Robson Bonnichsen and Marcella H. Sorg (Orono: Center for the Study of the First Americans, University of Maine, 1989), 395-412.

animal killed and butchered at the site. Both the people who made the artifacts and the animal may have been independently attracted to a location by the proximity of a spring or pond. In rare cases, a projectile point may be found imbedded in a skeleton (for example, *Manis* [Washington]).[120] Usually, however, evidence for dismembering the carcass and stripping meat off the bones is based on the presence of characteristic cut and per-cussion marks left by stone tools (for example, *Lubbock Lake* [western Texas]) (box 7.2). And if artifacts are altogether absent, taphonomic analy-sis is necessary simply to establish the presence of humans at the site (for example, *Lindsay* [Montana]).[121]

Modern humans dispersed across North America at a time when post-LGM climates were rapidly transforming the landscape. Retreating ice sheets and altered drainage systems created lakes and ponds across much of the continent, and many of the early sites are associated with them.[122] At *Hebior* (southern Wisconsin), about 90 percent of the skeleton of an adult male mammoth (*Mammuthus jeffersoni*) was found buried in pond depos-its along with two bifacial tools, a chopper, and a flake. Some of the bones (primarily the foot bones) exhibit cut-marks. At the nearby site of *Schaefer*, roughly 75 percent of the skeleton of an adult male mammoth was exca-vated from the margin of a former pond. Two blade-like stone artifacts were found with the bones, about 6 percent of which exhibit cut-marks (on foot bones and axial parts).[123] At *Page-Ladson* (northern Florida), mast-odon remains were found associated with a biface fragment and several flakes in sinkhole deposits near a small pond.[124]

The earliest known diagnostic artifacts (Clovis fluted points) are 13,400 to 13,300 years old, and their appearance coincides with an increase in the size and number of sites. Although the earlier sites are often designated "pre-Clovis," the label is premature because the small samples of artifacts in these sites lack any diagnostic forms.[125] Clovis points are distributed across the entire continent, and the pattern—unique to the North Ameri-can archaeological record—may reflect the initial spread of people whose characteristic artifacts became archaeologically visible only in later and larger sites. (The alternative and less credible explanation for their broad distribution is that the distinctive point style was adopted by many exist-ing groups over a wide area.)

The increase in the number and size of sites may be a product of grow-ing population density (analysis of the genetics of living Native Americans indicates rapid population growth during this interval). In addition to kill-butchery locations, some sites contain traces of extended habitation. The oldest dated Clovis points (ca. 13,400 years ago) are found at *Aubrey* (north-eastern Texas), where both a bison-kill site and a habitation area contain-ing thousands of artifacts were occupied near a spring-fed pond.[126] Younger

BOX 7.2
Taphonomy and the Interpretation of Archaeological Sites

The term *taphonomy* was proposed by the Russian zoologist I. A. Efremov in 1940 for the study of how living animals are transformed into part of the geologic record.[1] Taphonomy is critically important to the interpretation of archaeological sites that contain faunal remains: the bones and teeth (and soft parts, if preserved) of animals contain many clues to the processes by which they died and were buried. The clues may be found among the represented parts of the skeleton, degree of weathering, patterns of breakage, surficial marks (for example, tracks of carnivore teeth), age and sex of the animal, season of death, and other characteristics of the remains.[2]

By itself, the co-occurrence of animal remains and artifacts provides little information about the human past. Taphonomic analyses have played an especially important role in the interpretation of archaeological sites in North America that date to the initial period of settlement. Most of these sites represent locations where large mammals were killed and butchered or their carcasses were scavenged, and they typically contain few or no artifacts.

A classic example is Lubbock Lake (western Texas), where the remains of mammoths, camels, horses, and other vertebrates were found buried in low-energy stream deposits that date to around 13,100 to 12,100 years ago. Associated with the animal remains were stones and boulders too heavy to have been deposited by the stream. The remains were analyzed by taphonomist Eileen Johnson, who determined that the mammoths were represented by an adult and two juveniles, and that some of their bones exhibited breakage characteristic of heavy percussion by large rocks.[3] The tusks had been exposed to significant weathering before burial. Traces of sharp stone tools or flakes used for skinning and defleshing were identified on the mandibles and on some limb bones. Some traces of damage done by carnivores also were observed on the remains.

BOX FIGURE 7.2 Crescentic gouge mark, probably made by a stone scraper, on the surface of the shaft of a mammoth tibia from the site of Dent, identified by Jeffrey J. Saunders. (Photograph by the author [2008])

At *Dent* (northeastern Colorado), where the original Clovis point was found in 1932, Jeffrey Saunders reported diagnostic carcass-processing marks on the bones and teeth of an estimated 13 mammoths.[4] The marks include gouges and pits probably made by bone implements used as pry bars and wedges to dismember the heavy postcranial elements. Traces of stone tools include crescentic marks caused by scrapers (box figure 7.2).

1. I. A. Efremov, "Taphonomy: A New Branch of Paleontology," *Pan-American Geologist* 74 (1940): 81–93.
2. R. Lee Lyman, *Vertebrate Taphonomy* (Cambridge: Cambridge University Press, 1994).
3. Eileen Johnson, "Cultural Activities and Interactions," in *Lubbock Lake: Late Quaternary Studies on the Southern High Plains*, ed. Eileen Johnson (College Station: Texas A&M University Press, 1987), 120–158.
4. Jeffrey J. Saunders, "Processing Marks on Remains of *Mammuthus columbi* from the Dent Site, Colorado, in Light of Those from Clovis, New Mexico," in *Frontiers in Colorado Paleoindian Archaeology: From the Dent Site to the Rocky Mountains*, ed. Robert H. Brunswig and Bonnie L. Pitblado (Boulder: University Press of Colorado, 2007), 155–184.

sites (ca. 12,750 years ago) in the San Pedro Valley of southeastern Arizona include a habitation area and an associated bison-kill location (*Murray Springs*), along with nearby mammoth-kill sites along spring-fed drainages (figure 7.9).[127]

Most of the large-mammal taxa found in the early North American sites (mammoth, mastodon, gomphothere, ground sloth, camel, and horse) were extinct by 12,000 years ago (that is, before the end of the Younger Dryas cold interval).[128] As in Australia (chapter 5), it appears likely that human hunting played some role in these extinctions, but the impact of changing climates on NPP between 15,000 and 12,000 years ago probably was another factor. Unlike in Australia, there is abundant evidence of human predation on these taxa (that is, numerous kill sites dating to the first few millennia of human settlement). Moreover, the effects of the Younger Dryas on NPP varied from one region to another; in some places, climates were wetter during the Younger Dryas than in the preceding warm interval, presumably increasing local NPP and herbivore biomass.[129] In these areas, it would appear that human hunting was the decisive factor in the extinctions.

Radiocarbon dates from the *Paisley Caves* (southern Oregon) suggest that stemmed points in northwestern North America (Western Stemmed Tradition) are as old as the earliest Clovis points. New dates from Cave 5 bracket a stemmed-point fragment between 13,519 to 13,293 and 13,054 to

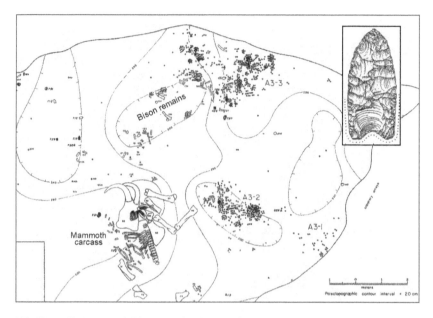

FIGURE 7.9 Clovis mammoth-kill site at and a Clovis fluted point (*inset*) from Murray Springs. (From E. Thomas Hemmings, "Buried Animal Kills and Processing Localities, Areas 1–5," in *Murray Springs: A Clovis Site with Multiple Activity Areas in the San Pedro Valley, Arizona*, ed. C. Vance Haynes and Bruce B. Huckell [Tucson: University of Arizona Press, 2007], 98, fig. 5.7; Bruce B. Huckell, "Clovis Lithic Technology: A View from the Upper San Pedro Valley," in *Murray Springs*, ed. Haynes and Huckell, 196, fig. 8.7*i*. Reprinted with permission from the University of Arizona Press)

12,864 years ago, comparable to the early dates from Aubrey.[130] The Northwest also contains traces of earlier settlement lacking diagnostic artifacts, but the genetic data—combined with the growing evidence that a viable interior route was unavailable until later—suggest that both stemmed and fluted (Clovis) points were developed by the same (diverse) population that left Beringia by way of the Northwest coast. Lineages represented by mtDNA haplogroups A and B were identified from human coprolites in older deposits at the Paisley Caves (see table 7.1; figure 7.10).[131]

The movement of people southward from Beringia along the Pacific coast also helps explain the presence of an established maritime economy and watercraft on the Channel Islands (southern California) by 13,000 years ago.[132] A voyage of at least 9 to 10 kilometers was required to reach the islands at the time. In addition to artifacts and associated faunal remains, *Arlington Springs* (Santa Rosa Island) yielded the oldest dated skeleton in the Western Hemisphere. The diet included marine mammals, fish, and sea birds.

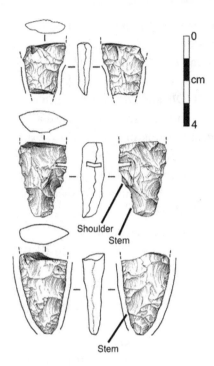

Shoulder
Stem

Stem

FIGURE 7.10 Western Stemmed points from
Paisley Cave 5. (From Dennis L. Jenkins et al.,
"Clovis Age Western Stemmed Projectile Points
and Human Coprolites at the Paisley Caves," *Science*
337 [2012]: 224, fig. 1. Copyright AAAS 2012)

Significantly, the occupants of the Channel Islands made stemmed points, as well as stone crescents, another early diagnostic artifact for western North America.

The Settlement of South America

Both genetic and archaeological evidence indicate that the initial dispersal in South America occurred along the Pacific coast. If it was a southward continuation of the coastal settlement of North America, the dating of the earliest South American sites suggests that the occupation of the Channel Islands was preceded by that of older coastal sites in North America (that is, more than 14,000 years old). And both the genetic and archaeological data suggest that movement down the coast of South America was rapid—completed within a millennium at most. In many respects, the initial settlement of South America appears similar to that of southern Asia, which also entailed the occupation of a rich tropical-coastal environment (chapter 5). The abundance of resources concentrated along a narrow linear habitat is likely to have promoted both steady population growth and rapid movement along the coastal corridor.

Three of the major Native American mtDNA haplogroups (B, C, and D) are widely distributed in South America among living Native Americans, but two subclades of maternal lineage D1 (D1g and D1j) that diverged early from their parent group—probably in Beringia—are concentrated along the Pacific coast.[133] Dated archaeological remains suggest that these sub-clades reached the southern tip of the continent by 13,000 years ago, although there is no available aDNA to confirm this date. As in southern Asia and Australia/New Guinea, some indigenous groups in southern South America appear to represent the original population.

The earliest dated archaeological sites contain substantial evidence of adaptation to coastal environments (table 7.4). At *Huaca Prieta* (Peru), which would have been at least 20 kilometers from the shore at the time, Tom Dillehay and his colleagues found occupations as early as 14,000 years ago containing remains of sea lion (*Otaria* sp.), fish (including sharks), crab, and shellfish, along with deer and various birds. The artifacts were confined to several dozen flake tools and flaking waste.[134] Somewhat younger (ca. 13,000 years old), *Quebrada Jaguay* (Peru) yielded an artifact assemblage of comparable poverty, but faunal remains com-pletely dominated by marine taxa—primarily fish and clams. This site was located only 7 to 8 kilometers from the coast at the time of occupation. The climate and biota were similar to those of the present day (figure 7.11).[135]

The paltry artifact assemblages of the earliest South American sites pro-vide even less insight into the technology of their makers than do the arti-facts from the early sites in North America. The coastal economy pursued by the initial inhabitants of South America would have demanded an array of complex instruments and devices for harvesting fish, possibly including watercraft, nets, leisters, and other gear. Birds may have been hunted with bolas, pitfall traps, and throwing darts.[136] Reduced mobility—often associ-ated with a coastal-maritime economy—could have encouraged semi-permanent dwellings (as among the Andaman Islanders [chapter 5]). A series of circular holes at Quebrada Jaguay may represent some form of structure, while a slightly younger level at the site contained specimens of cordage that may have been part of a fishing net.[137]

Surprisingly, the early South American sites of the coastal zone also include some examples similar to those of the North American interior.

TABLE 7.4

Archaeological Record of *Homo sapiens* in South America, Before 12,800 Years Ago

Site, Layer	Estimated Age (method)	Description	Comments
Monte Verde II (southern Chile)[a]	Ca. 14,260 years (radiocarbon)	Isolated artifacts; former structures	New dates (on seaweed) younger than previously reported dates
Huaca Prieta (Peru)[b]	14,090–13,350 years (radiocarbon)	42 stone artifacts include retouched flakes and utilized cobbles	Diverse array of fauna includes mammals, birds, fish, shellfish, and crabs
Arroyo Seco 2 (Argentina)[c]	Ca. 14,000–13,000 years (radiocarbon)	Unifacial marginally retouched artifacts (quartzite)	Extinct megafauna (horse, ground sloth); early eastern lowland occupation
Lapa do Boquete (eastern Brazil)[d]	Ca. 14,000–12,000 years (radiocarbon)	Artifacts include unifacial flake tools (Itaparica Tradition)	Early lowland occupation
Fell's Cave (southern Chile)[e]	Ca. 13,000 years (radiocarbon)	Artifacts include Fishtail points	Documents humans at southern tip of South America by 13,000 years ago
Quebrada Jaguay (Peru)[f]	Ca. 13,000 years (radiocarbon)	Artifacts include bifaces, flake tools, and debitage	Fauna dominated by fish and molluscs
Quebrada Santa Julia (southern Chile)[g]	Ca. 12,925 years (radiocarbon)	Artifacts include two bifaces, four knives, scraper, graver, and retouched flakes	Extinct megafauna (horse); kill-butchery site on lake margin; fluted-point blank
Piedra Museo, AEP-1 (Argentina)[h]	12,890 years (radiocarbon)	Isolated artifacts include Fishtail point	Rock shelter; extinct megafauna (horse, camel)
Cerro Tres Tetas 1 (Argentina)[i]	12,845 years (radiocarbon)	Artifacts include two cores, two bifaces, eight scrapers, and 474 debitage	Faunal remains sparse, include *Lama guanicoe*

(continued)

(continued)

Site, Layer	Estimated Age (method)	Description	Comments
Quebrada Maní (northern Chile)[j]	12,800–11,700 years (radiocarbon)	Artifacts include three cores, five biface fragments, scrapers, and debitage	Wooden artifacts include possible atlatl shaft; arid-zone occupation

Note: All radiocarbon dates are given in calibrated radiocarbon years before present.

[a]Tom C. Dillehay et al., "Monte Verde: Seaweed, Food, Medicine, and the Peopling of South America," Science 320 (2008): 784–786; Michael R. Waters and Thomas Wier Stafford Jr., "The First Americans: A Review of the Evidence for the Late-Pleistocene Peopling of the Americas," in Paleoamerican Odyssey, ed. Kelly E. Graf, Caroline V. Ketron, and Michael R. Waters (College Station: Texas A&M University Press, 2013), 541–560.

[b]Tom C. Dillehay et al., "A Late Pleistocene Human Presence at Huaca Prieta, Peru, and Early Pacific Coastal Adaptations," Quaternary Research 77 (2012): 418–423.

[c]James Steele and Gustavo Politis, "AMS [14]C Dating of Early Human Occupation of Southern South America," Journal of Archaeological Science 36 (2009): 419–429.

[d]Adriana Schmidt Dias and Lucas Bueno, "The Initial Colonization of South America Eastern Lowlands: Brazilian Archaeology Contributions to Settlement of America Models," in Paleoamerican Odyssey, ed. Graf, Ketron, and Waters, 339–357.

[e]Junius B. Bird, Travels and Archaeology in South Chile, ed. John Hyslop (Iowa City: University of Iowa Press, 1988).

[f]Daniel F. Sandweiss et al., "Quebrada Jaguay: Early South American Maritime Adaptations," Science 281 (1998): 1830–1832.

[g]Donald Jackson et al., "Initial Occupation of the Pacific Coast of Chile During Late Pleistocene Times," Current Anthropology 48 (2007): 725–731.

[h]Steele and Politis, "AMS [14]C Dating of Early Human Occupation of Southern South America."

[i]Rafael S. Paunero, "The Presence of a Pleistocene Colonizing Culture in La Maria Archaeological Locality: Casa del Minero 1, Argentina," in Where the South Winds Blow: Ancient Evidence for Paleo South Americans, ed. Laura Miotti, Monica Salemme, and Nora Flegenheimer (College Station: Texas A&M University Press, 2003), 127–132; Steele and Politis, "AMS [14]C Dating of Early Human Occupation of Southern South America."

[j]Claudio Latorre et al., "Late Pleistocene Human Occupation of the Hyperarid Core in the Atacama Desert, Northern Chile," Quaternary Science Reviews 77 (2013): 19–30.

Quebrada Santa Julia (northern Chile) dates to roughly the same time period as Quebrada Jaguay and is only a few kilometers from the modern coast. Quebrada Santa Julia, however, is more reminiscent of sites such as Hebior and Schaefer in Wisconsin. It was located on the margin of a lake and yielded a small number of stone artifacts (for example, a scraper, a broken pebble, and flakes) associated with bone fragments of extinct horse. In fact, Quebrada Santa Julia contained a fluted-point blank and other traces of fluted-point production.[138] Fluted points are known from other South American sites in this time range (ca. 13,000 years ago); their relationship to the fluted points in North America—which appeared at roughly the same date—is unclear.[139]

Quebrada Santa Julia is one of several South American sites that contain evidence for the hunting of extinct megafauna. South America

FIGURE 7.11 Map of South America, showing archaeological sites, dated to more than 12,000 years ago. (Base map adapted from en.wikipedia.org/ wiki/File:Blank Map-World6.svg)

witnessed a major extinction event following the initial settlement by humans, and, as in North America and Australia, the invaders are suspected to have played some role in it. Many of the extinctions of large mammals occurred between about 13,300 and 11,800 years ago (immediately preceding and during the Younger Dryas), including horses (*Equus* spp. and *Hippidion* spp.), ground sloths (*Eremotherium* spp., *Glossotherium* spp., and *Mylodon darwinii*), and the large ungulate *Toxodon*.[140] As elsewhere, the impact of changing climates on vegetation also must have been a factor in the extinctions, even in the lower latitudes (figure 7.12).[141]

The initial migration down the western coast of South America apparently was soon followed by movements across the Andes and into the

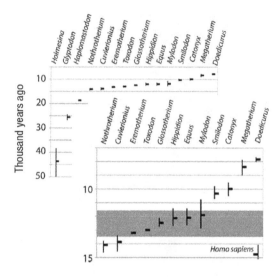

FIGURE 7.12 Last-appearance dates for South American megafauna, most of which cluster between roughly 13,500 and 11,800 years ago. (From Anthony D. Barnofsky and Emily L. Lindsey, "Timing of Quaternary Megafaunal Extinction in South America in Relation to Human Arrival and Climate Change," *Quaternary International* 217 [2010]: 13, fig. 2, with permission of Elsevier)

eastern lowlands. The deglaciation of mountain passes, allowing access to the eastern slope of the range, is thought to have occurred before 14,000 years ago.[142] According to Adriana Schmidt Dias and Lucas Bueno, the eastern lowlands were colonized in the same way as the North American interior, with a "pioneering phase" of low-density population across wide areas. Although several sites in the São Francisco Basin (eastern Brazil) yield dates older than 14,000 years, most of the early lowland occupations date to about 13,000 years ago or later.[143] The artifacts comprise thick flake scrapers, other unifacial flake tools, large utilized flakes, and other relatively nondiagnostic forms (*Itaparica Tradition*).

By 13,000 years ago, modern humans had occupied all unglaciated continental landmasses outside Antarctica (which remained undiscovered until 1775). As noted earlier, central and eastern Canada (as well as coastal Greenland) were settled after the glaciers retreated in the mid-Holocene. The colonization of areas in northwestern Europe covered by the Scandinavian ice sheet during the LGM had taken place earlier, following deglaciation at the end of the Pleistocene.[144] Dispersal among the islands of the central Pacific (Oceania) began after about 3,500 years ago with the spread of the *Lapita cultural complex* across the South Pacific. Subsequent movement northward into central and eastern Polynesia (less than 1,500 years ago) was facilitated by the development of the double canoe, which allowed travel across thousands of kilometers of open sea.[145]

NOTES

1. INFORMATION, COMPLEXITY, AND HUMAN EVOLUTION

1. Charles Darwin, *The Descent of Man and Selection in Relation to Sex* (London: Murray, 1871), 1:136.
2. Thomas Hobbes, *Leviathan, or the Matter, Forme & Power of a Common-wealth Ecclesiastical and Civill* (1651; London: Penguin, 1968), 81.
3. Eörs Szathmáry and John Maynard Smith, "The Major Evolutionary Transitions," *Nature* 374 (1995): 227–232. See also John Maynard Smith and Eörs Szathmáry, *The Major Transitions in Evolution* (Oxford: Oxford University Press, 1995); and John Maynard Smith and Eörs Szathmáry, *The Origins of Life: From the Birth of Life to the Origin of Language* (Oxford: Oxford University Press, 1999). On "emergence" in the context of complexity in evolution, see Stephen Jay Gould, *The Structure of Evolutionary Theory* (Cambridge, Mass.: Belknap Press of Harvard University Press, 2002), 618.
4. The "major transitions" in evolution proposed by Maynard Smith and Szathmáry in the 1990s stimulated much discussion and debate, some of it critical. See Brett Calcott and Kim Sterelny, eds., *The Major Transitions in Evolution Revisited* (Cambridge, Mass.: MIT Press, 2011). Szathmáry revealed that he and Maynard Smith had planned to publish a revised synthesis containing several "major transitions" that had been omitted from *The Major Transitions in Evolution*, including the immune and nervous systems, but the project was abandoned after Maynard Smith's death in 2004. See Eörs Szathmáry and Chrisantha Fernando, "Concluding Remarks," in *Major Transitions in Evolution Revisited*, ed. Calcott and Sterelny, 301–310. Recently, Szathmáry published a revised overview of the transitions: Eörs Szathmáry, "Toward Major Evolutionary Transitions Theory 2.0," *Proceedings of the National Academy of Sciences* 112 (2015): 10104–10111.

5. Andy Clark, *Being There: Putting Brain, Body, and World Together Again* (Cambridge, Mass.: MIT Press, 1998), and *Supersizing the Mind: Embodiment, Action, and Cognitive Extension* (New York: Oxford University Press, 2011), 44–60.
6. Daniel C. Dennett, *Consciousness Explained* (Boston: Little, Brown, 1991), 210.
7. Maynard Smith and Szathmáry, *Major Transitions in Evolution*, 61–78.
8. John Tyler Bonner, *The Evolution of Complexity by Means of Natural Selection* (Princeton, N.J.: Princeton University Press, 1988); Sean B. Carroll, *Endless Forms Most Beautiful: The New Science of Evo Devo and the Making of the Animal Kingdom* (New York: Norton, 2005).
9. Laura F. Landweber and Erik Winfree, eds., *Evolution as Computation: DIMACS Workshop, Princeton, January 1999* (Berlin: Springer, 2002); John E. Mayfield, *The Engine of Complexity: Evolution as Computation* (New York: Columbia University Press, 2013).
10. Charles Darwin, *On the Origin of Species by Means of Natural Selection* (London: Murray, 1859), 490.
11. Mayfield, *Engine of Complexity*.
12. See, for example, John Morgan Allman, *Evolving Brains* (New York: Scientific American Library, 1999), 3–6.
13. Marvin L. Minsky, *Computation: Finite and Infinite Machines* (Englewood Cliffs, N.J.: Prentice Hall, 1967).
14. Clark, *Being There*, and *Supersizing the Mind*, 44–60.
15. Dennett, *Consciousness Explained*, 177.
16. Marc Hauser, Noam Chomsky, and W. Tecumseh Fitch, "The Faculty of Language: What Is It, Who Has It, and How Did It Evolve?" *Science* 298 (2002): 1569–1579; George Boole, *An Investigation of the Laws of Thought, on Which Are Founded the Mathematical Theories of Logic and Probabilities* (1854; New York: Dover, 1958), 24.
17. John F. Hoffecker, "The Information Animal and the Super-Brain," *Journal of Archaeological Method and Theory* 20 (2013): 18–41. Richard Dawkins wrote, "Language . . . is the networking system by which brains (as they are called on this planet) exchange information with sufficient intimacy to allow the development of a cooperative technology" (*River Out of Eden: A Darwinian View of Life* [New York: Basic Books, 1995], 158).
18. Anthony F. C. Wallace, *The Social Context of Innovation: Bureaucrats, Families, and Heroes in the Early Industrial Revolution, as Foreseen in Bacon's* New Atlantis (Princeton, N.J.: Princeton University Press, 1982), 11–61.
19. J. G. D. Clark, "Neolithic Bows from Somerset, England, and the Prehistory of Archery in North-west Europe," *Proceedings of the Prehistoric Society* 29 (1963): 50–98.
20. Thomas D. Seeley compares the "cognitive entity" of the honeybee swarm with the brain of an individual primate, in *Honeybee Democracy* (Princeton, N.J.: Princeton University Press, 2010), 198–217.
21. Rolfe Landauer, "Information Is Physical," *Physics Today* 44 (1991): 23–29.
22. John Napier, *Hands*, rev. Russell H. Tuttle (Princeton, N.J.: Princeton University Press, 1993), 75–87.
23. John F. Hoffecker, *Landscape of the Mind: Human Evolution and the Archaeology of Thought* (New York: Columbia University Press, 2011), 42–52.

24. Jeffrey Laitman, "The Anatomy of Human Speech," *Natural History*, August 1984, 20–27; Philip Lieberman, *The Biology and Evolution of Language* (Cambridge, Mass.: Harvard University Press, 1984), 271–283.

25. Robert Seyfarth, Dorothy Cheney, and Peter Marler, "Vervet Monkey Alarm Calls: Semantic Communication in a Free-Ranging Primate," *Animal Behaviour* 28 (1980): 1070–1094. See also Marc D. Hauser, *The Evolution of Communication* (Cambridge, Mass.: MIT Press, 1996), 306–309.

26. Ralph Holloway, "Evolution of the Human Brain," in *Handbook of Human Symbolic Evolution*, ed. Andrew Lock and Charles R. Peters (Oxford: Clarendon Press, 1995), 74–125. See also Phillip V. Tobias, "The Brain of *Homo habilis*: A New Level of Organization in Cerebral Evolution," *Journal of Human Evolution* 16 (1987): 741–762; and Dean Falk, *Braindance: New Discoveries About Human Origins and Brain Evolution*, rev. ed. (Gainesville: University Press of Florida, 2004), 145–148.

27. Steven Pinker, *The Language Instinct: How the Mind Creates Language* (New York: Morrow, 1994), 295–301.

28. Tanya M. Smith et al., "Earliest Evidence of Modern Human Life History in North African Early *Homo sapiens*," *Proceedings of the National Academy of Sciences* 104 (2007): 6128–6133; Tanya M. Smith et al., "Dental Evidence for Ontogenetic Differences Between Modern Humans and Neanderthals," *Proceedings of the National Academy of Sciences* 107 (2010): 20923–20928.

29. Allman, *Evolving Brains*, 122–157.

30. Daniel E. Lieberman, *The Evolution of the Human Head* (Cambridge, Mass.: Belknap Press of Harvard University Press, 2011), 503–506.

31. Karin Isler and Carel P. van Schaik, "How Humans Evolved Large Brains: Comparative Evidence," *Evolutionary Anthropology* 23 (2014): 65–75.

32. For the definition of an evolutionarily stable (and unstable) strategy or phenotype, see John Maynard Smith, *Evolution and the Theory of Games* (Cambridge: Cambridge University Press, 1982), 10–11.

33. Robin I. M. Dunbar, "The Social Brain Hypothesis," *Evolutionary Anthropology* 6 (1998): 178–190.

34. For the definition of an evolutionary "arms race," see Richard Dawkins, *The Extended Phenotype: The Long Reach of the Gene* (Oxford: Oxford University Press, 1999). Potential constraints on human brain growth are discussed in Lieberman, *Evolution of the Human Head*, 221–223.

35. Christopher Dean et al., "Growth Processes in Teeth Distinguish Modern Humans from *Homo erectus* and Earlier Hominins," *Nature* 414 (2001): 628–631; H. Coqueugniot et al., "Early Brain Growth in *Homo erectus* and Implications for Cognitive Ability," *Nature* 431 (2004): 299–302.

36. Smith et al., "Earliest Evidence of Modern Human Life History in North African Early *Homo sapiens*"; Smith et al., "Dental Evidence for Ontogenetic Differences Between Modern Humans and Neanderthals."

37. Philipp Gunz et al., "A Uniquely Modern Human Pattern of Endocranial Development: Insights from a New Cranial Reconstruction of the Neandertal Newborn from Mezmaiskaya," *Journal of Human Evolution* 62 (2012): 300–313.

38. Quentin D. Atkinson, "Phonemic Diversity Supports a Serial Founder Effect Model of Language Expansion from Africa," *Science* 332 (2011): 346–349.

39. Richard G. Klein, *The Human Career: Human Biological and Cultural Origins*, 3rd ed. (Chicago: University of Chicago Press, 2009), 282–304. There is considerable anatomical variability in the African fossils in this time range assigned to *Homo sapiens*. See, for example, Günter Bräuer, "The Origin of Modern Anatomy: By Speciation or Intraspecific Evolution?" *Evolutionary Anthropology* 17 (2008): 22–37; and Philipp Gunz et al., "Early Modern Human Diversity Suggests Subdivided Population Structure and a Complex Out-of-Africa Scenario," *Proceedings of the National Academy of Sciences* 106 (2009): 6094–6098.

40. Lieberman, *Evolution of the Human Head*, 570, table 13.5.

41. G. David Poznik et al., "Sequencing Y Chromosomes Resolves Discrepancy in Time to Common Ancestor of Males versus Females," *Science* 341 (2013): 562–569.

42. J. R. Stewart and C. B. Stringer, "Human Evolution Out of Africa: The Role of Refugia and Climate Change," *Science* 335 (2012): 1317–1321.

43. For example, Klein, *Human Career*, 627.

44. Toomas Kivisild, "Maternal Ancestry and Population History from Whole Mitochondrial Genomes," *Investigative Genetics* 6 (2015).

45. Brenna M. Henn et al., "Hunter-Gatherer Genomic Diversity Suggests a Southern African Origin for Modern Humans," *Proceedings of the National Academy of Sciences* 108 (2011): 5154–5162.

46. See, for example, John J. Shea, "*Homo sapiens* Is as *Homo sapiens* Was," *Current Anthropology* 52 (2011): 1–15; and Paola Villa and Wil Roebroeks, "Neandertal Demise: An Archaeological Analysis of the Modern Human Superiority Complex," *PLoS ONE* 9 (2014): e96424

47. Jayne Wilkins et al., "Evidence for Early Hafted Hunting Technology," *Science* 338 (2012): 942–946; Sally McBrearty and Alison S. Brooks, "The Revolution That Wasn't: A New Interpretation of the Origin of Modern Human Behavior," *Journal of Human Evolution* 39 (2000): 453–563.

48. Johannes Krause et al., "The Derived FOXP2 Variant of Modern Humans Was Shared with Neandertals," *Current Biology* 17 (21) (2007): 1908–1912.

49. See, for example, Klein, *Human Career*, 643–649.

50. Kyle S. Brown et al., "Fire as an Engineering Tool of Early Modern Humans," *Science* 325 (2009): 859–862; John E. Yellen et al., "A Middle Stone Age Worked Bone Industry from Katanda, Upper Semliki Valley, Zaire," *Science* 268 (1995): 553–556.

51. Marian Vanhaeren et al., "Middle Paleolithic Shell Beads in Israel and Algeria," *Science* 312 (2006): 1785–1788.

52. Lyn Wadley, "Were Snares and Traps Used in the Middle Stone Age and Does It Matter? A Review and a Case Study from Sibudu, South Africa," *Journal of Human Evolution* 58 (2010): 179–192; Paola Villa et al., "Border Cave and the Beginning of the Later Stone Age in South Africa," *Proceedings of the National Academy of Sciences* 109 (2012): 13208–13213.

53. Klein wrote that "archaeology helps to explain why the expansion occurred only about 50 ka" (*Human Career*, 644).

54. See, for example, John F. Hoffecker, *Desolate Landscapes: Ice-Age Settlement of Eastern Europe* (New Brunswick, N.J.: Rutgers University Press, 2002), 158.

55. For discussions of hunter-gatherer diet and ecology in cold and dry environments, as well as complex technology, see Robert L. Kelly, *The Lifeways of Hunter-Gatherers: The Foraging Spectrum* (Cambridge: Cambridge University Press, 2013).

56. Derek Bickerton, *Language and Species* (Chicago: University of Chicago Press, 1990).

57. Ernst Mayr, *The Growth of Biological Thought: Diversity, Evolution, and Inheritance* (Cambridge, Mass.: Belknap Press of Harvard University Press, 1982), 520.

58. Darwin, *On the Origin of Species*, 45–46. *Coccus* is actually a genus (of beetles), and Darwin presumably was referring to variation within one species of *Coccus*.

59. According to Mayr, "Throughout his life Darwin believed in both soft and hard inheritance, changing his opinion only on the relative importance of the two" (*Growth of Biological Thought*, 689). The term "soft inheritance" refers to the belief that either the environment or "use versus disuse" could affect heritable characters in plants and animals.

60. Jacques Monod, *Chance and Necessity: An Essay of the Natural Philosophy of Modern Biology*, trans. Austryn Wainhouse (New York: Knopf, 1971), 118. Richard C. Lewontin wrote, "The essential new feature that Wright brought into the theory of evolution really was the role of random processes. Most important, he realized that random processes played a significant part in the process of speciation" ("Theoretical Population Genetics in the Evolutionary Synthesis," in *The Evolutionary Synthesis: Perspectives on the Unification of Biology*, ed. Ernst Mayr and William B. Provine [Cambridge, Mass.: Harvard University Press, 1980], 61). Lewontin noted the impact of a specific paper: Sewall Wright, "Evolution of Mendelian Populations," *Genetics* 16 (1931): 97–159.

61. Although the structure of DNA molecules (that is, double helix) was discovered in 1953, the complete genetic code (comprising 64 triplet codons for specific amino acids) was not deciphered until 1966 by Marshall Nirenberg and Har Gobind Khorana.

62. In addition to skeletal remains of anatomically modern humans discovered in association with Paleolithic artifacts and extinct animals such as mammoth (for example, the discovery in 1868 at Cro-Magnon [France]), the remains of a Neanderthal, discovered in Germany in 1856, were widely known and discussed as fossil evidence for human evolution.

63. Darwin wrote: "It is therefore probable that Africa was formerly inhabited by extinct apes closely allied to the gorilla and chimpanzee; and as these two species are now man's nearest allies, it is somewhat more probable that our early progenitors lived on the African continent than elsewhere" (*Descent of Man*, 1:199).

64. Ibid., 140–141.

65. Ibid., 139.

66. Ibid., 57–58.

67. W. Tecumseh Fitch found Darwin's theory of language evolution "still worthy of serious attention today" (*The Evolution of Language* [Cambridge: Cambridge University Press, 2010], 398).

68. Darwin, *Descent of Man*, 1:56–57.

69. Ibid., 57.

70. Ibid., 2–3.

71. Ibid, 240–250.

72. Charles Darwin, *The Descent of Man and Selection in Relation to Sex* (London: Murray, 1871), 2:355–384.

73. Richard C. Lewontin, "The Apportionment of Human Diversity," *Evolutionary Biology* 6 (1972): 381–398.
74. See, for example, Roger Lewin, *Bones of Contention: Controversies in the Search for Human Origins* (New York: Simon and Schuster, 1987), 47–62.
75. For an account of evolutionary biology during the years between the publication of *The Origin* and the evolutionary synthesis, see Mayr, *Growth of Biological Thought*, 535–570. He argues that part of the problem lay in the diversification of biology into many sub-fields in the late nineteenth century.
76. Michael H. Day, *Guide to Fossil Man: A Handbook of Human Palaeontology*, 3rd ed. (Chicago: University of Chicago Press, 1977), 284–324; Klein, *Human Career*, 282–304.
77. J. S. Weiner, K. P. Oakley, and W. E. Le Gros Clark, "The Solution of the Piltdown Problem," *Bulletin of the British Museum (Natural History), Geology* 2 (1953): 139–146.
78. Day, *Guide to Fossil Man*, 47–81.
79. Earnest Albert Hooton thought that the Broken Hill skull "with his wholly simian brow ridges and . . . small brain of inferior human pattern" might "not be more than a few thousand years old," but that "it is perfectly possible that an ancient and primitive type of man may have survived to comparatively recent times in this out of the way spot in south central Africa" (*Up from the Ape* [New York: Macmillan, 1937], 346, 339). While Hooton thought it unlikely that any modern humans were directly descended from Broken Hill, he wrote that "to me there is something about the face of this specimen which suggests a very low and bestial proto-Negroid," and that "some features . . . also recall the cranial form of the Australian native" (347).
80. Richard G. Klein, "Geological Antiquity of Rhodesian Man," *Nature* 244 (1973): 311–312.
81. Raymond A. Dart, "*Australopithecus africanus*: The Man-Ape of South Africa," *Nature* 115 (1925): 195–199.
82. Lewin, *Bones of Contention*, 47–84.
83. The classic example is W. J. Sollas, *Ancient Hunters and Their Modern Representatives* (London: Macmillan, 1911), which compared the Tasmanians to the Neanderthals.
84. For example, the Heidelberg mandible had been recovered from deposits containing an older faunal complex than that found in association with Neanderthal and early modern human remains, as described by Marcellin Boule, *Fossil Men: Elements of Human Palaeontology*, trans. Jessie Elliot Ritchie and James Ritchie (Edinburgh: Oliver and Boyd, Tweeddale Court, 1923), 147–150.
85. Carl C. Swisher III, Garniss H. Curtis, and Roger Lewin, *Java Man: How Two Geologists' Dramatic Discoveries Changed Our Understanding of the Evolutionary Path to Modern Humans* (New York: Scribner, 2000).
86. Some perspective on the challenge faced by paleoanthropologists during the early twentieth century is provided by Osbjorn M. Pearson, "Statistical and Biological Definitions of 'Anatomically Modern' Humans: Suggestions for a Unified Approach to Modern Morphology," *Evolutionary Anthropology* 17 (2008): 38–48. He describes the struggle to arrive at a systematic and quantitative definition of *Homo sapiens* anatomy during the past few decades.

87. Day, *Guide to Fossil Man*, 269.

88. Franz Weidenreich, *Apes, Giants, and Man* (Chicago: University of Chicago Press, 1946), 84–85. The potential pitfalls of relying on skeletal anatomy for human phylogenetic reconstruction are best illustrated by the recent controversy over the nearly complete skeleton found at Kennewick in the state of Washington in 1996 (dated to between 9,200 and 8,340 years ago). On the basis of the skeletal traits, the remains were variously attributed to European and North Pacific peoples. But analysis of ancient DNA revealed that the skeleton belonged to well-established Native American lineages, including mtDNA haplogroup X2a and Y-DNA haplogroup Q-M3. See Morten Rasmussen et al., "The Ancestry and Affiliations of Kennewick Man," *Nature* 523 (2015): 455–458.

89. C. Loring Brace, "The Roots of the Race Concept in American Physical Anthropology," in *A History of American Physical Anthropology, 1930–1980*, ed. Frank Spencer (New York: Academic Press, 1982), 11–29; Lewin, *Bones of Contention*, 305–308.

90. Henry Fairfield Osborn wrote in reference to the geographic races of Europe, East Asia, and Africa:

    The spiritual, moral, and physical characters which separate these three great human stocks are very profound and ancient. In the author's opinion these three primary stocks diverged from each other during the Age of Mammals, even before the beginning of the Pleistocene or Ice Age. . . . The Mongolian is somewhat less profoundly different from the Caucasian than is the Negro. The intelligence and morale of the Mongolian may fully reach the high Caucasian level. (*Man Rises to Parnassus: Critical Epochs in the Prehistory of Man*, 2nd ed. [Princeton, N.J.: Princeton University Press, 1928], 201–202)

91. Sir Arthur Keith, *The Antiquity of Man* (London: Williams and Norgate, 1915). Keith was particularly outspoken in his concerns about what he termed "deracialization," suggesting some years later that "this antipathy or race prejudice Nature has implanted within you for her own ends—the improvement of Mankind through race differentiation" (quoted in Brace, "Roots of the Race Concept in American Physical Anthropology," 13).

92. Harry Lionel Shapiro, *Peking Man* (New York: Simon and Schuster, 1974), 11–26. Although the prewar Zhoukoudian fossils were never recovered, high-quality casts had been made of all the remains.

93. Richard G. Klein, *F. Clark Howell, 1925–2007: A Biographical Memoir* (Washington, D.C.: National Academy of Sciences, 2013).

94. William Glen, *The Road to Jaramillo: Critical Years of the Revolution in Earth Science* (Stanford, Calif.: Stanford University Press, 1982). Methodological advances in geosciences with implications for paleoanthropology were not confined to chronometry. The entire framework for the Quaternary was revised on the basis of oxygen-isotope stratigraphy.

95. President Harry S Truman signed Executive Order 9981 on July 26, 1948, abolishing racial discrimination in the United States armed forces. A few days earlier, Governor Strom Thurmond of South Carolina had been nominated in Alabama for president on a segregationist platform.

96. Richard Leroy Hay, *Geology of the Olduvai Gorge: A Study of the Sedimentation in a Semiarid Basin* (Berkeley: University of California Press, 1976); Glen, *Road to Jaramillo*, 76–77.
97. L. S. B. Leakey, P. V. Tobias, and J. R. Napier, "A New Species of the Genus *Homo* from Olduvai Gorge," *Nature* 202 (1964): 7–9.
98. Robert Broom had discovered more australopith fossils in South Africa immediately after the Second World War (1947–1951).
99. Yves Coppens et al., eds., *Earliest Man and Environments in the Lake Rudolf Basin: Stratigraphy, Paleoecology, and Evolution* (Chicago: University of Chicago Press, 1976).
100. Carlton S. Coon concluded, "If Africa was the cradle of mankind, it was only an indifferent kindergarten. Europe and Asia were our principal schools" (*The Origin of Races* [New York: Knopf, 1962], 656).
101. For a detailed account of the reaction to Coon's book and related developments, see Erik Trinkaus and Pat Shipman, *The Neandertals: Of Skeletons, Scientists, and Scandal* (New York: Vintage Books, 1994), 312–324. A more balanced view is presented in Ian Tattersall and Rob DeSalle, *Race? Debunking a Scientific Myth* (College Station: Texas A&M University Press, 2011), 41–45.
102. Lawrence I. Grossman and Derek E. Wildman, *Morris Goodman, 1925–2010: A Biographical Memoir* (Washington, D.C.: National Academy of Sciences, 2014), 7; Morris Goodman, "Immunochemistry of the Primates and Primate Evolution," *Annals of the New York Academy of Sciences* 102 (1962): 219–234.
103. See, for example, David Pilbeam, *The Ascent of Man: An Introduction to Human Evolution* (New York: Macmillan, 1972), 85–99. Despite the paucity of fossil data, *Ramapithecus* was believed to have been bipedal and a tool-maker. See Lewin, *Bones of Contention*, 87–95.
104. Morris Goodman, "Evolution of the Immunologic Species Specificity of Human Serum Proteins," *Human Biology* 34 (1962):104–150.
105. Vincent M. Sarich and Allan C. Wilson, "Immunological Time Scale for Hominid Evolution," *Science* 158 (1967): 1200–1203.
106. Leonard Owen Greenfield, "On the Adaptive Pattern of 'Ramapithecus,'" *American Journal of Physical Anthropology* 50 (1979): 527–548. See also Lewin, *Bones of Contention*, 105–127.
107. See, for example, Philip W. Hedrick, *Genetics of Populations*, 4th ed. (Sudbury, Mass.: Jones and Bartlett, 2011), 295–315; and Richard C. Lewontin and John L. Hubby, "A Molecular Approach to the Study of Genic Heterozygosity in Natural Populations of *Drosophila pseudoobscura*," *Genetics* 54 (1966): 595–609.
108. Motoo Kimura, "Evolutionary Rate at the Molecular Level," *Nature* 217 (1968): 624–626.
109. Lewontin, "Apportionment of Human Diversity."
110. Decades later, a former student of Lewontin's emphasized the relevance of "The Apportionment of Human Diversity" to the study of the origins of modern humans. See Spencer Wells, *The Journey of Man: A Genetic Odyssey* (New York: Random House, 2002), 16–18.
111. Lewontin, "Apportionment of Human Diversity," 395. A decade later, Richard Lewontin mentioned "the recency of the spread of the human species", in *Human Diversity* (New York: Scientific American Books, 1982), 167.

112. Eric S. Lander et al., "Initial Sequencing and Analysis of the Human Genome," *Nature* 409 (2001): 860–921.
113. Nessa Carey, *Junk DNA: A Journey Through the Dark Matter of the Genome* (New York: Columbia University Press, 2015), 25–36.
114. John S. Mattick, "A New Paradigm for Developmental Biology," *Journal of Experimental Biology* 210 (2007): 1526–1547; John S. Mattick et al., "RNA Regulation of Epigenetic Processes," *BioEssays* 31 (2009): 51–59. In 2007, Mattick labeled the earlier belief that the non-protein-coding DNA has little or no function in evolution or development as "the biggest mistake in the history of molecular biology" ("New Paradigm for Developmental Biology," 1526).
115. Carroll, *Endless Forms Most Beautiful*, 6.
116. Nessa Carey, *The Epigenetics Revolution: How Modern Biology Is Rewriting Our Understanding of Genetics, Disease, and Inheritance* (New York: Columbia University Press, 2012).
117. John Maynard Smith wrote, "Around 1960, I conceived the idea that, using information theory, one could quantify evolution simultaneously at three levels—genetic, selective, and morphological" ("The Concept of Information in Biology," *Philosophy of Science* 67 [2000]: 185–186).
118. Claude E. Shannon, "A Mathematical Theory of Communication," *Bell System Technical Journal*, July 1948, 379–423. See also John R. Pierce, *An Introduction to Information Theory: Symbols, Signals and Noise*, 2nd ed. (New York: Dover, 1980); and Luciano Floridi, *Information: A Very Short Introduction* (New York: Oxford University Press, 2010).
119. Peter Godfrey-Smith, "Information in Biology," in *The Cambridge Companion to the Philosophy of Biology*, ed. David L. Hull and Michael Ruse (Cambridge: Cambridge University Press, 2007), 103–119. See also Philip C. Jackson Jr., *Introduction to Artificial Intelligence*, 2nd ed. (New York: Dover, 1985), 273–340.
120. Fred I. Dretske, *Knowledge and the Flow of Information* (Cambridge, Mass.: MIT Press, 1981), 45; Kim Sterelny, Kelly C. Smith, and Michael Dickison, "The Extended Replicator," *Biology and Philosophy* 19 (1996): 205–222.
121. Maynard Smith, "Concept of Information in Biology," 183; Ulrich E. Stegman, "The Arbitrariness of the Genetic Code," *Biology and Philosophy* 19 (2004): 205–222.
122. Maynard Smith, "Concept of Information in Biology," 192–193; Godfrey-Smith, "Information in Biology."
123. Landauer, "Information is Physical." The evolutionary biologist George C. Williams wrote, "Information can exist only as a material pattern, but the same information can be recorded by a variety of patterns in many different kinds of material" (*Natural Selection: Domains, Levels, and Challenges* [New York: Oxford University Press, 1992], 10). For a discussion of the distinction between "digital" and "analog," see Dretske, *Knowledge and the Flow of Information*, 135–141.
124. Dretske, *Knowledge and the Flow of Information*, 63–82.
125. See, for example, Melanie Mitchell, *Complexity: A Guided Tour* (Oxford: Oxford University Press, 2009), 169.
126. Rainer Feistal and Werner Ebeling wrote that "there is no symbolic information processing without life, and there is no life without symbolic information processing" (*Physics of Self-Organization and Evolution* [Weinheim: Wiley-VCH, 2011], 363).

127. Herbert A. Simon, "The Architecture of Complexity," *Proceedings of the American Philosophical Society* 106 (1962): 467–482.
128. See, for example, Gould, *Structure of Evolutionary Theory*, 595–744.
129. Mitchell, *Complexity*, 94–111.
130. Christos H. Papadimitriou and Kenneth Steiglitz, *Combinatorial Optimization: Algorithms and Complexity* (Mineola, N.Y.: Dover, 1998).
131. Christoph Adami, Charles Ofria, and Travis C. Collier, "Evolution of Biological Complexity," *Proceedings of the National Academy of Sciences* 97 (2000): 4463–4468.
132. See, for example, Christoph Adami, *Introduction to Artificial Life* (New York: Springer, 1998).
133. Christoph Adami, "The Use of Information Theory in Evolutionary Biology," *Annals of the New York Academy of Sciences* 1256 (2012): 49–65.
134. Mayr, *Growth of Biological Thought*, 520.
135. Robert J. Brooker, *Genetics: Analysis and Principles*, 2nd ed. (New York: McGraw-Hill, 2005), 292.
136. Landweber and Winfree, eds., *Evolution as Computation*; Mayfield, *Engine of Complexity*, 137–144. See also Richard C. Lewontin, "Evolution and the Theory of Games," *Journal of Theoretical Biology* 1 (1961): 382–403.
137. Adami, "Use of Information Theory in Evolutionary Biology," 54–61.
138. For example, Matthias Meyer and his colleagues reported a 400,000-year-old human mitogenome from Europe, in "A Mitochondrial Genome Sequence of a Hominin from Sima de los Huesos," *Nature* 505 (2014): 403–406.
139. Bonner, *Evolution of Complexity by Means of Natural Selection*; Daniel W. McShea, "Metazoan Complexity and Evolution: Is There a Trend?" *Evolution: International Journal of Organic Evolution* 50 (1996): 477–492.
140. Eörs Szathmáry, "Toward Major Evolutionary Transitions Theory 2.0," *Proceedings of the National Academy of Sciences* 112 (2015): 10104–10111.
141. Szathmáry and Maynard Smith, "Major Evolutionary Transitions." See also Maynard Smith and Szathmáry, *Major Transitions in Evolution*, and *Origins of Life*; and Szathmáry and Fernando, "Concluding Remarks."
142. McShea, "Metazoan Complexity and Evolution"; Danail Bonchev and Gregory A. Buck, "Quantitative Measures of Network Complexity," in *Complexity in Chemistry, Biology, and Ecology*, ed. Danail Bonchev and Dennis H. Rouvray (New York: Springer, 2005), 191–235. See also Mitchell, *Complexity*, 94–111.
143. Nick Lane, *The Vital Question: Energy, Evolution, and the Origins of Complex Life* (New York: Norton, 2015).
144. Bonner, *Evolution of Complexity by Means of Natural Selection*, 61–97; Maynard Smith and Szathmáry, *Major Transitions in Evolution*, 203–215; Mattick, "New Paradigm for Developmental Biology."
145. George C. Williams, *Sex and Evolution* (Princeton, N.J.: Princeton University Press, 1975); Maynard Smith and Szathmáry, *Origins of Life*, 79–93.
146. Dawkins, *Extended Phenotype*, 55–80. Leigh Van Valen characterized the "perpetual motion" of competition among life-forms as the "Red Queen's Hypothesis," in the classic paper "A New Evolutionary Law," *Evolutionary Theory* 1 (1973): 1–30.
147. Allman, *Evolving Brains*, 64–83.
148. Maynard Smith and Szathmáry, *Major Transitions in Evolution*, 121–145.

149. Gerald M. Edelman, *Neural Darwinism: The Theory of Neuronal Group Selection* (New York: Basic Books, 1987), 73–104; Bonner, *Evolution of Complexity by Means of Natural Selection*, 154–160; Mayfield, *Engine of Complexity*, 176–181.

150. William H. Calvin, "The Brain as a Darwin Machine," *Nature* 330 (1987): 33–34; Michael J. Noad et al., "Cultural Revolution in Whale Songs," *Nature* 408 (2000): 537. Another possible example is found among New Caledonian crows, which exhibit individual variations in making tools, as reported by Gavin R. Hunt et al., "Innovative Pandanus-Tool Folding by New Caledonian Crows," *Australian Journal of Zoology* 55 (2007): 291–298.

151. Calvin, "Brain as a Darwin Machine;" Mayfield, *Engine of Complexity*, 176–181.

152. As noted earlier, male humpback whale songs, which manifest individual creative variations, are an isolated exception to this pattern among nonhuman animals, according to Noad et al. "Cultural Revolution in Whale Songs."

153. Hoffecker, "Information Animal and the Super-Brain."

154. Dawkins, *River Out of Eden*, 158. Collective technological computation with genetic information is found among social insects—including ants, termites, and bees—some of which construct complex nests with hierarchically organized components. See, for example, Bert Hölldobler and E. O. Wilson, *The Super-Organism: The Beauty, Elegance, and Strangeness of Insect Societies* (New York: Norton, 2009), 338–339.

155. Francis Heylighen, "The Growth of Structural and Functional Complexity During Evolution," in *The Evolution of Complexity: The Violet Book of "Einstein Meets Magritte,"* ed. Francis Heylighen, Johan Bollen, and Alexander Riegler (Dordrecht: Kluwer, 1999), 17–44 .

156. Regna Darnell, "The Anthropological Concept of Culture at the End of the Boasian Century," *Social Analysis* 41 (1997): 42–54. See also Adam Kuper, *Culture: The Anthropologist's Account* (Cambridge, Mass.: Harvard University Press, 1999).

157. Hoffecker, *Landscape of the Mind*, 100–109.

158. See, for example, Gordon W. Hewes, "A History of Speculation on the Relation Between Tools and Language," in *Tools, Language and Cognition in Human Evolution*, ed. Kathleen R. Gibson and Tim Ingold (Cambridge: Cambridge University Press, 1993), 20–31; and Thomas Wynn and Frederick L. Coolidge, "Archeological Insights into Hominin Cognitive Evolution," *Evolutionary Anthropology* 25 (2016): 200–213.

2. MODERN HUMAN ORIGINS AND DISPERSAL

1. Richard E. Leakey provides a colorful account of the Omo Valley expedition in 1967, during the course of which he and his colleagues were attacked by a crocodile, in *One Life: An Autobiography* (Salem, Mass.: Salem House, 1983), 84–93.

2. Karl W. Butzer, "The Mursi, Nkalabong, and Kibish Formations, Lower Omo Basin, Ethiopia," in *Earliest Man and Environments in the Lake Rudolf Basin: Stratigraphy, Paleoecology, and Evolution*, ed. Yves Coppens et al. (Chicago: University of Chicago Press, 1976), 12–23. Member I of the Kibish Formation was dated by uranium-thorium method.

3. Richard G. Klein, "Geological Antiquity of Rhodesian Man," *Nature* 244 (1973): 311–312. For an updated discussion of the dating of the Broken Hill skull, see Chris Stringer, *Lone Survivors: How We Came to Be the Only Humans on Earth* (New York: Times Books, 2012), 253–258. The skull now appears to be somewhat younger than believed in recent years—only 200,000 years old.

4. Günter Bräuer, "The Evolution of Modern Humans: A Comparison of the African and Non-African Evidence," in *The Human Revolution: Behavioural and Biological Perspectives in the Origins of Modern Humans*, ed. Paul Mellars and Chris Stringer (Princeton, N.J.: Princeton University Press, 1989), 123–154; Richard G. Klein, *The Human Career: Human Biological and Cultural Origins*, 3rd ed. (Chicago: University of Chicago Press, 2009), 628, table 7.2.

5. Michael H. Day and Chris B. Stringer, "A Reconsideration of the Omo Kibish Remains and the *erectus–sapiens* Transition," in *L'Homo erectus et la place de l'homme de Tautavel parmi les hominides fossiles*, ed. Henry de Lumley (Nice: Centre National de la Recherche Scientifique, 1982), 814–846; Günter Bräuer, "Early Anatomically Modern Man in Africa and the Replacement of the Mediterranean and European Neandertals," in *L'Homo erectus et la place de l'homme de Tautavel parmi les hominides fossiles*, ed. Lumley, 112. Stringer has emphasized his use of multivariate analyses in the study of human fossils, underscoring the weakness of earlier research on human fossil anatomy—the biased selection of particular traits used as a basis for interpretations of evolutionary relationships among various fossils—in *Lone Survivors*, 19–20.

6. In their paper, Day and Stringer observed that "if we cannot provide a valid definition of the only living human species there is little hope of our recognising extinct taxa" ("Reconsideration of the Omo Kibish Remains and the *erectus-sapiens* Transition," 831).

7. Osbjorn M. Pearson, "Statistical and Biological Definitions of 'Anatomically Modern' Humans: Suggestions for a Unified Approach to Modern Morphology," *Evolutionary Anthropology* 17 (2008): 38–48.

8. See, for example, Richard G. Klein, *The Human Career: Human Biological and Cultural Origins* (Chicago: University of Chicago Press, 1989), 348–351; and Ian Tattersall and Jeffrey H. Schwartz, "The Morphological Distinctiveness of *Homo sapiens* and Its Recognition in the Fossil Record: Clarifying the Problem," *Evolutionary Anthropology* 17 (2008): 49–54.

9. Milford H. Wolpoff, "Multiregional Evolution: The Fossil Alternative to Eden," in *Human Revolution*, ed. Mellars and Stringer, 91.

10. See, for example, Milford H. Wolpoff, Wu Xin Zhi, and Alan G. Thorne, "Modern *Homo sapiens* Origins: A General Theory of Hominid Evolution Involving the Fossil Evidence from East Asia," in *The Origins of Modern Humans: A World Survey of the Fossil Evidence*, ed. Fred H. Smith and Frank Spencer (New York: Liss, 1984), 411–483.

11. Rebecca L. Cann, Mark Stoneking, and Allan C. Wilson, "Mitochondrial DNA and Human Evolution," *Nature* 325 (1987): 31–36; L. Vigilant et al., "African Populations and the Evolution of Human Mitochondrial DNA," *Science* 253 (1991): 1503–1507.

12. Richard C. Lewontin, "The Apportionment of Human Diversity," *Evolutionary Biology* 6 (1972): 381–398.

13. Alan R. Templeton, "Human Origins and the Analysis of Mitochondrial DNA Sequences," *Science* 255 (1992): 737, and "The 'Eve' Hypothesis: A Genetic Critique and Re-analysis," *American Anthropologist* 95 (1993): 51–72.
14. Satoshi Horai et al., "Recent African Origin of Modern Humans Revealed by Complete Sequences of Hominoid Mitochondrial DNAs," *Proceedings of the National Academy of Sciences* 92 (1995): 532–536. Another major mitochondrial DNA study was published in 2000: Max Ingman et al., "Mitochondrial Genome Variation and the Origin of Modern Humans," *Nature* 408 (2000): 708–713.
15. Sally McBrearty and Alison S. Brooks, "The Revolution That Wasn't: A New Interpretation of the Origin of Modern Human Behavior," *Journal of Human Evolution* 39 (2000): 453–563.
16. R. L. Dorit, H. Akashi, and W. Gilbert, "Absence of Polymorphism at the ZFY Locus in the Human Y Chromosome," *Science* 268 (1995): 1183–1185. See also Svante Pääbo, "The Y Chromosome and the Origin of All of Us (Men)," *Science* 268 (1995): 1141–1142.
17. Peter A. Underhill et al., "Y Chromosome Sequence Variation and the History of Human Populations," *Nature Genetics* 26 (2000): 358–361.
18. Matthias Krings et al., "Neanderthal DNA Sequences and the Origin of Modern Humans," *Cell* 90 (1997): 19–30.
19. Cidália Duarte et al., "The Early Upper Paleolithic Human Skeleton from the Abrigo do Lagar Velho (Portugal) and Modern Human Emergence in Iberia," *Proceedings of the National Academy of Sciences* 96 (1999): 7604–7609; João Zilhão et al., "The Peştera cu Oase People, Europe's Earliest Modern Humans," in *Rethinking the Human Revolution: New Behavioural and Biological Perspectives on the Origin and Dispersal of Modern Humans*, ed. Paul Mellars et al. (Cambridge: McDonald Institute for Archaeological Research, Cambridge University, 2007), 249–262.
20. An example is the Szeletian industry of central Europe, which contains artifact types associated with both Neanderthals and modern humans, and was attributed to "an acculturation process at the junction of Middle and Upper Paleolithic . . . most likely the creation of Neanderthal man" (P. Allsworth-Jones, "The Szeletian and the Stratigraphic Succession in Central Europe and Adjacent Areas: Main Trends, Recent Results, and Problems for Resolution," in *The Emergence of Modern Humans: An Archaeological Perspective*, ed. Paul Mellars [Edinburgh: Edinburgh University Press, 1990], 160–240).
21. Stanley H. Ambrose, "Chronology of the Later Stone Age and Food Production in East Africa," *Journal of Archaeological Science* 25 (1998): 377–392; John E. Yellen et al., "A Middle Stone Age Worked Bone Industry from Katanda, Upper Semliki Valley, Zaire," *Science* 268 (1995): 553–556.
22. Richard G. Klein, "Anatomy, Behavior, and Modern Human Origins," *Journal of World Prehistory* 9 (1995): 167–198. The first book-length synthesis of the global dispersal was Brian M. Fagan, *The Journey from Eden: The Peopling of Our World* (London: Thames and Hudson, 1990). This was followed a few years later by Christopher Stringer and Robin McKie, *African Exodus: The Origins of Modern Humanity* (New York: Holt, 1996). More recent syntheses include Stephen Oppenheimer, *The Real Eve: Modern Man's Journey Out of Africa* (New York: Basic Books, 2003); and Alice Roberts, *The Incredible Human Journey: The Story of How We Colonised the Planet* (London: Bloomsbury, 2009).

23. Christopher S. Henshilwood et al., "Emergence of Modern Human Behavior: Middle Stone Age Engravings from South Africa," *Science* 295 (2002): 1278–1280.
24. E. Watson et al., "Mitochondrial Footprints of Human Expansion in Africa," *American Journal of Human Genetics* 61 (1997): 691–704; Spencer Wells, *The Journey of Man: A Genetic Odyssey* (New York: Random House, 2002); Oppenheimer, *Real Eve*.
25. Georgi Hudjashov et al., "Revealing the Prehistoric Settlement of Australia by Y Chromosome and mtDNA Analysis," *Proceedings of the National Academy of Sciences* 104 (2007): 8726–8730; G. David Poznik et al., "Punctuated Bursts in Human Male Demography Inferred from 1,244 Worldwide Y-Chromosome Sequences," *Nature Genetics* 48 (2016): 593–599.
26. Phillip Endicott et al., "Evaluating the Mitochondrial Timescale of Human Evolution," *Trends in Ecology and Evolution* 24 (2009): 515–521.
27. Johannes Krause et al., "A Complete mtDNA Genome of an Early Modern Human from Kostenki, Russia," *Current Biology* 20 (2010): 231–236; Spencer Wells, *Deep Ancestry: Inside the Genographic Project* (Washington, D.C.: National Geographic Society, 2007).
28. Maanasa Raghavan et al., "Upper Palaeolithic Siberian Genome Reveals Dual Ancestry of Native Americans," *Nature* 505 (2013): 87–91.
29. Qiaomei Fu et al., "Genome Sequence of a 45,000-Year-Old Modern Human from Western Siberia," *Nature* 514 (2014): 445–449; Morten Rasmussen et al., "The Genome of a Late Pleistocene Human from a Clovis Burial Site in Western Montana," *Nature* 506 (2014): 225–229.
30. Cosimo Posth et al., "Pleistocene Mitochondrial Genomes Suggest a Single Major Dispersal of Non-Africans and a Late Glacial Population Turnover in Europe," *Current Biology* 26 (2016): 827–833.
31. Adrien Rieux et al., "Improved Calibration of the Human Mitochondrial Clock Using Ancient Genomes," *Molecular Biology and Evolution* 31 (2014): 2780–2792. A new technique (*collagen peptide mass fingerprinting*) that allows the rapid identification of taxonomic group on the basis of protein sequencing recently was applied to a sample of more than 2,000 unidentifiable bone fragments from Denisova Cave; one fragment was identified as a human bone, as reported in Samantha Brown et al., "Identification of a New Hominin Bone from Denisova Cave, Siberia Using Collagen Fingerprinting and Mitochondrial DNA Analysis," *Scientific Reports* 6 (2016): 23559. The wider application of this method has the potential to significantly increase the human aDNA database, because many sites, especially caves, are likely to contain small human bone fragments that have not or cannot be identified as such on the basis of anatomy.
32. Richard E. Green et al., "A Draft Sequence of the Neandertal Genome," *Science* 328 (2010): 710–722; David Reich et al., "Genetic History of an Archaic Hominin Group from Denisova Cave in Siberia," *Nature* 468 (2010): 1053–1060.
33. Svante Pääbo, *Neanderthal Man: In Search of Lost Genomes* (New York: Basic Books, 2014).
34. Qiaomei Fu et al., "The Genetic History of Ice Age Europe," *Nature* 534 (2016): 200–205.
35. David Reich et al., "Denisova Admixture and the First Modern Human Dispersals into Southeast Asia and Oceania," *American Journal of Human Genetics* 89 (2011): 516–528.

36. Susanna Sawyer et al., "Nuclear and Mitochondrial DNA Sequences from Two Denisovan Individuals," *Proceedings of the National Academy of Sciences* 112 (2015): 15696–15700.

37. Sydney Brenner et al. "Gene Expression Analysis by Massively Parallel Signature Sequencing (MPSS) on Microbead Arrays," *Nature Biotechnology* 18 (2000): 630–634.

38. James P. Noonan, "Neanderthal Genomics and the Evolution of Modern Humans," *Genome Research* 20 (2010): 547–553; Matthias Meyer et al., "A High-Coverage Genome Sequence from an Archaic Denisovan Individual," *Science* 338 (2012): 222–226; Federico Sánchez-Quinto and Carles Lalueza-Fox, "Almost 20 Years of Neanderthal Paleogenetics: Adaptation, Admixture, Diversity, Demography and Extinction," *Philosophical Transactions of the Royal Society B: Biological Sciences* 370 (2015): 20130374.

39. David Gokhman et al., "Reconstructing the DNA Methylation Maps of the Neandertal and Denisovan," *Science* 344 (2014): 523–527; Ludovic Orlando and Eske Willerslev, "An Epigenetic Window into the Past?" *Science* 345 (2014): 511–512. See also Nessa Carey, *Junk DNA: A Journey Through the Dark Matter of the Genome* (New York: Columbia University Press, 2015).

40. See, for example, Noah A. Rosenberg et al., "Genetic Structure of Human Populations," *Science* 298 (2002): 2381–2385; and Eran Elhaik et al., "Geographic Population Structure Analysis of Worldwide Human Populations Infers Their Biogeographical Origins," *Nature Communications* 5 (2014): 3513.

41. Charles Darwin, *On the Origin of Species by Means of Natural Selection* (London: Murray, 1859), 64.

42. Anders Eriksson et al., "Late Pleistocene Climate Change and the Global Expansion of Anatomically Modern Humans," *Proceedings of the National Academy of Sciences* 109 (2012): 16089–16094.

43. Huw S. Groucutt and Michael D. Petraglia, "The Prehistory of the Arabian Peninsula: Deserts, Dispersals, and Demography," *Evolutionary Anthropology* 21 (2012): 113–125.

44. Chris R. Stokes, Lev Tarasov, and Arthur S. Dyke, "Dynamics of the North American Ice Sheet Complex During Its Inception and Build-up to the Last Glacial Maximum," *Quaternary Science Reviews* 50 (2012): 86–104.

45. John F. Hoffecker et al., "Beringia and the Global Dispersal of Modern Humans," *Evolutionary Anthropology* 25 (2016): 64–78.

46. Ibid., 73–74.

47. Robert L. Kelly, "Colonization of New Land by Hunter-Gatherers," in *Colonization of Unfamiliar Landscapes: The Archaeology of Adaptation*, ed. Marcy Rockman and James Steele (London: Routledge, 2003), 44–58.

48. Benjamin Vernot et al., "Excavating Neandertal and Denisovan DNA from the Genomes of Melanesian Individuals," *Science* 352 (2016): 235–239.

49. Steven E. Churchill, *Thin on the Ground: Neandertal Biology, Archeology, and Ecology* (Hoboken, N.J.: Wiley-Blackwell, 2014), 347–360.

50. Corinne N. Simonti et al., "The Phenotypic Legacy of Admixture Between Modern Humans and Neandertals," *Science* 351 (2016): 737–741; Sriram Sankararaman et al., "The Combined Landscape of Denisovan and Neanderthal Ancestry in Present-Day Humans," *Current Biology* 26 (2016): 1241–1247.

51. Klein, *Human Career*, 3rd ed., 622–643.
52. Mary C. Stiner et al., "Paleolithic Population Growth Pulses Evidenced by Small Animal Exploitation," *Science* 283 (1999): 190–194.
53. Robert L. Kelly, *The Lifeways of Hunter-Gatherers: The Foraging Spectrum* (Cambridge: Cambridge University Press, 2013), 137–165.
54. See, for example, M. V. Anikovich et al., "Early Upper Paleolithic in Eastern Europe and Implications for the Dispersal of Modern Humans," *Science* 315 (2007): 223–226.
55. Paul Mellars, *The Neanderthal Legacy: An Archaeological Perspective from Western Europe* (Princeton, N.J.: Princeton University Press, 1996), 193–244.
56. Ian Buvit et al. report evidence of a spear-thrower from southern Siberia dating to more than 40,000 years ago, in "The Emergence of Modern Behavior in the Trans-Baikal, Russia," in *Emergence and Diversity of Modern Human Behavior in Paleolithic Asia*, ed. Yousuke Kaifu et al. (College Station: Texas A&M University Press, 2015), 490–505.
57. See, for example, Kelly, *Lifeways of Hunter-Gatherers*, 205–208; and Azar Gat, "Proving Communal Warfare Among Hunter-Gatherers: The Quasi-Rousseauan Error," *Evolutionary Anthropology* 24 (2015): 111–126.
58. Sankararaman et al., "Combined Landscape of Denisovan and Neanderthal Ancestry in Present-Day Humans, " 1242–1243.
59. Peter Hiscock, *Archaeology of Ancient Australia* (London: Routledge, 2008), 45–54.
60. Eyed needles are reported from Mezmaiskaya Cave, Layer 1C, in the northern Caucasus (L. V. Golovanova, V. B. Doronichev, and N. E. Cleghorn, "The Emergence of Bone-Working and Ornamental Art in the Caucasian Upper Paleolithic," *Antiquity* 84 [2010]: 299–320) and from Denisova Cave, Layer 11, in the Altai region (A. P. Derevianko, M. V. Shunkov, and S. V. Markin, *Dinamika Paleoliticheskikh Industrii v Afrike i Evrazii v Pozdnem Pleistotsene i Problema Formirovaniya Homo sapiens* [Novosibirsk: Institute of Archaeology and Ethnography SB RAS Press, 2014]).
61. Michael P. Richards et al., "Stable Isotope Evidence for Increasing Dietary Breadth in the European Mid-Upper Paleolithic," *Proceedings of the National Academy of Sciences* 98 (2001): 6528–6532; John F. Hoffecker, "Innovation and Technological Knowledge in the Upper Paleolithic of Northern Eurasia," *Evolutionary Anthropology* 14 (2005): 186–198; Fu et al., "Genome Sequence of a 45,000-Year-Old Modern Human from Western Siberia."
62. Anikovich et al., "Early Upper Paleolithic in Eastern Europe and Implications for the Dispersal of Modern Humans."
63. Michael R. Waters and Thomas Wier Stafford Jr., "The First Americans: A Review of the Evidence for the Late-Pleistocene Peopling of the Americas," in *Paleoamerican Odyssey*, ed. Kelly E. Graf, Caroline V. Ketron, and Michael R. Waters (College Station: Texas A&M University Press, 2013), 541–560.
64. Stokes , Tarasov, and Dyke, "Dynamics of the North American Ice Sheet Complex During Its Inception and Build-up to the Last Glacial Maximum."
65. Waters and Stafford, "First Americans."
66. Erika Tamm et al., "Beringian Standstill and Spread of Native American Founders," *PLoS ONE* 9 (2007): e829; Andrew Kitchen, Michael M. Miyamoto, and Connie J. Mulligan, "A Three-Stage Colonization Model for the Peopling of the

Americas," *PLoS ONE* 3 (2008): e1596; Hoffecker et al., "Beringia and the Global Dispersal of Modern Humans," 66–70.

67. Poznik et al., "Punctuated Bursts in Human Male Demography Inferred from 1,244 Worldwide Y-Chromosome Sequences," 595–596.

68. G. Richard Scott and Christy G. Turner II, *The Anthropology of Modern Human Teeth: Dental Morphology and Its Variation in Recent Human Populations* (Cambridge: Cambridge University Press, 1997), 304–307.

69. Posth et al., "Pleistocene Mitochondrial Genomes Suggest a Single Major Dispersal of Non-Africans and a Late Glacial Population Turnover in Europe," 829, fig. 2.

70. The high levels of NPP in Southeast and South Asia are reflected in the estimated population densities for various recent hunter-gatherers in this part of the world. For example, the *Semang* of Malaysia are estimated to have lived at five to 19 persons per 100 square kilometers and the *Birhor* of India at 22 persons per 100 square kilometers. By contrast, the population density of *Yukaghir* in Northeast Asia is estimated at only 0.5 person per 100 square kilometers. See Kelly, *Lifeways of Hunter-Gatherers*, 178–184, table 7-3.

71. Posth et al., "Pleistocene Mitochondrial Genomes Suggest a Single Major Dispersal of Non-Africans and a Late Glacial Population Turnover in Europe," 829, fig. 2.

72. Ibid., 828.

73. Andaine Seguin-Orlando et al., "Genomic Structure in Europeans Dating Back at Least 36,200 years," *Science* 346 (2014): 1113–1118; John F. Hoffecker et al., "Kostenki 1 and the Early Upper Paleolithic of Eastern Europe," *Journal of Archaeological Science: Reports* 5 (2016): 307–326.

74. Hoffecker et al., "Beringia and the Global Dispersal of Modern Humans," 73–74.

75. Tamm et al., "Beringian Standstill and Spread of Native American Founders," fig. 2.

76. Mark A. Sicoli and Gary Holton, "Linguistic Phylogenies Support Back-Migration from Beringia to Asia," *PLoS ONE* 9 (2014): e91722.

77. Manfred Kayser, "The Human Genetic History of Oceania: Near and Remote Views of Dispersal," *Current Biology* 20 (2010): R194–R201.

78. The bone fragment was a distal manual phalanx of a juvenile, excavated in 2008, as reported in Reich et al., "Genetic History of an Archaic Hominin Group from Denisova Cave in Siberia," 1053.

79. Fu et al., "Genome Sequence of a 45,000-Year-Old Modern Human from Western Siberia."

80. Shara E. Bailey, Timothy D. Weaver, and Jean-Jacques Hublin, "Who Made the Aurignacian and Other Early Upper Paleolithic Industries?" *Journal of Human Evolution* 57 (2009): 11–26.

81. On the presence of shovel-shaped incisors in modern Asian populations and *Homo erectus* fossils from East Asia, see, for example, Franz Weidenreich, *Apes, Giants, and Man* (Chicago: University of Chicago Press, 1946), 84; and Carlton S. Coon, *Origin of Races* (New York: Knopf), 454, fig. 64. For general comments on the use of selected skeletal traits to infer regional continuity, see Klein, *Human Career*, 3rd ed., 630.

82. Morten Rasmussen et al., "The Ancestry and Affiliations of Kennewick Man," *Nature* 523 (2015): 455–458.

83. Christopher Dean et al., "Growth Processes in Teeth Distinguish Modern Humans from *Homo erectus* and Earlier Hominins," *Nature* 414 (2001): 628–631; Tanya M. Smith et al., "Earliest Evidence of Modern Human Life History in North African Early *Homo sapiens*," *Proceedings of the National Academy of Sciences* 104 (2007): 6128–6133; Tanya M. Smith et al., "Dental Evidence for Ontogenetic Differences Between Modern Humans and Neanderthals," *Proceedings of the National Academy of Sciences* 107 (2010): 20923–20928.

84. Philipp Gunz et al., "A Uniquely Modern Human Pattern of Endocranial Development: Insights from a New Cranial Reconstruction of the Neandertal Newborn from Mezmaiskaya," *Journal of Human Evolution* 62 (2012): 300–313.

85. See, for example, overviews in Wells, *Deep Ancestry*; Klein, *Human Career*, 3rd ed., 631–638; and Stringer, *Lone Survivors*, 174–200.

86. Eva-Liis Loogväli et al., "Explaining the Imperfection of the Molecular Clock of Hominid Mitochondria," *PLoS ONE* 4 (2009): e8260.

87. Qiaomei Fu et al., "A Revised Timescale for Human Evolution Based on Ancient Mitochondrial Genomes," *Current Biology* 23 (2013): 553–559; Rieux et al., "Improved Calibration of the Human Mitochondrial Clock Using Ancient Genomes."

88. Jeffrey Rogers and Richard A. Gibbs, "Comparative Primate Genomics: Emerging Patterns of Genome Content and Dynamics," *Nature Reviews Genetics* 15 (2014): 347–359.

89. Ryosuke Kimura, "Human Migrations and Adaptations in Asia Inferred from Genomic Diversity," in *Emergence and Diversity of Modern Human Behavior in Paleolithic Asia*, ed. Kaifu et al., 34–50.

90. Seguin-Orlando et al., "Genomic Structure in Europeans Dating Back at Least 36,200 Years."

91. Most notably in the case of DNA from one of the burials at Lake Mungo in Australia, as reported in Hiscock, *Archaeology of Ancient Australia*, 87.

92. Colin I. Smith et al., "The Thermal History of Human Fossils and the Likelihood of Successful DNA Amplification," *Journal of Human Evolution* 45 (2003): 203–217.

93. As discussed in chapter 5, however, the absence of preserved aDNA in the lower latitudes does not preclude a role in reconstructing the dispersal of modern humans in southern Asia and Australia. The percentages of Denisovan genes (which can be estimated only with reference to Denisovan aDNA recovered from Siberia) in the living peoples of the southwest Pacific and Australia provide important clues to the initial dispersal in this part of the world.

94. Kirsten Ziesemer et al., "Challenging Environments: Ancient DNA Research in the Circum-Caribbean," in *Abstracts of the SAA 81st Annual Meeting* (Orlando, Fla.) (Washington, D.C.: Society for American Archaeology, 2016), 496.

95. Henshilwood et al., "Emergence of Modern Human Behavior"; Marian Vanhaeran et al., "Middle Paleolithic Shell Beads in Israel and Algeria," *Science* 312 (2006): 1785–1788.

96. Yellen et al., "Middle Stone Age Worked Bone Industry from Katanda"; Sarah Wurz, "Technological Trends in the Middle Stone Age of South Africa Between MIS 7 and MIS 3," *Current Anthropology* 54 (2013): S305–S319.

97. Kyle S. Brown et al., "Fire as an Engineering Tool of Early Modern Humans," *Science* 325 (2009): 859–862; Christopher S. Henshilwood et al., "A 100,000-Year-Old Ochre-Processing Workshop at Blombos Cave, South Africa," *Science* 334 (2011): 219–222.

98. John F. Hoffecker, "The Early Upper Paleolithic of Eastern Europe Reconsidered," *Evolutionary Anthropology* 20 (2011): 24–39.

99. Michael Petraglia et al., "Middle Paleolithic Assemblages from the Indian Subcontinent Before and After the Toba Super-Eruption," *Science* 317 (2007): 114–116; Paul Mellars et al., "Genetic and Archaeological Perspectives on the Initial Modern Human Colonization of Southern Asia," *Proceedings of the National Academy of Sciences* 110 (2013): 10699–10704.

100. Ludovic Slimak et al., "Late Mousterian Persistence near the Arctic Circle," *Science* 332 (2011): 841–845.

101. John F. Hoffecker, "The Spread of Modern Humans in Europe," *Proceedings of the National Academy of Sciences* 106 (2009): 16040–16045.

102. L. V. Golovanova et al., "Significance of Ecological Factors in the Middle to Upper Paleolithic Transition," *Current Anthropology* 51 (2010): 655–691.

103. Day and Stringer, "Reconsideration of the Omo Kibish Remains and the *erectus-sapiens* Transition;" Klein, *Human Career*, 3rd ed., 622–626; Ian Tattersall, "Human Origins: Out of Africa," *Proceedings of the National Academy of Sciences* 106 (2009): 16018–16021.

104. Pearson, "Statistical and Biological Definitions of 'Anatomically Modern' Humans"; Klein, *Human Career*, 3rd ed., 622–626; Tattersall, "Human Origins."

105. Klein, *Human Career*, 3rd ed., 465–481; Stringer, *Lone Survivors*, 251.

106. See, for example, Günter Bräuer, "The Origin of Modern Anatomy: By Speciation or Intraspecific Evolution?" *Evolutionary Anthropology* 17 (2008): 22–37.

107. Scott and Turner, *Anthropology of Modern Human Teeth*.

108. See, for example, Klein, *Human Career*, 3rd ed., 505–512.

109. Ofer Bar-Yosef, "Upper Pleistocene Cultural Stratigraphy in Southwest Asia," in *The Emergence of Modern Humans: Biocultural Adaptations in the Later Pleistocene*, ed. Erik Trinkaus (Cambridge: Cambridge University Press, 1989), 154–180; Ofer Bar-Yosef and Bernard Vandermeersch, "Modern Humans in the Levant," *Scientific American*, April 1993, 64–70; Christopher Stringer and Clive Gamble, *In Search of the Neanderthals: Solving the Puzzle of Human Origins* (New York: Thames and Hudson, 1993), 96–122.

110. Jane Qiu, "The Forgotten Continent," *Nature* 535 (2016): 218–220.

111. Florent Détroit et al., "Upper Pleistocene *Homo sapiens* from the Tabon Cave (Palawan, the Philippines): Description and Dating of New Discoveries," *Comptes Rendus Paleovol* 3 (2004): 705–712; Israel Hershkovitz et al., "Levantine Cranium from Manot Cave (Israel) Foreshadows the First European Modern Humans," *Nature* 520 (2015): 216–219.

112. Robin Dennell and Michael D. Petraglia, "The Dispersal of *Homo sapiens* Across Southern Asia: How Early, How Often, How Complex?" *Quaternary Science Reviews* 47 (2012): 15–22; Guanjun Shen et al., "U-Series Dating of Liujiang Hominid Site in Guangxi, Southern China," *Journal of Human Evolution* 43 (2002): 817–829.

113. Armand Salvador Mijares et al., "New Evidence for a 67,000-Year-Old Human Presence at Callao Cave, Luzon, Philippines," *Journal of Human Evolution* 59 (2010): 123–132.
114. Stefano Benazzi et al., "Early Dispersal of Modern Humans in Europe and Implications for Neanderthal Behaviour," *Nature* 479 (2011): 525–528; Fu et al., "Genome Sequence of a 45,000-Year-Old Modern Human from Western Siberia."
115. Anikovich et al., "Early Upper Paleolithic in Eastern Europe and Implications for the Dispersal of Modern Humans."
116. Hiscock, *Archaeology of Ancient Australia*, 84–91.
117. Klein, *Human Career*, 3rd ed., 620–622, table 7.1; Gary Haynes, *The Early Settlement of North America: The Clovis Era* (Cambridge: Cambridge University Press, 2002), 16–18.
118. Christopher J. Bae et al., "Modern Human Teeth from Late Pleistocene Luna Cave (Guangxi, China)," *Quaternary International* 354 (2014): 169–183; Yanjun Cai et al., "The Age of Human Remains and Associated Fauna from Zhiren Cave in Guangxi, Southern China," *Quaternary International* 434 (2017): 84–91.
119. Mellars et al., "Genetic and Archaeological Perspectives on the Initial Modern Human Colonization of Southern Asia."
120. John H. Relethford and Henry C. Harpending, "Craniometric Variation, Genetic Theory, and Modern Human Origins," *American Journal of Physical Anthropology* 95 (1994): 249–270; Joel D. Irish, "Ancestral Dental Traits in Recent Sub-Saharan Africans and the Origins of Modern Humans," *Journal of Human Evolution* 34 (1998): 81–98.
121. Richards et al., "Stable Isotope Evidence for Increasing Dietary Breadth in the European Mid-Upper Paleolithic"; Fu et al., "Genome Sequence of a 45,000-Year-Old Modern Human from Western Siberia," 445.
122. John F. Hoffecker, *Desolate Landscapes: Ice-Age Settlement in Eastern Europe* (New Brunswick, N.J.: Rutgers University Press, 2002), 155–158.
123. Erik Trinkaus et al., "An Early Modern Human from the Peştera cu Oase, Romania," *Proceedings of the National Academy of Sciences* 100 (2003): 11231–11236.
124. Qiaomei Fu et al., "An Early Modern Human from Romania with a Recent Neanderthal Ancestor," *Nature* 524 (2105): 216–219.
125. Noonan, "Neanderthal Genomics and the Evolution of Modern Humans"; Gokhman et al., "Reconstructing the DNA Methylation Maps of the Neandertal and Denisovan"; Orlando and Willerslev, "Epigenetic Window into the Past?"
126. Green et al., "Draft Sequence of the Neandertal Genome"; Reich et al., "Genetic History of an Archaic Hominin Group from Denisova Cave in Siberia"; Sánchez-Quinto and Lalueza-Fox, "Almost 20 Years of Neanderthal Paleogenetics."
127. Johannes Krause et al., "The Derived FOXP2 Variant of Modern Humans Was Shared with Neandertals," *Current Biology* 17 (2007): 1908–1912; Green et al., "Draft Sequence of the Neandertal Genome," 710.
128. Katherine S. Pollard et al., "Forces Shaping the Fastest Evolving Regions in the Human Genome," *PLoS Genetics* 2 (2006): e168; Hernán A. Burbano et al., "Analysis of Human Accelerated DNA Regions Using Archaic Hominin Genomes," *PLoS ONE* 7 (2012): e32877. In general, the changes in the HARs precede the Neanderthal–modern human split; however, analysis of a complete Neanderthal

genome indicated that there are "51 positions in 45 HARs where Neanderthals carry the ancestral version whereas all known present-day humans carry the derived version" (Green et al., "Draft Sequence of the Neandertal Genome," 715).

129. Rieux et al., "Improved Calibration of the Human Mitochondrial Clock Using Ancient Genomes," 2787; Poznik et al., "Punctuated Bursts in Human Male Demography Inferred from 1,244 Worldwide Y-Chromosome Sequences," 594.

130. John H. Relethford, "Genetics and Modern Human Origins," *Evolutionary Anthropology* 4 (1995): 53–63; Hua Liu et al., "A Geographically Explicit Genetic Model of Worldwide Human-Settlement History," *American Journal of Human Genetics* 79 (2006): 230–237.

131. Peter Forster, "Ice Ages and the Mitochondrial DNA Chronology of Human Dispersals: A Review," *Philosophical Transactions of the Royal Society B: Biological Sciences* (2004): 255–264; Wells, *Deep Ancestry*.

132. Doron M. Behar et al., "A 'Copernican' Reassessment of the Human Mitochondrial DNA Tree from Its Root," *American Journal of Human Genetics* 90 (2012): 675–684; Fu et al., "Revised Timescale for Human Evolution Based on Ancient Mitochondrial Genomes"; Stringer, *Lone Survivors*, 178–180; Rieux et al., "Improved Calibration of the Human Mitochondrial Clock Using Ancient Genomes," 2787.

133. Wei Wei et al., "A Calibrated Human Y-Chromosomal Phylogeny Based on Resequencing," *Genome Research* 23 (2013): 388–395; Poznik et al., "Punctuated Bursts in Human Male Demography Inferred from 1,244 Worldwide Y-Chromosome Sequences," fig. 2.

134. Rosa Fregel et al., "Carriers of Mitochondrial DNA Macrohaplogroup N Lineages Reached Australia Around 50,000 Years Ago Following a Northern Asian Route," *PLoS ONE* 10 (2015): e0129839.

135. Posth et al., "Pleistocene Mitochondrial Genomes Suggest a Single Major Dispersal of Non-Africans and a Late Glacial Population Turnover in Europe," 828–829.

136. Tamm et al., "Beringian Standstill and Spread of Native American Founders."

137. Hudjashov et al., "Revealing the Prehistoric Settlement of Australia by Y Chromosome and mtDNA Analysis; Wells, *Deep Ancestry*, 175–227; Peter A. Underhill et al., "The Phylogeography of Y Chromosome Binary Haplotypes and the Origins of Modern Human Populations," *Annals of Human Genetics* 65 (2001): 43–62; Dennis H. O'Rourke and Jennifer A. Raff, "Human Genetic History of the Americas," *Current Biology* 20 (2010): R202–R207; Poznik et al., "Punctuated Bursts in Human Male Demography Inferred from 1,244 Worldwide Y-Chromosome Sequences."

138. Eske Willerslev et al., "Diverse Plant and Animal Genetic Records from Holocene and Pleistocene Sediments," *Science* 300 (2003): 791–795.

139. Jim Allen and James F. O'Connell, "Both Half Right: Updating the Evidence for Dating the First Arrivals in Sahul," *Australian Archaeology* 79 (2015): 86–108.

140. S. Benazzi et al., "The Makers of the Protoaurignacian and Implications for Neandertal Extinction," *Science* 348 (2015): 793–796.

141. Posth et al., "Pleistocene Mitochondrial Genomes Suggest a Single Major Dispersal of Non-Africans and a Late Glacial Population Turnover in Europe," 828–829.

142. Qiaomei Fu et al., "DNA Analysis of an Early Modern Human from Tianyuan Cave, China," *Proceedings of the National Academy of Sciences* 110 (2013): 2223–2227.

143. Dennis L. Jenkins et al., "Geochronology, Archaeological Context, and DNA at the Paisley Caves," in *Paleoamerican Odyssey*, ed. Graf, Ketron, and Waters, 485–510; Frederika A. Kaestle and David Glenn Smith, "Ancient Mitochondrial DNA Evidence for Prehistoric Population Movements: The Numic Expansion," *American Journal of Physical Anthropology* 115 (2001): 1–12.

144. David Glenn Smith et al., "Mitochondrial DNA Haplogroups of Paleoamericans in North America," in *Paleoamerican Origins: Beyond Clovis*, ed. Robson Bonnichsen et al. (College Station: Texas A&M University Press, 2005), 243–254; Rasmussen et al., "Genome of a Late Pleistocene Human from a Clovis Burial Site in Western Montana"; Rasmussen et al., "Ancestry and Affiliations of Kennewick Man."

145. Liu et al., "Geographically Explicit Genetic Model of Worldwide Human-Settlement History," 234. This estimate was derived from a study of 52 populations analyzed at 783 autosomal DNA "microsatellite" markers. The microsatellites are blocks of DNA that contain repeats of the same sequence of base-pairs and appear to be selectively neutral (that is, mutations or copying errors will not be selected out of the genome).

146. Stringer, *Lone Survivors*, 183–189.

147. Mark Stoneking, "Genetic Evidence Concerning the Origins and Dispersals of Modern Humans," in *Causes and Consequences of Human Migration: An Evolutionary Perspective*, ed. Michael H. Crawford and Benjamin C. Campbell (Cambridge: Cambridge University Press, 2012), 12.

148. Reich et al., "Genetic History of an Archaic Hominin Group from Denisova Cave in Siberia"; Vania Yotova et al., "An X-Linked Haplotype of Neandertal Origin Is Present Among All Non-African Populations," *Molecular Biology and Evolution* 28 (2011): 1957–1962; Pääbo, *Neanderthal Man*.

149. Marek Kohn and Steven J. Mithen, "Handaxes: Products of Sexual Selection?" *Antiquity* 73 (1999): 518–526; April Nowell and Melanie Lee Chang, "The Case Against Sexual Selection as an Explanation of Handaxe Morphology," *PaleoAnthropology* (2009): 77–88.

150. McBrearty and Brooks, "Revolution That Wasn't."

151. Golovanova et al., "Significance of Ecological Factors in the Middle to Upper Paleolithic Transition"; Feng Li et al., "The Development of Upper Palaeolithic China: New Results from the Shuidonggou Site," *Antiquity* 87 (2013): 368–383.

152. Lyn Wadley, "Were Snares and Traps Used in the Middle Stone Age and Does It Matter? A Review and a Case Study from Sibudu, South Africa," *Journal of Human Evolution* 58 (2020): 179–192.

153. Paola Villa et al., "Border Cave and the Beginning of the Later Stone Age in South Africa," *Proceedings of the National Academy of Sciences* 109 (2012): 13208–13213.

154. Vanhaeran et al., "Middle Paleolithic Shell Beads in Israel and Algeria."

155. Christopher S. Henshilwood, "Fully Symbolic *sapiens* Behaviour: Innovation in the Middle Stone Age at Blombos Cave, South Africa," in *Rethinking the Human Revolution*, ed. Mellars et al., 123–132; Pierre-Jean Texier et al., "A Howiesons Poort Tradition of Engraving Ostrich Eggshell Containers Dated to 60,000 Years Ago at Diepkloof Rock Shelter, South Africa," *Proceedings of the National Academy of Sciences* 107 (2010): 6180–6185.

156. Yellen et al., "Middle Stone Age Worked Bone Industry from Katanda."

157. Pamela R. Willoughby, *The Evolution of Modern Humans in Africa: A Comprehensive Guide* (Lanham: Altamira Press, 2007), 280–317; Wurz, "Technological Trends in the Middle Stone Age of South Africa Between MIS 7 and MIS 3," S309–S311.
158. McBrearty and Brooks, "Revolution that Wasn't," 510–513; Wadley, "Were Snares and Traps Used in the Middle Stone Age and Does It Matter?" A general pattern of regional variability in artifact style—expressed in stone points and small bifacial points—was observed some years ago in the MSA by J. Desmond Clark. The pattern seems to anticipate the regional variability in artifact style that is characteristic of modern humans, and is not apparent in the archaeological record of Eurasia until the dispersal of *Homo sapiens* out of Africa. Given the increasingly early dates on the beginning of the MSA (more than 250,000 years ago), this may be the earliest indirect evidence for modern cognitive faculties See J. Desmond Clark, "African and Asian Perspectives on the Origin of Modern Humans," in *The Origin of Modern Humans and the Impact of Chronometric Dating*, ed. Martin Jim Aitken, Chris B. Stringer, and Paul A. Mellars (Princeton, N.J.: Princeton University Press, 1993), 148–178.
159. Anikovich et al., "Early Upper Paleolithic in Eastern Europe and Implications for the Dispersal of Modern Humans"; M. Aubert et al., "Pleistocene Cave Art from Sulawesi, Indonesia," *Nature* 514 (2014): 223–227.
160. James F. O'Connell and Jim Allen, "Dating the Colonization of Sahul (Pleistocene Australia–New Guinea): A Review of Recent Research," *Journal of Archaeological Science* 31 (2004): 835–853; John F. Hoffecker and Scott A. Elias, *Human Ecology of Beringia* (New York: Columbia University Press, 2007).
161. Jiří A. Svoboda and Ofer Bar-Yosef, eds., *Stránská skála: Origins of the Upper Paleolithic in the Brno Basin, Moravia, Czech Republic* (Cambridge, Mass.: Peabody Museum of Archaeology and Ethnology, 2003); Hoffecker, "Early Upper Paleolithic of Eastern Europe Reconsidered," 32–35.
162. Klein, *Human Career*, 3rd ed., 654–656; O. Bar-Yosef and J.-G. Bordes, "Who Were the Makers of the Châtelperronian Culture?" *Journal of Human Evolution* 59 (2010): 586–593; J.-G. Bordes and N. Teyssandier, "The Upper Paleolithic Nature of the Châtelperronian in Southwestern France: Archeostratigraphic and Lithic Evidence," *Quaternary International* 259 (2012): 95–101.
163. Bailey, Weaver, and Hublin, "Who Made the Aurignacian and Other Early Upper Paleolithic Industries?"; Hoffecker, "Spread of Modern Humans in Europe,"16041.
164. Anikovich et al., "Early Upper Paleolithic in Eastern Europe and Implications for the Dispersal of Modern Humans."
165. Sue O'Connor, "Crossing the Wallace Line: The Maritime Skills of the Earliest Colonists in the Wallacean Archipelago," in *Emergence and Diversity of Modern Human Behavior in Paleolithic Asia*, ed. Kaifu et al., 214–224.
166. Hoffecker, "Innovation and Technological Knowledge in the Upper Paleolithic of Northern Eurasia."
167. Anikovich et al., "Early Upper Paleolithic in Eastern Europe and Implications for the Dispersal of Modern Humans."
168. Pelayo Correa and M. Blanca Piazuelo, "Evolutionary History of the *Helicobacter pylori* Genome: Implications for Gastric Carcinogenesis," *Gut Liver* 6 (2012): 21–28.
169. Ralf Kittler, Manfred Kayser, and Mark Stoneking, "Molecular Evolution of *Pediculus humanus* and the Origin of Clothing," *Current Biology* 13 (2003): 1414–1417.

170. Quentin D. Atkinson, "Phonemic Diversity Supports a Serial Founder Effect Model of Language Expansion from Africa," *Science* 332 (2011): 346–349. See also Luigi Luca Cavalli-Sforza and Francesco Cavalli-Sforza, *The Great Human Diasporas: The History of Diversity and Evolution* (Reading, Mass.: Addison-Wesley, 1995), 164–202.

3. AN EVOLUTIONARY CONTEXT FOR *HOMO SAPIENS*

1. Simon Singh, *The Code Book: The Science of Secrecy from Ancient Egypt to Quantum Cryptography* (New York: Anchor Books, 1999), 124–142.
2. Ibid., 143–189.
3. Patricia K. Theodosopoulos and Theodore V. Theodosopoulos, "Evolution at the Edge of Chaos: A Paradigm for the Maturation of the Humoral Immune Response," in *Evolution as Computation: DIMACS Workshop, Princeton, January 1999*, ed. Laura F. Landweber and Eric Winfree (Berlin: Springer, 2002), 41–66; John E. Mayfield, *The Engine of Complexity: Evolution as Computation* (New York: Columbia University Press, 2013), 181–189.
4. Christoph Adami, "The Use of Information Theory in Evolutionary Biology," *Annals of the New York Academy of Sciences* 1256 (2011): 49–65.
5. James A. Shapiro, "Genome System Architecture and Natural Genetic Engineering," in *Evolution as Computation*, ed. Landweber and Winfree, 1–14; Mayfield, *Engine of Complexity*, 91–144.
6. Gerald M. Edelman, *Neural Darwinism: The Theory of Neuronal Group Selection* (New York: Basic Books, 1987).
7. Stanislas Dehaene, *Reading in the Brain: The New Science of How We Read* (New York: Penguin, 2009), 42–46.
8. The distributed character of visual and auditory word recognition in the brain is confirmed by the pattern of neuronal activation revealed by brain imaging data. See J. Hart Jr. et al., "Neural Substrates of Orthographic Lexical Access as Demonstrated by Functional Brain Imaging," *Neuropsychiatry, Neuropsychology and Behavioral Neurology* 13 (2000): 1–7; and Angela D. Friederici, "The Brain Basis of Language Processing: From Structure to Function," *Physiological Reviews* 91 (2011): 1357–1392.
9. See, for example, Milford H. Wolpoff, *Paleo-Anthropology* (New York: Knopf, 1980); G. E. Kennedy, *Paleo-Anthropology* (New York: McGraw-Hill, 1980); and Bernard Campbell, *Humankind Emerging*, 3rd ed. (Boston: Little, Brown, 1982).
10. John Maynard Smith and Eörs Szathmáry, *The Origins of Life: From the Birth of Life to the Origin of Language* (Oxford: Oxford University Press, 1999).
11. John Morgan Allman observes: "Brains exist because the distribution of resources necessary for survival and the hazards that threaten survival vary in space and time. There would be little need for a nervous system in an immobile organism or an organism that lived in regular and predictable surroundings" (*Evolving Brains* [New York: Scientific American Library, 1999], 2).
12. See, for example, R. Michael Mulligan, "Signaling in Plants," *Proceedings of the National Academy of Sciences* 94 (1997): 2793–2795.
13. Sean B. Carroll, *Endless Forms Most Beautiful: The New Science of Evo Devo and the Making of the Animal Kingdom* (New York: Norton, 2005), 140–145.

14. Larry W. Swanson, *Brain Architecture: Understanding the Basic Plan*, 2nd ed. (New York: Oxford University Press, 2012), 43–49.

15. Adami, "Use of Information Theory in Evolutionary Biology," 56–61; Mayfield, *Engine of Complexity*, 250–254.

16. Allman, *Evolving Brains*, 73–83. The cephalopods also never evolved myelin or something comparable to insulate axons and allow the development of a densely interconnected neuronal network.

17. On the evolution of adaptive immunity in the vertebrates, see Charles A. Janeway et al., *Immunobiology: The Immune System in Health and Disease*, 5th ed. (New York: Garland, 2001). Mayfield wrote: "The immune response really does illustrate evolution in the body" (*Engine of Complexity*, 187). For a comparison of the immune system with the vertebrate brain, see Edelman, *Neural Darwinism*, 17.

18. Georg F. Striedter, *Principles of Brain Evolution* (Sunderland, Mass.: Sinauer, 2005), 268–287. Striedter suggests that processing high-frequency sounds may have driven brain expansion in early mammals (264).

19. Allman, *Evolving Brains*, 107–119. Ryan Lister et al. report new research on the role of epigenetics in mammalian brain development, in "Global Epigenomic Reconfiguration During Mammalian Brain Development," *Science* 341 (2013): 1237905.

20. John G. Fleagle, *Primate Adaptation and Evolution*, 3rd ed. (San Diego, Calif.: Academic Press, 2013).

21. Allman, *Evolving Brains*, 152–153; Striedter, *Principles of Brain Evolution*, 307.

22. David Marr, *Vision: A Computational Investigation into the Human Representation and Processing of Visual Information* (San Francisco: Freeman, 1982).

23. Oliver G. Selfridge, "Pandemonium: A Paradigm for Learning," in *Mechanisation of Thought Processes: Proceedings of a Symposium Held at the National Physical Laboratory on 24th, 25th, 26th and 27th November 1958* (London: Her Majesty's Stationary Office, 1959), 511–530. Although never mentioned in the paper, it is widely assumed that the "demons" in the model allude to "Maxwell's demon" (see box 1.1). See also Philip C. Jackson Jr., *Introduction to Artificial Intelligence*, 2nd ed. (New York: Dover, 1985), 169–212.

24. Dehaene, *Reading in the Brain*, 41–51; James L. McClelland and David E. Rumelhart, "An Interactive Activation Model of Context Effects in Letter Perception: Part 1. An Account of Basic Findings," *Psychological Review* 88 (1981): 375–407. Selfridge's model is similar to the more recently proposed "Bayesian brain" concept, as noted in Mayfield, *Engine of Complexity*, 245–246.

25. Marc Hauser, Noam Chomsky, and W. Tecumseh Fitch, "The Faculty of Language: What Is It, Who Has It, and How Did It Evolve?" *Science* 298 (2002): 1569–1579. See also Michael C. Corballis, *The Recursive Mind: The Origins of Human Language, Thought, and Civilization* (Princeton, N.J.: Princeton University Press, 2011).

26. Michael J. Noad et al., "Cultural Revolution in Whale Songs," *Nature* 408 (2000): 537. Another possible example is found among New Caledonian crows, which reportedly exhibit individual variations in making tools. See Gavin R. Hunt et al., "Innovative Pandanus-Tool Folding by New Caledonian Crows," *Australian Journal of Zoology* 55 (2007): 291–298.

27. Creative recombination of neuronal information in the brain may be most clearly manifest in dreams, which appear to be widespread among mammals, if not

other vertebrates, although it remains unclear whether nonhuman mammals generate creative dreams or simply replay recently stored visual memories. See Kenway Louie and Matthew A. Wilson, "Temporally Structured Replay of Awake Hippocampal Ensemble Activity During Rapid Eye Movement Sleep," *Neuron* 29 (2001): 145–156.

28. Chris Stringer and Peter Andrews, *The Complete World of Human Evolution*, 2nd ed. (London: Thames and Hudson, 2011), 84–88.

29. Allman, *Evolving Brains*, 143–146.

30. Este Armstrong, "Relative Brain Size in Monkeys and Prosimians," *American Journal of Physical Anthropology* 66 (1985): 263–273. As a group, primates have brains that are about 2.3 times larger, relative to body weight, than those of non-primates. See Allman, *Evolving Brains*, 161.

31. Robin I. M. Dunbar, "The Social Brain Hypothesis," *Evolutionary Anthropology* 6 (1998): 178–190.

32. See, for example, Striedter, *Principles of Brain Evolution*, 304–305.

33. Evelyn McGuinness, Dave Sivertsen, and John M. Allman, "Organization of the Face Representation in Macaque Motor Cortex," *Journal of Comparative Neurology* 193 (1980): 591–608.

34. Robert Seyfarth, Dorothy Cheney, and Peter Marler, "Vervet Monkey Alarm Calls: Semantic Communication in a Free-Ranging Primate," *Animal Behaviour* 28 (1980): 1070–1094. See also Marc D. Hauser, *The Evolution of Communication* (Cambridge, Mass.: MIT Press, 1996), 306–309.

35. Claudia Fichtel, "Predation," in *The Evolution of Primate Societies*, ed. John C. Mitani et al. (Chicago: University of Chicago Press, 2012), table 8.2.

36. Nancy J. Stevens et al., "Palaeontological Evidence for an Oligocene Divergence Between Old World Monkeys and Apes," *Nature* 497 (2013): 611–614.

37. Estimates of the time of divergence between Old World monkeys and apes based on molecular data fall between 25 and 30 million years ago. See M. E. Steiper and N. M. Young, "Timing Primate Evolution: Lessons from the Discordance Between Molecular and Paleontological Evidence," *Evolutionary Anthropology* 17 (2008): 179–188.

38. Alan Walker and Pat Shipman, *The Ape in the Tree: An Intellectual and Natural History of* Proconsul (Cambridge, Mass.: Belknap Press of Harvard University Press, 2005); Stringer and Andrews, *Complete World of Human Evolution*, 88–89.

39. Salvador Moyà-Solà and Meike Köhler, "A *Dryopithecus* Skeleton and the Origin of Great-Ape Locomotion," *Nature* 379 (1996): 156–159. The more recently discovered *Graecopithecus* (also known as *Ouranopithecus*) may represent a more likely common ancestor of living apes and humans. See Stringer and Andrews, *Complete World of Human Evolution*, 110–113.

40. John Napier, *Hands*, rev. Russell H. Tuttle (Princeton, N.J.: Princeton University Press, 1993), 75–87.

41. Charles Darwin, *The Descent of Man and Selection in Relation to Sex* (London: Murray, 1871), 1:51–53; Jane Goodall, "My Life Among the Wild Chimpanzees," *National Geographic Magazine*, August 1963, 272–308.

42. William McGrew, *Chimpanzee Material Culture: Implications for Human Evolution* (Cambridge: Cambridge University Press, 1992). See also Michael Tomasello and Josep Call, *Primate Cognition* (New York: Oxford University Press, 1997),

71–80; and William McGrew, *The Cultured Chimpanzee: Reflections on Cultural Primatology* (Cambridge: Cambridge University Press, 2004), 103–130.

43. Sue Savage-Rumbaugh and Roger Lewin, *Kanzi: The Ape at the Brink of the Human Mind* (New York: Wiley, 1994), 135–150; Sue Savage-Rumbaugh, Stuart G. Shanker, and Talbot J. Taylor, *Apes, Language, and the Human Mind* (New York: Oxford University Press, 1998), 14–18.

44. N. Toth et al., "Pan the Toolmaker: Investigations into the Stone-Tool Making and Tool-Using Capabilities of a Bonobo (*Pan paniscus*)," *Journal of Archaeological Science* 20 (1993): 81–91; Savage-Rumbaugh and Lewin, *Kanzi*, 211–221.

45. C. Hobaiter and R. W. Byrne, "The Gestural Repertoire of the Wild Chimpanzee," *Animal Cognition* 14 (2011): 745–767.

46. See, for example, Steven Pinker, *The Language Instinct: How the Mind Creates Language* (New York: Morrow, 1994), 343–351.

47. Leslie Aiello and Christopher Dean, *An Introduction to Human Evolutionary Anatomy* (London: Academic Press, 1990), 232–243.

48. Esther Clarke, Ulrich H. Reichard, and Klaus Zuberbühler, "The Syntax and Meaning of Wild Gibbon Songs," *PLoS ONE* 1 (2006) 1: e73.

49. Robert C. Berwick et al., "Songs to Syntax: The Linguistics of Birdsong," *Trends in Cognitive Science* 15 (2011): 113–121; Robert C. Berwick et al., "A Bird's Eye View of Human Language Evolution," *Frontiers in Evolutionary Neuroscience* 4 (2012): 1–25.

50. Richard G. Klein, *The Human Career: Human Biological and Cultural Origins*, 3rd ed. (Chicago: University of Chicago Press, 2009), 271–274.

51. Richard Wrangham, *Catching Fire: How Cooking Made Us Human* (New York: Basic Books, 2009); Daniel E. Lieberman, *The Evolution of the Human Head* (Cambridge, Mass.: Belknap Press of Harvard University Press, 2011).

52. Dunbar, "Social Brain Hypothesis"; Clive Gamble, John Gowlett, and Robin Dunbar, *Thinking Big: How the Evolution of Social Life Shaped the Human Mind* (London: Thames and Hudson, 2014).

53. Karin Isler and Carel P. van Schaik, "How Humans Evolved Large Brains: Comparative Evidence," *Evolutionary Anthropology* 23 (2014): 65–75.

54. Lieberman, *Evolution of the Human Head*, 221–223.

55. Vincent M. Sarich and Allan C. Wilson; "Immunological Time Scale for Hominid Evolution," *Science* 158 (1967): 1200–1203. A recent estimate of TMRCA for chimpanzees and humans, based on recalibrated mtDNA substitution rates, is 5.4 to 2.9 million years ago (ca. 4.1 million years ago). See Adrien Rieux et al., "Improved Calibration of the Human Mitochondrial Clock Using Ancient Genomes," *Molecular Biology and Evolution* 31 (2014): 2780–2792.

56. Klein, *Human Career*, 131–278. See also Stringer and Andrews, *Complete World of Human Evolution*, 114–129.

57. Klein, *Human Career*, 131–234; Rob DeSalle and Ian Tattersall, *The Brain: Big Bangs, Behaviors, and Beliefs* (New Haven, Conn.: Yale University Press, 2012), 267–272.

58. Mary W. Marzke, "Joint Function and Grips of the *Australopithecus afarensis* Hand, with Special Reference to the Region of the Capitate," *Journal of Human Evolution* 12 (1983): 197–211; Napier, *Hands*, 87–88.

59. Vernon B. Mountcastle, *The Sensory Hand: Neural Mechanisms of Somatic Sensation* (Cambridge, Mass.: Harvard University Press, 2005).

60. John F. Hoffecker, *Landscape of the Mind: Human Evolution and the Archaeology of Thought* (New York: Columbia University Press, 2011), 47–52.
61. Sonia Harmand et al., "3.3-Million-Year-Old Stone Tools from Lomekwi 3, West Turkana, Kenya," *Nature* 521 (2015): 310–315.
62. Jean de Heinzelin et al., "Environment and Behavior of 2.5-Million-Year-Old Bouri Hominids," *Science* 284 (1999): 625–629.
63. Matt Sponheimer and Julia A. Lee-Thorp, "Isotopic Evidence for the Diet of an Early Hominid *Australopithecus africanus*," *Science* 283 (1999): 368–370; Jonathan G. Wynn et al., "Diet of *Australopithecus afarensis* from the Pliocene Hadar Formation, Ethiopia," *Proceedings of the National Academy of Sciences* 110 (2013): 10495–10500; Thure E. Cerling et al., "Stable-Isotope-Based Diet Reconstructions of Turkana Basin Hominins," *Proceedings of the National Academy of Sciences* 110 (2013): 10501–10506. Some of the carbon-isotope evidence for grasses and sedges also may reflect the consumption of animals that feed on them.
64. Greg Laden and Richard Wrangham, "The Rise of the Hominids as an Adaptive Shift in Fallback Foods: Plant Underground Storage Organs (USOs) and Australopith Origins," *Journal of Human Evolution* 49 (2005): 482–498; Frank W. Marlowe, *The Hadza: Hunter-Gatherers of Tanzania* (Berkeley: University of California Press, 2010), 101–131. The consumption of tubers has implications for hand use and technology, because they must be dug out of a hard ground surface with a stick.
65. Amanda G. Henry et al., "The Diet of *Australopithecus sediba*," *Nature* 487 (2012): 90–93.
66. Meave G. Leakey et al., "New Fossils from Koobi Fora in Northern Kenya Confirm Taxonomic Diversity in Early *Homo*," *Nature* 488 (2012): 201–204.
67. Brian Villmoare et al., "Early *Homo* at 2.8 Ma from Ledi-Geraru, Afar, Ethiopia," *Science* 347 (2015): 1352–1355. The remains from Ledi-Geraru consist of "the left side of an adult mandibular corpus that preserves the partial or complete crowns and roots of the canine, both premolars, and all three molars" (1352).
68. Klein, *Human Career*, 234–271.
69. Ralph Holloway, "Evolution of the Human Brain," in *Handbook of Human Symbolic Evolution*, ed. Andrew Lock and Charles R. Peters (Oxford: Clarendon Press, 1995), 74–125. See also Phillip V. Tobias, "The Brain of *Homo habilis*: A New Level of Organization in Cerebral Evolution," *Journal of Human Evolution* 16 (1987): 741–762; and Dean Falk, *Braindance: New Discoveries About Human Origins and Brain Evolution*, rev. ed. (Gainesville: University Press of Florida, 2004), 145–148.
70. Alan Walker and Pat Shipman, *The Wisdom of the Bones: In Search of Human Origins* (New York: Knopf, 1996), 242–246.
71. Napier, *Hands*, 88–89; Randall L. Susman, "Hand of *Paranthropus robustus* from Member 1, Swartkrans: Fossil Evidence for Tool Behavior," *Science* 240 (1988): 781–784.
72. Lieberman, *Evolution of the Human Head*, 503–506.
73. Henry T. Bunn and Ellen M. Kroll, "Systematic Butchery by Plio/Pleistocene Hominids at Olduvai Gorge, Tanzania," *Current Anthropology* 27 (1986): 431–452; M. Domínguez-Rodrigo and T. R. Pickering, "Early Hominid Hunting and Scavenging: A Zooarchaeological Review," *Evolutionary Anthropology* 12 (2003): 275–282.

74. Joseph V. Ferraro et al., "Earliest Archaeological Evidence of Persistent Hominin Carnivory," *PLoS ONE* 8 (2013): e62174.
75. Cerling et al., "Stable-Isotope-Based Diet Reconstructions of Turkana Basin Hominins"; Matt Sponheimer and Darna L. Dufour, "Increased Dietary Breadth in Early Hominin Evolution: Revisiting Arguments and Evidence with a Focus on Biogeochemical Contributions," in *The Evolution of Hominid Diets: Integrating Approaches to the Study of Palaeolithic Subsistence*, ed. Jean-Jacques Hublin and Michael P. Richards (Berlin: Springer, 2009), 229–240.
76. Leslie C. Aiello and Peter Wheeler, "The Expensive-Tissue Hypothesis: The Brain and the Digestive System in Human and Primate Evolution," *Current Anthropology* 36 (1995): 199–221.
77. Robert L. Kelly, *The Lifeways of Hunter-Gatherers: The Foraging Spectrum* (Cambridge: Cambridge University Press, 2013).
78. Kathy Schick and Nicholas Toth, "An Overview of the Oldowan Industrial Complex: The Sites and the Nature of Their Evidence," in *The Oldowan: Case Studies into the Earliest Stone Age*, ed. Nicholas Toth and Kathy Schick (Gosport, Ind.: Stone Age Press, 2006), 3–42.
79. Toth et al., "Pan the Toolmaker."
80. Wrangham, *Catching Fire*, 83–103. For a discussion of the archaeological evidence for fire in this time range, see Klein, *Human Career*, 261–262.
81. Thomas W. Plummer et al., "Oldest Evidence of Toolmaking Hominins in a Grassland-Dominated Ecosystem," *PLoS ONE* 4 (2009): e7199.
82. Thure E. Cerling, "Development of Grasslands and Savannas in East Africa During the Neogene," *Palaeogeography, Palaeoclimatology, Palaeoecology* 97 (1992): 241–247.
83. Klein, *Human Career*, 346–357.
84. Glynn Isaac, "The Food-Sharing Behavior of Protohuman Hominids," *Scientific American*, April 1978, 90–108.
85. See, for example, M. Domínguez-Rodrigo et al., "New Excavations at the FLK *Zinjanthropus* Site and Its Surrounding Landscape and Their Behavioral Implications," *Quaternary Research* 74 (2010): 315–332.
86. Ferraro et al., "Earliest Archaeological Evidence of Persistent Hominin Carnivory."
87. Domínguez-Rodrigo et al., "New Excavations at the FLK *Zinjanthropus* Site and Its Surrounding Landscape and Their Behavioral Implications."
88. Thomas Plummer, "Flaked Stones and Old Bones: Biological and Cultural Evolution at the Dawn of Technology," *Yearbook of Physical Anthropology* 47 (2004): 118–164.
89. Henry S. Horn, "The Adaptive Significance of Colonial Nesting in the Brewer's Blackbird (*Euphagus cyanocephalus*)," *Ecology* 49 (1968): 682–694.
90. Although other factors may affect sexual dimorphism, body-size dimorphism in the anthropoids is attributed to sexual selection. See J. Michael Plavcan, "Understanding Dimorphism as a Function of Changes in Male and Female Traits," *Evolutionary Anthropology* 20 (2011): 143–155.
91. Several researchers have argued that the extended post-menopausal life span of modern humans evolved at this time (ca. 1.8 million years ago) because of its contribution to child rearing ("grandmother hypothesis"). See, for example,

378   3. AN EVOLUTIONARY CONTEXT FOR *HOMO SAPIENS*

K. Hawkes et al., "Grandmothering, Menopause, and the Evolution of Human Life Histories," *Proceedings of the National Academy of Sciences* 95 (1998): 1336–1339. It has been difficult to confirm extended life expectancy in early *Homo* remains, however; paleoanthropologists have had more success estimating maturation rates in early human fossils. See Klein, *Human Career*, 424–426.

92. Bernard Chapais, *Primeval Kinship: How Pair-Bonding Gave Birth to Human Society* (Cambridge, Mass.: Harvard University Press, 2008).

93. There is a piece of evidence from late *Homo* times, however, that provides a starting point. An analysis of aDNA from the remains of a dozen Neanderthals found together in the cave of El Sidrón (northern Spain) revealed three adult males of the same maternal linage, but adult females from three different maternal lineages. See Carles Lalueza-Fox et al., "Genetic Evidence for Patrilocal Mating Behavior Among Neandertal Groups," *Proceedings of the National Academy of Sciences* 108 (2011): 250–253. The pattern indicates a social group based on male kinship and exogamous mating—although not necessarily *reciprocal*—with females from other groups, as in modern humans. It seems improbable that exogamous mate exchange evolved independently in Neanderthals and modern humans, and it may be assumed that the pattern was inherited from their most recent common ancestor, which would extend it back to at least 500,000 years ago. See Phillip Endicott, Simon Ho, and Chris Stringer, "Using Genetic Evidence to Evaluate Four Palaeoanthropological Hypotheses for the Timing of Neanderthal and Modern Human Origins," *Journal of Human Evolution* 59 (2010): 87–95. Equally significant is that analysis of the remains from El Sidrón suggests that the low mean coefficient of relationship found in modern human foraging peoples has some time depth in human evolution.

94. Thomas D. Seeley, *Honeybee Ecology: A Study of Adaptation in Social Life* (Princeton, N.J.: Princeton University Press, 1985).

95. Madeleine Beekman and Jie Bin Lew, "Foraging in Honeybees—When Does It Pay to Dance?" *Behavioral Ecology* 19 (2007): 255–261.

96. Lieberman, *Evolution of the Human Head*, 564–567.

97. Suzana Herculano-Houzel, "The Human Brain in Numbers: A Linearly Scaled-Up Primate Brain," *Frontiers in Human Neuroscience* 3 (2009): 1–11.

98. Isler and van Schaik, "How Humans Evolved Large Brains."

99. Lieberman, *Evolution of the Human Head*, 203–223. See also Allman, *Evolving Brains*, 160–168.

100. Wrangham, *Catching Fire*, 55–81. "The major factor," as Wrangham notes, is "denaturation of the food proteins, induced by heat . . . causing the molecule to open up. . . . [D]enatured proteins are more digestible because their open structure exposes them to the action of digestive enzymes" (65).

101. Aiello and Wheeler, "Expensive-Tissue Hypothesis"; Isler and van Schaik, "How Humans Evolved Large Brains," 68–69.

102. Richard Dawkins, *The Ancestor's Tale: A Pilgrimage to the Dawn of Evolution* (Boston: Houghton Mifflin, 2004), 270–273.

103. Matt Ridley, *The Red Queen: Sex and the Evolution of Human Nature* (New York: HarperCollins, 1993).

104. Dunbar, "Social Brain Hypothesis," 184–185.

105. Nick Lane, *The Vital Question: Energy, Evolution, and the Origins of Complex Life* (New York: Norton, 2015), 158.

106. In the 1940s, Sir Arthur Keith proposed a "cerebral rubicon" of 750 cc for brain volume as part of the anatomical definition of humans.

107. Seeley, *Honeybee Democracy*, 198–217.

108. Nicholas Toth, "The Oldowan Reassessed: A Close Look at Early Stone Artifacts," *Journal of Archaeological Science* 12 (1985): 101–120.

109. Christopher J. Lepre et al., "An Earlier Origin for the Acheulian," *Nature* 477 (2011): 82–85.

110. See, for example, John McNabb, Francesca Binyon, and Lee Hazelwood, "The Large Cutting Tools from the South African Acheulean and the Question of Social Traditions," *Current Anthropology* 45(2004): 653–677; and John A. J. Gowlett, "The Elements of Design Form in Acheulian Bifaces: Modes, Modalities, Rules and Language," in *Axe Age: Acheulian Tool-making from Quarry to Discard*, ed. Naama Goren-Inbar and Gonen Sharon (London: Equinox, 2006), 203–221. Raymond Corbey et al. recently suggested that the design of the hand ax was at least partially based on genetic information, in "The Acheulean Handaxe: More Like a Bird's Song Than a Beatles' Tune?" *Evolutionary Anthropology* 25 (2016): 6–19.

111. John A. J. Gowlett, "Mental Abilities of Early Man: A Look at Some Hard Evidence," in *Hominid Evolution and Community Ecology: Prehistoric Human Adaptation in Biological Perspective*, ed. Robert Foley (New York: Academic Press, 1984), 167–192; Kathy Schick and Nicholas Toth, *Making Silent Stones Speak: Human Evolution and the Dawn of Technology* (New York: Simon and Schuster, 1993), 237–245; Jacques Pelegrin, "Cognition and the Emergence of Language: A Contribution from Lithic Technology," in *Cognitive Archaeology and Human Evolution*, ed. Sophie A. de Beaune, Frederick L. Coolidge, and Thomas Wynn (Cambridge: Cambridge University Press, 2009), 95–108.

112. Schick and Toth, *Making Silent Stones Speak*, 258–260; Michael Pitts and Mark Roberts, *Fairweather Eden: Life in Britain Half a Million Years Ago as Revealed by the Excavations at Boxgrove* (London: Century, 1998), 285–294.

113. Arthur J. Jelinek, "The Lower Paleolithic: Current Evidence and Interpretation," *Annual Review of Anthropology* 6 (1977): 11–32.

114. Marek Kohn and Steven J. Mithen, "Handaxes: Products of Sexual Selection?" *Antiquity* 73 (1999): 518–526.

115. April Nowell and Melanie Lee Chang, "The Case Against Sexual Selection as an Explanation of Handaxe Morphology," *PaleoAnthropology* (2009): 77–88.

116. John Maynard Smith and Eörs Szathmáry, *The Major Transitions in Evolution* (Oxford: Oxford University Press, 1995); Eörs Szathmáry and John Maynard Smith, "The Major Evolutionary Transitions," *Nature* 374 (1995): 227–232.

117. Dunbar, "Social Brain Hypothesis"; Gamble, Gowlett, and Dunbar, *Thinking Big*.

4. RECENT AFRICAN ORIGIN

1. Jean Itard, *The Memorandum and Report on Victor de l'Aveyron* (1806). See also Jean-Didier Vincent and Pierre-Marie Lledo, *The Custom-Made Brain: Cerebral Plasticity, Regeneration, and Enhancement*, trans. Laurence Garey (New York: Columbia University Press, 2014), 6–7.

2. Steven Pinker, *The Language Instinct: How the Mind Creates Language* (New York: Morrow, 1994).
3. Russ Rymer, Genie: A Scientific Tragedy (New York: Harper Perennial, 1993). See also Derek Bickerton, *Language and Species* (Chicago: University of Chicago Press, 1990), 115–118.
4. Nessa Carey, *The Epigenetics Revolution: How Modern Biology Is Rewriting Our Understanding of Genetics, Disease, and Inheritance* (New York: Columbia University Press, 2012).
5. Vincent and Lledo, *Custom-Made Brain*, 5–8.
6. Andy Clark, *Being There: Putting Brain, Body, and World Together Again* (Cambridge, Mass.: MIT Press, 1998).
7. Charles Darwin wrote: "A complex train of thought can no more be carried on without the aid of words, whether spoken or silent, than a long calculation without the use of figures or algebra" (*The Descent of Man and Selection in Relation to Sex* [London: Murray, 1871], 1:57–58).
8. Daniel C. Dennett, *Consciousness Explained* (Boston: Little, Brown,, 1991), 210.
9. Richard Dawkins and John R. Krebs, "Animal Signals: Information or Manipulation?" in *Behavioural Ecology: An Evolutionary Approach*, ed. John R. Krebs and Nicholas B. Davies (Oxford: Blackwell Scientific, 1978), 282–309.
10. Noam Chomsky, *On Nature and Language* (Cambridge: Cambridge University Press, 2002), 107.
11. Herbert A. Simon, "The Architecture of Complexity," *Proceedings of the American Philosophical Society* 106 (1962): 467–482.
12. Patricia M. Greenfield, "Language, Tools, and Brain: The Ontogeny and Phylogeny of Hierarchically Organized Sequential Behavior," *Behavioral and Brain Sciences* 14 (1991): 531–595.
13. Almerindo E. Ojeda, *A Computational Introduction to Linguistics: Describing Language in Plain Prolog* (Stanford, Calif.: CSLI Publications, 2013).
14. Marc D. Hauser, Noam Chomsky, and W. Tecumseh Fitch, "The Faculty of Language: What Is It, Who Has It, and How Did It Evolve?" *Science* 298 (2002): 1569–1579.
15. Christos H. Papadimitriou, *Computational Complexity* (Reading, Mass.: Addison-Wesley, 1994), 66–67.
16. René Descartes, *Philosophical Works*, trans. Elizabeth S. Haldane and G. R. T. Ross (New York: Dover, 1955).
17. Michael J. Noad et al., "Cultural Revolution in Whale Songs," *Nature* 408 (2000): 537. See also Roger Payne, *Among Whales* (New York: Delta, 1995), 141–167.
18. Gerald M. Edelman, *Neural Darwinism: The Theory of Neuronal Group Selection* (New York: Basic Books, 1987).
19. Merlin Donald, *Origins of the Modern Mind: Three Stages in the Evolution of Culture and Cognition* (Cambridge, Mass.: Harvard University Press, 1991).
20. Jean Piaget, *The Origins of Intelligence in Children*, trans. Margaret Cook (New York: International Universities Press, 1952), and *The Language and Thought of the Child* (New York: New American Library, 1974).
21. L. S. Vygotsky, *Mind in Society: The Development of Higher Psychological Processes* (Cambridge, Mass.: Harvard University Press, 1978), and *Thought and Language*, rev. and ed. Alex Kozulin (Cambridge, Mass.: MIT Press, 1986).

22. Richard G. Klein, *The Human Career: Human Biological and Cultural Origins*, 3rd ed. (Chicago: University of Chicago Press, 2009), 424–426.

23. Daniel E. Lieberman, *The Evolution of the Human Head* (Cambridge, Mass.: Harvard University Press, 2011).

24. John E. Mayfield, *The Engine of Complexity: Evolution as Computation* (New York: Columbia University Press, 2013), 177.

25. G. Philip Rightmire, "Human Evolution in the Middle Pleistocene: The Role of *Homo heidelbergensis*," *Evolutionary Anthropology* 6 (1998): 218–227. Some paleo-anthropologists prefer to assign the Broken Hill skull to the taxon *Homo rhodesiensis* and all its European descendants to *H. neanderthalensis*.

26. The estimated coalescence ages of living maternal and paternal lineages now appear to significantly postdate the appearance of modern humans in the fossil record, however, and apparently represent a later evolutionary event or a genetic bottleneck (or simply low effective population size) in the African population. See Timothy D. Weaver, "Did a Discrete Event 200,000–100,000 Years Ago Produce Modern Humans?" *Journal of Human Evolution* 63 (2012): 121–126.

27. Christopher Dean et al., "Growth Processes in Teeth Distinguish Modern Humans from *Homo erectus* and Earlier Hominins," *Nature* 414 (2001): 628–631; H. Coqueugniot et al., "Early Brain Growth in *Homo erectus* and Implications for Cognitive Ability," *Nature* 431 (2004): 299–302.

28. Fernando V. Ramirez Rozzi and José Maria Bermudez de Castro, "Surprisingly Rapid Growth in Neanderthals," *Nature* 428 (2004): 936–939.

29. Hennady P. Shulha et al., "Human-Specific Histone Methylation Signatures at Transcription Start Sites in Prefrontal Neurons," *PLoS Biology* 10 (2012): e1001427; Irene Hernando-Herraez et al., "Dynamics of DNA Methylation in Recent Human and Great Ape Evolution," *PLoS Genetics* 9 (2013): e1003763. Neanderthal and Denisovan genomes can be compared with those of living and recent modern humans, but aDNA is not available for the African ancestors of modern humans because of preservation factors (chapter 2).

30. Tanya M. Smith et al., "Earliest Evidence of Modern Human Life History in North African Early *Homo sapiens*," *Proceedings of the National Academy of Sciences* 104 (2007): 6128–6133; Tanya M. Smith et al., "Dental Evidence for Ontogenetic Differences Between Modern Humans and Neanderthals," *Proceedings of the National Academy of Sciences* 107 (2010): 20923–20928; April Nowell, "Childhood, Play and the Evolution of Cultural Capacity in Neanderthals and Modern Humans," in *The Nature of Culture*, ed. Miriam N. Haidle, Nicolas J. Conard, and Michael Bolus (Dordrecht: Springer, 2016), 87–98.

31. Philipp Gunz et al., "A Uniquely Modern Human Pattern of Endocranial Development: Insights from a New Cranial Reconstruction of the Neandertal Newborn from Mezmaiskaya," *Journal of Human Evolution* 62 (2012): 300–313. At present, no Denisovan crania are available for comparison, because this taxon is known from only teeth, small bone fragments, and aDNA.

32. Bruce E. Murdoch, "The Cerebellum and Language: Historical Perspective and Review," *Cortex* 46 (2009): 858–868.

33. Chris Stringer, "The Status of *Homo heidelbergensis* (Schoetensack 1908)," *Evolutionary Anthropology* 21 (2012): 101–107.

34. See, for example, Klein, *Human Career*, 308–311.

35. The earliest human aDNA recovered to date is estimated to be around 400,000 years old, from Sima de los Huesos at Atapuerca (Spain). See Matthias Meyer et al., "A Mitochondrial Genome Sequence of a Hominin from Sima de los Huesos," *Nature* 505 (2014): 403–406.

36. East Asian fossils considered possible representatives of *Homo heidelbergensis* include Dali, Jinniushan, and Yunxian (all from China), as noted in Stringer, "Status of *Homo heidelbergensis*," 103, table 2.

37. Ibid., 102, table 1; G. Philip Rightmire, *The Evolution of* Homo erectus: *Comparative Anatomical Studies of an Extinct Human Species* (Cambridge: Cambridge University Press, 1990), 204–233.

38. Lieberman, *Evolution of the Human Head*, 564–567.

39. Ralph L. Holloway et al., "Evolution of the Brain in Humans—Paleoneurology," in *Encyclopedia of Neuroscience*, ed. Marc D. Binder, Nobutaka Hirokawa, and Uwe Windhorst (New York: Springer, 2009), 1332, table 4; Kenneth L. Beals, Courtland L. Smith, and Stephen M. Dodd, "Brain Size, Cranial Morphology, Climate, and Time Machines," *Current Anthropology* 25 (1984): 301–330.

40. Lieberman, *Evolution of the Human Head*, 564–567.

41. Dean et al., "Growth Processes in Teeth Distinguish Modern Humans from *Homo erectus* and Earlier Hominins."

42. Coqueugniot et al., "Early Brain Growth in *Homo erectus* and Implications for Cognitive Ability."

43. Ramirez Rozzi and Bermudez de Castro, "Surprisingly Rapid Gowth in Neanderthals."

44. Klein, *Human Career*, 581–584.

45. Rozzi and Bermudez de Castro, "Surprisingly Rapid Growth in Neanderthals," 939.

46. Klein, *Human Career*, 414–415.

47. Ibid., 416–420.

48. Matt I. Pope and Mark B. Roberts, "Observations on the Relationship Between Palaeolithic Individuals and Artifact Scatters at the Middle Pleistocene Site of Boxgrove, UK," in *The Hominid Individual in Context: Archaeological Excavations of Lower and Middle Palaeolithic Landscapes, Locales and Artefacts*, ed. Clive Gamble and Martin Porr (London: Routledge, 2005), 81–97.

49. Paul Mellars, *The Neanderthal Legacy: An Archaeological Perspective from Western Europe* (Princeton, N.J.: Princeton University Press, 1996); John F. Hoffecker, *Desolate Landscapes: Ice-Age Settlement of Eastern Europe* (New Brunswick, N.J.: Rutgers University Press, 2002).

50. Hervé Bocherens et al., "Isotopic Evidence for Diet and Subsistence Pattern of the Saint Cesaire I Neanderthal: Review and Use of a Multi-Source Mixing Model," *Journal of Human Evolution* 49 (2005): 71–87; Yuichi I. Naito et al., "Ecological Niche of Neanderthals from Spy Cave Revealed by Nitrogen Isotopes of Individual Amino Acids in Collagen," *Journal of Human Evolution* 93 (2016): 82–90.

51. Robert L. Kelly, *The Lifeways of Hunter-Gatherers: The Foraging Spectrum* (Cambridge: Cambridge University Press, 2013).

52. Gavin R. Hunt, "Manufacture and Use of Hook-Tools by New Caledonian Crows," *Nature* 379 (1996): 249–251.

53. Jayne Wilkins et al., "Evidence for Early Hafted Hunting Technology," *Science* 338 (2012): 942–946.

54. Ibid.

55. See, for example, Kathy Schick and Nicholas Toth, *Making Silent Stones Speak: Human Evolution and the Dawn of Technology* (New York: Simon and Schuster, 1993), 292–293.

56. Miriam Noël Haidle, "How to Think a Simple Spear," in *Cognitive Archaeology and Human Evolution*, ed. Sophie A. de Beaune, Frank L. Coolidge, and Thomas Wynn (Cambridge: Cambridge University Press, 2009), 57–73.

57. Naama Goren-Inbar et al., "Evidence of Hominin Control of Fire at Gesher Benot Ya'aqov, Israel," *Science* 304 (2004): 725–727; Dennis M. Sandgathe et al., "On the Role of Fire in Neandertal Adaptations in Western Europe: Evidence from Pech de l'Aze IV and Roc de Marsal, France," *PaleoAnthropology* 2011 (2011): 216–242.

58. An analysis of aDNA from the remains of a dozen Neanderthals found together in the cave of El Sidrón (northern Spain) revealed three adult males of the same maternal linage, but adult females from three different maternal lineages. See Carles Lalueza-Fox et al., "Genetic Evidence for Patrilocal Mating Behavior Among Neandertal Groups," *Proceedings of the National Academy of Sciences* 108 (2011): 250–253.

59. J. Desmond Clark, "African and Asian Perspectives on the Origin of Modern Humans," in *The Origin of Modern Humans and the Impact of Chronometric Dating*, ed. Martin Jim Aitken, Chris B. Stringer, and Paul A. Mellars (Princeton, N.J.: Princeton University Press, 1993), 148–178; Sally McBrearty and Alison S. Brooks, "The Revolution That Wasn't: A New Interpretation of the Origin of Modern Human Behavior," *Journal of Human Evolution* 39 (2000): 453–563.

60. *Homo heidelbergensis* and its relatives also may have been translating and transmitting information with mineral pigment in the form of drawing or painting. Red-ocher "crayons" exhibiting use-wear facets are dated as early as 300,000 to 250,000 years ago in sub-Saharan Africa. See Ian Watts, "The Origin of Symbolic Culture," in *The Evolution of Culture*, ed. Robin Dunbar, Chris Knight, and Camilla Power (New Brunswick, N.J.: Rutgers University Press, 1999), 113–146. An alternative explanation is that the mineral pigment was used to preserve hides or was mixed with other materials to produce an adhesive. Both arguments have implications for the increased complexity of computation with information and/or materials.

61. Timothy D. Weaver, "The Meaning of Neandertal Skeletal Morphology," *Proceedings of the National Academy of Sciences* 106 (2009): 16028–16033.

62. A rough equivalence between modern humans and Neanderthals with respect to cognition and behavior has been asserted in several recent publications, such as Paola Villa and Wil Roebroeks, "Neandertal Demise: An Archaeological Analysis of the Modern Human Superiority Complex," *PLoS ONE* 9 (2014): e96424. Evidence both for structurally and functionally complex technology—such as traps, bows and arrows, tailored clothing, and watercraft—and for representational art is absent in the Neanderthal archaeological record, however.

63. Shulha et al., "Human-Specific Histone Methylation Signatures at Transcription Start Sites in Prefrontal Neurons."

64. John A. Capra et al., "Many Human Accelerated Regions Are Developmental Enhancers," *Philosophical Transactions of the Royal Society B: Biological Sciences* 368 (2013): 20130025; Federico Sánchez-Quinto and Carles Lalueza-Fox, "Almost

20 Years of Neanderthal Palaeogenetics: Adaptation, Admixture, Diversity, Demography, and Extinction," *Philosophical Transactions of the Royal Society B: Biological Sciences* 370 (2015): 20130374.

65. Maxime Aubert et al., "Confirmation of a Late Middle Pleistocene Age for the Omo Kibish 1 Cranium by Direct Uranium-Series Dating," *Journal of Human Evolution* 63 (2012): 704–710.

66. Klein, *Human Career*, 465–481; Chris Stringer suggests that several major African fossils dating to around 300,000 to 140,000 years ago and usually assigned to *Homo sapiens* are not actually modern human—Omo II, Eliye Springs, Laetoli H. 18, and Singa—in *Lone Survivors: How We Came to Be the Only Humans on Earth* (New York: Times Books, 2012), 251. See also Philipp Gunz et al., "Early Modern Human Diversity Suggests Subdivided Population Structure and a Complex Out-of-Africa Scenario," *Proceedings of the National Academy of Sciences* 106 (2009): 6094–6098.

67. See, for example, Klein, *Human Career*, 465–481.

68. Paul Mellars, "Why Did Modern Human Populations Disperse from Africa ca. 60,000 Years Ago? A New Model," *Proceedings of the National Academy of Sciences* 103 (2006): 9381–9386.

69. Stringer, *Lone Survivors*, 44–47.

70. Wu Liu et al., "The Earliest Unequivocally Modern Humans in Southern China," *Nature* 526 (2015): 696–700; Yanjun Cai et al., "The Age of Human Remains and Associated Fauna from Zhiren Cave in Guangxi, Southern China," *Quaternary International* 434 (2017): 84–91.

71. Robin Dennell and Michael D. Petraglia, "The Dispersal of *Homo sapiens* Across Southern Asia: How Early, How Often, How Complex?" *Quaternary Science Reviews* 47 (2012): 15–22.

72. Genetic traces of an expansion of *Homo sapiens* into Eurasia around 100,000 years ago may be present in Neanderthal aDNA from the Altai Mountains in southwestern Siberia. See Martin Kuhlwilm et al., "Ancient Gene Flow from Early Modern Humans into Eastern Neanderthals," *Nature* 530 (2016): 429–433.

73. See, for example, Osbjorn M. Pearson, "Integration of the Genetic, Anatomical and Archaeological Data for the African Origin of Modern Humans: Problems and Prospects," in *African Genesis: Perspectives on Hominin Evolution*, ed. Sally C. Reynolds and Andrew Gallagher (Cambridge: Cambridge University Press, 2012), 423–448.

74. Curtis Marean et al., "Early Use of Marine Resources and Pigment in South Africa During the Middle Pleistocene," *Nature* 449 (2007): 905–908.

75. John E. Yellen et al., "A Middle Stone Age Worked Bone Industry from Katanda, Upper Semliki Valley, Zaire," *Science* 268 (1995): 553–556.

76. Marian Vanhaeren et al., "Middle Paleolithic Shell Beads in Israel and Algeria," *Science* 312 (2006): 1785–1788.

77. Francisco d'Errico et al., "Archaeological Evidence for the Emergence of Language, Symbolism, and Music—An Alternative Multidisciplinary Perspective," *Journal of World Prehistory* 17 (2003): 1–70.

78. The earliest dated examples of visual art that exhibits three or more hierarchical levels of organization actually date (radiocarbon) to about 35,000 calibrated years before the present (for example, Hohle Fels Cave [Germany]). See Nicolas J.

Conard, "A Female Figurine from the Basal Aurignacian of Hohle Fels Cave in Southwestern Germany," *Nature* 459 (2009): 248–252.

79. See, for example, John J. Shea, "*Homo sapiens* Is as *Homo sapiens* Was," *Current Anthropology* 52 (2011): 1–15.

80. Richard G. Klein, "Anatomy, Behavior, and Modern Human Origins," *Journal of World Prehistory* 9 (1995): 167–198. See also Klein, *Human Career*, 397–398; and Mellars, "Why Did Modern Human Populations Disperse from Africa ca. 60,000 Years Ago?"

81. Günter Bräuer, "The Origin of Modern Anatomy: By Speciation or Intraspecific Evolution?" *Evolutionary Anthropology* 17 (2008): 22–37. See also Osbjorn M. Pearson, "Statistical and Biological Definitions of 'Anatomically Modern' Humans: Suggestions for a Unified Approach to Modern Morphology," *Evolutionary Anthropology* 17 (2008): 38–48; and Ian Tattersall and Jeffrey H. Schwartz, "The Morphological Distinctiveness of *Homo sapiens* and Its Recognition in the Fossil Record: Clarifying the Problem," *Evolutionary Anthropology* 17 (2008): 49–54. The pattern of pronounced anatomical variability also is evident in a more recent sample of African remains that date to between about 25,000 and 20,000 years ago, as reported in I. Crevecoeur et al., "Late Stone Age Human Remains from Ishango (Democratic Republic of Congo): New Insights on Late Pleistocene Modern Human Diversity in Africa," *Journal of Human Evolution* 96 (2016): 35–57.

82. Weaver, "Did a Discrete Event 200,000–100,000 Years Ago Produce Modern Humans?" 122–123.

83. McBrearty and Brooks, "Revolution That Wasn't."

84. Sylvian Soriano, Paola Villa, and Lyn Wadley, "Blade Technology and Tool Forms in the Middle Stone Age of South Africa: The Howiesons Poort and Post-Howiesons Poort at Rose Cottage Cave," *Journal of Archaeological Science* 34 (2007): 681–703; Paola Villa et al., "The Howiesons Poort and MSA III at Klasies River Main Site, Cave 1A," *Journal of Archaeological Science* 37 (2010): 630–635; Paola Villa et al., "Border Cave and the Beginning of the Later Stone Age in South Africa," *Proceedings of the National Academy of Sciences* 109 (2012): 13208–13213. On climate change in southern Africa during this period, see D. H. Urrego et al., "Increased Aridity in Southwestern Africa During the Warmest Periods of the Last Interglacial," *Climate of the Past* 11 (2015): 1417–1431.

85. Alex Mackay, Brian A. Stewart, and Brian M. Chase, "Coalescence and Fragmentation in the Late Pleistocene Archaeology of Southernmost Africa," *Journal of Human Evolution* 72 (2014): 26–51.

86. Smith et al., "Earliest Evidence of Modern Human Life History in North African Early *Homo sapiens*."

87. Murdoch, "Cerebellum and Language."

88. Lyn Wadley, "Were Snares and Traps Used in the Middle Stone Age and Does It Matter? A Review and a Case Study from Sibudu, South Africa," *Journal of Human Evolution* 58 (2010): 179–192; Aurore Val, Paloma de la Peña, and Lyn Wadley, "Direct Evidence for Human Exploitation of Birds in the Middle Stone Age of South Africa: The Example of Sibudu Cave, KwaZulu-Natal," *Journal of Human Evolution* 99 (2016): 1–17. See also Richard G. Klein, "Stone Age Predation on Small African Bovids," *South African Archaeological Bulletin* 36 (1981): 55–65.

89. See, for example, Mellars, "Why Did Modern Human Populations Disperse from Africa ca. 60,000 Years Ago?" 9384–9385; and Quentin D. Atkinson, Russell D. Gray, and Alexei J. Drummond, "Bayesian Coalescent Inference of Major Human Mitochondrial DNA Haplogroup Expansions in Africa," *Proceedings of the Royal Society B: Biological Sciences* 276 (2009): 367–373.

90. PingHsun Hsieh et al., "Model-Based Analyses of Whole-Genome Data Reveal a Complex Evolutionary History Involving Archaic Introgression in Central African Pygmies," *Genome Research* 26 (2016): 291–300.

91. Klein, *Human Career,* 647–660.

92. Lieberman, *Evolution of the Human Head,* 564–575.

93. Klein, *Human Career,* 625–626.

94. Daniel E. Lieberman, Brandeis M. McBratney, and Gail Krovitz, "The Evolution and Development of Cranial Form in *Homo sapiens,*" *Proceedings of the National Academy of Sciences* 99 (2002): 1134–1139.

95. Elkhonon Goldberg, *The New Executive Brain: Frontal Lobes in a Complex World* (New York: Oxford University Press, 2009); Frederick Coolidge and Thomas Wynn, "Working Memory, Its Executive Functions, and the Emergence of Modern Thinking," *Cambridge Archaeological Journal* 15 (2005): 5–26. See also Frederick L. Coolidge and Thomas Wynn, *The Rise of* Homo sapiens: *The Evolution of Modern Thinking* (Malden, Mass.: Wiley-Blackwell, 2009).

96. Shulha et al., "Human-Specific Histone Methylation Signatures at Transcription Start Sites in Prefrontal Neurons."

97. Pearson, "Statistical and Biological Definitions of 'Anatomically Modern' Humans," 44–45.

98. Zaneta M. Thayer and Seth D. Dobson, "Sexual Dimorphism in Chin Shape: Implications for Adaptive Hypotheses," *American Journal of Physical Anthropology* 143 (2010): 417–425.

99. Smith et al., "Earliest Evidence of Modern Human Life History in North African Early *Homo sapiens.*"

100. Pearson, "Statistical and Biological Definitions of 'Anatomically Modern' Humans," 41–44.

101. Stephen Jay Gould, *Ontogeny and Phylogeny* (Cambridge, Mass.: Belknap Press of Harvard University Press, 1977), 352–404.

102. Gunz et al., "Uniquely Modern Human Pattern of Endocranial Development."

103. Carey, *Epigenetics Revolution*; Vincent and Lledo, *Custom-Made Brain,* 5–8.

104. Hauser, Chomsky, and Fitch, "Faculty of Language," 1569.

105. Clark, *Being There.*

106. Ojeda, *Computational Introduction to Linguistics.*

107. Noam Chomsky, "Three Models for the Description of Language," *IRE Transactions on Information Theory* 2 (1956): 113–124, and "On Certain Formal Properties of Grammars," *Information and Control* 2 (1959): 137–167.

108. Chomsky, *On Nature and Language.*

109. Papadimitriou, *Computational Complexity,* 66.

110. Dennett, *Consciousness Explained,* 210. See also Andy Clark, *Surfing Uncertainty: Prediction, Action, and the Embodied Mind* (Oxford: Oxford University Press, 2016).

111. d'Errico et al., "Archaeological Evidence for the Emergence of Language, Symbolism, and Music."

112. Quentin D. Atkinson, "Phonemic Diversity Supports a Serial Founder Effect Model of Language Expansion from Africa," *Science* 332 (2011): 346–349.

113. John F. Hoffecker, *Landscape of the Mind: Human Evolution and the Archaeology of Thought* (New York: Columbia University Press, 2011), 101–102.

114. Pierre-Jean Texier et al., "A Howiesons Poort Tradition of Engraving Ostrich Eggshell Containers Dated to 60,000 Years Ago at Diepkloof Rock Shelter, South Africa," *Proceedings of the National Academy of Sciences* 107 (2010): 6180–6185.

115. Yellen et al., "Middle Stone Age Worked Bone Industry from Katanda"; McBrearty and Brooks, "Revolution That Wasn't."

116. McBrearty and Brooks, "Revolution That Wasn't," 510–513.

117. Mackay, Stewart, and Chase, "Coalescence and Fragmentation in the Late Pleistocene Archaeology of Southernmost Africa"; David J. Nash et al., "Going the Distance: Mapping Mobility in the Kalahari Desert During the Middle Stone Age Through Multi-Site Geochemical Provenancing of Silcrete Artefacts," *Journal of Human Evolution* 96 (2016): 113–133.

118. Syntactic language is "acquired" through exposure to spoken language during a critical period between the ages of approximately five and 12 years. Language acquisition is part of the process of brain growth and development, and is inextricably tied to the significantly delayed maturation that evolved in modern humans. The fossil record indicates that a comparable period of language acquisition did not exist in early *Homo*, because a 1.8-million-year-old cranium from an infant (estimated 1–1.5 years in age) from Southeast Asia exhibits a brain that already is 72 to 84 percent of its adult size (on a par with the timing of brain growth in chimpanzees). See Coqueugniot et al., "Early Brain Growth in *Homo erectus* and Implications for Cognitive Ability."

119. Greenfield observed that between the ages of 11 and 36 months, the child acquires a capacity for manipulating objects with "increasing hierarchical complexity" ("Language, Tools, and Brain," 532). By contrast, a juvenile chimpanzee never attains a comparable level of complexity in the manipulation of objects.

120. Eörs Szathmáry and John Maynard Smith, "The Major Evolutionary Transitions," *Nature* 374 (1995): 227–232. See also John Maynard Smith and Eörs Szathmáry, *The Major Transitions in Evolution* (Oxford: Oxford University Press, 1995), and *The Origins of Life: From the Birth of Life to the Origin of Language* (Oxford: Oxford University Press, 1999).

121. Clark, *Being There*; Andy Clark, *Supersizing the Mind: Embodiment, Action, and Cognitive Extension* (New York: Oxford University Press, 2011), 44–60; Dennett, *Consciousness Explained*, 210.

122. Francesco d'Errico et al., "Early Evidence of San Material Culture Represented by Organic Artifacts from Border Cave, South Africa," *Proceedings of the National Academy of Sciences* 109 (2012): 13214–13219.

123. John F. Hoffecker, "The Information Animal and the Super-Brain," *Journal of Archaeological Method and Theory* 20 (2013): 18–41.

124. Adam Kuper, *Culture: The Anthropologist' Account* (Cambridge, Mass.: Harvard University Press, 1999).

125. See, for example, Dean Falk, *Braindance: New Discoveries About Human Origins and Brain Evolution*, rev. ed. (Gainesville: University Press of Florida, 2004), 145–148.

126. Thomas D. Seeley compared the "cognitive entity" of the honeybee swarm with the brain of an individual primate, in *Honeybee Democracy* (Princeton, N.J.: Princeton University Press, 2010), 198–217.

127. Hoffecker, "Information Animal and the Super-Brain," 21, table 1.

128. John von Neumann, *Theory of Self-Reproducing Automata*, ed. Arthur W. Burks (Urbana: University of Illinois Press, 1966).

129. See, for example, Mellars, "Why Did Modern Human Populations Disperse from Africa ca. 60,000 Years Ago?" 9384–9385; Richard G. Klein et al., "The Ysterfontein 1 Middle Stone Age Site, South Africa, and Early Human Exploitation of Coastal Resources," *Proceedings of the National Academy of Sciences* 101 (2004): 5708–5715; and Klein, *Human Career*, 647–660.

130. Sarah A. Tishkoff et al., "The Genetic Structure and History of Africans and African Americans," *Science* 324 (2009): 1035–1043.

131. Brenna M. Henn et al., "Hunter-Gatherer Genomic Diversity Suggests a Southern African Origin for Modern Humans," *Proceedings of the National Academy of Sciences* 108 (2011): 5154–5162.

132. Doron M. Behar et al., "The Dawn of Human Matrilineal Diversity," *American Journal of Human Genetics* 82 (2008): 1130–1140.

133. Colin I. Smith et al., "The Thermal History of Human Fossils and the Likelihood of Successful DNA Amplification," *Journal of Human Evolution* 45 (2003): 203–217.

134. Tishkoff et al., "Genetic Structure and History of Africans and African Americans," 1035.

135. Ibid., 1037–1039.

136. Ibid., 1041.

137. Henn et al., "Hunter-Gatherer Genomic Diversity Suggests a Southern African Origin for Modern Humans," 5154.

138. Behar et al., "Dawn of Human Matrilineal Diversity," 1136. The southern African speakers of click languages represent clades L0d and L0k and include the L0d1 lineage.

139. Brenna Mariah Henn, "Inferring Modern Human Migration Patterns Within Africa Using Calibrated Mitochondrial and Y-Chromosomal DNA" (Ph.D. diss., Stanford University, 2009), 30.

140. Tishkoff et al., "Genetic Structure and History of Africans and African Americans," 1041–1042.

141. Karima Fadhlaoui-Zid et al., "Genome-Wide and Paternal Diversity Reveal a Recent Origin of Human Populations in North Africa," *PLoS ONE* 8 (2013): e80293.

142. Tishkoff et al., "Genetic Structure and History of Africans and African Americans," 1041.

143. Brigitte Pakendorf, Koen Bostoen, and Cesare de Filippo, "Molecular Perspectives on the Bantu Expansion: A Synthesis," *Language Dynamics and Change* 1 (2011): 50–88. The genetic analyses of Bantu-speaking people confirm that the spread of the language was effected by population movements, which are better represented by Y-DNA (paternal lineages).

144. Klein, *Human Career*, 505–512.

145. Mae Goder-Goldberger, Natalia Gubenko, and Erella Hovers, " 'Diffusion with Modifications': Nubian Assemblages in the Central Negev Highlands of Israel and

Their Implications for Middle Paleolithic Inter-Regional Interactions," *Quaternary International* 408 (2016): 121–139.

146. Kuhlwilm et al., "Ancient Gene Flow from Early Modern Humans into Eastern Neanderthals." See also Richard E. Green et al., "A Draft Sequence of the Neandertal Genome," *Science* 328 (2010): 710–722.

147. It is worth noting that the living representatives of the oldest maternal and paternal lineages (for example, Bushmen) exhibit the same cognitive faculties as other living peoples. The complexity of the click languages is comparable to that of other syntactic languages, and artifacts include mechanical devices with multiple components (for example, spring-pole snare comprising seven "techno-units"). See Wendell H. Oswalt, *An Anthropological Analysis of Food-Getting Technology* (New York: Wiley, 1976), 238.

148. R. N. E. Barton et al., "OSL Dating of the Aterian Levels at Dar es-Soltan I (Rabat, Morocco) and Implications for the Dispersal of Modern *Homo sapiens*," *Quaternary Science Reviews* 28 (2009): 1914–1931.

149. Philip Van Peer, "The Nile Corridor and the Out-of-Africa Model: An Examination of the Archaeological Record," *Current Anthropology* 39 (1998): S115–S140; P. M. Vermeersch et al., "A Middle Palaeolithic Burial of a Modern Human at Taramsa Hill, Egypt," *Antiquity* 72 (1998): 475–484.

150. Zenobia Jacobs et al., "Single-Grain OSL Dating at La Grotte des Contrebandiers ('Smugglers Cave'), Morocco: Improved Age Constraints for the Middle Paleolithic Levels," *Journal of Archaeological Science* 38 (2011): 3611–3643; Zenobia Jacobs et al., "Single-Grain OSL Chronologies for Middle Palaeolithic Deposits at El Mnasra and El Harhoura 2, Morocco: Implications for Late Pleistocene Human-Environment Interactions Along the Atlantic Coast of Northwest Africa," *Journal of Human Evolution* 62 (2012): 377–394; Katerina Douka et al., "The Chronostratigraphy of the Haua Fteah Cave (Cyrenaica, Northeast Libya)," *Journal of Human Evolution* 66 (2014): 39–63.

151. Qiaomei Fu et al., "Genome Sequence of a 45,000-Year-Old Modern Human from Western Siberia," *Nature* 514 (2014): 445–449.

152. Marean et al., "Early Use of Marine Resources and Pigment in South Africa During the Middle Pleistocene," 905–906.

153. Soriano, Villa, and Wadley, "Blade Technology and Tool Forms in the Middle Stone Age of South Africa"; Villa et al., "Howiesons Poort and MSA III at Klasies River Main Site."

154. McBrearty and Brooks, "Revolution That Wasn't," 513–517.

155. Kyle S. Brown et al., "Fire as an Engineering Tool of Early Modern Humans," *Science* 325 (2009): 859–862.

156. McBrearty and Brooks, "Revolution That Wasn't," 510–513; Klein, *Human Career*, 562–563.

157. Yellen et al., "Middle Stone Age Worked Bone Industry from Katanda," 554–555.

158. Miguel Cortés-Sánchez et al., "Earliest Known Use of Marine Resources by Neanderthals," *PLoS ONE* 6 (2011): e24026.

159. Richard G. Klein and Teresa E. Steele, "Archaeological Shellfish Size and Later Human Evolution in Africa," *Proceedings of the National Academy of Sciences* 110 (2013): 10910–10915.

160. Wadley, "Were Snares and Traps Used in the Middle Stone Age and Does It Matter?";
     Villa et al., "Border Cave and the Beginning of the Later Stone Age in South Africa."
161. Texier, et al., "Howiesons Poort Tradition of Engraving Ostrich Eggshell Con-
     tainers," 6183; Klein suggests that the container fragments, which occur in the
     uppermost part of the early occupations (MSA), may have been introduced by
     later occupants of the site (LSA), in *Human Career*, 535.
162. McBrearty and Brooks, "Revolution That Wasn't," 516; Pamela R. Willoughby,
     The Evolution of Modern Humans in Africa: A Comprehensive Guide (Lanham,
     Md.: Altamira Press, 2007), 245.

5. GLOBAL DISPERSAL: SOUTHERN ASIA
AND AUSTRALIA

1. Phillip Endicott et al., "The Genetic Origin of the Andaman Islanders," *American
   Journal of Human Genetics* 72 (2003): 178–184; Vincent Macaulay et al., "Single,
   Rapid Coastal Settlement of Asia Revealed by Analysis of Complete Mitochondrial
   Genomes," *Science* 308 (2005): 1034–1036; Vikrant Kumar et al., "Y-Chromosome
   Evidence Suggests a Common Paternal Heritage of Austro-Asiatic Populations,"
   BMC Evolutionary Biology 7 (2007): 47; Morten Rasmussen et al., "An Aboriginal
   Australian Genome Reveals Separate Human Dispersals into Asia," *Science* 334
   (2011): 94–98.
2. Huw S. Groucutt and Michael D. Petraglia, "The Prehistory of the Arabian Peninsula: Deserts, Dispersals, and Demography," *Evolutionary Anthropology* 21
   (2012): 113–125.
3. High marine productivity characterizes the entire coastal route. See Paul Mellars
   et al., "Genetic and Archaeological Perspectives on the Initial Modern Human
   Colonization of Southern Asia," *Proceedings of the National Academy of Sciences*
   110 (2013): 10699–10704.
4. See, for example, Sheela Athreya, "Modern Human Emergence in South Asia:
   A Review of the Fossil and Genetic Evidence," in *Emergence and Diversity of
   Modern Human Behavior in Paleolithic Asia*, ed. Yousuke Kaifu et al. (College
   Station: Texas A&M University Press, 2015), 61–79.
5. In the early 1990s, Luigi L. Cavalli-Sforza, Paolo Menozzi, and Alberto Piazza
   wrote: "Use of rafts and boats must have been common because around 55,000 or
   60,000 years ago *H[omo] s[apiens] s[apiens]* passed several sea tracts to reach Aus-
   tralia. The development of navigational skills may have helped Africans to reach
   Southeast Asia (and finally Australia) along the southern coasts of Asia" ("Demic
   Expansions and Human Evolution," *Science* 259 [1993]: 639–646).
6. Svante Pääbo, *Neanderthal Man: In Search of Lost Genomes* (New York: Basic
   Books, 2014).
7. Rosa Fregel et al., "Carriers of Mitochondrial DNA Macrohaplogroup N Lineages
   Reached Australia Around 50,000 Years Ago Following a Northern Asian Route,"
   PLoS ONE 10 (2015): e0129839.
8. Groucutt and Petraglia, "Prehistory of the Arabian Peninsula."
9. In "Carriers of Mitochondrial DNA Macrohaplogroup N Lineages Reached
   Australia Around 50,000 Years Ago," Fregel et al. suggested, on the basis of the

distribution of mtDNA haplogroup N, that Australia was initially settled by modern humans who expanded across northern Eurasia—not southern Asia. Additional supporting evidence for a Levantine route, based on a comparative analysis of Egyptian and Ethiopian genomes, is reported in Luca Pagani et al., "Tracing the Route of Modern Humans Out of Africa by Using 225 Human Genome Sequences from Ethiopians and Egyptians," *American Journal of Human Genetics* 96 (2015): 986–991.

10. Pääbo, *Neanderthal Man.*

11. Sue O'Connor, Rintaro Ono, and Chris Clarkson, "Pelagic Fishing at 42,000 BP and the Maritime Skills of Modern Humans," *Science* 334 (2011): 1117–1121.

12. Armand Salvador Mijares et al., "New Evidence for a 67,000-Year-Old Human Presence at Callao Cave, Luzon, Philippines," *Journal of Human Evolution* 59 (2010): 123–132; Robin Dennell and Michael D. Petraglia, "The Dispersal of *Homo sapiens* Across Southern Asia: How Early, How Often, How Complex?" *Quaternary Science Reviews* 47 (2012): 15–22.

13. See, for example, Amanuel Beyin, "Upper Pleistocene Human Dispersals Out of Africa: A Review of the Current State of the Debate," *International Journal of Evolutionary Biology* (2011): 615094; Tim Appenzeller, "Eastern Odyssey," *Nature* 485 (2012): 24–26; and Dennell and Petraglia, "Dispersal of *Homo sapiens* Across Southern Asia."

14. Glenn R. Summerhayes et al., "Human Adaptation and Plant Use in Highland New Guinea 49,000 to 44,000 Years Ago," *Science* 330 (2010): 78–81; Jim Allen and James F. O'Connell, "Both Half Right: Updating the Evidence for Dating First Human Arrivals in Sahul," *Australian Archaeology* 79 (2014): 86–108; Dennell and Petraglia, "Dispersal of *Homo sapiens* Across Southern Asia," 17.

15. Michael Petraglia et al., "Middle Paleolithic Assemblages from the Indian Subcontinent Before and After the Toba Super-Eruption," *Science* 317 (2007): 114–116.

16. Rasmussen et al., "Aboriginal Australian Genome Reveals Separate Human Dispersals into Asia"; Qiaomei Fu et al., "A Revised Timescale for Human Evolution Based on Ancient Mitochondrial Genomes," *Current Biology* 23 (2013): 553–559; Adrien Rieux et al., "Improved Calibration of the Human Mitochondrial Clock Using Ancient Genomes," *Molecular Biology and Evolution* 31 (2014): 2780–2792; G. David Poznik et al., "Punctuated Bursts in Human Male Demography Inferred from 1,244 Worldwide Y-Chromosome Sequences," *Nature Genetics* 48 (2016): 593–599.

17. As discussed in chapter 2, lineages may diverge with or without the migration of some portion of the population to another geographic area. If the maternal and/or paternal lineages now associated with Eurasia diverged from their African source populations before the latter moved into Eurasia, however, their African descendants are absent (that is, the defining mutations of the Eurasian lineages are absent) in the modern African population.

18. Guanjun Shen et al., "U-Series Dating of Liujiang Hominid Site in Guangxi, Southern China," *Journal of Human Evolution* 43 (2002): 817–829.

19. Christopher J. Bae et al., "Modern Human Teeth from Late Pleistocene Luna Cave (Guangxi, China)," *Quaternary International* 354 (2014): 169–183; Wu Liu et al., "The Earliest Unequivocally Modern Humans in Southern China," *Nature* 526 (2015): 696–700; Yanjun Cai et al., "The Age of Human Remains and Associated

Fauna from Zhiren Cave in Guangxi, Southern China," *Quaternary International* 434 (2017): 84–91.

20. Jane Qiu, "The Forgotten Continent," *Nature* 535 (2016): 218–220.

21. At the southern end of the Red Sea, the southwestern tip of the Arabian Peninsula is less than 30 kilometers from the northeastern coast of Africa (Eritrea and Djibouti). Moreover, this narrow strait (known as the Bab al Mandab) contains a number of small islands that limit the maximum required water crossing to 18 kilometers. During periods of lowered sea level, the distance would have been reduced further See Groucutt and Petraglia, "Prehistory of the Arabian Peninsula," 114.

22. Michael D. Petraglia and Jeffrey I. Rose, eds., *The Evolution of Human Populations in Arabia: Paleoenvironments, Prehistory, and Genetics* (Dordrecht: Springer, 2009).

23. Jeffrey I. Rose, "New Light on Human Prehistory in the Arabo-Persian Gulf Oasis," *Current Anthropology* 51 (2010): 849–883; Andrew Lawler, "In Search of Green Arabia," *Science* 345 (2014): 994–997. The floor of the Persian Gulf averages about 40 meters below the modern sea level and would have been exposed repeatedly during the late Pleistocene.

24. Jeffrey I. Rose et al., "The Nubian Complex of Dhofar, Oman: An African Middle Stone Age Industry in Southern Arabia," *PLoS ONE* 6 (2011): e28239; Groucutt and Petraglia, "Prehistory of the Arabian Peninsula," 114–115.

25. Khaled K. Abu-Amero et al., "Mitochondrial DNA Structure in the Arabian Peninsula," *BMC Evolutionary Biology* 8 (2008): 45; Khaled K. Abu-Amero et al., "Saudi Arabian Y-Chromosome Diversity and Its Relationship with Nearby Regions," *BMC Evolutionary Biology* 10 (2009): 59–68; Fregel et al., "Carriers of Mitochondrial DNA Macrohaplogroup N Lineages Reached Australia Around 50,000 Years Ago."

26. Rose et al., "Nubian Complex of Dhofar, Oman."

27. Simon J. Armitage et al., "The Southern Route 'Out of Africa': Evidence for an Early Expansion of Modern Humans into Arabia," *Science* 331 (2011): 453–456.

28. Appenzeller, "Eastern Odyssey"; Lawler, "In Search of Green Arabia."

29. Carlos A. Fernandes, "Bayesian Coalescent Inference from Mitochondrial DNA Variation of the Colonization Time of Arabia by the Hamadryas Baboon (*Papio hamadryas hamadryas*)," in *Evolution of Human Populations in Arabia*, ed. Petraglia and Rose, 89–102.

30. Mae Goder-Goldberger, Natalia Gubenko, and Erella Hovers, "'Diffusion with Modifications': Nubian Assemblages in the Central Negev Highlands of Israel and Their Implications for Middle Paleolithic Inter-Regional Interactions," *Quaternary International* 408 (2016): 121–139.

31. Partha P. Majumder, "The Human Genetic History of South Asia," *Current Biology* 20 (2010): R184–R187.

32. Fu et al., "Revised Timescale for Human Evolution Based on Ancient Mitochondrial Genomes," 555, table 3.

33. Sanghamitra Sahoo et al., "A Prehistory of Indian Y Chromosomes: Evaluating Demic Diffusion Scenarios," *Proceedings of the National Academy of Sciences* 103 (2006): 843–848; Poznik et al., "Punctuated Bursts in Human Male Demography Inferred from 1,244 Worldwide Y-Chromosome Sequences," 596.

34. Michael Bamshad et al., "Genetic Evidence on the Origins of Indian Caste Populations," *Genome Research* 11 (2003): 994–1004.
35. Common Y-chromosome haplogroups in India include R-M17, J-M172, R-M124, and L-M20. See Richard Cordaux et al., "Independent Origins of Indian Caste and Tribal Paternal Lineages," *Current Biology* 14 (2004): 231–235.
36. David Reich et al., "Reconstructing Indian Population History," *Nature* 461 (2009): 489–494.
37. Majumder, "Human Genetic History of South Asia," R185.
38. Analabha Basu et al., "Ethnic India: A Genomic View, with Special Reference to Peopling and Structure," *Genome Research* 13 (2003): 2277–2290; Mait Metspalu et al., "Most of the Extant mtDNA Boundaries in South and Southwest Asia Were Likely Shaped During the Initial Settlement of Eurasia by Anatomically Modern Humans," *BMC Genetics* 5 (2004): 26.
39. Endicott et al., "Genetic Origin of the Andaman Islanders."
40. Kumar et al., "Y-Chromosome Evidence Suggests a Common Paternal Heritage of Austro-Asiatic Populations."
41. Analabha Basu, Neeta Sarkar-Roy, and Partha P. Majumder, "Genomic Reconstruction of the History of Extant Populations of India Reveals Five Distinct Ancestral Components and a Complex Structure," *Proceedings of the National Academy of Sciences* 113 (2016): 1594–1599.
42. Basu, Sarkar-Roy, and Majumder concluded that endogamy had been established abruptly about 70 generations, or about 1,500 years, ago in India, in ibid.
43. David Bulbeck, "Craniodental Affinities of Southeast Asia's 'Negritos' and the Concordance with Their Genetic Affinities," *Human Biology* 85 (2013): 95–134; Jat T. Stock, "The Skeletal Phenotype of 'Negritos' from the Andaman Islands and Philippines Relative to Global Variation Among Hunter-Gatherers," *Human Biology* 85 (2013): 67–94.
44. Zarine Cooper, *Archaeology and History: Early Settlements in the Andaman Islands* (Oxford: Oxford University Press, 2002); Endicott et al., "Genetic Origin of the Andaman Islanders."
45. Endicott et al. found that 50 percent of their sample of aDNA extracted from the remains of 12 Andaman Islanders belonged to the mtDNA M2 sub-lineage (estimated coalescence age of 63,000 years), while the other half of the sample represented the younger M4 lineage (estimated coalescence age of 32,00 years), in "Genetic Origin of the Andaman Islanders," 181. Kumar et al. reported that a sample of people from the neighboring Nicobar Islands were exclusively Y-DNA haplogroup O-M95, in "Y-Chromosome Evidence Suggests a Common Paternal Heritage of Austro-Asiatic Populations," fig. 2.
46. The Andaman Islanders are unlikely to be directly descended from the earliest modern human population in Southeast Asia, however, because they lack any traces of genetic material from the Denisovans. See David Reich et al., "Denisova Admixture and the First Modern Human Dispersals into Southeast Asia and Oceania," *American Journal of Human Genetics* 89 (2011): 516–528.
47. A. R. Radcliffe-Brown, *The Andaman Islanders* (Cambridge: Cambridge University Press, 1922), 18. The Andaman Islanders appear to have lived at one of the highest population densities of any tropical foraging people recorded in ethnography.

See Robert L. Kelly, *The Lifeways of Hunter-Gathereers: The Foraging Spectrum* (Cambridge: Cambridge University Press, 2013), 178–184, table 7-3.

48. Only two other foraging peoples are reported to have lacked fire-making technology: the Mbuti (Pygmies) in central Africa and the original inhabitants of Tasmania.

49. See, for example, Dennell and Petraglia, "Dispersal of *Homo sapiens* Across Southern Asia"; Appenzeller, "Eastern Odyssey"; and Mellars et al., "Genetic and Archaeological Perspectives on the Initial Modern Human Colonization of Southern Asia."

50. Mijares et al., "New Evidence for a 67,000-Year-Old Human Presence at Callao Cave."

51. *Homo floresiensis* (nicknamed "hobbit") is a small non-modern-human fossil taxon discovered in 2003 on the island of Flores (ca. 480 kilometers east of Java) in Southeast Asia. Its relationship to modern humans and other fossil humans is unclear, and many controversies have surrounded the interpretation of the skeletal remains. See, for example, Chris Stringer and Peter Andrews, *The Complete World of Human Evolution*, 2nd ed. (London: Thames and Hudson, 2011), 174–175.

52. Armand Salvador Mijares observed that "it is difficult to determine whether the Callao specimen belongs to *H. sapiens* or to a more archaic form of *Homo*" ("Human Emergence and Adaptation to an Island Environment in the Philippine Paleolithic," in *Emergence and Diversity of Modern Human Behavior in Paleolithic Asia*, ed. Kaifu et al., 178).

53. Shen et al., "U-Series Dating of Liujiang Hominid Site in Guangxi."

54. Dennell and Petraglia, "Dispersal of *Homo sapiens* Across Southern Asia," 19.

55. Cai et al., "Age of Human Remains and Associated Fauna from Zhiren Cave in Guangxi."

56. Wu Liu et al., "Human Remains from Zhirendong, South China, and Modern Human Emergence in East Asia," *Proceedings of the National Academy of Sciences* 107 (2010): 19201–19206.

57. Bae et al., "Modern Human Teeth from Late Pleistocene Luna Cave."

58. Kenneth A. R. Kennedy, "The Deep Skull of Niah: An Assessment of Twenty Years of Speculation Concerning Its Evolutionary Significance," *Asian Perspectives* 20 (1977): 32–50.

59. Phillip J. Habgood, "The Origin of Anatomically Modern Humans in Australasia," in *The Human Revolution: Behavioural and Biological Perspectives on the Origins of Modern Humans*, ed. Paul Mellars and Chris Stringer (Princeton, N.J.: Princeton University Press, 1989), 245–273; Graeme Barker et al., "The 'Human Revolution' in Lowland Tropical Southeast Asia: The Antiquity and Behavior of Anatomically Modern Humans at Niah Cave (Sarawak, Borneo)," *Journal of Human Evolution* 52 (2007): 243–261.

60. D. R. Brothwell, "Upper Pleistocene Human Skull from Niah Caves," *Sarawak Museum Journal* 9 (1960): 323–349; Kennedy, "Deep Skull of Niah," 35; Habgood, "Origin of Anatomically Modern Humans in Australasia," 264–265; Barker et al., "'Human Revolution' in Lowland Tropical Southeast Asia," 251.

61. Fabrice Demeter et al., "Anatomically Modern Human in Southeast Asia (Laos) by 46 ka," *Proceedings of the National Academy of Sciences* 109 (2012): 14375–14380.

62. Florent Détroit et al., "Upper Pleistocene *Homo sapiens* from the Tabon Cave (Palawan, the Philippines): Description and Dating of New Discoveries," *Comptes Rendus Paleovol* 3 (2004): 705–712.

63. Habgood, "Origin of Anatomically Modern Humans in Australasia," 264–266; Détroit et al., "Upper Pleistocene *Homo sapiens* from the Tabon Cave," 711.

64. Athreya, "Modern Human Emergence in South Asia," 68–70.

65. Kenneth A. R. Kennedy and Siran U. Deraniyagala, "Fossil Remains of 28,000-Year-Old Hominids from Sri Lanka," *Current Anthropology* 30 (1989): 394–399; Nimal Perera et al., "People of the Ancient Rainforest: Late Pleistocene Foragers at the Batadomba-lena Rockshelter, Sri Lanka," *Journal of Human Evolution* 61 (2011): 254–269.

66. Kenneth A. R. Kennedy et al., "Upper Pleistocene Fossil Hominids from Sri Lanka," *American Journal of Physical Anthropology* 72 (1987): 441–461.

67. Philip J. Habgood and Natalie R. Franklin, "The Revolution That Didn't Arrive: A Review of Pleistocene Sahul," *Journal of Human Evolution* 55 (2008): 187–222; Peter Hiscock, "Cultural Diversification and the Global Dispersion of *Homo sapiens*," in *Emergence and Diversity of Modern Human Behavior in Paleolithic Asia*, ed. Kaifu et al., 225–236.

68. Petraglia et al., "Middle Paleolithic Assemblages from the Indian Subcontinent Before and After the Toba Super-Eruption."

69. See, for example, Ravi Korisettar, "Antiquity of Modern Humans and Behavioral Modernity in the Indian Subcontinent," in *Emergence and Diversity of Modern Human Behavior in Paleolithic Asia*, ed. Kaifu et al., 80–93.

70. Mellars et al., "Genetic and Archaeological Perspectives on the Initial Modern Human Colonization of Southern Asia."

71. Robin W. Dennell et al., "A 45,000-Year-Old Open-air Paleolithic Site at Riwat, Northern Pakistan," *Journal of Field Archaeology* 19 (1992): 17–33.

72. Sheila Mishra, Naveen Chauhan, and Ashkok K. Singhvi, "Continuity of Microblade Technology in the Indian Subcontinent Since 45 ka: Implications for the Dispersal of Modern Humans," *PLoS ONE* 8 (2013): e69280.

73. Chris Clarkson et al., "The Oldest and Longest Enduring Microlithic Sequence in India: 35 000 Years of Modern Human Occupation and Change at the Jwalapuram Locality 9 Rockshelter," *Antiquity* 83 (2009): 326–348; Mellars et al., "Genetic and Archaeological Perspectives on the Initial Modern Human Colonization of Southern Asia," 10701–10703.

74. Peter Bellwood, "Southeast Asia Before History," in *The Cambridge History of Southeast Asia*, vol. 1, part 1, *From Early Times to c. 1500*, ed. Nicholas Tarling (Cambridge: Cambridge University Press, 1999), 55–136.

75. Good-quality stone (for example, chert) was available in some parts of Southeast Asia, however, and is reported abundant in artifact assemblages from East Timor and the Talaud Islands (see table 5.3).

76. Earlier reports of bone points and spatulae from Niah Great Cave in Borneo (Tom Harrisson and Lord Medway, "A First Classification of Prehistoric Bone and Tooth Artifacts Based on Material from Niah Great Cave," *Asian Perspectives* 6 [1962]: 219–229; Bellwood, "Southeast Asia Before History," 83) apparently refer to artifacts in a younger context at this site (Barker et al., "'Human Revolution' in Lowland Tropical Southeast Asia," 255–256).

77. Sue O'Connor, Gail Robertson, and K. P. Aplin, "Are Osseous Artefacts a Window to Perishable Material Culture? Implications of an Unusually Complex Bone Tool from the Late Pleistocene of Timor," *Journal of Human Evolution* 67 (2014): 108–119.

78. Michelle C. Langley, Sue O'Connor, and Elena Piotto, "42,000-Year-Old Worked and Pigment-Stained *Nautilus* Shell from Jerimalai (Timor-Leste): Evidence for an Early Coastal Adaptation in ISEA," *Journal of Human Evolution* 97 (2016): 1–16.

79. M. Aubert et al., "Pleistocene Cave Art from Sulawesi, Indonesia," *Nature* 514 (2014): 223–227.

80. Commenting on the implications of the faunal remains from a roughly 40,000-year-old shelter on East Timor, Sue O'Connor wrote, "Although the lithic assemblage from Jerimalai appears exceeding simple, the faunal remains show that the people living in this region were skilled in the acquisition of marine resources that must have required the manufacture of sophisticated organic technology" ("Crossing the Wallace Line: The Maritime Skills of the Earliest Colonists in the Wallacean Archipelago," in *Emergence and Diversity of Modern Human Behavior in Paleolithic Asia*, ed. Kaifu et al., 218).

81. Radcliffe-Brown, *Andaman Islanders*, 485–492.

82. O'Connor, Ono, and Clarkson, "Pelagic Fishing at 42,000 BP and the Maritime Skills of Modern Humans"; O'Connor, "Crossing the Wallace Line," 217–220. The inshore species included parrot fish, unicorn fish, trevallies, triggerfish, snapper, emperors, and groupers. For a critique of pelagic fishing in this context, see Atholl Anderson, "The Antiquity of Sustained Offshore Fishing," *Antiquity* 87 (2013): 879–895.

83. Radcliffe-Brown, *Andaman Islanders*, 417–418. A fish hook (made of shell) was recovered from a younger level (23,000–16,000 years ago) at Jerimalai Shelter, as reported in O'Connor, "Crossing the Wallace Line," 217–220.

84. Rintaro Ono, Santoso Soegondho, and Minoru Yoneda, "Changing Marine Exploitation During Late Pleistocene in Northern Wallacea: Shell Remains from Leang Sarru Rockshelter in Talaud Islands," *Asian Perspectives* 48 (2009): 318–341. The reason for the absence of fish remains at this site is unclear, but may be related to preservation (323).

85. Tim Reynolds and Graeme Barker wrote, "The lack of specific targeting of any particular age group is unlike modern hunting practices . . . and suggests that traps or snares may have been used" ("Reconstructing Late Pleistocene Climates, Landscapes, and Human Activities in Northern Borneo from Excavations in the Niah Caves," in *Emergence and Diversity of Modern Human Behavior in Paleolithic Asia*, ed. Kaifu et al., 150). See also Barker et al., "'Human Revolution' in Lowland Tropical Southeast Asia," 255–256.

86. Ann Gibbons, "Five Matings for Moderns, Neandertals," *Science* 351 (2016): 1250–1251.

87. Pääbo, *Neanderthal Man*, 188.

88. Fregel et al., "Carriers of Mitochondrial DNA Macrohaplogroup N Lineages Reached Australia Around 50,000 Years Ago."

89. Hallam L. Movius Jr. wrote:

> [T]he archaeological, or palaeo-ethnological, material very definitely indicates that as early as Lower Palaeolithic times Southern and Eastern Asia

as a whole was a region of cultural retardation. Therefore, it seems very unlikely that this vast area could ever have played a vital and dynamic role in early human evolution, although very primitive forms of Early Man apparently persisted there long after types of a comparable stage of physical evolution had become extinct elsewhere. ("The Lower Palaeolithic Cultures of Southern and Eastern Asia," *Transactions of the American Philosophical Society* 38 [1948]: 411)

90. C. C. Swisher III et al., "Latest *Homo erectus* of Java: Potential Contemporaneity with *Homo sapiens* in Southeast Asia," *Science* 274 (1996): 1870–1874.
91. See, for example, Stringer and Andrews, *Complete World of Human Evolution*, 174–175.
92. David Reich et al., "Genetic History of an Archaic Hominin Group from Denisova Cave in Siberia," *Nature* 468 (2010): 1053–1060.
93. Reich et al., "Denisova Admixture and the First Modern Human Dispersals into Southeast Asia and Oceania."
94. Ibid., 522; Benjamin Vernot et al., "Excavating Neandertal and Denisovan DNA from the Genomes of Melanesian Individuals," *Science* 352 (2016): 235–239.
95. Morris Goodman, "Evolution of the Immunologic Species Specificity of Human Serum Proteins," *Human Biology* 34 (1962): 104–150.
96. Susanna Sawyer et al., "Nuclear and Mitochondrial DNA Sequences from Two Denisovan Individuals," *Proceedings of the National Academy of Sciences* 112 (2015): 15696–15700.
97. Qiu, "Forgotten Continent."
98. Athreya, "Modern Human Emergence in South Asia," 67–68. Current dating of the specimen from Narmada is based on ESR dating of an associated mammal tooth; earlier estimates based on U-series dating and biostratigraphy were significantly older (more than 200,000 years ago).
99. Chris Stringer, "The Status of *Homo heidelbergensis* (Schoetensack 1908)," *Evolutionary Anthropology* 21 (2012): 101–107.
100. Hannah V. A. James, "The Emergence of Modern Human Behavior in South Asia: A Review of the Current Evidence and Discussion of Its Possible Implications," in *The Evolution and History of Human Populations in South Asia: Interdisciplinary Studies in Archaeology, Biological Anthropology, Linguistics and Genetics*, ed. Michael D. Petraglia and Bridget Allchin (Dordrecht: Springer, 2007), 201–227.
101. Kurt Lambeck, Yusuke Yokoyama, and Tony Purcell, "Into and Out of the Last Glacial Maximum: Sea-Level Change During Oxygen-Isotope Stages 3 and 2," *Quaternary Science Reviews* 21 (2002): 343–360. Readers are urged to consult the interactive graphic of sea-level history and early settlement of Sahul designed by Matthew Coller, with the Monash School of Geography and Environmental Science, "SahulTime," http://sahultime.monash.edu.au/. Despite lower sea levels and exposed coastal-shelf areas, most of the water distances involved in crossing from Southeast Asia to Sahul were similar to those of the present day. See Jim Allen and James F. O'Connell, "Getting from Sunda to Sahul," in *Islands of Inquiry: Colonisation, Seafaring and the Archaeology of Maritime Landscapes*, ed. Geoffrey Clark, Foss Leach, and Sue O'Connor (Canberra: ANU E Press, 2008), 31–46.

102. Peter Hiscock, *Archaeology of Ancient Australia* (London: Routledge, 2008), 102–161.
103. Georgi Hudjashov et al., "Revealing the Prehistoric Settlement of Australia by Y Chromosome and mtDNA Analysis," *Proceedings of the National Academy of Sciences* 104 (2007): 8726–8730; Manfred Kayser, "The Human Genetic History of Oceania: Near and Remote Views of Dispersal," *Current Biology* 20 (2010): R194–R201.
104. Anders Bergström et al., "Deep Roots for Aboriginal Australian Y Chromosomes," *Current Biology* 26 (2016): 809–813.
105. Kayser, "Human Genetic History of Oceania."
106. Rasmussen et al., "Aboriginal Australian Genome Reveals Separate Human Dispersals into Asia," 96. Papuans and Aboriginal Australians also cluster together on the basis of genetic diversity in the human bacterial parasite *Helicobacter pylori*, according to Kayser, "Human Genetic History of Oceania," R195.
107. James M. Bowler et al., "Pleistocene Human Remains from Australia: A Living Site and Human Cremation from Lake Mungo, Western New South Wales," *World Archaeology* 2 (1970): 39–60; Richard G. Klein, *The Human Career: Human Biological and Cultural Origins*, 3rd ed. (Chicago: University of Chicago Press, 2009), 717–719.
108. Gregory J. Adcock et al., "Mitochondrial DNA Sequences in Ancient Australians: Implications for Modern Human Origins," *Proceedings of the National Academy of Sciences* 98 (2001): 537–542; Alan Cooper et al., "Human Origins and Ancient Human DNA," *Science* 292 (2001): 1656–1657. Recently, another specimen (WLH4) yielded a complete mitochondrial genome (assigned to mtDNA subclade S2), but it is relatively young—only a few thousand years old. See Tim H. Heupink et al., "Ancient mtDNA Sequences from the First Australians Revisited," *Proceedings of the National Academy of Sciences* 113 (2016): 6892–6897.
109. Habgood, "Origin of Anatomically Modern Humans in Australasia," 266–268; Hiscock, *Archaeology of Ancient Australia*, 82–101.
110. See, for example, M. I. Bird et al., "Radiocarbon Dating of 'Old' Charcoal Using a Wet Oxidation, Stepped-Combustion Procedure," *Radiocarbon* 41 (1999): 127–140. See also the summary discussion of recent developments in radiocarbon dating in Allen and O'Connell, "Both Half Right."
111. Peter Hiscock, "Occupying New Lands: Global Migrations and Cultural Diversification with Particular Reference to Australia," in *Paleoamerican Odyssey*, ed. Kelly E. Graf, Caroline V. Ketron, and Michael R. Waters (College Station: Texas A&M University Press, 2013), 3–11.
112. For example, Phillip Endicott et al., "Evaluating the Mitochondrial Timescale of Human Evolution," *Trends in Ecology and Evolution* 24 (2009): 515–521.
113. Rasmussen et al., "Aboriginal Australian Genome Reveals Separate Human Dispersals into Asia," 97; Rieux et al., "Improved Calibration of the Human Mitochondrial Clock Using Ancient Genomes."
114. Hiscock, *Archaeology of Ancient Australia*, 84–91; Allen and O'Connell estimate slightly less than 50,000 years ago, primarily on the basis of radiocarbon dating, in "Both Half Right." New evidence for somewhat earlier colonization was reported recently in the form of burned eggshell fragments—apparently, human

food debris—from multiple sites in two regions of Australia that are dated as early as 54,000 years ago. See Gifford Miller et al., "Human Predation Contributed to the Extinction of the Australian Megafaunal Bird *Genyornis newtoni* ~47 ka," *Nature Communications* 7 (2016): 10496.

115. Summerhayes et al., "Human Adaptation and Plant Use in Highland New Guinea 49,000 to 44,000 Years Ago." Waisted axes also are known from coral terraces on the *Huon Peninsula* (northeastern New Guinea) in buried contexts that are at least 44,000 years old. See Les Groube, John Chappell, and David Price, "A 40,000-Year-Old Human Occupation Site at Huon Peninsula, Papua New Guinea," *Nature* 324 (1986): 453–455; and James F. O'Connell and Jim Allen, "Dating the Colonization of Sahul (Pleistocene Australia–New Guinea): A Review of Recent Research," *Journal of Archaeological Science* 31 (2004): 838–840.

116. Among the early sites, *Devil's Lair*, located a few kilometers from the modern coast of southwestern Australia and dated to almost 50,000 years ago, may be nearest to the former coast, but it contains no evidence of a coastal economy. See Chris S. M. Turney et al., "Early Occupation at Devil's Lair, Southwestern Australia 50,000 Years Ago," *Quaternary Research* 55 (2001): 3–13. Hiscock currently estimates "landfall by *H. sapiens* on the northern coastline of Sahul at or slightly before 55,000 ± 5000 years BP . . . and entering Tasmania at 41,000 cal BP" ("Cultural Diversification and the Global Dispersion of *Homo sapiens*," 228).

117. James F. O'Connell, Jim Allen, and Kristen Hawkes note the evidence for deep-sea fishing before 40,000 years ago, as well as for open sea voyages of at least 140 kilometers before 30,000 years ago, in "Pleistocene Sahul and the Origins of Seafaring," in *The Global Origins and Development of Seafaring*, ed. Atholl Anderson, James H. Barrett, and Katherine V. Boyle (Cambridge: McDonald Institute for Archaeological Research, Cambridge University, 2010), 57–68.

118. J. B. Birdsell, "The Recalibration of a Paradigm for the First Peopling of Greater Australia," in *Sunda and Sahul: Prehistoric Studies in Southeast Asia, Melanesia, and Australia*, ed. J. Allen, J. Golson, and R. Jones (London: Academic Press, 1977), 113–167; Allen and O'Connell, "Getting from Sunda to Sahul," 37–38.

119. Hiscock, *Archaeology of Ancient Australia*, 102–128.

120. Peter Hiscock et al., "World's Earliest Ground-Edge Axe Production Coincides with Human Colonisation of Australia," *Australian Archaeology* 82 (2016): 2–11.

121. Wendell H. Oswalt, *An Anthropological Analysis of Food-Getting Technology* (New York: Wiley, 1976), 245–246.

122. Ibid., 236–237.

123. The peoples of the Western Desert, including the Aranda, supported some of the lowest population densities known among foragers (0.01–0.03 person per square kilometer). See Kelly, *Lifeways of Hunter-Gatherers*, 178–184, table 7-3.

124. R. A. Gould, *Living Archaeology* (Cambridge: Cambridge University Press, 1980), 60–87.

125. The technology of the Western Desert peoples also included little in the way of shelter or clothing, despite nighttime temperatures as low as 0°C, to which they famously acclimatized themselves. See G. A. Harrison et al., *Human Biology: An Introduction to Human Evolution, Variation, Growth and Ecology*, 3rd ed. (Oxford: Oxford University Press, 1988), 430–431.

126. Oswalt counted only 11 artifacts related to subsistence among the Tasmanians, in *Anthropological Analysis of Food-Getting Technology*, 263–264. The most complex artifact was a baited bird blind composed of four parts.
127. Gifford H. Miller et al., "Ecosystem Collapse in Pleistocene Australia and a Human Role in Megafaunal Extinction," *Science* 309 (2005): 287–290.
128. Another factor in the visibility of megafauna kill sites in North America is that the prey animals were social species and often were killed in groups or herds, creating large bone beds.
129. Judith Field and John Dodson, "Late Pleistocene Megafauna and Archaeology from Cuddie Springs, South-eastern Australia," *Proceedings of the Prehistoric Society* 65 (1999): 275–301; Judith H. Field, "Trampling Through the Pleistocene: Does Taphonomy Matter?" *Australian Archaeology* 63 (2006): 9–20.
130. Hiscock, *Archaeology of Ancient Australia*, 72–75. Significant climate changes took place at the time humans invaded Sahul, and their impacts on plant and animal life could account for at least some of the extinctions of megafauna. See, for example, Stephen Wroe et al., "Climate Change Frames Debate over the Extinction of Megafauna in Sahul (Pleistocene Australia-New Guinea)," *Proceedings of the National Academy of Sciences* 110 (2013): 8777–8781. See also James F. O'Connell and Jim Allen, "The Process, Biotic Impact, and Global Implications of the Human Colonization of Sahul About 47,000 Years Ago," *Journal of Archaeological Science* 56 (2015): 73–84.
131. Miller et al., "Human Predation Contributed to the Extinction of the Australian Megafaunal Bird *Genyornis newtoni*."
132. Susan Rule et al., "The Aftermath of Megafaunal Extinction: Ecosystem Transformation in Pleistocene Australia," *Science* 335 (2012): 1483–1486.
133. O'Connell and Allen, argue that modern humans dispersed initially into favorable habitats and only later began to occupy less productive environments, in "Process, Biotic Impact, and Global Implications of the Human Colonization of Sahul About 47,000 Years Ago."

6. GLOBAL DISPERSAL: NORTHERN EURASIA

1. Qiaomei Fu et al., "Genome Sequence of a 45,000-Year-Old Modern Human from Western Siberia," *Nature* 514 (2014): 445–449.
2. Adrien Rieux et al., "Improved Calibration of the Human Mitochondrial Clock Using Ancient Genomes," *Molecular Biology and Evolution* 31 (2014): 2780–2792; G. David Poznik et al., "Punctuated Bursts in Human Male Demography Inferred from 1,244 Worldwide Y-Chromosome Sequences," *Nature Genetics* 48 (2016): 593–599.
3. Fu et al., "Genome Sequence of a 45,000-Year-Old Modern Human from Western Siberia," 446; Qiaomei Fu et al., "A Revised Timescale for Human Evolution Based on Ancient Mitochondrial Genomes," *Current Biology* 23 (2013): 553–559.
4. Fu et al., "Genome Sequence of a 45,000-Year-Old Modern Human from Western Siberia," 445.
5. Ibid., 447–448.

6. Iosif Lazaridis et al., "Ancient Human Genomes Suggest Three Ancestral Populations for Present-Day Europeans," *Nature* 513 (2014): 409–413.

7. As noted in chapter 2, aDNA recently has been recovered from human remains in the lower latitudes that are at least a few thousand years old, most notably from the Willandra Lakes region in Australia. See Tim H. Heupink et al., "Ancient mtDNA Sequences from the First Australians Revisited," *Proceedings of the National Academy of Sciences* 113 (2016): 6892–6897. It is not clear if aDNA can be extracted from significantly older remains found in a warm-climate setting.

8. It may be significant that the "near modern" representatives of *Homo sapiens*, who occupied the Levant roughly 120,000 to 90,000 years ago and possibly parts of southern Asia during the same period, may have been unable to inhabit the colder regions of northern Eurasia. While the presence of the Neanderthals may have excluded them from places like Europe and the Altai area, most of Siberia apparently was uninhabited by other forms of *Homo*.

9. Pavel Pavlov, John Inge Svendsen, and Svein Indrelid, "Human Presence in the European Arctic Nearly 40,000 Years Ago," *Nature* 413 (2001): 64–67. Recently, new evidence for the presence of modern humans above the Arctic Circle in Siberia before 45,000 calibrated years before the present was reported in Vladimir V. Pitulko et al., "Early Human Presence in the Arctic: Evidence from 45,000-Year-Old Mammoth Remains," *Science* 351 (2016): 260–263.

10. John F. Hoffecker, *A Prehistory of the North: Human Settlement of the Higher Latitudes* (New Brunswick, N.J.: Rutgers University Press, 2005), 77–81.

11. For global temperature and precipitation data, see www.weatherbase.com, and for estimates of net primary productivity, see, for example, O. W. Archibold, *Ecology of World Vegetation* (London: Chapman & Hall, 1995), 1–14.

12. Several major sites on the East European Plain widely believed to represent Neanderthal occupations now appear to be too young to be firmly or even plausibly ascribed to Neanderthals (for example, *Betovo* on the Desna River near Bryansk), while other sites containing Levallois blades and points—widely assumed in the past to represent Neanderthal occupations—now appear more likely to be those of modern humans (for example, *Shlyakh* near Volgograd).

13. The presence of Neanderthals and Denisovans in the Altai region provides a clue to the principal barrier to the settlement of the colder and drier parts of Eurasia by pre-modern humans. Today, the area experiences low winter temperatures (January mean of –15°C in Barnaul) but somewhat higher net primary productivity than surrounding areas, suggesting that scarcity of resources—not low temperature—was the key factor.

14. For a summary discussion of foraging adaptations to cold environments, see John F. Hoffecker, *Desolate Landscapes: Ice-Age Settlement of Eastern Europe* (New Brunswick, N.J.: Rutgers University Press, 2002), 6–12.

15. Wendell H. Oswalt, *An Anthropological Analysis of Food-Getting Technology* (New York: Wiley, 1976), 181–190.

16. Robin Torrence argued that the increased complexity of technology among foragers in cold environments was a function of greater time constraints, in "Time Budgeting and Hunter-Gatherer Technology," in *Hunter-Gatherer Economy in*

Prehistory: A European Perspective, ed. Geoff Bailey (Cambridge: Cambridge University Press, 1983), 11–22.

17. Wendell H. Oswalt, "Technological Complexity: The Polar Eskimos and the Tareumiut," *Arctic Anthropology* 24 (1987): 82–98. It should be noted that Oswalt calculated the number of parts for some subsistants that had not been included in the earlier study (for example, Tareumiut umiak = 35 technounits, or components), which exceeded the complexity of toggle-headed harpoons.

18. Anecdotal examples of heavy dependence on snaring and trapping illustrate how critical they may be in specific temporal/spatial settings. For example, Richard K. Nelson described how Chalkyitsik families in northern interior Alaska had subsisted primarily on meat from hares during a winter when other sources of food were unexpectedly scarce: "A few families probably consumed more hare meat during 1969–70 than any other food, largely because other game species such as fish, moose, and bear were scarce at that time. . . . In good times [hares] are a supplement, in hard times a lifesaver" (*Hunters of the Northern Forest: Designs for Survival Among the Alaskan Kutchin* [Chicago: University of Chicago Press, 1973], 142).

19. The Andaman Islanders, who formerly lacked the technology for making fire, expended considerable time and resources on maintaining constant fires in their camps and transporting fire on journeys. According to A. R. Radcliffe-Brown, "Fires were and still are carefully kept alive in the village, and carefully carried when travelling. Every hunting party carries its fire with it. The natives are very skillful in selecting wood that will smoulder for a long time without going out and without breaking into flame" (*The Andaman Islanders* [Cambridge: Cambridge University Press, 1922], 472).

20. Lyn Wadley, "Were Snares and Traps Used in the Middle Stone Age and Does It Matter? A Review and a Case Study from Sibudu, South Africa," *Journal of Human Evolution* 58 (2010): 179–192; Paola Villa et al., "Border Cave and the Beginning of the Later Stone Age in South Africa," *Proceedings of the National Academy of Sciences* 109 (2012): 13208–13213.

21. Oswalt, *Anthropological Analysis of Food-Getting Technology*.

22. Richard B. Lee, *The !Kung San: Men, Women, and Work in a Foraging Society* (Cambridge: Cambridge University Press, 1979), 141–142.

23. John F. Hoffecker, "Innovation and Technological Knowledge in the Upper Paleolithic of Northern Eurasia," *Evolutionary Anthropology* 14 (2005): 186–198.

24. Michael P. Richards et al., "Stable Isotope Evidence for Increasing Dietary Breadth in the European Mid-Upper Paleolithic," *Proceedings of the National Academy of Sciences* 98 (2001): 6528–6532. Noting a spike in background ^{15}N values in Europe that appears to be related to increased aridity during a cold phase (HE4), Hervé Bocherens, Dorothée G. Drucker, and Stéphanie Madelaine caution against the assumption that increased ^{15}N values in human bone are necessarily related to changes in diet, in "Evidence for a ^{15}N Positive Excursion in Terrestrial Foodwebs at the Middle to Upper Palaeolithic Transition in South-Western France: Implications for Early Modern Human Palaeodiet and Palaeoenvironment," *Journal of Human Evolution* 69 (2014): 31–43. However, both the Ust'-Ishim bone and younger finds (ca. 35,000–30,000 years ago) that yielded high ^{15}N values date to warm periods (GI 12 and G 8) associated with reduced aridity.

25. John J. Shea and Matthew L. Sisk, "Complex Projectile Technology and *Homo sapiens* Dispersal into Western Eurasia," *PaleoAnthropology* 2010 (2010): 100–122; Katsuhiro Sano, "Evidence for the Use of Bow-and-Arrow Technology by the First Modern Humans in the Japanese Islands," *Journal of Archaeological Science: Reports* 10 (2016): 130–141.

26. Domesticated dogs, which are listed in table 6.4, represent a potentially important "technology" (or biotechnology) related to subsistence, but their presence before 15,000 years ago remains problematic. See, for example, Laurent A. F. Frantz et al., "Genomic and Archaeological Evidence Suggests a Dual Origin of Domestic Dogs," *Science* 352 (2016): 1228–1231.

27. Oswalt, "Technological Complexity," 87–90. For example, among the Tareumiut of northern Alaska, the number of parts that make up the following items of clothing are in parentheses: dress jacket of women (30), rain shirt of men (28), dress jacket of men (18), dress pants of women (16), and outer-skin jacket of men (14). Oswalt estimated a total of 215 components (or technounits) for all clothing and accessories (for example, belts and snow goggles) (89, table 2).

28. Ian Gilligan, "The Prehistoric Development of Clothing: Archaeological Implications of a Thermal Model," *Journal of Archaeological Theory and Method* 17 (2010): 15–80.

29. Eyed needles are reported from Mezmaiskaya Cave, Layer 1C, in the northern Caucasus (L. V. Golovanova, V. B. Doronichev, and N. E. Cleghorn, "The Emergence of Bone-Working and Ornamental Art in the Caucasian Upper Paleolithic," *Antiquity* 84 [2010]: 299–320), and from Denisova Cave, Layer 11, in the Altai region (A. P. Derevianko, M. V. Shunkov, and S. V. Markin, *Dinamika Paleoliticheskikh Industrii v Afrike i Evrazii v Pozdnem Pleistotsene i Problema Formirovaniya Homo sapiens* [Novosibirsk: Institute of Archaeology and Ethnography, SB RAS Press, 2014]).

30. V. N. Stepanchuk, "The Archaic to True Upper Paleolithic Interface: The Case of Mira in the Middle Dnieper Area," *Eurasian Prehistory* 3 (2005): 23–41.

31. Evidence of hand-operated drills is reported from Kostenki 17, Layer II (P. I. Boriskovskii, *Ocherki po Paleolitu Basseina Dona*, Materialy i Issledovaniya po Arkheologii SSSR 121 [Moscow: Nauka, 1963]), and from Denisova Cave, Layer 11 (Derevianko, Shunkov, and Markin, *Dinamika Paleoliticheskikh Industrii v Afrike i Evrazii v Pozdnem Pleistotsene*, 180–184).

32. Buvit et al., "Emergence of Modern Behavior in the Trans-Baikal," 494–497, table 33.1. The site of *Sungir'* (northern Russia), for example, which dates to about 35,000 years ago, contains several small pits thought by the excavator to represent storage facilities. See O. N. Bader, *Sungir' Verkhnepaleoliticheskaya Stoyanka* (Moscow: Nauka, 1978), 67–113. Larger and deeper pits, which are common in later sites on the East European Plain—such as Kostenki 1, Layer I, and Avdeevo—may have been used primarily for cold storage of bone fuel, which must be kept fresh to retain its flammability. See Hoffecker, *Desolate Landscapes*, 226.

33. Alex Mackay, Brian A. Stewart, and Brian M. Chase, "Coalescence and Fragmentation in the Late Pleistocene Archaeology of Southernmost Africa," *Journal of Human Evolution* 72 (2014): 26–51.

34. M. V. Anikovich et al., "Early Upper Paleolithic in Eastern Europe and Implications for the Dispersal of Modern Humans," *Science* 315 (2007): 223–226; Evgeny

P. Rybin, "Middle and Upper Paleolithic Interactions and the Emergence of 'Modern Behavior' in Southern Siberia and Mongolia," in *Emergence and Diversity of Modern Human Behavior in Paleolithic Asia*, ed. Kaifu et al., 470–489.

35. John F. Hoffecker et al., "Evidence for Kill-Butchery Events of Early Upper Paleolithic Age at Kostenki, Russia," *Journal of Archaeological Science* 37 (2010): 1073–1089.

36. Fu et al., "Genome Sequence of a 45,000-Year-Old Modern Human from Western Siberia," 448. The earlier estimate was published in Richard E. Green et al., "A Draft Sequence of the Neandertal Genome," *Science* 328 (2010): 710–722.

37. March Haber et al., "Genome-Wide Diversity in the Levant Reveals Recent Structuring by Culture," *PLoS Genetics* 9 (2013): e1003316. This study analyzed more than 500,000 genome-wide single nucleotide polymorphisms (SNPs) from a sample of 1,341, compared with samples from 48 populations worldwide.

38. Lluís Quintana-Murci et al., "Where West Meets East: The Complex mtDNA Landscape of the Southwest and Central Asian Corridor," *American Journal of Human Genetics* 74 (2004): 827–845; David Comas et al., "Admixture, Migrations, and Dispersals in Central Asia: Evidence from Maternal DNA Lineages," *European Journal of Human Genetics* 12 (2004): 495–504; Danielle A. Badro et al., "Y-Chromosome and mtDNA Genetics Reveal Significant Contrasts in Affinities of Modern Middle Eastern Populations with European and African Populations," *PLoS ONE* 8 (2013): e54616.

39. Israel Hershkovitz et al., "Levantine Cranium from Manot Cave (Israel) Foreshadows the First European Modern Humans," *Nature* 520 (2015): 216–219.

40. Ibid., 217.

41. Ibid., 218.

42. In addition to the fossil remains of *Homo sapiens* described in the text, isolated human teeth (ca. 10) were recovered from both the IUP and the Early Ahmarian layers at Üçağızlı Cave (southern Turkey) and tentatively assigned to modern humans. See Steven L. Kuhn et al., "The Early Upper Paleolithic Occupations at Üçağızlı Cave (Hatay, Turkey)," *Journal of Human Evolution* 56 (2009): 87–113.

43. Katerina Douka et al., "Chronology of Ksar Akil (Lebanon) and Implications for the Colonization of Europe by Anatomically Modern Humans," *PLoS ONE* 8 (2013): 372931. "Egbert" is thought to represent a juvenile (roughly seven to nine years old) and possibly a female; the cranium exhibits a modern morphology.

44. Margolein D. Bosch et al., "New Chronology for Ksâr ʿAkil (Lebanon) Supports Levantine Route of Modern Human Dispersal into Europe," *Proceedings of the National Academy of Sciences* 112 (2015): 7683–7688. The new dates have been critiqued on methodological grounds in Katerina Douka, Thomas F. G. Higham, and Christopher A. Bergman, "Statistical and Archaeological Errors Invalidate the Proposed Chronology for the Site of Ksar Akil," *Proceedings of the National Academy of Sciences* 112 (2015): E7034.

45. J. Franklin Ewing, "A Probable Neanderthaloid from Ksar ʿAkil, Lebanon," *American Journal of Physical Anthropology* 21 (1963): 101–104; M. Metni, "A Re-Examination of a Proposed Neandertal Maxilla from Ksar ʿAkil Rock Shelter, Antelias, Lebanon," *American Journal of Physical Anthropology*, suppl. 28 (1999): 202.

46. Douka et al., "Chronology of Ksar Akil (Lebanon) and Implications for the Colonization of Europe"; Bosch et al., "New Chronology for Ksâr ʿAkil (Lebanon) Supports Levantine Route of Modern Human Dispersal into Europe."

47. R. Kh. Suleimanov, *Statisticheskoe Izuchenie Kul'tury Grota Obi-Rakhmat* (Tashkent: Fan, 1972).
48. Andrei I. Krivoshapkin, Yaroslav V. Kuzmin, and A. J. Timothy Jull, "Chronology of the Obi-Rakhmat Grotto (Uzbekistan): First Results on the Dating and Problems of the Paleolithic Key Site in Central Asia," *Radiocarbon* 52 (2010): 549–554; Michelle M. Glantz, "The History of Hominin Occupation of Central Asia in Review," in *Asian Paleoanthropology: From Africa to China and Beyond*, ed. Christopher J. Norton and David R. Braun (Dordrecht: Springer, 2010), 101–112.
49. Andrei. I. Krivoshapkin, A. A. Anoikan, and P. Jeffrey Brantingham, "The Lithic Industry of Obi-Rakhmat Grotto, Uzbekistan," *Bulletin of the Indo-Pacific Association* 26 (2007): 5–19.
50. Shara Bailey et al., "The Affinity of the Dental Remains from Obi-Rakhmat Grotto, Uzbekistan," *Journal of Human Evolution* 55 (2008): 238–248; Michelle Glantz et al., "New Hominin Remains from Uzbekistan," *Journal of Human Evolution* 55 (2008): 223–237; Glantz, "History of Hominin Occupation of Central Asia," 108.
51. Glantz et al., "New Hominin Remains from Uzbekistan," 225.
52. See, for example, Janusz K. Kozlowski, "The Significance of Blade Technologies in the Period 50–35 kya BP for the Middle–Upper Palaeolithic Transition in Central and Eastern Europe," in *Rethinking the Human Revolution: New Behavioural and Biological Perspectives on the Origin and Dispersal of Modern Humans*, ed. Paul Mellars et al. (Cambridge: McDonald Institute for Archaeological Research, Cambridge University, 2007), 317–328.
53. Jiří A. Svoboda, "On Modern Human Penetration into Northern Eurasia: The Multiple Advances Hypothesis," in *Rethinking the Human Revolution*, ed. Mellars et al., 329–339; John F. Hoffecker, "The Spread of Modern Humans in Europe," *Proceedings of the National Academy of Sciences* 106 (2009): 16040–16045.
54. See, for example, Doron M. Behar et al., "A 'Copernican' Reassessment of the Human Mitochondrial DNA Tree from Its Root," *American Journal of Human Genetics* 90 (2012): 680.
55. Anna Belfer-Cohen and Ofer Bar-Yosef, "The Levantine Aurignacian: 60 Years of Research," in *Dorothy Garrod and the Progress of the Palaeolithic*, ed. William Davies and Ruth Charles (Barnsley: Oxbow Books, 1999), 119–134. Another important figure in the development of the chronology of the Middle and Upper Paleolithic of the Levant was René Neuville (1899–1952).
56. D. A. E. Garrod, "The Mugharet el Emireh in Lower Galilee: Type-Station of the Emiran Industry," *Journal of the Royal Anthropological Institute* 85 (1955): 141–162.
57. Anthony E. Marks and C. Reid Ferring, "The Early Upper Paleolithic of the Levant," in *The Early Upper Paleolithic: Evidence from Europe and the Near East*, ed. J. F. Hoffecker and C. A. Wolf, BAR International Series 437 (Oxford: British Archaeological Reports, 1988), 43–72.
58. Steven L. Kuhn, Mary C. Stiner, and Erksin Güleç, "Initial Upper Palaeolithic in South-Central Turkey and Its Regional Context: A Preliminary Report," *Antiquity* 73 (1999): 505–517.
59. Ofer Bar-Yosef, "The Middle and Early Upper Paleolithic in Southwest Asia and Neighboring Regions," in *The Geography of Neandertals and Modern Humans in*

Europe and the Greater Mediterranean, ed. Ofer Bar-Yosef and David Pilbeam (Cambridge, Mass.: Peabody Museum of Archaeology and Ethnology, 2000), 107–156.

60. Marks and Ferring, "Early Upper Paleolithic of the Levant," 48–56.

61. The mean estimated age of the cranium from Manot Cave (ca. 55,000 years ago) exceeds current dates on the earliest IUP assemblages in the Levant.

62. The most commonly cited example of local variation in the IUP of the Levant is the heavy production of chamfered pieces (*chanfreins*) with rounded edges at Ksâr 'Akil (also popular in North Africa at roughly the same time). See Bar-Yosef, "Middle and Early Upper Paleolithic in Southwest Asia and Neighboring Regions," 124.

63. Ibid., 127.

64. Steven L. Kuhn et al., "Ornaments of the Earliest Upper Paleolithic: New Insights from the Levant," *Proceedings of the National Academy of Sciences* 98 (2001): 7641–7646. Some freshwater gastropods also are present at Üçağızlı Cave (Layer F).

65. The dating of the lowest Early Ahmarian levels at Ksâr 'Akil and Üçağızlı Cave is comparable and is consistent with the dating of the earliest related sites (Proto-Aurignacian) in southern Europe.

66. Bruce G. Trigger, *A History of Archaeological Thought* (Cambridge: Cambridge University Press, 1989).

67. Cosimo Posth et al., "Pleistocene Mitochondrial Genomes Suggest a Single Major Dispersal of Non-Africans and a Late Glacial Population Turnover in Europe," *Current Biology* 26 (2016): 827–833.

68. Qiaomei Fu et al., "An Early Modern Human from Romania with a Recent Neanderthal Ancestor," *Nature* 524 (2015): 216–219; S. Benazzi et al., "The Makers of the Protoaurignacian and Implications for Neandertal Extinction," *Science* 348 (2015): 793–796. Maternal lineage mtDNA haplogroup M has been identified in slightly younger remains from *Goyet Cave* (Belgium). See Posth et al., "Pleistocene Mitochondrial Genomes Suggest a Single Major Dispersal of Non-Africans."

69. Pedro Soares et al., "The Archaeogenetics of Europe," *Current Biology* 20 (2010): R174–R183; Boris Malyarchuk et al., "The Peopling of Europe from the Mitochondrial Haplogroup U5 Perspective," *PLoS ONE* 5 (2010): e10285.

70. Andaine Seguin-Orlando et al., "Genomic Structure in Europeans Dating Back at Least 36,200 Years," *Science* 346 (2014): 1113–1118.

71. Soares et al., "Archaeogenetics of Europe."

72. Poznik et al., "Punctuated Bursts in Human Male Demography Inferred from 1,244 Worldwide Y-Chromosome Sequences."

73. Svoboda, "On Modern Human Penetration into Northern Eurasia," 331–334.

74. B. Ginter et al., eds., *Temnata Cave: Excavations in Karlukovo Karst Area, Bulgaria* (Krakow: Jagiellonian University Press, 2000); Jiří A. Svoboda and Ofer Bar-Yosef, eds., *Stránská skála: Origins of the Upper Paleolithic in the Brno Basin, Moravia, Czech Republic* (Cambridge, Mass.: Peabody Museum of Archaeology and Ethnology, 2003); Tsenka Tsanova and Jean-Guillaume Bordes, "Contribution au débat sur l'origine de l'Aurignacien: Principaux résultats d'une étude technologique de l'industrie lithique de la couche 11 de Bacho Kiro," in *The Humanized Mineral World: Towards Social and Symbolic Evaluation of Prehistoric Technologies in Southeastern Europe*, ed. Tsoni Tsonev and Emmanuela Montagnari Kokelj, ERAUL 103 (Liege: University of Liege, 2003), 41–50.

75. A bone perforator and a bone pendant were recovered from Layer 11 at Bacho Kiro. See Janusz K. Kozlowski et al., "Upper Paleolithic Assemblages," in *Excavation in the Bacho Kiro Cave (Bulgaria): Final Report*, ed. Janusz K. Kozlowski (Warsaw: Państwowe Wydawnictwo Naukowe, 1982), 141, fig. 5; and Steven L. Kuhn and Nicolas Zwyns, "Rethinking the Initial Upper Paleolithic," *Quaternary International* 347 (2014): 29–38.

76. P. Allsworth-Jones, "The Szeletian and the Stratigraphic Succession in Central Europe and Adjacent Areas: Main Trends, Recent Results, and Problems for Resolution," in *The Emergence of Modern Humans: An Archaeological Perspective*, ed. Paul Mellars (Edinburgh: Edinburgh University Press, 1990), 160–242.

77. Brian Adams suggests that the Szeletian may represent "an activity facies" of the local early Upper Paleolithic (which he identifies as "Aurignacian"), in "Gulyás Archaeology: The Szeletian and the Middle to Upper Palaeolithic Transition in Hungary and Central Europe," in *New Approaches to the Study of Early Upper Paleolithic 'Transitional' Industries in Western Eurasia: Transitions Great and Small*, ed. Julien Riel-Salvatore and Geoffrey A. Clark (Oxford: Archaeopress, 2007), 91–110. He notes that the perceived "archaic" appearance of the Szeletian may be explained in part by post-depositional processes and the low quality of the raw materials used at these sites, in Brian Adams, "The Impact of Lithic Raw Material Quality and Post-Depositional Processes on Cultural/Chronological Classification: The Hungarian Szeletian Case," in *Lithic Materials and Paleolithic Societies*, ed. Brian Adams and Brook S. Blades (Oxford: Blackwell, 2009), 248–255.

78. D. Richter, G. Tostevin, and P. Škrdla, "Bohunician Technology and Thermoluminescence Dating of the Type Locality of Brno-Bohunice (Czech Republic)," *Journal of Human Evolution* 55 (2008): 871–885.

79. In "Rethinking the Initial Upper Paleolithic," Kuhn and Zwyns observe that calibrated radiocarbon dates on the IUP in general (that is, across northern Eurasia) range between 47,000 and 32,000 years ago.

80. Elżbieta Gleń and Krzysztof Kaczanowski, "Human Remains," in *Excavation in the Bacho Kiro Cave (Bulgaria)*, ed. Kozlowski, 75, pl. I.

81. Shara E. Bailey, Timothy D. Weaver, and Jean-Jacques Hublin, "Who Made the Aurignacian and Other Early Upper Paleolithic Industries?" *Journal of Human Evolution* 57 (2009): 11–26.

82. A modern human maxilla from *Kent's Cavern* (England) is now estimated to be 44,200 to 41,500 years old and possibly too early for other industries associated with modern humans in Europe. See Tom Higham et al., "The Earliest Evidence for Anatomically Modern Humans in Northwestern Europe," *Nature* 479 (2011): 521–524. Following similar reasoning, Fu et al. suggested that the 45,000-year-old modern human bone from Ust'-Ishim in western Siberia probably was linked to the appearance of IUP-like assemblages in Northern Asia, in "Genome Sequence of a 45,000-Year-Old Modern Human from Western Siberia," 449.

83. Vadim Y. Cohen and Vadim N. Stepanchuk, "Late Middle and Early Upper Paleolithic Evidence from the East European Plain and Caucasus: A New Look at Variability, Interactions, and Transitions," *Journal of World Prehistory* 13 (1999): 265–319; L. Meignen et al., "Koulichivka and Its Place in the Middle-Upper Paleolithic Transition in Eastern Europe," in *The Early Upper Paleolithic Beyond*

Western Europe, ed. P. Jeffrey Brantingham, Steven L. Kuhn, and Kristopher W. Kerry (Berkeley: University of California Press, 2004), 50–63.

84. P. E. Nehoroshev, *Tekhnologicheskii metod izucheniya pervobytnogo rasshchepleniya kamnya srednego paleolita* (St. Petersburg: Evropeiskii Dom, 1999), and "Rezul'taty Datirovaniya Stoyanki Shlyakh," *Rossiiskaya arkheologiya* 3 (2006): 21–30.

85. John F. Hoffecker, "The Early Upper Paleolithic of Eastern Europe Reconsidered," *Evolutionary Anthropology* 20 (2011): 24–39.

86. V. M. Kharitonov, "Iskopaemye Gominidy Severnogo Kavkaza," in *Drevneishii Kavkaz: Perekrestok Azii i Evropy*, ed. S. A. Vasil'ev and A. V. Larionova (St. Petersburg: IIMK Russian Academy of Sciences, 2013), 79–80. The teeth recovered from Monasheskaya Cave exhibit relatively thick enamel, which is associated with modern humans.

87. Meignen et al., "Koulichivka and Its Place in the Middle-Upper Paleolithic Transition in Eastern Euroipe"; Svoboda and Bar-Yosef, eds., *Stránská skála*. The high percentage of quarry/workshop locations among the IUP sites has yet to be explained.

88. If the Levallois blades and points in Layer 4 at Molodova I in the Dniester Valley (western Ukraine) represent an IUP occupation, there may be traces of artificial shelters in the form of patterned arrangements of mammoth bones. See A. P. Chernysh, "Mnogosloinaya Paleoliticheskaya Stoyanka Molodova I," in *Molodova I: Unikal'noe Must'erskoe Poselenie na Srednem Dnestre*, ed. G. I. Goretskii and I. K. Ivanova (Moscow: Nauka, 1982), 6–102.

89. Francesco G. Fedele, Biagio Giaccio, and Irka Hajdas, "Timescales and Cultural Process at 40,000 BP in the Light of the Campanian Ignimbrite Eruption, Western Eurasia," *Journal of Human Evolution* 55 (2008): 834–857.

90. Steven L. Kuhn and Amilcare Bietti, "The Late Middle and Early Upper Paleolithic in Italy," in *Geography of Neandertals and Modern Humans in Europe and the Greater Mediterranean*, ed. Bar-Yosef and Pilbeam, 49–76.

91. Benazzi et al., "Makers of the Protoaurignacian and Implications for Neandertal Extinction"; Fu et al., "Genome Sequence of a 45,000-Year-Old Modern Human from Western Siberia."

92. See, for example, Paul Mellars, "Archeology and the Dispersal of Modern Humans in Europe: Deconstructing the 'Aurignacian,'" *Evolutionary Anthropology* 15 (2006): 167–182.

93. Katerina Douka et al., "On the Chronology of the Uluzzian," *Journal of Human Evolution* 68 (2014): 1–13.

94. Marco Peresani, Emanuela Cristiani, and Matteo Romandini, "The Uluzzian Technology of Grotta di Fumane and Its Implication for Reconstructing Cultural Dynamics in the Middle–Upper Palaeolithic Transition of Western Eurasia," *Journal of Human Evolution* 91 (2016): 36–56.

95. Stefano Benazzi et al., "Early Dispersal of Modern Humans in Europe and Implications for Neanderthal Behaviour," *Nature* 479 (2011): 525–528.

96. Alan Houghton Broderick, *Father of Prehistory: The Abbé Henri Breuil: His Life and Times* (New York: Morrow, 1963), 53–73.

97. Thomas Higham et al., "Testing Models for the Beginnings of the Aurignacian and the Advent of Figurative Art and Music: The Radiocarbon Chronology of

Geißenklösterle," *Journal of Human Evolution* 62 (2012): 664–676; Philip R. Nigst et al., "Early Modern Human Settlement of Europe North of the Alps Occurred 43,500 Years Ago in a Cold Steppe-Type Environment," *Proceedings of the National Academy of Sciences* 111 (2014): 14394–14399.

98. Daniel S. Adler, Anna Belfer-Cohen, and Ofer Bar-Yosef, "Between a Rock and a Hard Place: Neanderthal–Modern Human Interactions in the Southern Caucasus," in *When Neanderthals and Modern Humans Met*, ed. Nicholas J. Conard (Tübingen: Krems, 2006), 165–187; Golovanova, Doronichev, and Cleghorn, "Emergence of Bone-Working and Ornamental Art in the Caucasian Upper Paleolithic." Although the layer (Layer 1C) containing the eyed-needle fragment in Mezmaiskaya Cave has been dated to around 38,000 to 37,000 years ago, the application of more rigorous pretreatment methods to old radiocarbon samples indicates that the dates probably are too young by several thousand years. See Katerina Douka, Thomas Higham, and Andrey Sinitsyn, "The Influence of Pretreatment Chemistry on the Radiocarbon Dating of Campanian Ignimbrite–Aged Charcoal from Kostenki 14 (Russia)," *Quaternary Research* 73 (2010): 583–587.

99. John F. Hoffecker et al., "From the Bay of Naples to the River Don: The Campanian Ignimbrite Eruption and the Middle to Upper Paleolithic Transition in Eastern Europe," *Journal of Human Evolution* 55 (2008): 858–870.

100. Hoffecker et al., "Evidence for Kill-Butchery Events of Early Upper Paleolithic Age at Kostenki," 1087–1088.

101. Kill-butchery sites that contain artifacts similar to those at the Kostenki locations where large mammals were killed and butchered are especially common in North America.

102. Hoffecker et al., "Evidence for Kill-Butchery Events of Early Upper Paleolithic Age at Kostenki," 1081–1083.

103. A. A. Sinitsyn, "Nizhnie Kul'turnye Sloi Kostenok 14 (Markina Gora) (Raskopki 1998–2001 gg.)," in *Kostenki v Kontekste Paleolita Evrazii*, ed. A. A. Sinitsyn, V. Ya. Sergin, and J. F. Hoffecker (St. Petersburg: Russian Academy of Sciences, 2002), 219–236.

104. Ibid., 230, fig. 9.

105. N. K. Vereshchagin and I. E. Kuz'mina, "Ostatki Mlekopitayushchikh iz Paleoliticheskikh Stoyanok na Donu i Verkhnei Desne," *Trudy Zoologicheskogo Instituta AN SSSR* 72 (1977): 77–110. The author examined the hare remains from Layer IV at Kostenki 14 in 2009.

106. The harvesting of hares and other small mammals by recent foraging peoples in northern environments sometimes provided a critical food supply during winters when other foods were scarce. See, for example, Nelson, *Hunters of the Northern Forest*, 142.

107. Vereshchagin and Kuz'mina, "Ostatki Mlekopitayushchikh iz Paleoliticheskikh Stoyanok na Donu i Verkhnei Desne," 107. A total of 34 bird bones were reported from Layer IV at Kostenki 14 (excavations of A. N. Rogachev in 1953 and 1954).

108. Richards et al., "Stable Isotope Evidence for Increasing Dietary Breadth in the European Mid-Upper Paleolithic," 6530.

109. Pavlov, Svendsen, and Indrelid, "Human Presence in the European Arctic Nearly 40,000 Years Ago."

110. Vance T. Holliday et al., "Geoarchaeology of the Kostenki-Borshchevo Sites, Don River Valley, Russia," *Geoarchaeology: An International Journal* 22 (2007): 181–228. The CI tephra is up to 20 centimeters thick at Borshchevo 5 (213, table X).

111. Fedele, Giaccio, and Hajdas, "Timescales and Cultural Process at 40,000 BP in the Light of the Campanian Ignimbrite Eruption," 838–840.

112. Hoffecker et al., "From the Bay of Naples to the River Don," 866–867. Typical Aurignacian artifacts are reported from Layer X at Molodova 5 on the Dniester River in western Ukraine, which dates to roughly the same time period. See Hoffecker, "Early Upper Paleolithic of Eastern Europe Reconsidered," 28–29; and Seguin-Orlando et al., "Genomic Structure in Europeans Dating Back at Least 36,200 Years."

113. Paul Mellars and Jennifer C. French, "Tenfold Population Increase in Western Europe at the Neanderthal-to-Modern Human Transition," *Nature* 333 (2011): 623–627.

114. Malyarchuk et al., "Peopling of Europe from the Mitochondrial Haplogroup U5 Perspective."

115. Ted Goebel, "The Overland Dispersal of Modern Humans to Eastern Asia," in *Emergence and Diversity of Modern Human Behavior in Paleolithic Asia*, ed. Kaifu et al., 437–452.

116. Rybin, "Middle and Upper Paleolithic Interactions and the Emergence of 'Modern Behavior' in Southern Siberia and Mongolia."

117. Qiaomei Fu et al., "DNA Analysis of an Early Modern Human from Tianyuan Cave, China," *Proceedings of the National Academy of Sciences* 110 (2013): 2223–2227.

118. Rybin, "Middle and Upper Paleolithic Interactions and the Emergence of 'Modern Behavior' in Southern Siberia and Mongolia," 476–480.

119. Derevianko, Shunkov, and Markin, *Dinamika Paleoliticheskikh Industrii v Afrike i Evrazii v Pozdnem Pleistotsene*, 180–183.

120. David Reich et al., "Genetic History of an Archaic Hominin Group from Denisova Cave in Siberia," *Nature* 468 (2010): 1053–1060.

121. L. V. Lbova, *Paleolit Severnoi Zoni Zapadnogo Zabaikal'ya* (Ulan-Ude: Buryat Science Center, 2000); Buvit et al., "Emergence of Modern Behavior in the Trans-Baikal." The site of Kamenka also reportedly contains at least one pit, which may be the earliest evidence for storage in the archaeological record of modern humans.

122. Rybin, "Middle and Upper Paleolithic Interactions and the Emergence of 'Modern Behavior' in Southern Siberia and Mongolia," 476. The IUP layer at Ushlep 6 has yielded dates on bone of more than 40,000 years ago.

123. Fu et al., "Genome Sequence of a 45,000-Year-Old Modern Human from Western Siberia," p. 445.

124. Rybin, "Middle and Upper Paleolithic Interactions and the Emergence of 'Modern Behavior' in Southern Siberia and Mongolia," 481.

125. Feng Li et al., "The Development of Upper Palaeolithic China: New Results from the Shuidonggou Site," *Antiquity* 87 (2013): 368–383.

126. Fu et al., "DNA Analysis of an Early Modern Human from Tianyuan Cave," 2224.

127. Rybin, "Middle and Upper Paleolithic Interactions and the Emergence of 'Modern Behavior' in Southern Siberia and Mongolia," 477.

128. Buvit et al., "Emergence of Modern Behavior in the Trans-Baikal," 494, table 33.1.

129. Robert L. Kelly, *The Lifeways of Hunter-Gatherers: The Foraging Spectrum* (Cambridge: Cambridge University Press, 2013), 103–104.
130. Masami Izuho and Yosuke Kaifu, "The Appearance and Characteristics of the Early Upper Paleolithic in the Japanese Archipelago," in *Emergence and Diversity of Modern Human Behavior in Paleolithic Asia*, ed. Kaifu et al., 289–313.
131. In 2016, new evidence—in the form of traces of butchery on a mammoth carcass—for the presence of modern humans above the Arctic Circle in Siberia before 45,000 calibrated years before the present was reported in Pitulko et al., "Early Human Presence in the Arctic."
132. This is the Ust'-Ishim bone from the Irtysh River in western Siberia, reported in Fu et al., "Genome Sequence of a 45,000-Year-Old Modern Human from Western Siberia."
133. V. V. Pitulko et al., "The Yana RHS Site: Humans in the Arctic Before the Last Glacial Maximum," *Science* 303 (2004): 52–56.
134. Vladimir Pitulko et al., "Human Habitation in Arctic Western Beringia Prior to the LGM," in *Paleoamerican Odyssey*, ed. Kelly E. Graf, Caroline V. Ketron, and Michael R. Waters (College Station: Texas A&M University Press, 2013), 13–44.
135. Ibid., 21–36; V. V. Pitulko et al., "The Oldest Art of the Eurasian Arctic: Personal Ornaments and Symbolic Objects from Yana RHS, Arctic Siberia," *Antiquity* 86 (2012): 642–659.
136. See, for example, John F. Hoffecker and Scott A. Elias, *Human Ecology of Beringia* (New York: Columbia University Press, 2007), 92–94.
137. Pitulko et al., "Human Habitation in Arctic Western Beringia Prior to the LGM," 22–25. Other large mammals represented in the Yana River sites include horse, mammoth, wolf, and arctic fox. Earlier reports on these sites emphasized the presence of bird remains.
138. Erika Tamm et al., "Beringian Standstill and Spread of Native American Founders," *PLoS ONE* 9 (2007): e829; Andrew Kitchen, Michael M. Miyamoto, and Connie J. Mulligan, "A Three-Stage Colonization Model for the Peopling of the Americas," *PLoS ONE* 3 (2008): e1596; John F. Hoffecker, Scott A. Elias, and Dennis H. O'Rourke, "Out of Beringia?" *Science* 343 (2014): 979–980.

7. GLOBAL DISPERSAL: BERINGIA AND THE AMERICAS

1. Kurt Lambeck et al., "Sea Level and Global Ice Volumes from the Last Glacial Maximum to the Holocene," *Proceedings of the National Academy of Sciences* 111 (2014): 15296–15303.
2. Chris R. Stokes, Lev Tarasov, and Arthur S. Dyke, "Dynamics of the North American Ice Sheet Complex During Its Inception and Build-up to the Last Glacial Maximum," *Quaternary Science Reviews* 50 (2012): 86–104.
3. Lambeck et al., "Sea Level and Global Ice Volumes from the Last Glacial Maximum to the Holocene," 15301. As the authors note, the estimate of –134 meters for maximum fall in sea level during the LGM, compared with that of the present day, is greater than the frequently cited estimate of –125 meters.
4. M. Claussen et al., "Impact of CO_2 and Climate on Last Glacial Maximum Vegetation—A Factor of Separation," *Biogeosciences* 10 (2013): 3593–3604. During

the millennia preceding the LGM, climates were cooler and drier than those of the present day (and the LGM was preceded by a cold event between roughly 32,000 and 30,000 years ago). However, climates during the LGM often are estimated with reference to those of the present day.

5. Ibid., 3597–3598. Claussen et al. developed estimates of preindustrial climate and net primary productivity for comparison with LGM conditions.

6. See, for example, Lotfi Cherni et al., "Post-Last Glacial Maximum Expansion from Iberia to North Africa Revealed by Fine Characterization of mtDNA H Haplogroup in Tunisia," *American Journal of Physical Anthropology* 139 (2009): 253–260; Viktor Černý et al., "Internal Diversification of Mitochondrial Haplogroup R0a Reveals Post-Last Glacial Maximum Demographic Expansions in South Arabia," *Molecular Biology and Evolution* 28 (2010): 71–78; and Xiaoyun Cai et al., "Human Migration Through Bottlenecks from Southeast Asia into East Asia During Last Glacial Maximum Revealed by Y Chromosomes," *PLoS ONE* 6 (2011): e24282.

7. See, for example, Colin Renfrew, "Archaeogenetics—Towards a 'New Synthesis'?" *Current Biology* 20 (2010): R162–R165; Pedro Soares et al., "The Archaeogenetics of Europe," *Current Biology* 20 (2010): R174–R183; and Miroslava Derenko et al., "Complete Mitochondrial DNA Analysis of Eastern Eurasian Haplogroups Rarely Found in Populations of Northern Asia and Eastern Europe," *PLoS ONE* 7 (2012): e32179.

8. See, for example, Ted Goebel, "The Pleistocene Colonization of Siberia and Peopling of the Americas: An Ecological Approach," *Evolutionary Anthropology* 8 (1999): 208–227; Huw S. Groucutt and Michael D. Petraglia, "The Prehistory of the Arabian Peninsula: Deserts, Dispersals, and Demography," *Evolutionary Anthropology* 21 (2012): 113–125; and Ian Buvit et al., "Last Glacial Maximum Human Abandonment of the Transbaikal," *PaleoAmerica* 1 (2015): 374–376. Major stratified sites in northern Eurasia that exhibit an occupation hiatus that corresponds to the LGM include Solutré (eastern France) and Molodova 5 (western Ukraine). Surprising evidence for a human presence on the central East European Plain during the LGM emerged in 2016 at Kostenki (Russia), where two radiocarbon dates on charcoal fragments recovered from a hearth (primarily comprising burned bone and bone ash) were roughly 25,350 to 24,100 calibrated years before the present.

9. See, for example, Peter Hiscock, *Archaeology of Ancient Australia* (London: Routledge, 2008), 60–61; and Alan Williams et al., "Human Refugia in Australia During the Last Glacial Maximum and Terminal Pleistocene: A Geospatial Analysis of the 25–12 ka Australian Archaeological Record," *Journal of Archaeological Science* 40 (2013): 4612–4625. Although radiocarbon dates—typically on bone—from the colder and drier regions of Eurasia do sometimes fall into the LGM time range, they appear likely have been contaminated by younger carbon. Most such dates were processed decades ago without current pretreatment methods. A common pattern is that occupations reliably dated to a period predating the LGM, such as *Sungir'* (northern Russia), also yield much younger dates on bone or teeth that often fall in the LGM time range.

10. John F. Hoffecker, *A Prehistory of the North: Human Settlement of the Higher Latitudes* (New Brunswick, N.J.: Rutgers University Press, 2005), 80–81.

11. Isabelle Théry-Parisot et al., "The Use of Bone as Fuel During the Palaeolithic, Experimental Study of Bone Combustible Properties," in *The Zooarchaeology of Fats, Oils, Milk and Dairying*, ed. J. Mulville and A. K. Outram (Oxford: Oxbow Books, 2005), 50–59.

12. An estimate of the mean TMRCA for the five major maternal lineages, calibrated with aDNA, is 27,380 years ago, although three of the mtDNA lineages yielded coalescence estimates of more than 30,000 years ago. See Adrien Rieux et al., "Improved Calibration of the Human Mitochondrial Clock Using Ancient Genomes," *Molecular Biology and Evolution* 31 (2014): 2780–2792. The paternal lineages have consistently yielded younger coalescence dates (for example, Y-DNA haplogroup Q-M3 estimated at about 22,000 years ago). See Matthew C. Dulik et al., "Mitochondrial DNA and Y Chromosome Variation Provides Evidence for a Recent Common Ancestry Between Native Americans and Indigenous Altaians," *American Journal of Human Genetics* 90 (2012): 229–246. More recent mtDNA and whole-genome analyses (also using aDNA) have produced estimates of roughly 25,000 to 23,000 years ago. See Maanasa Raghavan et al., "Genomic Evidence for the Pleistocene and Recent Population History of Native Americans," *Science* 349 (2015): aab3884; and Bastien Llamas et al., "Ancient Mitochondrial DNA Provides High-Resolution Time Scale of the Peopling of the Americas," *Science Advances* 2 (2016): e1501385.

13. Erika Tamm et al., "Beringian Standstill and Spread of Native American Founders," *PLoS ONE* 9 (2007): e829.

14. John F. Hoffecker, Scott A. Elias, and Dennis H. O'Rourke, "Out of Beringia?" *Science* 343 (2014): 979–980.

15. Scott A. Elias and Brian Crocker, "The Bering Land Bridge: A Moisture Barrier to the Dispersal of Steppe-Tundra Biota?" *Quaternary Science Reviews* 27 (2008): 2473–2483.

16. Boris A. Yurtsev, "The Pleistocene Tundra-Steppe and the Productivity Paradox: The Landscape Approach," *Quaternary Science Reviews* 20 (2001): 165–174. See also Hoffecker, Elias, and O'Rourke, "Out of Beringia?"

17. Hoffecker, Elias, and O'Rourke, "Out of Beringia?"

18. John F. Hoffecker, *Landscape of the Mind: Human Evolution and the Archaeology of Thought* (New York: Columbia University Press, 2011), 128–137.

19. John F. Hoffecker and Scott A. Elias, *Human Ecology of Beringia* (New York: Columbia University Press, 2007), 132–161.

20. Aurelio Marangoni, David Caramelli, and Giorgio Manzi, "*Homo sapiens* in the Americas: Overview of the Earliest Human Expansion in the New World," *Journal of Anthropological Sciences* 92 (2014): 79–97; Llamas et al., "Ancient Mitochondrial DNA Provides High-Resolution Time Scale of the Peopling of the Americas."

21. New research on the "ice-free corridor" in western Canada indicates that it probably did not support a viable habitat for humans throughout its entire length until about 12,600 years ago. See Mikkel W. Pedersen et al., "Postglacial Viability and Colonization in North America's Ice-Free Corridor," *Nature* 537 (2016): 45–49.

22. Ted Goebel, Michael R. Waters, and Dennis H. O'Rourke, "The Late Pleistocene Dispersal of Modern Humans in the Americas," *Science* 319 (2008): 1497–1502.

23. Martin Bodner et al., "Rapid Coastal Spread of First Americans: Novel Insights from South America's Southern Cone Mitochondrial Genomes," *Genome Research*

22 (2012): 811–820; Goebel, Waters, and O'Rourke, "Late Pleistocene Dispersal of Modern Humans in the Americas," 1499.

24. Richard C. Lewontin, "The Apportionment of Human Diversity," *Evolutionary Biology* 6 (1972): 381–398.

25. Dennis H. O'Rourke and Jennifer A. Raff, "Human Genetic History of the Americas," *Current Biology* 20 (2010): R202–R207.

26. See, for example, Joanna Nichols, "Linguistic Diversity and the First Settlement of the New World," *Language* 66 (1990): 475–521; and Lyle Campbell, *American Indian Languages: The Historical Linguistics of Native America* (New York: Oxford University Press, 1997).

27. Sija Wang et al., "Genetic Variation and Population Structure in Native Americans," *PLoS Genetics* 3 (2007): 2049–2067.

28. Llamas et al., "Ancient Mitochondrial DNA Provides High-Resolution Time Scale of the Peopling of the Americas."

29. Joseph H. Greenberg, Christy G. Turner II, and Stephen L. Zegura, "The Settlement of the Americas: A Comparison of the Linguistic, Dental, and Genetic Evidence," *Current Anthropology* 27 (1986): 477–497.

30. See comments in ibid; and Nichols, "Linguistic Diversity and the First Settlement of the New World."

31. Emöke J. E. Szathmary and Nancy S. Ossenberg, "Are the Biological Differences Between North American Indians and Eskimos Truly Profound?" *Current Anthropology* 19 (1978): 673–701; Emöke J. E. Szathmary, "Genetics of Aboriginal North Americans," *Evolutionary Anthropology* 1 (1993): 202–220.

32. O'Rourke and Raff, "Human Genetic History of the Americas," R202–R203. The major Native American mtDNA haplogroups were defined before the others—in the early 1990s—which is why they were assigned the initial letters of the alphabet.

33. Jennifer A. Raff et al., "Ancient DNA Perspectives on American Colonization and Population History," *American Journal of Physical Anthropology* 146 (2011): 503–514; Miroslava V. Derenko et al., "The Presence of Mitochondrial Haplogroup X in Altaians from South Siberia," *American Journal of Human Genetics* 69 (2001): 237–241. Although the higher frequencies of mtDNA haplogroup X in western Eurasia (and North Africa) have encouraged speculation about migrations to the Americas from Europe during the LGM (for example, Dennis J. Stanford and Bruce Bradley, *Across Atlantic Ice: The Origin of America's Clovis Culture* [Berkeley: University of California Press, 2012]), the Native American sublineage (mtDNA X2a) is not found in western Eurasia (Jennifer A. Raff and Deborah A. Bolnick, "Does Mitochondrial Haplogroup X Indicate Ancient Trans-Atlantic Migration to the Americas? A Critical Re-evaluation," *PaleoAmerica* 1 [2015]: 297–304).

34. Morten Rasmussen et al., "The Ancestry and Affiliations of Kennewick Man," *Nature* 523 (2015): 455–458.

35. Stephen L. Zegura et al., "High-Resolution SNPs and Microsatellite Haplotypes Point to a Single, Recent Entry of Native American Y Chromosomes into the Americas," *Molecular Biology and Evolution* 21 (2004): 164–175; Dulik et al., "Mitochondrial DNA and Y Chromosome Variation Provides Evidence for a Recent Common Ancestry Between Native Americans and Indigenous Altaians."

36. O'Rourke and Raff, "Human Genetic History of the Americas," R204.

37. P. Forster et al., "Origin and Evolution of Native American mtDNA Variation: A Reappraisal," *American Journal of Human Genetics* 59 (1996): 935–945.
38. Sandro L. Bonatto and Francisco M. Salzano, "A Single and Early Migration for the Peopling of the Americas Supported by Mitochondrial DNA Sequence Data," *Proceedings of the National Academy of Sciences* 94 (1997): 1870. This paper was influenced by earlier publications, including Szathmary and Ossenberg, "Are the Biological Differences Between North American Indians and Eskimos Truly Profound?"
39. Tamm et al., "Beringian Standstill and Spread of Native American Founders."
40. Ibid.
41. Andrew Kitchen, Michael M. Miyamoto, and Connie J. Mulligan, "A Three-Stage Colonization Model for the Peopling of the Americas," *PLoS ONE* 3 (2008): e1596.
42. John F. Hoffecker et al., "Beringia and the Global Dispersal of Modern Humans," *Evolutionary Anthropology* 25 (2016): 64–78.
43. Ibid., 73–74.
44. David Reich et al., "Reconstructing Native American Population History," *Nature* 488 (2012): 370–374.
45. An allele at a microsatellite locus on chromosome 9 (D9S1120) exhibits a pattern similar to that of the mtDNA data; that is, it is unique to Native Americans but found in all groups, including Aleut-Inuit and Na-Dené. See K. B. Schroeder et al., "A Private Allele Ubiquitous in the Americas," *Biology Letters* 3 (2007): 218–223.
46. Hoffecker et al., "Beringia and the Global Dispersal of Modern Humans," 68–69; G. Richard Scott et al., "Sinodonty, Sundadonty, and the Beringian Standstill Model: Issues of Timing and Migrations into the New World," *Quaternary International* (2016): in press.
47. The identification of one or more of the ancestral Native American mtDNA haplogroups (listed in Tamm et al., "Beringian Standstill and Spread of Native American Founders") in aDNA recovered from human remains in Beringia that date to the LGM (or, possibly, aDNA recovered from sediment associated with a human occupation in the same spatial-temporal setting) seems to be the most effective way to confirm the Beringian Standstill Hypothesis.
48. Alessandro Achilli et al., "Reconciling Migration Models to the Americas with the Variation of North American Native Mitogenomes," *Proceedings of the National Academy of Sciences* 110 (2013): 14308–14313.
49. Matthew C. Dulik et al., "Y-Chromosome Analysis Reveals Genetic Divergence and New Founding Native Lineages in Athapaskan- and Eskimoan-Speaking Populations," *Proceedings of the National Academy of Sciences* 109 (2012): 8471–8476.
50. Mark A. Sicoli and Gary Holton, "Linguistic Phylogenies Support Back-Migration from Beringia to Asia," *PLoS ONE* 9 (2014): e91722.
51. Lauriane Bourgeon, Ariane Burke, and Thomas Higham, "Earliest Human Presence in North America Dated to the Last Glacial Maximum: New Radiocarbon Dates from Bluefish Caves, Canada," *PLoS ONE* 12 (2017): e0169486.
52. Hoffecker et al., "Beringia and the Global Dispersal of Modern Humans."
53. Eric Hultén, *Outline of the History of Arctic and Boreal Biota During the Quaternary Period* (New York: Cramer, 1937).
54. David M. Hopkins, "Introduction," in *The Bering Land Bridge*, ed. David M. Hopkins (Stanford, Calif.: Stanford University Press, 1967), 1–6.

55. José de Acosta, *Historia natural y moral de las Indias*, ed. J. Mangan; trans. Frances Lopez-Morillas (1590; Durham, N.C.: Duke University Press, 2002).
56. Hopkins, "Introduction," 1–3. See also Hoffecker and Elias, *Human Ecology of Beringia*, 2–3.
57. David M. Hopkins defined Beringia as "western Alaska, Northeastern Siberia, and the shallow parts of the Bering and Chukchi Seas" ("Preface," in *Bering Land Bridge*, ed. Hopkins, vii).
58. Paul A. Colinvaux, "Quaternary Vegetational History of Arctic Alaska," in *Bering Land Bridge*, ed. Hopkins, 207–231.
59. R. Dale Guthrie, "Paleoecology of the Large-Mammal Community in Interior Alaska During the Late Pleistocene," *American Midland Naturalist* 79 (1968): 346–463. See also R. Dale Guthrie, "Mammals of the Mammoth Steppe as Paleoenvironmental Indicators," in *Paleoecology of Beringia*, ed. David M. Hopkins et al. (New York: Academic Press, 1982), 307–326.
60. David M. Hopkins, P. A. Smith, and J. V. Matthews Jr., "Dated Wood from Alaska and the Yukon: Implications for Forest Refugia in Beringia," *Quaternary Research* 15 (1981): 217–249.
61. Frederick Hadleigh West, *The Archaeology of Beringia* (New York: Columbia University Press, 1981); William R. Powers and John F. Hoffecker, "Late Pleistocene Settlement in the Nenana Valley, Central Alaska," *American Antiquity* 54 (1989): 263–287. The Russian phytogeographer Boris Yurtsev proposed the term *Megaberingia*, based on the distribution of key modern plant species, for an even larger area that extended into the Lena Basin and portions of northwestern Canada, in "Problems of the Late Cenozoic Paleogeography of Beringia in Light of Phytogeographic Evidence," in *Beringia in the Cenozoic Era*, ed. V. L. Kontrimavichus (New Delhi: Amerind, 1984), 129–153.
62. See, for example, Brian Fagan, *The First North Americans: An Archaeological Journey* (London: Thames and Hudson, 2011), 21–24.
63. Scott A. Elias et al., "Life and Times of the Bering Land Bridge," *Nature* 382 (1996): 60–63.
64. R. Dale Guthrie, "Origin and Causes of the Mammoth Steppe: A Story of Cloud Cover, Woolly Mammoth Tooth Pits, Buckles, and Inside-Out Beringia," *Quaternary Science Reviews* 20 (2001): 549–574; Elias and Crocker, "Bering Land Bridge."
65. Hoffecker and Elias, *Human Ecology of Beringia*, 1–10.
66. David M. Hopkins, "Cenozoic History of the Bering Land Bridge," *Science* 129 (1959): 1519–1528.
67. Yurtsev, "Pleistocene Tundra-Steppe and the Productivity Paradox."
68. Linda B. Brubaker et al., "Beringia as a Glacial Refugium for Boreal Trees and Shrubs: New Perspectives from Mapped Pollen Data," *Journal of Biogeography* 32 (2005): 833–848; Rachel Westbrook et al., "Evidence for Glacial Refugium in Central Beringia" (poster presented at the annual meeting of the Geological Society of America, Charlotte, N.C., November 4–7, 2012). See also Patricia M. Anderson and Anatoly V. Lozhkin, "Late Quaternary Vegetation of Chukotka (Northeast Russia): Implications for Glacial and Holocene Environments of Beringia," *Quaternary Science Reviews* 107 (2015): 112–128.

69. Vladimir Pitulko et al., "Human Habitation in Arctic Western Beringia Prior to the LGM," in *Paleoamerican Odyssey*, ed. Kelly E. Graf, Caroline V. Ketron, and Michael R. Waters (College Station: Texas A&M University Press, 2013), 13–44.

70. Vladimir V. Pitulko et al., "Early Human Presence in the Arctic: Evidence from 45,000-Year-Old Mammoth Remains," *Science* 351 (2016): 260–263.

71. Carol Gelvin-Reymiller et al., "Technical Aspects of a Worked Proboscidean Tusk from Inmachuk River, Seward Peninsula, Alaska," *Journal of Archaeological Science* 33 (2006): 1088–1094. As the authors note, there are other examples of worked mammoth tusk from recent contexts in northern Alaska.

72. Stokes, Tarasov, and Dyke, "Dynamics of the North American Ice Sheet Complex During Its Inception and Build-up to the Last Glacial Maximum," 91–93.

73. Kelin Zhuang and John R. Giardino, "Ocean Cooling Pattern at the Last Glacial Maximum," *Advances in Meteorology* 2012 (2012): e213743.

74. Jed O. Kaplan, "Geophysical Applications of Vegetation Modeling" (Ph.D. diss., Lund University, 2001). See also Thomas Alerstam et al., "A Polar System of Intercontinental Bird Migration," *Proceedings of the Royal Society B: Biological Sciences* 274 (2007): 2523–2530.

75. Elias and Crocker, "Bering Land Bridge," 2476, fig. 2.

76. Bas van Geel et al., "The Ecological Implications of a Yakutian Mammoth's Last Meal," *Quaternary Research* 69 (2008): 361–376.

77. Jacques Cinq-Mars, "La place des grottes du Poisson-Bleu dans la prehistoire béringienne," *Revista de Arquelogía Americana* 1 (1990): 9–32; Lauriane Bourgeon, "Bluefish Cave II (Yukon Territory, Canada): Taphonomic Study of a Bone Assemblage," *PaleoAmerica* 1 (2015): 105–108.

78. Bourgeon, Burke, and Higham, "Earliest Human Presence in North America Dated to the Last Glacial Maximum."

79. Connie J. Mulligan and Andrew Kitchen, "Three-Stage Colonization Model for the Peopling of the Americas," in *Paleoamerican Odyssey*, ed. Graf, Ketron, and Waters, 171–181.

80. See, for example, R. N. E. Barton et al., "The Late Upper Palaeolithic Occupation of the Moroccan Northwest Maghreb During the Last Glacial Maximum," *African Archaeological Review* 22 (2005): 77–100; and Hiscock, *Archaeology of Australia*, 58–61.

81. Soares et al., "Archaeogenetics of Europe," R177–R178.

82. Ted Goebel, "The 'Microblade Adaptation' and Recolonization of Siberia During the Late Upper Pleistocene," in *Thinking Small: Global Perspectives on Microlithization*, ed. Robert G. Elston and Steven L. Kuhn (Arlington, Va.: American Anthropological Association, 2002), 117–131.

83. Y. A. Mochanov, *Drevneishie Etapy Zaseleniya Chelovekom Severo-Vostochnoi Azii* (Novosibirsk: Nauka, 1977); Kelly E. Graf, "Siberian Odyssey," in *Paleoamerican Odyssey*, ed. Graf, Ketron, and Waters, 65–80. The Yubetsu technique entails the production of microblade cores from small bifaces through a sequence of steps that are described in Richard E. Morlan, "Technological Characteristics of Some Wedge-Shaped Cores in Northwestern North America and Northeast Asia," *Asian Perspectives* 19 (1976): 96–106.

84. Ben A. Potter, Charles E. Holmes, and David R. Yesner, "Technology and Economy Among the Earliest Prehistoric Foragers in Interior Eastern Beringia," in *Paleoamerican Odyssey*, ed. Graf, Ketron, and Waters, 81–103.

85. Théry-Parisot et al., "Use of Bone as Fuel During the Paleolithic."

86. N. N. Dikov, *Drevnie Kul'tury Severo-Vostochnoi Azii* (Moscow: Nauka, 1979).

87. Powers and Hoffecker, "Late Pleistocene Settlement in the Nenana Valley."

88. Vladimir V. Pitulko, Aleksandr E. Basilyan, and Elena Y. Pavlova, "The Berelekh 'Graveyard': New Chronological and Stratigraphical Data from the 2009 Field Season," *Geoarchaeology: An International Journal* 29 (2014): 277–299.

89. N. N. Dikov, "The Ushki Sites, Kamchatka Peninsula," in *American Beginnings: The Prehistory and Paleoecology of Beringia*, ed. Frederick Hadleigh West (Chicago: University of Chicago Press, 1996), 244–250; Ted Goebel, Michael R. Waters, and Margarita Dikova, "The Archaeology of Ushki Lake, Kamchatka, and the Pleistocene Peopling of the Americas," *Science* 301 (2003): 501–505.

90. Powers and Hoffecker, "Late Pleistocene Settlement in the Nenana Valley"; John F. Hoffecker, W. Roger Powers, and Ted Goebel, "The Colonization of Beringia and the Peopling of the New World," *Science* 259 (1993): 46–53; Kelly E. Graf et al., "Dry Creek Revisited: New Excavations, Radiocarbon Dates, and Site Formation Inform on the Peopling of Eastern Beringia," *American Antiquity* 80 (2015): 671–694.

91. Nancy H. Bigelow and Wm. Roger Powers, "Climate, Vegetation, and Archaeology 14,000–9000 Cal Yr B.P. in Central Alaska," *Arctic Anthropology* 38 (2001): 171–195.

92. Hoffecker and Elias, *Human Ecology of Beringia*, 211–212.

93. Tamm et al., "Beringian Standstill and Spread of Native American Founders."

94. Mulligan and Kitchen, "Three-Stage Colonization Model for the Peopling of the Americas."

95. Tamm et al., "Beringian Standstill and Spread of Native American Founders."

96. Beringian sites that contain microblade assemblages of Younger Dryas age overlying Nenana Complex (or stemmed-point assemblages) include Dry Creek, Moose Creek, Little John, and Ushki I and V.

97. Morten Rasmussen et al., "Ancient Human Genome Sequence of an Extinct Palaeo-Eskimo," *Nature* 463 (2010): 757–762.

98. John F. Hoffecker, "Assemblage Variability in Beringia: The Mesa Factor," in *From the Yenisei to the Yukon: Interpreting Lithic Assemblage Variability in Late Pleistocene/Early Holocene Beringia*, ed. Ted Goebel and Ian Buvit (College Station: Texas A&M University Press, 2011), 165–178.

99. Justin C. Tackney et al., "Two Contemporaneous Mitogenomes from Terminal Pleistocene Burials in Eastern Beringia," *Proceedings of the National Academy of Sciences* 112 (2015): 13833–13838. One of the burials (USR1) was found to be closely related to an individual of the living Arara people in Brazil (13836).

100. Ibid., 13837.

101. Although the burials at Upward Sun River are assigned to the microblade industry, their affiliation with it is ambiguous. See Ben A. Potter et al., "New Insights into Eastern Beringian Mortuary Behavior: A Terminal Pleistocene Double Infant Burial at Upward Sun River," *Proceedings of the National Academy of Sciences* 111 (2014): 17060–17065. The associated artifacts lack diagnostic elements of the microblade industry, but include lanceolate points conceivably related to the point types

typical of the North American Plains that were found in Alaska and dated to the Younger Dryas.

102. Tamm et al., "Beringian Standstill and Spread of Native American Founders"; Phillip Endicott et al., "Evaluating the Mitochondrial Timescale of Human Evolution," *Trends in Ecology and Evolution* 24 (2009): 515–521. See also Llamas et al., "Ancient Mitochondrial DNA Provides High-Resolution Time Scale of the Peopling of the Americas."

103. Arthur S. Dyke concluded that a narrow ice-free corridor was present by roughly 15,200 to 14,400 calibrated years before the present, in "An Outline of North American Deglaciation with Emphasis on Central and Northern Canada," in *Quaternary Glaciations—Extent and Chronology*, vol. 2, part 2, *North America*, ed. Jürgen Ehlers and Philip L. Gibbard (Amsterdam: Elsevier, 2004), 388. Recently published research on the environments of the ice-free corridor reveals that a viable habitat for humans may not have existed until about 12,600 years ago, which significantly postdates the appearance of occupations south of the retreating ice sheets. See Pedersen et al., "Postglacial Viability and Colonization in North America's Ice-Free Corridor."

104. For a brief overview of the dating and contents of the earliest sites, see Michael R. Waters and Thomas Wier Stafford Jr., "The First Americans: A Review of the Evidence for the Late-Pleistocene Peopling of the Americas," in *Paleoamerican Odyssey*, ed. Graf, Ketron, and Waters, 541–560.

105. Robert L. Kelly and Lawrence C. Todd, "Coming into the Country: Early Paleoindian Hunting and Mobility," *American Antiquity* 53 (1988): 231–244.

106. Adriana Schmidt Dias and Lucas Bueno, "The Initial Colonization of South America Eastern Lowlands: Brazilian Archaeology Contributions to Settlement of America Models," in *Paleoamerican Odyssey*, ed. Graf, Ketron, and Waters, 339–357.

107. See, for example, Fagan, *First North Americans*.

108. Much of the Northwest coast may have been deglaciated before the opening of an interior ice-free corridor, encouraging speculation that it was the initial route of migration from Beringia to mid-latitude North America. See, for example, Carol A. S. Mandryk et al., "Late Quaternary Paleoenvironments of Northwestern North America: Implications for Inland Versus Coastal Migration Routes," *Quaternary Science Reviews* 20 (2001): 301–314.

109. Heiner Josenhans et al., "Early Humans and Rapidly Changing Holocene Sea Levels in the Queen Charlotte Islands–Hecate Strait, British Columbia, Canada," *Science* 277 (1997): 71–74.

110. Quentin Mackie et al., "Locating Pleistocene-Age Submerged Archaeological Sites on the Northwest Coast: Current Status of Research and Future Directions," in *Paleoamerican Odyssey*, ed. Graf, Ketron, and Waters, 133–147.

111. Daryl Fedje et al., "Function, Variability, and Interpretation of Archaeological Assemblages at the Pleistocene/Holocene Transition in Haida Gwaii," in *From the Yenisei to the Yukon*, ed. Goebel and Buvit, 323–342.

112. Mackie et al., "Locating Pleistocene-Age Submerged Archaeological Sites on the Northwest Coast," 138–141.

113. E. James Dixon, "Human Colonization of the Americas: Timing, Technology and Process," *Quaternary Science Reviews* 20 (2001): 277–299.

114. Michael R. Waters et al., "Late Pleistocene Horse and Camel Hunting at the Southern Margin of the Ice-Free Corridor: Reassessing the Age of Wally's Beach, Canada," *Proceedings of the National Academy of Sciences* 112 (2015): 4263–4267.
115. John W. Ives et al., "Vectors, Vestiges, and Valhallas—Rethinking the Corridor," in *Paleoamerican Odyssey*, ed. Graf, Ketron, and Waters, 149–169.
116. Ugo A. Perego et al., "Distinctive Paleo-Indian Migration Routes from Beringia Marked by Two Rare mtDNA Haplogroups," *Current Biology* 19 (2009): 1–8.
117. Morten Rasmussen et al., "The Genome of a Late Pleistocene Human from a Clovis Burial Site in Western Montana," *Nature* 506 (2014): 225–229.
118. Rasmussen et al., "Ancestry and Affiliations of Kennewick Man." See also Raff and Bolnick, "Does Mitochondrial Haplogroup X Indicate Ancient Trans-Atlantic Migration to the Americas?"
119. Pedersen et al., "Postglacial Viability and Colonization in North America's Ice-Free Corridor."
120. Some archaeologists have suggested that the bone fragment embedded in the Manis mastodon skeleton may not be a projectile point. See, for example, Gary Haynes, *The Early Settlement of North America: The Clovis Era* (Cambridge: Cambridge University Press, 2002), 72.
121. Kathryn E. Krasinski, "Broken Bones and Cut Marks: Taphonomic Analyses and Implications for the Peopling of North America" (Ph.D. diss., University of Nevada, Reno, 2010). In addition to taphonomic analysis of the bones, the Lindsay mammoth remains were associated with large sandstone blocks that probably were brought to the site by humans.
122. Vance T. Holliday and D. Shane Miller, "The Clovis Landscape," in *Paleoamerican Odyssey*, ed. Graf, Ketron, and Waters, 221–245.
123. Eileen Johnson, "Along the Ice Margin—The Cultural Taphonomy of Late Pleistocene Mammoth in Southeastern Wisconsin (USA)," *Quaternary International* 169–170 (2007): 64–83.
124. Jessi J. Halligan et al., "Pre-Clovis Occupation 14,550 Years Ago at the Page-Ladson Site, Florida, and the Peopling of the Americas," *Science Advances* 2 (2016): e1600375.
125. See, for example, Daniel J. Joyce, "Pre-Clovis Megafauna Butchery Sites in the Western Great Lakes Region, USA," in *Paleoamerican Odyssey*, ed. Graf, Ketron, and Waters, 467–483. An exception to the lack of diagnostic artifacts in older contexts may be a bifacial point recovered from the Meadowcroft Rockshelter (Pennsylvania) that is thought to predate 14,000 years, although similar points are thus far unknown from well-dated contexts in this time range.
126. C. Reid Ferring, *The Archaeology and Paleoecology of the Aubrey Clovis Site (41DN479) Denton County, Texas* (Denton: Department of Geography, University of North Texas, 2001).
127. C. Vance Haynes and Bruce B. Huckell, eds., *Murray Springs: A Clovis Site with Multiple Activity Areas in the San Pedro Valley, Arizona* (Tucson: University of Arizona Press, 2007).
128. Nicole M. Waguespack, "Pleistocene Extinctions: The State of Evidence and the Structure of Debate," in *Paleoamerican Odyssey*, ed. Graf, Ketron, and Waters, 311–319.

129. Jesse A. M. Ballenger et al., "Evidence for Younger Dryas Global Climate Oscillation and Human Response in the American Southwest," *Quaternary International* 242 (2011): 502–519.
130. Dennis L. Jenkins et al., "Clovis Age Western Stemmed Projectile Points and Human Coprolites at the Paisley Caves," *Science* 337 (2012): 223–228.
131. Dennis L. Jenkins et al., "Geochronology, Archaeological Context, and DNA at the Paisley Caves," in *Paleoamerican Odyssey*, ed. Graf, Ketron, and Waters, 485–510.
132. Jon M. Erlandson et al., "Paleoindian Seafaring, Maritime Technologies, and Coastal Foraging on California's Channel Islands," *Science* 331 (2011): 1181–1185.
133. Bodner et al., "Rapid Coastal Spread of First Americans."
134. Tom D. Dillehay et al., "A Late Pleistocene Human Presence at Huaca Prieta, Peru, and Early Pacific Coastal Adaptations," *Quaternary Research* 77 (2012): 418–423.
135. Daniel H. Sandweiss et al., "Quebrada Jaguay: Early South American Maritime Adaptations," *Science* 281 (1998): 1830–1832.
136. For a description of the food-getting technology of the Aymara, who lived in the Lake Titicaca area of Bolivia and Peru, see, for example, Wendell H. Oswalt, *An Anthropological Analysis of Food-Getting Technology* (New York: Wiley, 1976), 271–274.
137. Sandweiss et al., "Quebrada Jaguay."
138. Donald Jackson et al., "Initial Occupation of the Pacific Coast of Chile During Late Pleistocene Times," *Current Anthropology* 48 (2007): 725–731.
139. See, for example, William J. Mayer-Oakes, "Early Man in the Andes," *Scientific American*, May 1963, 117–128; and Goebel, Waters, and O'Rourke, "Late Pleistocene Dispersal of Modern Humans in the Americas," 1499. Fluted points in South America are stemmed fluted points (fluted Fishtail points) that may have some connection with Western Stemmed points in North America (and possibly even those of Beringia). The most parsimonious explanation for the appearance of fluted points in North and South America is that they were produced by people with a common heritage. Because the stemmed fluted points of South America are closely linked to lineages that dispersed along the Pacific coast (for example, mtDNA subclade D1), it is conceivable that the Clovis fluted points were made by related lineages that dispersed into the North American interior (and not by people who migrated through the ice-free corridor directly from Beringia, which apparently was not available as a migration route until after the Clovis fluted points were present in mid-latitude North America).
140. Anthony D. Barnofsky and Emily L. Lindsey, "Timing of Quaternary Megafaunal Extinction in South America in Relation to Human Arrival and Climate Change," *Quaternary International* 217 (2010): 10–29.
141. See, for example, Alberto L. Cione, Eduardo P. Tonni, and Leopoldo Soibelzon, "The Broken Zig-Zag: Late Cenozoic Large Mammal and Tortoise Extinction in South America," *Revista del Museo Argentino de Ciencias Naturales*, n.s., 5 (2003): 1–19.
142. Bodner et al., "Rapid Coastal Spread of First Americans," 816.
143. Dias and Bueno, "Initial Colonization of South America Eastern Lowlands."
144. Hoffecker, *Prehistory of the North*, 120–134.
145. Peter Bellwood, *First Migrants: Ancient Migration in Global Perspective* (Oxford: Wiley Blackwell, 2013), 194–197.

BIBLIOGRAPHY

Abu-Amero, Khaled K., Ali Hellani, Ana M. González, José M. Larruga, Vicente M. Cabrera, and Peter A. Underhill. "Saudi Arabian Y-Chromosome Diversity and Its Relationship with Nearby Regions." *BMC Evolutionary Biology* 10 (2009): 59–68.

Abu-Amero, Khaled K., José M. Larruga, Vicente M. Cabrera, and Ana M. González. "Mitochondrial DNA Structure in the Arabian Peninsula." *BMC Evolutionary Biology* 8 (2008): 45.

Achilli, Alessandro, Ugo A. Perego, Hovirag Lancioni, Anna Olivieri, Francesca Gandini, Baharak Hooshiar Kashani, Vincenza Battaglia, Viola Grugni, Norman Angerhofer, Mary P. Rogers, et al. "Reconciling Migration Models to the Americas with the Variation of North American Native Mitogenomes." *Proceedings of the National Academy of Sciences* 110 (2013): 14308–14313.

Acosta, José de. *Historia natural y moral de las Indias*. 1590. Edited by J. Mangan. Translated by Frances Lopez-Morillas. Durham, N.C.: Duke University Press, 2002.

Adami, Christoph. *Introduction to Artificial Life*. New York: Springer, 1998.

Adami, Christoph. "The Use of Information Theory in Evolutionary Biology." *Annals of the New York Academy of Sciences* 1256 (2011): 49–65.

Adami, Christoph, Charles Ofria, and Travis C. Collier. "Evolution of Biological Complexity." *Proceedings of the National Academy of Sciences* 97 (2000): 4463–4468.

Adams, Brian. "Gulyás Archaeology: The Szeletian and the Middle to Upper Palaeolithic Transition in Hungary and Central Europe." In *New Approaches to the Study of Early Upper Paleolithic 'Transitional' Industries in Western Eurasia: Transitions Great and Small*, edited by Julien Riel-Salvatore and Geoffrey A. Clark, 91–110. Oxford: Archaeopress, 2007.

Adams, Brian. "The Impact of Lithic Raw Material Quality and Post-Depositional Processes on Cultural/Chronological Classification: The Hungarian Szeletian Case." In *Lithic Materials and Paleolithic Societies*, edited by Brian Adams and Brook S. Blades, 248–255. Oxford: Blackwell, 2009.

Adcock, Gregory J., Elizabeth S. Dennis, Simon Easteal, Gavin A. Huttley, Lars S. Jermlin, W. James Peacock, and Alan Thorne. "Mitochondrial DNA Sequences in Ancient Australians: Implications for Modern Human Origins." *Proceedings of the National Academy of Sciences* 98 (2001): 537–542.

Adler, Daniel S., Anna Belfer-Cohen, and Ofer Bar-Yosef. "Between a Rock and a Hard Place: Neanderthal–Modern Human Interactions in the Southern Caucasus." In *When Neanderthals and Modern Humans Met*, edited by Nicholas J. Conard, 165–187. Tübingen: Krems, 2006.

Aiello, Leslie, and Christopher Dean. *An Introduction to Human Evolutionary Anatomy*. London: Academic Press, 1990.

Aiello, Leslie C., and Peter Wheeler. "The Expensive-Tissue Hypothesis: The Brain and the Digestive System in Human and Primate Evolurion." *Current Anthropology* 36 (1995): 199–221.

Alerstam, Thomas, Johan Bäckman, Gudmundur A. Gudmundsson, Anders Hedenström, Sara S. Henningsson, Håkan Karlsson, Mikael Rosé, and Roine Strandberg. "A Polar System of Intercontinental Bird Migration." *Proceedings of the Royal Society B: Biological Sciences* 274 (2007): 2523–2530.

Allen, Jim, and James F. O'Connell. "Both Half Right: Updating the Evidence for Dating First Human Arrivals in Sahul." *Australian Archaeology* 79 (2014): 86–108.

Allen, Jim, and James F. O'Connell. "Getting from Sunda to Sahul." In *Islands of Inquiry: Colonisation, Seafaring and the Archaeology of Maritime Landscapes*, edited by Geoffrey Clark, Foss Leach, and Sue O'Connor, 31–46. Canberra: ANU E Press, 2008.

Allman, John Morgan. *Evolving Brains*. New York: Scientific American Library, 1999.

Allsworth-Jones, P. "The Szeletian and the Stratigraphic Succession in Central Europe and Adjacent Areas: Main Trends, Recent Results, and Problems for Resolution." In *The Emergence of Modern Humans: An Archaeological Perspective*, edited by Paul Mellars, 160–240. Edinburgh: Edinburgh University Press, 1990.

Ambrose, Stanley H. "Chronology of the Later Stone Age and Food Production in East Africa." *Journal of Archaeological Science* 25 (1998): 377–392.

Ambrose, Stanley H. "Coevolution of Composite-Tool Technology, Constructive Memory, and Language: Implications for the Evolution of Modern Human Behavior." *Current Anthropology* 51 (2010): S135–S147.

Ambrose, Stanley H. "Late Pleistocene Human Population Bottlenecks, Volcanic Winter, and Differentiation of Modern Humans." *Journal of Human Evolution* 34 (1998): 623–651.

Ambrose, Stanley H. "Paleolithic Technology and Human Evolution." *Science* 291 (2001): 1748–1753.

Anderson, Patricia M., and Anatoly V. Lozhkin. "Late Quaternary Vegetation of Chukotka (Northeast Russia): Implications for Glacial and Holocene Environments of Beringia." *Quaternary Science Reviews* 107 (2015): 112–128.

Anikovich, M. V., V. V. Popov, and N. I. Platonova. *Paleolit Kostenkovsko-Borshchevskogo Raiona v Kontekste Verkhnego Paleolita Evropy*. St. Petersburg: Russian Academy of Sciences, 2008.

Anikovich, M. V., A. A. Sinitsyn, J. F. Hoffecker, V. T. Holliday, V. V. Popov, S. N. Lisitsyn, S. L. Forman, G. M. Levkovskaya, G. A. Pospelova, I. E. Kuz'mina, et al. "Early

Upper Paleolithic in Eastern Europe and Implications for the Dispersal of Modern Humans." *Science* 315 (2007): 223–226.

Appenzeller, Tim. "Eastern Odyssey." *Nature* 485 (2012): 24–26.

Archibold, O. W. *Ecology of World Vegetation.* London: Chapman & Hall, 1995.

Armitage, Simon J., Sabah A. Jasmin, Anthony E. Marks, Adrian G. Parker, Vitaly I. Usik, and Hans-Peter Uerpmann. "The Southern Route 'Out of Africa': Evidence for an Early Expansion of Modern Humans into Arabia." *Science* 331 (2011): 453–456.

Armstrong, Este. "Relative Brain Size in Monkeys and Prosimians." *American Journal of Physical Anthropology* 66 (1985): 263–273.

Athreya, Sheela. "Modern Human Emergence in South Asia: A Review of the Fossil and Genetic Evidence." In *Emergence and Diversity of Modern Human Behavior in Paleolithic Asia,* edited by Yousuke Kaifu, Masami Izuho, Ted Goebel, Hiroyuki Sato, and Akira Ono, 61–79. College Station: Texas A&M University Press, 2015.

Atkinson, Quentin D. "Phonemic Diversity Supports a Serial Founder Effect Model of Language Expansion from Africa." *Science* 332 (2011): 346–349.

Atkinson, Quentin D., Russell D. Gray, and Alexei J. Drummond. "Bayesian Coalescent Inference of Major Human Mitochondrial DNA Haplogroup Expansions in Africa." *Proceedings of the Royal Society B: Biological Sciences* 276 (2009): 367–373.

Aubert, M., A. Brumm, M. Ramli, T. Sutikna, E. W. Saptomo, B. Hakim, M. J. Morwood, G. D. van den Bergh, L. Kinsley, and A. Dosseto. "Pleistocene Cave Art from Sulawesi, Indonesia." *Nature* 514 (2014): 223–227.

Aubert, Maxime, Alistair W. G. Pike, Chris Stringer, Antonis Bartsiokas, Les Kinsley, Stephen Eggins, Michael Day, and Rainer Grün. "Confirmation of a Late Middle Pleistocene Age for the Omo Kibish 1 Cranium by Direct Uranium-Series Dating." *Journal of Human Evolution* 63 (2012): 704–710.

Babbage, Charles. "On a Method of Expressing by Signs the Action of Machinery." *Philosophical Transactions of the Royal Society* 116 (1826): 250–265.

Bader, O. N. *Sungir' Verkhnepaleoliticheskaya Stoyanka.* Moscow: Nauka, 1978.

Bader, O. N., and N. O. Bader. "Verkhnepaleoliticheskoe Poselenie Sungir'." In *Homo Sungirensis: Verkhnepaleoliticheskii Chelovek: Ekologicheskie i Evolyutsionnye Aspekty Issledovaniya,* 21–29. Moscow: Nauchnyi Mir, 2000.

Badro, Danielle A., Bouchra Douaihy, Marc Haber, Sonia C. Youhanna, Angélique Salloum, Michella Ghassibe-Sabbagh, Brian Johnsrud, Georges Khazen, Elizabeth Matisoo-Smith, David F. Soria-Hernanz, et al. "Y-Chromosome and mtDNA Genetics Reveal Significant Contrasts in Affinities of Modern Middle Eastern Populations with European and African Populations." *PLoS ONE* 8 (2013): e54616.

Bae, Christopher J., Wei Wang, Jianxin Zhao, Shengming Huang, Feng Tian, and Guanjun Shen. "Modern Human Teeth from Late Pleistocene Luna Cave (Guangxi, China)." *Quaternary International* 354 (2014): 169–183.

Bailey, Shara, Michelle Glantz, Timothy D. Weaver, and Bence Viola. "The Affinity of the Dental Remains from Obi-Rakhmat Grotto, Uzbekistan." *Journal of Human Evolution* 55 (2008): 238–248.

Bailey, Shara E., Timothy D. Weaver, and Jean-Jacques Hublin. "Who Made the Aurignacian and Other Early Upper Paleolithic Industries?" *Journal of Human Evolution* 57 (2009): 11–26.

Ballenger, Jesse A. M., Vance T. Holliday, Andrew L. Kowler, William T. Reitze, Mary M. Prasciunas, D. Shane Miller, and Jason D. Windingstad. "Evidence for Younger Dryas Global Climate Oscillation and Human Response in the American Southwest." *Quaternary International* 242 (2011): 502–519.

Balme, Jane. "Excavations Revealing 40,000 Years of Occupation at Mimbi Caves, South Central Kimberley, Western Australia." *Australian Archaeology* 51 (2000): 1–5.

Bamshad, Michael, Toomas Kivisild, W. Scott Watkins, Mary E. Dixon, Chris E. Ricker, Baskara B. Rao, J. Mastan Naidu, B. V. Ravi Prasad, P. Govinda Reddy, Arani Rasanayagam, et al. "Genetic Evidence on the Origins of Indian Caste Populations." *Genome Research* 11 (2003): 994–1004.

Barker, Graeme, Huw Barton, Michael Bird, Patrick Daly, Ipoi Datan, Alan Dykes, Lucy Farr, David Gilbertson, Barbara Harrisson, Chris Hunt, et al. "The 'Human Revolution' in Lowland Tropical Southeast Asia: The Antiquity and Behavior of Anatomically Modern Humans at Niah Cave (Sarawak, Borneo)." *Journal of Human Evolution* 52 (2007): 243–261.

Barnofsky, Anthony D., and Emily L. Lindsey. "Timing of Quaternary Megafaunal Extinction in South America in Relation to Human Arrival and Climate Change." *Quaternary International* 217 (2010): 10–29.

Bartolomei, Giorgio, Alberto Broglio, Pier Francesco Cassoli, Lanfredo Castelletti, Laura Cattani, Mauro Cremaschi, Giacomo Giacobini, Giancarla Malerba, Alfio Maspero, Marco Peresani, et al. "La Grotte de Fumane: Un site aurignacien au pied des Alpes." *Preistoria Alpina* 28 (1992): 131–179.

Barton, R. N. E., A. Bouzouggar, S. N. Collcutt, R. Gale, T. F. G. Higham, L. T. Humphrey, S. Parfitt, E. Rhodes, C. B. Stringer, and F. Malek. "The Late Upper Palaeolithic Occupation of the Moroccan Northwest Maghreb During the Last Glacial Maximum." *African Archaeological Review* 22 (2005): 77–100.

Barton, R. N. E., A. Bouzouggar, S. N. Collcutt, J.-L. Schwenninger, and L. Clark-Balzan. "OSL Dating of the Aterian Levels at Dar es-Soltan I (Rabat, Morocco) and Implications for the Dispersal of Modern *Homo sapiens*." *Quaternary Science Reviews* 28 (2009): 1914–1931.

Bar-Yosef, Ofer. "The Middle and Early Upper Paleolithic in Southwest Asia and Neighboring Regions." In *The Geography of Neandertals and Modern Humans in Europe and the Greater Mediterranean*, edited by Ofer Bar-Yosef and David Pilbeam, 107–156. Cambridge, Mass.: Peabody Museum of Archaeology and Ethnology, 2000.

Bar-Yosef, Ofer. "Upper Pleistocene Cultural Stratigraphy in Southwest Asia." In *The Emergence of Modern Humans: Biocultural Adaptations in the Later Pleistocene*, edited by Erik Trinkaus, 154–180. Cambridge: Cambridge University Press, 1989.

Bar-Yosef, O., and J.-G. Bordes. "Who Were the Makers of the Châtelperronian Culture?" *Journal of Human Evolution* 59 (2010): 586–593.

Bar-Yosef, Ofer, and Bernard Vandermeersch. "Modern Humans in the Levant." *Scientific American*, April 1993, 64–70.

Basu, Analabha, Namita Mukherjee, Sangita Roy, Sanghamitra Sengupta, Sanat Banerjee, Madan Chakraborty, Badal Dey, Monami Roy, Bidyut Roy, Nitai P. Bhattacharyya, et al. "Ethnic India: A Genomic View, with Special Reference to Peopling and Structure." *Genome Research* 13 (2003): 2277–2290.

Basu, Analabha, Neeta Sarkar-Roy, and Partha P. Majumder. "Genomic Reconstruction of the History of Extant Populations of India Reveals Five Distinct Ancestral Components and a Complex Structure." *Proceedings of the National Academy of Sciences* 113 (2016): 1594–1599.

Beals, Kenneth L., Courtland L. Smith, and Stephen M. Dodd. "Brain Size, Cranial Morphology, Climate, and Time Machines." *Current Anthropology* 25 (1984): 301–330.

Beaumont, Will, Robert Beverly, John Southon, and R. E. Taylor. "Bone Preparation at the KCCAMS Laboratory." *Nuclear Instruments and Methods B* 268 (2010): 906–909.

Beekman, Madeleine, and Jie Bin Lew. "Foraging in Honeybees—When Does It Pay to Dance?" *Behavioral Ecology* 19 (2007): 255–261.

Behar, Doron M., Mannis van Oven, Saharon Rosset, Mait Metspalu, Eva-Liis Loogväli, Nuno M. Silva, Toomas Kivisild, Antonio Torroni, and Richard Villems. "A 'Copernican' Reassessment of the Human Mitochondrial DNA Tree from Its Root." *American Journal of Human Genetics* 90 (2012): 675–684.

Behar, Doron M., Richard Villems, Himla Soodyall, Jason Blue-Smith, Luisa Pereira, Ene Metspalu, Rosaria Scozzari, Heeran Makkan, Shay Tzur, David Comas, et al. "The Dawn of Human Matrilineal Diversity." *American Journal of Human Genetics* 82 (2008): 1130–1140.

Belfer-Cohen, Anna, and Ofer Bar-Yosef. "The Levantine Aurignacian: 60 Years of Research." In *Dorothy Garrod and the Progress of the Palaeolithic*, edited by William Davies and Ruth Charles, 119–134. Barnsley: Oxbow Books, 1999.

Bellwood, Peter. *First Migrants: Ancient Migration in Global Perspective.* Oxford: Wiley-Blackwell, 2013.

Bellwood, Peter. "Southeast Asia Before History." In *The Cambridge History of Southeast Asia.* Vol. 1, part 1, *From Early Times to c. 1500*, edited by Nicholas Tarling, 55–136. Cambridge: Cambridge University Press, 1999.

Benazzi, Stefano, Katerina Douka, Cinzia Fornai, Catherine C. Bauer, Ottmar Kullmer, Jiří Svoboda, Ildikó Pap, Francesco Mallegni, Priscilla Bayle, Michael Coquerelle, et al. "Early Dispersal of Modern Humans in Europe and Implications for Neanderthal Behaviour." *Nature* 479 (2011): 525–528.

Benazzi, S., V. Slon, S. Talamo, F. Negrino, M. Peresani, S. E. Bailey, S. Sawyer, D. Panetta, G. Vicino, E. Starnini, et al. "The Makers of the Protoaurignacian and Implications for Neandertal Extinction." *Science* 348 (2015): 793–796.

Bergström, Anders, Nano Nagle, Yuan Chen, Shane McCarthy, Martin O. Pollard, Qasim Ayub, Stephen Wilcox, Leah Wilcox, Roland A. H. van Oorschot, Peter McAllister, et al. "Deep Roots for Aboriginal Australian Y Chromosomes." *Current Biology* 26 (2016): 809–813.

Berwick, Robert C., Gabriel J. L. Beckers, Kazou Okanoya, and Johan J. Bolhuis. "A Bird's Eye View of Human Language Evolution." *Frontiers in Evolutionary Neuroscience* 4 (2012): 1–25.

Berwick, Robert C., Kazou Okanoya, Gabriel J. L. Beckers, and Johan J. Bolhuis. "Songs to Syntax: The Linguistics of Birdsong." *Trends in Cognitive Science* 15 (2011): 113–121.

Bettinger, Robert L. *Hunter-Gatherer Foraging: Five Simple Models.* Clinton Corners, N.Y.: Werner , 2009.

Beyin, Amanuel. "Upper Pleistocene Human Dispersals Out of Africa: A Review of the Current State of the Debate." *International Journal of Evolutionary Biology* (2011): 615094.

Bickerton, Derek. *Language and Species*. Chicago: University of Chicago Press, 1990.

Bigelow, Nancy H. and Wm. Roger Powers. "Climate, Vegetation, and Archaeology 14,000–9000 Cal Yr B.P. in Central Alaska." *Arctic Anthropology* 38 (2001): 171–195.

Bird, Junius B. *Travels and Archaeology in South Chile*. Edited by John Hyslop. Iowa City: University of Iowa Press, 1988.

Bird, M. I., L. K. Ayliffe, L. K. Fifeld, C. S. M. Turney, R. G. Creswell, T. T. Barrows, and B. David. "Radiocarbon Dating of 'Old' Charcoal Using a Wet Oxidation, Stepped-Combustion Procedure." *Radiocarbon* 41 (1999): 127–140.

Birdsell, J. B. "The Recalibration of a Paradigm for the First Peopling of Greater Australia." In *Sunda and Sahul: Prehistoric Studies in Southeast Asia, Melanesia, and Australia*, edited by J. Allen, J. Golson, and R. Jones, 113–167. London: Academic Press, 1977.

Blackwell, B. A. "Electron Spin Resonance Dating." In *Dating Methods for Quaternary Deposits* edited by N. W. Rutter and N. R. Catto, 209–268. St. John's, Newfoundland: Geological Association of Canada, 1995.

Bocherens, Hervé, Dorothée G. Drucker, Daniel Billiou, Marylène Patou-Mathis, and Bernard Vandermeersch. "Isotopic Evidence for Diet and Subsistence Pattern of the Saint Cesaire I Neanderthal: Review and Use of a Multi-Source Mixing Model." *Journal of Human Evolution* 49 (2005): 71–87.

Bocherens, Hervé, Dorothée G. Drucker, and Stéphanie Madelaine. "Evidence for a ^{15}N Positive Excursion in Terrestrial Foodwebs at the Middle to Upper Palaeolithic Transition in South-Western France: Implications for Early Modern Human Palaeodiet and Palaeoenvironment." *Journal of Human Evolution* 69 (2014): 31–43.

Bodner, Martin, Ugo A. Perego, Gabriela Huber, Liane Fendt, Alexander W. Röck, Bettina Zimmermann, Anna Olivieri, Alberto Gómez-Carballa, Hovirag Lancioni, Norman Angerhofer, et al. "Rapid Coastal Spread of First Americans: Novel Insights from South America's Southern Cone Mitochondrial Genomes." *Genome Research* 22 (2012): 811–820.

Bonatto, Sandro L., and Francisco M. Salzano. "A Single and Early Migration for the Peopling of the Americas Supported by Mitochondrial DNA Sequence Data." *Proceedings of the National Academy of Sciences* 94 (1997): 1866–1871.

Bonchev, Danail, and Gregory A. Buck. "Quantitative Measures of Network Complexity." In *Complexity in Chemistry, Biology, and Ecology*, edited by Danail Bonchev and Dennis H. Rouvray, 191–235. New York: Springer, 2005.

Bonner, John Tyler. *The Evolution of Complexity by Means of Natural Selection*. Princeton, N.J.: Princeton University Press, 1988.

Boole, George. *An Investigation of the Laws of Thought, on Which Are Founded the Mathematical Theories of Logic and Probabilities*. 1854. New York: Dover, 1958.

Boriskovskii, P. I. *Ocherki po Paleolitu Basseina Dona*. Materialy i Issledovaniya po Arkheologii SSSR 121. Moscow: Nauka, 1963.

Bosch, Margolein D., Marcello A. Mannino, Amy L. Prendergast, Tamsin C. O'Connell, Beatrice Demarchi, Sheila M. Taylor, Laura Niven, Johannes van der Plicht, and Jean-Jacques Hublin. "New Chronology for Ksâr 'Akil (Lebanon) Supports

Levantine Route of Modern Human Dispersal into Europe." *Proceedings of the National Academy of Sciences* 112 (2015): 7683–7688.

Boule, Marcellin. *Fossil Men: Elements of Human Palaeontology.* Translated by Jessie Elliot Ritchie and James Ritchie. Edinburgh: Oliver and Boyd, Tweeddale Court, 1923.

Bourgeon, Lauriane. "Bluefish Cave II (Yukon Territory, Canada): Taphonomic Study of a Bone Assemblage." *PaleoAmerica* 1 (2015): 105–108.

Bourgeon, Lauriane, Ariane Burke, and Thomas Higham. "Earliest Human Presence in North America Dated to the Last Glacial Maximum: New Radiocarbon Dates from Bluefish Caves, Canada." *PLoS ONE* 12 (2017): e0169486.

Bouzouggar, Abdeljalil, Nick Barton, Marian Vanhaeren, Francesco d'Errico, Simon Collcutt, Tom Higham, Edward Hodge, Simon Parfitt, Edward Rhodes, Jean-Luc Schwenninger, et al. "82,000-Year-Old Shell Beads from North Africa and Implications for the Origins of Modern Human Behavior." *Proceedings of the National Academy of Sciences* 104 (2007): 9964–9969.

Bowler, James M., Harvey Johnston, Jon M. Olley, John R. Prescott, Richard G. Roberts, Wilfred Shawcross, and Nigel A. Spooner. "New Ages for Human Occupation and Climatic Change at Lake Mungo, Australia." *Nature* 421 (2003): 837–840.

Bowler, James M., Rhys Jones, Harry Allen, and A. G. Thorne. "Pleistocene Human Remains from Australia: A Living Site and Human Cremation from Lake Mungo, Western New South Wales." *World Archaeology* 2 (1970): 39–60.

Brace, C. Loring. "The Roots of the Race Concept in American Physical Anthropology." In *A History of American Physical Anthropology, 1930–1980,* edited by Frank Spencer, 11–29. New York: Academic Press, 1982.

Braha, Dan, and Oded Maimon. "The Measurement of a Design Structural and Functional Complexity." *IEEE Transactions on Systems, Man, and Cybernetics Part A: Systems and Humans* 28 (1998): 527–535.

Bräuer, Günter. "Early Anatomically Modern Man in Africa and the Replacement of the Mediterranean and European Neandertals." In *L'Homo erectus et la place de l'homme de Tautavel parmi les hominides fossiles,* edited by Henry de Lumley, 112. Nice: Centre National de la Recherche Scientifique, 1982.

Bräuer, Günter. "The Evolution of Modern Humans: A Comparison of the African and Non-African Evidence." In *The Human Revolution: Behavioural and Biological Perspectives on the Origins of Modern Humans,* edited by Paul Mellars and Chris Stringer, 123–154. Princeton, N.J.: Princeton University Press, 1989.

Bräuer, Günter. "The Origin of Modern Anatomy: By Speciation or Intraspecific Evolution?" *Evolutionary Anthropology* 17 (2008): 22–37.

Brenner, Sydney, Maria Johnson, John Bridgham, George Golda, Danid H. Lloyd, David Johnson, Shujun Luo, Sarah McCurdy, Michael Foy, Mark Ewan, et al. "Gene Expression Analysis by Massively Parallel Signature Sequencing (MPSS) on Microbead Arrays." *Nature Biotechnology* 18 (2000): 630–634.

Broderick, Alan Houghton. *Father of Prehistory: The Abbé Henri Breuil: His Life and Times.* New York: Morrow, 1963.

Brooker, Robert J. *Genetics: Analysis and Principles.* 2nd ed. New York: McGraw-Hill, 2005.

Brothwell, D. R. "Upper Pleistocene Human Skull from Niah Caves." *Sarawak Museum Journal* 9 (1960): 323–349.

Brown, Kyle S., Curtis W. Marean, Andy I. R. Herries, Zenobia Jacobs, Chantal Tribolo, David Braun, David L. Roberts, Michael C. Meyer, and Jocelyn Bernatchez. "Fire as an Engineering Tool of Early Modern Humans." *Science* 325 (2009): 859–862.

Brown, Samantha, Thomas Higham, Viviane Slon, Svante Pääbo, Matthias Meyer, Katerina Douka, Fiona Brock, Daniel Comeskey, Noemi Procopio, Michael Shunkov, et al. "Identification of a New Hominin Bone from Denisova Cave, Siberia Using Collagen Fingerprinting and Mitochondrial DNA Analysis." *Scientific Reports* 6 (2016): 23559.

Brubaker, Linda B., Patricia M. Anderson, Mary E. Edwards, and Anatoly V. Lozhkin. "Beringia as a Glacial Refugium for Boreal Trees and Shrubs: New Perspectives from Mapped Pollen Data." *Journal of Biogeography* 32 (2005): 833–848.

Bulbeck, David. "Craniodental Affinities of Southeast Asia's 'Negritos' and the Concordance with Their Genetic Affinities." *Human Biology* 85 (2013): 95–134.

Bunn, Henry T., and Ellen M. Kroll. "Systematic Butchery by Plio/Pleistocene Hominids at Olduvai Gorge, Tanzania." *Current Anthropology* 27 (1986): 431–452.

Burbano, Hernán A., Richard E. Green, Tomislav Maricic, Carles Lalueza-Fox, Marco de la Rasilla, Antonio Rosas, Janet Kelso, Katherine S. Pollard, Michael Lachmann, and Svante Pääbo. "Analysis of Human Accelerated DNA Regions Using Archaic Hominin Genomes." *PLoS ONE* 7 (2012): e32877.

Bush, Michael E., C. Owen Lovejoy, Donald C. Johanson, and Yves Coppens. "Hominid Carpal, Metacarpal, and Phalangeal Bones Recovered from the Hadar Formation: 1974–1977 Collections." *American Journal of Physical Anthropology* 57 (1982): 651–677.

Butzer, Karl W. "The Mursi, Nkalabong, and Kibish Formations, Lower Omo Basin, Ethiopia." In *Earliest Man and Environments in the Lake Rudolf Basin: Stratigraphy, Paleoecology, and Evolution*, edited by Yves Coppens, F. Clark Howell, Glynn Ll. Isaac, and Richard E. F. Leakey, 12–23. Chicago: University of Chicago Press, 1976.

Buvit, Ian, Masami Izuho, Karissa Terry, Alan Carter, Mikhail V. Konstantinov, and Aleksandr V. Konstantinov. "Last Glacial Maximum Human Abandonment of the Transbaikal." *PaleoAmerica* 1 (2015): 374–376.

Buvit, Ian, Karisa Terry, Masami Izuho, and Mikhail V. Konstantinov. "The Emergence of Modern Behavior in the Trans-Baikal, Russia." In *Emergence and Diversity of Modern Human Behavior in Paleolithic Asia*, edited by Yousuke Kaifu, Masami Izuho, Ted Goebel, Hiroyuki Sato, and Akira Ono, 490–505. College Station: Texas A&M University Press, 2015.

Cai, Xiaoyun, Zhendong Qin, Bo Wen, Shuhua Xu, Yi Wang, Yan Lu, Lanhai Wei, Chuanchao Wang, Shilin Li, Xingqiu Huang, et al. "Human Migration Through Bottlenecks from Southeast Asia into East Asia During Last Glacial Maximum Revealed by Y Chromosomes." *PLoS ONE* 6 (2011): e24282.

Cai, Yanjun, Xiaoke Qiang, Xulong Wang, Changzhu Jin, Yuan Wang, Yingqi Zhang, Erik Trinkaus, and Zhisheng An. "The Age of Human Remains and Associated Fauna from Zhiren Cave in Guangxi, Southern China." *Quaternary International* 434 (2017): 84–91.

Calcott, Brett, and Kim Sterelny, eds. *The Major Transitions in Evolution Revisited.* Cambridge, Mass.: MIT Press, 2011.

Calvin, William H. "The Brain as a Darwin Machine." *Nature* 330 (1987): 33–34.

Campbell, Bernard. *Humankind Emerging*. 3rd ed. Boston: Little, Brown, 1982.

Campbell, Lyle. *American Indian Languages: The Historical Linguistics of Native America*. New York: Oxford University Press, 1997.

Cann, Rebecca L., Mark Stoneking, and Allan C. Wilson. "Mitochondrial DNA and Human Evolution." *Nature* 325 (1987): 31–36.

Capra, John A., Genevieve D. Erwin, Gabriel McKinsey, John L. R. Rubenstein, and Katherine S. Pollard. "Many Human Accelerated Regions Are Developmental Enhancers." *Philosophical Transactions of the Royal Society B: Biological Sciences* 368 (2013): 20130025.

Caramelli, David, Lucio Milani, Stefania Vai, Alessandra Modi, Elena Pecchioli, Matteo Girardi, Elena Pilli, Martina Lari, Barbara Lippi, Annamaria Ronchitelli, et al. "A 28,000 Years Old Cro-Magnon mtDNA Sequence Differs from All Potentially Contaminating Modern Sequences." *PLoS ONE* 3 (2008): e2700.

Carey, Nessa. *The Epigenetics Revolution: How Modern Biology Is Rewriting Our Understanding of Genetics, Disease, and Inheritance*. New York: Columbia University Press, 2012.

Carey, Nessa. *Junk DNA: A Journey Through the Dark Matter of the Genome*. New York: Columbia University Press, 2015.

Carroll, Sean B. *Endless Forms Most Beautiful: The New Science of Evo Devo and the Making of the Animal Kingdom*. New York: Norton, 2005.

Cavalli-Sforza, Luigi Luca, and Francesco Cavalli-Sforza. *The Great Human Diasporas: The History of Diversity and Evolution*. Reading, Mass.: Addison-Wesley, 1995.

Cavalli-Sforza, Luigi L., Paolo Menozzi, and Alberto Piazza. "Demic Expansions and Human Evolution." *Science* 259 (1993): 639–646.

Cerling, Thure E. "Development of Grasslands and Savannas in East Africa During the Neogene." *Palaeogeography, Palaeoclimatology, Palaeoecology* 97 (1992): 241–247.

Cerling, Thure E., Fredrick Kyalo Manthi, Emma N. Mbua, Louise N. Leakey, Meave G. Leakey, Richard E. Leakey, Francis H. Brown, Frederick E. Grine, John A. Hart, Prince Kaleme, et al. "Stable-Isotope-Based Diet Reconstructions of Turkana Basin Hominins." *Proceedings of the National Academy of Sciences* 110 (2013): 10501–10506.

Černý, Viktor, Connie J. Mulligan, Verónica Fernandes, Nuno M. Silva, Farida Alshamali, Amy Non, Nourdin Harich, Lotfi Cherni, Amel Ben Ammar El Gaaied, Ali Al-Meeri, et al. "Internal Diversification of Mitochondrial Haplogroup R0a Reveals Post-Last Glacial Maximum Demographic Expansions in South Arabia." *Molecular Biology and Evolution* 28 (2010): 71–78.

Ceruzzi, Paul E. *Computing: A Concise History*. Cambridge, Mass.: MIT Press, 2012.

Chapais, Bernard. *Primeval Kinship: How Pair-Bonding Gave Birth to Human Society*. Cambridge, Mass.: Harvard University Press, 2008.

Chatters, James C., Douglas J. Kennett, Yemane Asmerom, Brian M. Kemp, Victor Polyak, Alberto Nava Blank, Patricia A. Beddows, Eduard Reinhardt, Joaquin Arroyo-Cabrales, Deborah A. Bolnick, et al. "Late Pleistocene Human Skeleton and mtDNA Link Paleoamericans and Modern Native Americans." *Science* 344 (2014): 750–754.

Cherni, Lotfi, Verónica Fernandes, Joana B. Pereira, Marta D. Costa, Ana Goios, Sabeh Frigi, Besma Yacoubi-Loueslati, Mohamed Ben Amor, Abdelhakim Slama, António Amorim, et al. "Post-Last Glacial Maximum Expansion from Iberia to North Africa

Revealed by Fine Characterization of mtDNA H Haplogroup in Tunisia." *American Journal of Physical Anthropology* 139 (2009): 253–260.

Chernysh, A. P. "Mnogosloinaya Paleoliticheskaya Stoyanka Molodova I." In *Molodova I: Unikal'noe Must'erskoe Poselenie na Srednem Dnestre*, edited by G. I. Goretskii and I. K. Ivanova, 6–102. Moscow: Nauka, 1982.

Chomsky, Noam. *Language and Mind*. 3rd ed. Cambridge: Cambridge University Press, 2006.

Chomsky, Noam. "On Certain Formal Properties of Grammars." *Information and Control* 2 (1959): 137–167.

Chomsky, Noam. *On Nature and Language*. Cambridge: Cambridge University Press, 2002.

Chomsky, Noam. "Three Models for the Description of Language." *IRE Transactions on Information Theory* 2 (1956): 113–124.

Cinq-Mars, Jacques. "La place des grottes du Poission-Bleu dans la prehistoire béringi-enne." *Revista de Arqueología Americana* 1 (1990): 9–32.

Cione, Alberto L., Eduardo P. Tonni, and Leopoldo Soibelzon. "The Broken Zig-Zag: Late Cenozoic Large Mammal and Tortoise Extinction in South America." *Revista del Museo Argentino de Ciencias Naturales*, n.s., 5 (2003): 1–19.

Clark, Andy. *Being There: Putting Brain, Body, and World Together Again*. Cambridge, Mass.: MIT Press, 1998.

Clark, Andy. *Supersizing the Mind: Embodiment, Action, and Cognitive Extension*. New York: Oxford University Press, 2011.

Clark, Andy. *Surfing Uncertainty: Prediction, Action, and the Embodied Mind*. Oxford: Oxford University Press, 2016.

Clark, J. Desmond. "African and Asian Perspectives on the Origin of Modern Humans." In *The Origin of Modern Humans and the Impact of Chronometric Dating*, edited by Martin Jim Aitken, Chris B. Stringer, and Paul A. Mellars, 148–178. Princeton, N.J.: Princeton University Press, 1993.

Clark, J. G. D. "Neolithic Bows from Somerset, England, and the Prehistory of Archery in North-west Europe." *Proceedings of the Prehistoric Society* 29 (1963): 50–98.

Clarke, Esther, Ulrich H. Reichard, and Klaus Zuberbühler. "The Syntax and Meaning of Wild Gibbon Songs." *PLoS ONE* 1 (2006): e73.

Clarkson, Chris, Michael Petraglia, Ravi Korisettar, Michael Haslam, Nicole Boivin, Alison Crowther, Peter Ditchfield, Dorian Fuller, Preston Miracle, Clair Harris, et al. "The Oldest and Longest Enduring Microlithic Sequence in India: 35 000 Years of Modern Human Occupation and Change at the Jwalapuram Locality 9 Rockshelter." *Antiquity* 83 (2009): 326–348.

Claussen, M., K. Selent, V. Brovkin, T. Raddatz, and V. Gayler. "Impact of CO_2 and Climate on Last Glacial Maximum Vegetation—A Factor of Separation." *Biogeosciences* 10 (2013): 3593–3604.

Cohen, Vadim Y., and Vadim N. Stepanchuk. "Late Middle and Early Upper Paleolithic Evidence from the East European Plain and Caucasus: A New Look at Variability, Interactions, and Transitions." *Journal of World Prehistory* 13 (1999): 265–319.

Colinvaux, Paul A. "Quaternary Vegetational History of Arctic Alaska." In *The Bering Land Bridge*, edited by David M. Hopkins, 207–231. Stanford, Calif.: Stanford University Press, 1967.

Collard, Mark, Michael Kemery, and Samantha Banks. "Causes of Toolkit Variation Among Hunter-Gatherers: A Test of Four Competing Hypotheses." *Canadian Journal of Archaeology* 29 (2005): 1–19.

Comas, David, Stéphanie Plaza, R. Spencer Wells, Nadira Yuldaseva, Oscar Lao, Francesc Calafell, and Jaume Bertranpetit. "Admixture, Migrations, and Dispersals in Central Asia: Evidence from Maternal DNA Lineages." *European Journal of Human Genetics* 12 (2004): 495–504.

Conard, Nicholas J. "A Female Figurine from the Basal Aurignacian of Hohle Fels Cave in Southwestern Germany." *Nature* 459 (2009): 248–252.

Coolidge, Frederick L., and Thomas Wynn. *The Rise of* Homo sapiens: *The Evolution of Modern Thinking.* Malden, Mass.: Wiley-Blackwell, 2009.

Coolidge, Frederick L., and Thomas Wynn. "Working Memory, Its Executive Functions, and the Emergence of Modern Thinking." *Cambridge Archaeological Journal* 15 (2005): 5–26.

Coon, Carlton S. *The Origin of Races.* New York: Knopf, 1962.

Cooper, Alan, Andrew Rambaut, Vincent Macaulay, Eske Willerslev, Anders J. Hansen, and Chris Stringer. "Human Origins and Ancient Human DNA." *Science* 292 (2001): 1656–1657.

Cooper, Zarine. *Archaeology and History: Early Settlements in the Andaman Islands.* Oxford: Oxford University Press, 2002.

Coppens, Yves, F. Clark Howell, Glynn Ll. Isaac, and Richard E. F. Leakey, eds. *Earliest Man and Environments in the Lake Rudolf Basin: Stratigraphy, Paleoecology, and Evolution.* Chicago: University of Chicago Press, 1976.

Coqueugniot, H., J.-J. Hublin, F. Veillon, F. Houët, and T. Jacobs, "Early Brain Growth in *Homo erectus* and Implications for Cognitive Ability." *Nature* 431 (2004): 299–302.

Corballis, Michael C. *The Recursive Mind: The Origins of Human Language, Thought, and Civilization.* Princeton, N.J.: Princeton University Press, 2011.

Corbey, Raymond, Adam Jagich, Krist Vaesen, and Mark Collard, "The Acheulean Handaxe: More Like a Bird's Song Than a Beatles' Tune?" *Evolutionary Anthropology* 25 (2016): 6–19.

Cordaux, Richard, Robert Aunger, Gillian Bentley, Ivane Nasidze, S. M. Sirajuddin, and Mark Stoneking. "Independent Origins of Indian Caste and Tribal Paternal Lineages." *Current Biology* 14 (2004): 231–235.

Correa, Pelayo, and M. Blanca Piazuelo. "Evolutionary History of the *Helicobacter pylori* Genome: Implications for Gastric Carcinogenesis." *Gut Liver* 6 (2012): 21–28.

Cortés-Sánchez, Miguel, Arturo Morales-Muñiz, María D. Simón-Vallejo, María C. Lozano-Francisco, José L. Vera-Peláez, Clive Finlayson, Joaquín Rodríguez-Vidal, Antonio Delgado-Huertas, Francisco J. Jiménez-Espejo, Francisca Martínez-Ruiz, et al. "Earliest Known Use of Marine Resources by Neanderthals." *PLoS ONE* 6 (2011): e24026.

Cosgrove, Richard. "Late Pleistocene Behavioural Variation and Time Trends: The Case from Tasmania." *Archaeology of Oceania* 30 (1995): 83–104.

Crevecoeur, I., A. Brooks, I. Ribot, E. Cornelissen, and P. Sernal. "Late Stone Age Human Remains from Ishango (Democratic Republic of Congo): New Insights on Late Pleistocene Modern Human Diversity in Africa." *Journal of Human Evolution* 96 (2016): 35–57.

Dahlberg, Albert A. "Materials for the Establishment of Standards for Classification of Tooth Characters, Attributes, and Techniques in Morphological Studies of the Dentition." Zollar Laboratory of Dental Anthropology, University of Chicago. Mimeo, 1956.

Dannemann, Michael, Aida M. Andrés, and Janet Kelso. "Introgression of Neandertal- and Denisovan-like Haplotypes Contributes to Adaptive Variation in Human Toll-like Receptors." *American Journal of Human Genetics* 98 (2016): 22–33.

Darnell, Regna. "The Anthropological Concept of Culture at the End of the Boasian Century." *Social Analysis* 41 (1997): 42–54.

Dart, Raymond A. "*Australopithecus africanus*: The Man-Ape of South Africa." *Nature* 115 (1925): 195–199.

Darwin, Charles. *The Descent of Man and Selection in Relation to Sex*. 2 vols. London: Murray, 1871.

Darwin, Charles. *On the Origin of Species by Means of Natural Selection*. London: Murray, 1859.

David, Bruno, Jean-Michel Geneste, Ray L. Whear, Jean-Jacques Delannoy, Margaret Katherine, R. G. Gunn, Christopher Clarkson, Hugues Plisson, Preston Lee, Fiona Petchey, et al. "Nawarla Gabarnmang, a 45,180 ± 910 cal BP Site in Jawoyn Country, Southwest Arnhem Land Plateau." *Australian Archaeology* 73 (2011): 73–77.

Dawkins, Richard. *The Ancestor's Tale: A Pilgrimage to the Dawn of Evolution*. Boston: Houghton Mifflin, 2004.

Dawkins, Richard. *The Extended Phenotype: The Long Reach of the Gene*. Oxford: Oxford University Press, 1999.

Dawkins, Richard. *River Out of Eden: A Darwinian View of Life*. New York: Basic Books, 1995.

Dawkins, Richard. *The Selfish Gene*. New York: Oxford University Press, 1976.

Dawkins, Richard, and John R. Krebs. "Animal Signals: Information or Manipulation?" In *Behavioural Ecology: An Evolutionary Approach*, edited by John R. Krebs and Nicholas B. Davies, 282–309. Oxford: Blackwell Scientific, 1978.

Day, Michael H. *Guide to Fossil Man: A Handbook of Human Palaeontology*. 3rd ed. Chicago: University of Chicago Press, 1977.

Day, Michael H., and Chris B. Stringer. "A Reconsideration of the Omo Kibish Remains and the *erectus–sapiens* Transition." In *L'Homo erectus et la place de l'homme de Tautavel parmi les hominides fossiles*, edited by Henry de Lumley, 814–846. Nice: Centre National de la Recherche Scientifique, 1982.

Dean, Christopher, Meave G. Leakey, Donald Reid, Friedemann Schrenk, Gary T. Schwartz, Christopher Stringer, and Alan Walker. "Growth Processes in Teeth Distinguish Modern Humans from *Homo erectus* and Earlier Hominins." *Nature* 414 (2001): 628–631.

Deetz, James. *In Small Things Forgotten: The Archaeology of Early American Life*. Garden City, N.Y.: Anchor Books, 1977.

Deetz, James. *Invitation to Archaeology*. Garden City, N.Y.: Natural History Press, 1967.

de Heinzelin, Jean, J. Desmond Clark, Tim White, William Hart, Paul Renne, Giday WoldeGabriel, Yonas Beyene, and Elisabeth Vrba. "Environment and Behavior of 2.5-Million-Year-Old Bouri Hominids." *Science* 284 (1999): 625–629.

Demeter, Fabrice, Laura L. Shackelford, Anne-Marie Bacon, Philippe Duringer, Kira Westaway, Thongsa Sayavongkhamdy, José Braga, Phonephanh Sichanthongtip,

Phimmasaeng Khamdalavong, Jean-Luc Ponche, et al. "Anatomically Modern Human in Southeast Asia (Laos) by 46 ka." *Proceedings of the National Academy of Sciences* 109 (2012): 14375–14380.

Demidenko, Yuri E., and Pierre Noiret. "Radiocarbon Dates for the Siuren I Sequence." In *Siuren I Rock-Shelter: From Late Middle Paleolithic and Early Upper Paleolithic to Epi-Paleolithic in Crimea*, edited by Yuri E. Demidenko, Marcel Otte, and Pierre Noiret, 49–53. ERAUL 129 Liege: University of Liege, 2012.

Dennell, Robin, and Michael D. Petraglia. "The Dispersal of *Homo sapiens* Across Southern Asia: How Early, How Often, How Complex?" *Quaternary Science Reviews* 47 (2012): 15–22.

Dennell, Robin W., Helen M. Rendell, Mohammad Halim, and Eddie Moth. "A 45,000-Year-Old Open-air Paleolithic Site at Riwat, Northern Pakistan." *Journal of Field Archaeology* 19 (1992): 17–33.

Dennett, Daniel C. *Consciousness Explained*. Boston: Little, Brown, 1991.

Derenko, Miroslava V., Tomasz Grzybowski, Boris A. Malyarchuk, Jakub Czarny, Danutia Miścicka-Sliwka, and Ilia A. Zakharov. "The Presence of Mitochondrial Haplogroup X in Altaians from South Siberia." *American Journal of Human Genetics* 69 (2001): 237–241.

Derenko, Miroslava, Boris Malyarchuk, Galina Denisova, Maria Perkova, Urszula Rogalla, Tomasz Grzybowski, Elza Khusnutdinova, Irina Dambueva, and Ilia Zakharov. "Complete Mitochondrial DNA Analysis of Eastern Eurasian Haplogroups Rarely Found in Populations of Northern Asia and Eastern Europe." *PLoS ONE* 7 (2012): e32179.

Derevianko, A. P., M. V. Shunkov, and S. V. Markin. *Dinamika Paleoliticheskikh Industrii v Afrike i Evrazii v Pozdnem Pleistotsene i Problema Formirovaniya Homo sapiens*. Novosibirsk: Institute of Archaeology and Ethnography, SB RAS Press, 2014.

d'Errico, Francesco. "Palaeolithic Origins of Artificial Memory Systems: An Evolutionary Perspective." In *Cognition and Material Culture: The Archaeology of Symbolic Storage*, edited by Colin Renfrew and Chris Scarre, 19–50. Cambridge: McDonald Institute for Archaeological Research, Cambridge University, 1998.

d'Errico, Francesco, and Lucinda Blackwell. "Earliest Evidence of Personal Ornaments Associated with Burial: The *Conus* Shells from Border Cave." *Journal of Human Evolution* 93 (2016): 91–108.

d'Errico, Francesco, Lucinda Backwell, Paola Villa, Ilaria Degano, Jeannette J. Lucejko, Marion K. Bamford, Thomas F. G. Higham, Maria Perla Colombini, and Peter B. Beaumont. "Early Evidence of San Material Culture Represented by Organic Artifacts from Border Cave, South Africa." *Proceedings of the National Academy of Sciences* 109 (2012): 13214–13219.

d'Errico, Francesco, Christopher Henshilwood, Graeme Lawson, Marian Vanhaeren, Anne-Marie Tillier, Marie Soressi, Frédérique Bresson, Bruno Maureille, April Nowell, Joseba Lakarra, et al. "Archaeological Evidence for the Emergence of Language, Symbolism, and Music—An Alternative Multidisciplinary Perspective." *Journal of World Prehistory* 17 (2003): 1–70.

DeSalle, Rob, and Ian Tattersall, *The Brain: Big Bangs, Behaviors, and Beliefs*. New Haven, Conn.: Yale University Press, 2012.

Descartes, René. *Philosophical Works*. Translated by Elizabeth S. Haldane and G. R. T. Ross. New York: Dover, 1955.

Détroit, Florent, Eusebio Dizon, Christophe Falguères, Sébastien Hameau, Wilfredo Ronquillo, and François Sémah. "Upper Pleistocene *Homo sapiens* from the Tabon Cave (Palawan, the Philippines): Description and Dating of New Discoveries." *Comptes Rendus Paleovol* 3 (2004): 705–712.

Dias, Adriana Schmidt, and Lucas Bueno. "The Initial Colonization of South America Eastern Lowlands: Brazilian Archaeology Contributions to Settlement of America Models." In *Paleoamerican Odyssey*, edited by Kelly E. Graf, Caroline V. Ketron, and Michael R. Waters, 339–357. College Station: Texas A&M University Press, 2013.

Dikov, N. N. *Arkheologicheskie Pamyatniki Kamchatki, Chukotki i Verkhnei Kolymy.* Moscow: Nauka, 1977.

Dikov, N. N. *Drevnie Kul'tury Severo-Vostochnoi Azii.* Moscow: Nauka, 1979.

Dikov, N. N. "The Ushki Sites, Kamchatka Peninsula." In *American Beginnings: The Prehistory and Paleoecology of Beringia*, edited by Frederick Hadleigh West, 244–250. Chicago: University of Chicago Press, 1996.

Dillehay, Tom D., Duccio Bonavia, Steve L. Goodbred Jr., Mario Pino, Victor Vásquez, and Teresa Rosales Tham. "A Late Pleistocene Human Presence at Huaca Prieta, Peru, and Early Pacific Coastal Adaptations." *Quaternary Research* 77 (2012): 418–423.

Dillehay, Tom D., C. Ramírez, M. Pino, M. B. Collins, J. Rossen, and J. B. Pino-Navarro. "Monte Verde: Seaweed, Food, Medicine, and the Peopling of South America." *Science* 320 (2008): 784–786.

Dixon, E. James. "Human Colonization of the Americas: Timing, Technology and Process." *Quaternary Science Reviews* 20 (2001): 277–299.

Domínguez-Rodrigo, M., H. T. Bunn, A. Z. P. Mabulla, G. M. Ashley, F. Diez-Martin, D. Barboni, M. E. Prendergast, J. Yravedra, R. Barba, A. Sánchez, et al. "New Excavations at the FLK *Zinjanthropus* Site and Its Surrounding Landscape and Their Behavioral Implications." *Quaternary Research* 74 (2010): 315–332.

Domínguez-Rodrigo, M., and T. R. Pickering. "Early Hominid Hunting and Scavenging: A Zooarchaeological Review." *Evolutionary Anthropology* 12 (2003): 275–282.

Donald, Merlin. *Origins of the Modern Mind: Three Stages in the Evolution of Culture and Cognition.* Cambridge, Mass.: Harvard University Press, 1991.

Dorit, R. L., H. Akashi, and W. Gilbert. "Absence of Polymorphism at the ZFY Locus in the Human Y Chromosome." *Science* 268 (1995): 1183–1185.

Douka, Katerina, Christopher A. Bergman, Robert E. M. Hedges, Frank P. Wesselingh, and Thomas F. G. Higham. "Chronology of Ksar Akil (Lebanon) and Implications for the Colonization of Europe by Anatomically Modern Humans." *PLoS ONE* 8 (2013): 372931.

Douka, Katerina, Thomas F. G. Higham, and Christopher A. Bergman. "Statistical and Archaeological Errors Invalidate the Proposed Chronology for the Site of Ksar Akil." *Proceedings of the National Academy of Sciences* 112 (2015): E7034.

Douka, Katerina, Thomas Higham, and Andrey Sinitsyn. "The Influence of Pretreatment Chemistry on the Radiocarbon Dating of Campanian Ignimbrite–Aged Charcoal from Kostenki 14 (Russia)." *Quaternary Research* 73 (2010): 583–587.

Douka, Katerina, Thomas F. G. Higham, Rachel Wood, Paolo Boscato, Paolo Gambassini, Panagiotis Karkanas, Marco Peresani, and Anna Maria Ronchitelli. "On the Chronology of the Uluzzian." *Journal of Human Evolution* 68 (2014): 1–13.

Douka, Katerina, Zenobia Jacobs, Christine Lane, Rainer Grün, Lucy Farr, Chris Hunt, Robyn H. Inglis, Tim Reynolds, Paul Albert, Maxime Aubert, et al. "The Chronostratigraphy of the Haua Fteah Cave (Cyrenaica, Northeast Libya)." *Journal of Human Evolution* 66 (2014): 39–63.

Dretske, Fred I. *Knowledge and the Flow of Information.* Cambridge, Mass.: MIT Press, 1981.

Duarte, Cidália, João Mauricio, Paul B. Pettitt, Pedro Souto, Erik Trinkaus, Hans van der Plicht, and João Zilhão. "The Early Upper Paleolithic Human Skeleton from the Abrigo do Lagar Velho (Portugal) and Modern Human Emergence in Iberia." *Proceedings of the National Academy of Sciences* 96 (1999): 7604–7609.

Dulik, Matthew C., Amanda C. Owings, Jill B. Gaieski, Miguel G. Vilar, Alestine Andre, Crystal Lennie, Mary Adele Mackenzie, Ingrid Kritsch, Sharon Snowshoe, Ruth Wright, et al. "Y-Chromosome Analysis Reveals Genetic Divergence and New Founding Native Lineages in Athapaskan- and Eskimoan-Speaking Populations." *Proceedings of the National Academy of Sciences* 109 (2012): 8471–8476.

Dulik, Matthew C., Sergey I. Zhadanov, Ludmila P. Osipova, Ayken Askapuli, Lydia Gau, Omer Gokcumen, Samara Rubinstein, and Theodore G. Schurr. "Mitochondrial DNA and Y Chromosome Variation Provides Evidence for a Recent Common Ancestry Between Native Americans and Indigenous Altaians." *American Journal of Human Genetics* 90 (2012): 229–246.

Dunbar, James S. "Paleoindian Archaeology." In *First Floridians and Last Mastodons: The Page-Ladson Site in the Aucilla River,* edited by S. David Webb, 403–435. Dordrecht: Springer, 2006.

Dunbar, Robin. *Grooming, Gossip, and the Evolution of Language.* Cambridge, Mass.: Harvard University Press, 1996.

Dunbar, Robin I. M. "The Social Brain Hypothesis." *Evolutionary Anthropology* 6 (1998): 178–190.

Dyke, Arthur S. "An Outline of North American Deglaciation with Emphasis on Central and Northern Canada." In *Quaternary Glaciations—Extent and Chronology.* Vol. 2, part 2, *North America,* edited by Jürgen Ehlers and Philip L. Gibbard, 373–424. Amsterdam: Elsevier, 2004.

Dyson, George. *Turing's Cathedral: The Origins of the Digital Universe.* New York: Pantheon, 2012.

Easton, Norman Alexander, Glen R. Mackay, Patricia Bernice Young, Peter Schnurr, and David R. Yesner. "Chindadn in Canada? Emergent Evidence of the Pleistocene Transition in Southeast Beringia as Revealed by the Little John Site, Yukon." In *From the Yenisei to the Yukon: Interpreting Lithic Assemblage Variability in Late Pleistocene/Early Holocene Beringia,* edited by Ted Goebel and Ian Buvit, 289–307. College Station: Texas A&M University Press, 2011.

Edelman, Gerald M. *Neural Darwinism: The Theory of Neuronal Group Selection.* New York: Basic Books, 1987.

Edwards, A. W. F. "Human Genetic Diversity: Lewontin's Fallacy." *BioEssays* 25 (2003): 798–801.

Efremov, I. A. "Taphonomy: A New Branch of Paleontology." *Pan-American Geologist* 74 (1940): 81–93.

Elena, Santiago F., Vaughn S. Cooper, and Richard E. Lenski. "Punctuated Evolution Caused by Selection of Rare Beneficial Mutations." *Science* 272 (1996): 1802–1804.

Elhaik, Eran, Tatiana Tatarinova, Dmitri Chebotarev, Ignazio S. Piras, Carla Maria Calò, Antonella De Montis, Manuela Atzori, Monica Marini, Sergio Tofanelli, Paolo Francalacci, et al. "Geographic Population Structure Analysis of Worldwide Human Populations Infers Their Biogeographical Origins." *Nature Communications* 5 (2014): 3513.

Elias, Scott A., and Brian Crocker, "The Bering Land Bridge: A Moisture Barrier to the Dispersal of Steppe-Tundra Biota?" *Quaternary Science Reviews* 27 (2008): 2473–2483.

Elias, Scott A., Susan K. Short, C. Hans Nelson, and Hilary H. Birks. "Life and Times of the Bering Land Bridge." *Nature* 382 (1996): 60–63.

Emmorey, Karen, Sonya Mehta, and Thomas J. Grabowski. "The Neural Correlates of Sign Versus Word Production." *NeuroImage* 36 (2007): 202–208.

Enard, Wolfgang, Molly Przeworski, Simon E. Fisher, Cecilia S. L. Lai, Victor Wiebe, Takashi Kitano, Anthony P. Monaco, and Svante Pääbo. "Molecular Evolution of *FOXP2*, a Gene Involved in Speech and Language." *Nature* 418 (2002): 869–872.

Endicott, Phillip, M. Thomas P. Gilbert, Chris Stringer, Carles Lalueza-Fox, Eske Willerslev, Anders J. Hansen, and Alan Cooper. "The Genetic Origin of the Andaman Islanders." *American Journal of Human Genetics* 72 (2003): 178–184.

Endicott, Phillip, Simon Y. W. Ho, Mait Metspalu, and Chris Stringer. "Evaluating the Mitochondrial Timescale of Human Evolution." *Trends in Ecology and Evolution* 24 (2009): 515–521.

Endicott, Phillip, Simon Ho, and Chris Stringer. "Using Genetic Evidence to Evaluate Four Palaeoanthropological Hypotheses for the Timing of Neanderthal and Modern Human Origins." *Journal of Human Evolution* 59 (2010): 87–95.

Eriksson, Anders, Lia Betti, Andrew D. Friend, Stephen J. Lycett, Joy S. Singarayer, Noreen von Cramon-Taubadel, Paul J. Valdes, Francois Balloux, and Andrea Manica. "Late Pleistocene Climate Change and the Global Expansion of Anatomically Modern Humans." *Proceedings of the National Academy of Sciences* 109 (2012): 16089–16094.

Erlandson, Jon M., Madonna L. Moss, and Matthew Des Lauriers. "Life on the Edge: Early Maritime Cultures of the Pacific Coast of North America." *Quaternary Science Reviews* 27 (2008): 2232–2245.

Erlandson, Jon M., Torben C. Rick, Todd J. Braje, Molly Casperson, Brendan Culleton, Brian Fulfrost, Tracy Garcia, Daniel A. Guthrie, Nicholas Jew, Douglas J. Kennett, et al. "Paleoindian Seafaring, Maritime Technologies, and Coastal Foraging on California's Channel Islands." *Science* 331 (2011): 1181–1185.

Ewing, J. Franklin. "A Probable Neanderthaloid from Ksar 'Akil, Lebanon." *American Journal of Physical Anthropology* 21 (1963): 101–104.

Fadhlaoui-Zid, Karima, Marc Haber, Begoña Martínez-Cruz, Pierre Zalloua, Amel Benammar Elgaaied, and David Comas. "Genome-Wide and Paternal Diversity Reveal a Recent Origin of Human Populations in North Africa." *PLoS ONE* 8 (2013): e80293.

Fagan, Brian. *The First North Americans: An Archaeological Journey*. London: Thames and Hudson, 2011.

Fagan, Brian M. *The Journey from Eden: The Peopling of Our World*. London: Thames and Hudson, 1990.

Falk, Dean. *Braindance: New Discoveries About Human Origins and Brain Evolution*. Rev. ed. Gainesville: University Press of Florida, 2004.

Fedele, Francesco G., Biagio Giaccio, and Irka Hajdas. "Timescales and Cultural Process at 40,000 BP in the Light of the Campanian Ignimbrite Eruption, Western Eurasia." *Journal of Human Evolution* 55 (2008): 834–857.

Fedje, Daryl, Quentin Mackie, Nicole Smith, and Duncan McLaren. "Function, Variability, and Interpretation of Archaeological Assemblages at the Pleistocene/Holocene Transition in Haida Gwaii." In *From the Yenisei to the Yukon: Interpreting Lithic Assemblage Variability in Late Pleistocene/Early Holocene Beringia*, edited by Ted Goebel and Ian Buvit, 323–342. College Station: Texas A&M University Press, 2011.

Feistal, Rainer, and Werner Ebeling. *Physics of Self-Organization and Evolution*. Weinheim: Wiley-VCH, 2011.

Fernandes, Carlos A. "Bayesian Coalescent Inference from Mitochondrial DNA Variation of the Colonization Time of Arabia by the Hamadryas Baboon (*Papio hamadryas hamadryas*)." In *The Evolution of Human Populations in Arabia: Paleoenvironments, Prehistory, and Genetics*, edited by Michael D. Petraglia and Jeffrey I. Rose, 89–102. Dordrecht: Springer, 2009.

Ferraro, Joseph V., Thomas W. Plummer, Briana L. Pobiner, James S. Oliver, Laura C. Bishop, David R. Braun, Peter W. Ditchfield, John W. Seaman III, Katie M. Binetti, John W. Seaman Jr., et al. "Earliest Archaeological Evidence of Persistent Hominin Carnivory." *PLoS ONE* 8 (2013): e62174.

Ferring, C. Reid. *The Archaeology and Paleoecology of the Aubrey Clovis Site (41DN479) Denton County, Texas*. Denton: Department of Geography, University of North Texas, 2001.

Field, Judith H. "Trampling Through the Pleistocene: Does Taphonomy Matter?" *Australian Archaeology* 63 (2006): 9–20.

Field, Judith, and John Dodson. "Late Pleistocene Megafauna and Archaeology from Cuddie Springs, South-eastern Australia." *Proceedings of the Prehistoric Society* 65 (1999): 275–301.

Fifield, L. K., M. I. Bird, C. S. M. Turney, P. A. Hausladen, G. M. Santos, and M. L. di Tada. "Radiocarbon Dating of the Human Occupation of Australia Prior to 40 ka BP: Successes and Pitfalls." *Radiocarbon* 43 (2001): 1139–1145.

Fitch, W. Tecumseh. *The Evolution of Language*. Cambridge: Cambridge University Press, 2010.

Fitch, W. Tecumseh. "The Evolution of Speech: A Comparative Review." *Trends in Cognitive Sciences* 4 (2000): 258–267.

Fleagle, John G. *Primate Adaptation and Evolution*. 3rd ed. San Diego, Calif.: Academic Press, 2013.

Floridi, Luciano. *Information: A Very Short Introduction*. New York: Oxford University Press, 2010.

Forde, C. Daryl. *Habitat, Economy and Society: A Geographical Introduction to Ethnology*. London: Methuen, 1934.

Forman, Steven L., James Pierson, and Kenneth Lepper. "Luminescence Geochronology." In *Quaternary Geochronology: Methods and Applications*, edited by Jay Stratton Noller, Janet M. Sowers, and William R. Lettis, 157–176. Washington, D.C.: American Geophysical Union, 2000.

Forster, Peter. "Ice Ages and the Mitochondrial DNA Chronology of Human Dispersals: A Review." *Philosophical Transactions of the Royal Society B: Biological Sciences*
(2004): 255–264.

Forster, P., R. Harding, A. Torroni, and H.-J. Bandelt. "Origin and Evolution of Native
American mtDNA Variation: A Reappraisal." *American Journal of Human Genetics* 59 (1996): 935–945.

Francalacci, Paolo, Laura Morelli, Andrea Angius, Riccardo Berutti, Frederic Rei-
nier, Rossano Atzeni, Rosella Pilu, Fabio Busonero, Andrea Maschio, Ilenia Zara,
et al. "Low-Pass DNA Sequencing of 1200 Sardinians Reconstructs European
Y-Chromosome Phylogeny." *Science* 341 (2013): 565–569.

Frantz, Laurent A. F., Victoria E. Mullin, Maud Pionnier-Capitan, Ophélie Lebrasseur,
Morgane Ollivier, Angela Perri, Anna Linderholm, Valeria Mattiangeli, Matthew
D. Teasdale, Evangelos A. Dimopoulos, et al. "Genomic and Archaeological Evi-
dence Suggests a Dual Origin of Domestic Dogs." *Science* 352 (2016): 1228–1231.

Fregel, Rosa, Vicente Cabrera, Jose M. Larruga, Khaled K. Abu-Amero, and Ana M.
Gonzalez. "Carriers of Mitochondrial DNA Macrohaplogroup N Lineages Reached
Australia Around 50,000 Years Ago Following a Northern Asian Route." *PLoS ONE*
10 (2105): e0129839.

Friederici, Angela D. "The Brain Basis of Language Processing: From Structure to
Function." *Physiological Reviews* 91 (2011): 1357–1392.

Fu, Qiaomei, Mateja Hajdinjak, Oana Teodora Moldovan, Silviu Constantin, Swapan
Mallick, Pontus Skoglund, Nick Patterson, Nadin Rohland, Iosif Lazaridis, Birgit
Nickel, et al. "An Early Modern Human from Romania with a Recent Neanderthal
Ancestor." *Nature* 524 (2105): 216–219.

Fu, Qiaomei, Heng Li, Priya Moorjani, Flora Jay, Sergey M. Slepchenko, Aleksei A.
Bondarev, Philip L. F. Johnson, Ayinuer Aximu-Petri, Kay Prüfer, Cesare de Filippo,
et al. "Genome Sequence of a 45,000-Year-Old Modern Human from Western Siberia." *Nature* 514 (2014): 445–449.

Fu, Qiaomei, Matthias Meyer, Xing Gao, Udo Stenzel, Hernán A. Burbano, Janet Kelso,
and Svante Pääbo. "DNA Analysis of an Early Modern Human from Tianyuan Cave,
China." *Proceedings of the National Academy of Sciences* 110 (2013): 2223–2227.

Fu, Qiaomei, Alissa Mittnik, Philip L. F. Johnson, Kirsten Bos, Martina Lari, Ruth Bol-
longino, Chengkai Sun, Liane Giemsch, Ralf Schmitz, Joachim Burger, et al. "A
Revised Timescale for Human Evolution Based on Ancient Mitochondrial
Genomes." *Current Biology* 23 (2013): 553–559.

Fu, Qiaomei, Cosimo Posth, Mateja Hajdinjak, Martin Petr, Swapan Mallick, Daniel
Fernandes, Anja Furtwängler, Wolfgang Haak, Matthias Meyer, Alissa Mittnik, et al.
"The Genetic History of Ice Age Europe." *Nature* 534 (2016): 200–205.

Gambassini, P., ed. *Il Paleolitico di Castelcivita: Culture e ambiente.* Naples: Electa,
1997.

Gamble, Clive, John Gowlett, and Robin Dunbar. *Thinking Big: How the Evolution of
Social Life Shaped the Human Mind.* London: Thames and Hudson, 2014.

Garrod, D. A. E. "The Mugharet el Emireh in Lower Galilee: Type-Station of the
Emiran Industry." *Journal of the Royal Anthropological Institute* 85 (1955): 141–162.

Gat, Azar, "Proving Communal Warfare Among Hunter-Gatherers: The Quasi-
Rousseauan Error." *Evolutionary Anthropology* 24 (2015): 111–126.

Gatlin, Lila L. *Information Theory and the Living System*. New York: Columbia University Press, 1972.

Gelvin-Reymiller, Carol, Joshua D. Reuther, Ben A. Potter, and Peter M. Bowers. "Technical Aspects of a Worked Proboscidean Tusk from Inmachuk River, Seward Peninsula, Alaska." *Journal of Archaeological Science* 33 (2006): 1088–1094.

Gerasimova, M. M., S. N. Astakhov, and A. A. Velichko. *Paleoliticheskii Chelovek, Ego Material'naya Kul'tura i Prirodnaya Sreda Obitaniya*. St. Petersburg: Nestor-Istoriya, 2007.

Germonpré, Mietje, Mikhail V. Sablin, Rhiannon E. Stevens, Robert E. M. Hedges, Michael Hofreiter, Mathias Stiller, and Viviane R. Després. "Fossil Dogs and Wolves from Palaeolithic Sites in Belgium, the Ukraine, and Russia: Osteometry, Ancient DNA, and Stable Isotopes." *Journal of Archaeological Science* 36 (2009): 473–490.

Gertner, Jon. *The Idea Factory: Bell Labs and the Great Age of American Innovation*. New York: Penguin, 2012.

Gibbons, Ann. "Ancient DNA Pinpoints Paleolithic Liaison in Europe." *Science* 348 (2015): 847.

Gibbons, Ann. "Five Matings for Moderns, Neandertals." *Science* 351 (2016): 1250–1251.

Gilbert, M. Thomas P., Dennis L. Jenkins, Anders Götherstrom, Nuria Naveran, Juan J. Sanchez, Michael Hofreiter, Philip Francis Thomsen, Jonas Binladen, Thomas F. G. Higham, Robert M. Yohe II, et al. "DNA from Pre-Clovis Human Coprolites in Oregon, North America." *Science* 320 (2008): 786–789.

Gilligan, Ian. "The Prehistoric Development of Clothing: Archaeological Implications of a Thermal Model." *Journal of Archaeological Theory and Method* 17 (2010): 15–80.

Ginter, B., J. K. Kozlowski, J. Guadelli, and H. Laville, eds. *Temnata Cave: Excavations in Karlukovo Karst Area, Bulgaria*. Krakow: Jagiellonian University Press, 2000.

Glantz, Michelle M. "The History of Hominin Occupation of Central Asia in Review." In *Asian Paleoanthropology: From Africa to China and Beyond*, edited by Christopher J. Norton and David R. Braun, 101–112. Dordrecht: Springer, 2010.

Glantz, Michelle, Bence Viola, Patrick Wrinn, Tatiana Chikisheva, Anatoly Derevianko, Andrei Krivoshapkin, Uktur Islamov, Rustam Suleimanov, and Terrence Ritzman. "New Hominin Remains from Uzbekistan." *Journal of Human Evolution* 55 (2008): 223–237.

Gleick, James. *The Information: A History, a Theory, a Flood*. New York: Pantheon, 2011.

Gleń, Elżbieta, and Krzysztof Kaczanowski. "Human Remains." In *Excavation in the Bacho Kiro Cave (Bulgaria): Final Report*, edited by Janusz K. Kozlowski, 75–79. Warsaw: Państwowe Wydawnictwo Naukowe, 1982.

Glen, William. *The Road to Jaramillo: Critical Years of the Revolution in Earth Science*. Stanford, Calif.: Stanford University Press, 1982.

Goder-Goldberger, Mae, Natalia Gubenko, and Erella Hovers. "'Diffusion with Modifications': Nubian Assemblages in the Central Negev Highlands of Israel and Their Implications for Middle Paleolithic Inter-Regional Interactions." *Quaternary International* 408 (2016): 121–139.

Godfrey-Smith, Peter. "Information in Biology." In *The Cambridge Companion to the Philosophy of Biology*, edited by David L. Hull and Michael Ruse, 103–119. Cambridge: Cambridge University Press, 2007.

Goebel, Ted. "The Early Upper Paleolithic of Siberia." In *The Early Upper Paleolithic Beyond Western Europe*, edited by P. Jeffrey Brantingham, Steven L. Kuhn, and Kristopher W. Kerry, 162–195. Berkeley: University of California Press, 2004.

Goebel, Ted. "The 'Microblade Adaptation' and Recolonization of Siberia During the Late Upper Pleistocene." In *Thinking Small: Global Perspectives on Microlithization*, edited by Robert G. Elston and Steven L. Kuhn, 117–131. Arlington, Va.: American Anthropological Association, 2002.

Goebel, Ted. "The Overland Dispersal of Modern Humans to Eastern Asia." In *Emergence and Diversity of Modern Human Behavior in Paleolithic Asia*, edited by Yousuke Kaifu, Masami Izuho, Ted Goebel, Hiroyuki Sato, and Akira Ono, 437–452. College Station: Texas A&M University Press, 2015.

Goebel, Ted. "The Pleistocene Colonization of Siberia and Peopling of the Americas: An Ecological Approach." *Evolutionary Anthropology* 8 (1999): 208–227.

Goebel, Ted, Anatoli P. Derevianko, and Valerii T. Petrin. "Dating the Middle-to-Upper-Paleolithic Transition at Kara-Bom." *Current Anthropology* 34 (1993): 452–458.

Goebel, Ted, W. Roger Powers, Nancy H. Bigelow, and Andrew S. Higgs. "Walker Road." In *American Beginnings: The Prehistory and Palaeoecology of Beringia*, edited by Frederick Hadleigh West, 356–362. Chicago: University of Chicago Press, 1996.

Goebel, Ted, Michael R. Waters, and Margarita Dikova. "The Archaeology of Ushki Lake, Kamchatka, and the Pleistocene Peopling of the Americas." *Science* 301 (2003): 501–505.

Goebel, Ted, Michael R. Waters, and Dennis H. O'Rourke. "The Late Pleistocene Dispersal of Modern Humans in the Americas." *Science* 319 (2008): 1497–1502.

Gokhman, David, Eitan Lavi, Kay Prüfer, Mario F. Fraga, José A. Riancho, Janet Kelso, Svante Pääbo, Eran Meshorer, and Liran Carmel. "Reconstructing the DNA Methylation Maps of the Neandertal and Denisovan." *Science* 344 (2014): 523–527.

Goldberg, Elkhonon. *The New Executive Brain: Frontal Lobes in a Complex World*. New York: Oxford University Press, 2009.

Golovanova, L. V., V. B. Doronichev, and N. E. Cleghorn. "The Emergence of Bone-Working and Ornamental Art in the Caucasian Upper Paleolithic." *Antiquity* 84 (2010): 299–320.

Golovanova, L. V., V. B. Doronichev, N. E. Cleghorn, M. A. Kulkova, T. V. Sapelko, and M. S. Shackley. "Significance of Ecological Factors in the Middle to Upper Paleolithic Transition." *Current Anthropology* 51 (2010): 655–691.

Goodall, Jane. "My Life Among the Wild Chimpanzees." *National Geographic Magazine*, August 1963, 272–308.

Goodman, Morris. "Evolution of the Immunologic Species Specificity of Human Serum Proteins." *Human Biology* 34 (1962): 104–150.

Goodman, Morris. "Immunochemistry of the Primates and Primate Evolution." *Annals of the New York Academy of Sciences* 102 (1962): 219–234.

Goren-Inbar, Naama, Nira Alperson, Mordechai E. Kislev, Orit Simchoni, Yoel Melamed, Adi Ben-Nun, and Ella Werker. "Evidence of Hominin Control of Fire at Gesher Benot Ya'aqov, Israel." *Science* 304 (2004): 725–727.

Gould, James L., and Carol Grant Gould. *The Honey Bee*. New York: Scientific American Library, 1995.

Gould, R. A. *Living Archaeology*. Cambridge: Cambridge University Press, 1980.

Gould, Stephen Jay. *Ontogeny and Phylogeny*. Cambridge, Mass.: Belknap Press of Harvard University Press, 1977.

Gould, Stephen Jay. *The Structure of Evolutionary Theory*. Cambridge, Mass.: Belknap Press of Harvard University Press, 2002.

Gould, Stephen Jay. *Wonderful Life: The Burgess Shale and the Nature of History*. New York: Norton, 1989.

Gowlett, John A. J. "The Elements of Design Form in Acheulian Bifaces: Modes, Modalities, Rules and Language." In *Axe Age: Acheulian Tool-making from Quarry to Discard*, edited by Naama Goren-Inbar and Gonen Sharon, 203–221. London: Equinox, 2006.

Gowlett, John A. J. "Mental Abilities of Early Man: A Look at Some Hard Evidence." In *Hominid Evolution and Community Ecology: Prehistoric Human Adaptation in Biological Perspective*, edited by Robert Foley, 167–192. New York: Academic Press, 1984.

Graf, Kelly E. "Siberian Odyssey." In *Paleoamerican Odyssey*, edited by Kelly E. Graf, Caroline V. Ketron, and Michael R. Waters, 65–80. College Station: Texas A&M University Press, 2013.

Graf, Kelly E., Lyndsay M. DiPietro, Kathryn E. Krasinski, Angela K. Gore, Heather L. Smith, Brendan J. Culleton, Douglas J. Kennett, and David Rhode. "Dry Creek Revisited: New Excavations, Radiocarbon Dates, and Site Formation Inform on the Peopling of Eastern Beringia." *American Antiquity* 80 (2015): 671–694.

Green, Richard E., Johannes Krause, Adrian W. Briggs, Tomislav Maricic, Udo Stenzel, Martin Kircher, Nick Patterson, Heng Li, Weiwei Zhai, Markus Hsi-Yang Fritz, et al. "A Draft Sequence of the Neandertal Genome." *Science* 328 (2010): 710–722.

Greenberg, Joseph H., Christy G. Turner II, and Stephen L. Zegura. "The Settlement of the Americas: A Comparison of the Linguistic, Dental, and Genetic Evidence." *Current Anthropology* 27 (1986): 477–497.

Greenfield, Leonard Owen. "On the Adaptive Pattern of 'Ramapithecus.'" *American Journal of Physical Anthropology* 50 (1979): 527–548.

Greenfield, Patricia M. "Language, Tools, and Brain: The Ontogeny and Phylogeny of Hierarchically Organized Sequential Behavior." *Behavioral and Brain Sciences* 14 (1991): 531–595.

Gregory, William King, and Milo Hellman. *The Dentition of* Dryopithecus *and the Origin of Man*. Anthropological Papers of the American Museum of Natural History 28. New York: American Museum of Natural History, 1926.

Grossman, Lawrence I., and Derek E. Wildman. *Morris Goodman, 1925–2010: A Biographical Memoir*. Washington, D.C.: National Academy of Sciences, 2014.

Groube, Les, John Chappell, and David Price. "A 40,000-Year-Old Human Occupation Site at Huon Peninsula, Papua New Guinea." *Nature* 324 (1986): 453–455.

Groucutt, Huw S., and Michael D. Petraglia. "The Prehistory of the Arabian Peninsula: Deserts, Dispersals, and Demography." *Evolutionary Anthropology* 21 (2012): 113–125.

Gunz, Philipp, Fred L. Bookstein, Philipp Mitteroecker, Andrea Stadlmayr, Horst Seidler, and Gerhard W. Weber. "Early Modern Human Diversity Suggests Subdivided Population Structure and a Complex Out-of-Africa Scenario." *Proceedings of the National Academy of Sciences* 106 (2009): 6094–6098.

Gunz, Philipp, Simon Neubauer, Lubov Golovanova, Vladimir Doronichev, Bruno Maureille, and Jean-Jacques Hublin. "A Uniquely Modern Human Pattern of Endocranial Development: Insights from a New Cranial Reconstruction of the Neandertal Newborn from Mezmaiskaya." *Journal of Human Evolution* 62 (2012): 300–313.

Guthrie, R. Dale. "Mammals of the Mammoth Steppe as Paleoenvironmental Indicators." In *Paleoecology of Beringia*, edited by David M. Hopkins, John V. Matthews Jr., Charles E. Schweger, and Steven B. Young, 307–326. New York: Academic Press, 1982.

Guthrie, R. Dale. "Origin and Causes of the Mammoth Steppe: A Story of Cloud Cover, Woolly Mammoth Tooth Pits, Buckles, and Inside-Out Beringia." *Quaternary Science Reviews* 20 (2001): 549–574.

Guthrie, R. Dale. "Paleoecology of the Large-Mammal Community in Interior Alaska During the Late Pleistocene." *American Midland Naturalist* 79 (1968): 346–463.

Haber, March, Dominique Gauguier, Sonia Youhanna, Nick Patterson, Priya Moorjani, Laura R. Botigué, Daniel E. Platt, Elizabeth Matisoo-Smith, David F. Soria-Hernanz, R. Spencer Wells, et al. "Genome-Wide Diversity in the Levant Reveals Recent Structuring by Culture." *PLoS Genetics* 9 (2013): e1003316.

Habgood, Philip J. "The Origin of Anatomically Modern Humans in Australasia." In *The Human Revolution: Behavioural and Biological Perspectives on the Origins of Modern Humans*, edited by Paul Mellars and Chris Stringer, 245–273. Princeton, N.J.: Princeton University Press, 1989.

Habgood, Philip J., and Natalie R. Franklin. "The Revolution That Didn't Arrive: A Review of Pleistocene Sahul." *Journal of Human Evolution* 55 (2008): 187–222.

Haidle, Miriam Noël. "How to Think a Simple Spear." In *Cognitive Archaeology and Human Evolution*, edited by Sophie A. de Beaune, Frank L. Coolidge, and Thomas Wynn, 57–73. Cambridge: Cambridge University Press, 2009.

Halligan, Jessi J., Michael R. Waters, Angelina Perrotti, Ivy J. Owens, Joshua M. Feinberg, Mark D. Bourne, Brendan Fenerty, Barbara Winsborough, David Carlson, Daniel C. Fisher, et al. "Pre-Clovis Occupation 14,550 Years Ago at the Page-Ladson Site, Florida, and the Peopling of the Americas." *Science Advances* 2 (2016): e1600375.

Hammer, Michael F., August E. Woerner, Fernando L. Mendez, Joseph C. Watkins, and Jeffrey D. Wall. "Genetic Evidence for Archaic Admixture in Africa." *Proceedings of the National Academy of Sciences* 108 (2011): 15123–15128.

Hanihara, K. "Mongoloid Dental Complex in the Permanent Dentition." In *Proceedings of the VIIIth International Congress of Anthropological and Ethnological Sciences*, 1:298–300. Tokyo: Science Council of Japan, 1968.

Hannus, L. Adrien. "Flaked Mammoth Bone from the Lange/Ferguson Site, White River Badlands Area, South Dakota." In *Bone Modification*, edited by Robson Bonnichsen and Marcella H. Sorg, 395–412. Orono: Center for the Study of the First Americans, University of Maine, 1989.

Harmand, Sonia, Jason E. Lewis, Craig S. Feibel, Christopher J. Lepre, Sandrine Prat, Arnaud Lenoble, Xavier Boës, Rhonda L. Quinn, Michel Brenet, Adrian Arroyo, et al. "3.3-Million-Year-Old Stone Tools from Lomekwi 3, West Turkana, Kenya." *Nature* 521 (2015): 310–315.

Harris, Kelley, and Rasmus Nielsen. "The Genetic Cost of Neanderthal Introgression." *Genetics* 203 (2016): 881–891.

Harrison, G. A., J. M. Tanner, D. R. Pilbeam, and P. T. Baker. *Human Biology: An Intro-duction to Human Evolution, Variation, Growth, and Adaptability.* 3rd ed.. Oxford: Oxford University Press, 1988.

Harrisson, Tom, and Lord Medway. "A First Classification of Prehistoric Bone and Tooth Artifacts Based on Material from Niah Great Cave." *Asian Perspectives* 6 (1962): 219–229.

Hart, J., Jr., M. A. Kraut, S. Kremen, B. Soher, and B. Gordon. "Neural Substrates of Orthographic Lexical Access as Demonstrated by Functional Brain Imaging." *Neu-ropsychiatry, Neuropsychology, and Behavioral Neurology* 13 (2000): 1–7.

Hartley, Ralph V. L. "Transmission of Information." *Bell System Technical Journal,* July 1928, 535.

Hartmanis, Juris, and Richard E. Stearns. "On the Computational Complexity of Algo-rithms." *Transactions of the American Mathematical Society* 177 (1965): 285–305.

Haslam, Michael, Chris Clarkson, Michael Petraglia, Ravi Korisettar, Sacha Jones, C. Shipton, Peter Ditchfield, and Stanley H. Ambrose. "The 74,000 BP Toba Super-eruption and Southern Indian Hominins: Archaeology, Lithic Technology, and Environments at Jwalapuram Locality 3." *Journal of Archaeological Science* 37 (2010): 3370–3384.

Hauser, Marc D. *The Evolution of Communication.* Cambridge, Mass.: MIT Press, 1996.

Hauser, Marc, Noam Chomsky, and W. Tecumseh Fitch. "The Faculty of Language: What Is It, Who Has It, and How Did It Evolve?" *Science* 298 (2002): 1569–1579.

Hawkes, K., J. F. O'Connell, N. G. Blurton Jones, H. Alvarez, and E. L. Charnov. "Grandmothering, Menopause, and the Evolution of Human Life Histories." *Pro-ceedings of the National Academy of Sciences* 95 (1998): 1336–1339.

Hawkey, D. E. "Out of Asia: Dental Evidence for Affinities and Microevolution of Early Populations from India/Sri Lanka." Ph.D. diss., Arizona State University, 1998.

Hay, Richard Leroy. *Geology of the Olduvai Gorge: A Study of the Sedimentation in a Semiarid Basin.* Berkeley: University of California Press, 1976.

Haynes, C. Vance, and Bruce B. Huckell, eds. *Murray Springs: A Clovis Site with Mul-tiple Activity Areas in the San Pedro Valley, Arizona.* Tucson: University of Arizona Press, 2007.

Haynes, Gary. *The Early Settlement of North America: The Clovis Era.* Cambridge: Cambridge University Press, 2002.

Hazelwood, Lee, and James Steele. "Colonizing New Landscapes: Archaeological Detectability of the First Phase." In *Colonization of Unfamiliar Landscapes: The Archaeology of Adaptation,* edited by Marcy Rockman and JamesSteele, 203–221. London: Routledge, 2003.

Hedrick, Philip W. *Genetics of Populations.* 4th ed. Sudbury, Mass.: Jones and Bartlett, 2011.

Henn, Brenna Mariah. "Inferring Modern Human Migration Patterns Within Africa Using Calibrated Mitochondrial and Y-Chromosomal DNA." Ph.D. diss., Stanford University, 2009.

Henn, Brenna M., Christopher R. Gignoux, Matthew Jobin, Julie M. Granka, J. M. Macpherson, Jeffrey M. Kidd, Laura Rodríguez-Botigué, Sohini Ramachandran, Lawrence Hon, Abra Brisbin, et al. "Hunter-Gatherer Genomic Diversity Suggests a Southern African Origin for Modern Humans." *Proceedings of the National Acad-emy of Sciences* 108 (2011): 5154–5162.

Henry, Amanda G., Peter S. Ungar, Benjamin H. Passey, Matt Sponheimer, Lloyd Rossouw, Marion Bamford, Paul Sandberg, Darryl J. de Ruiter, and Lee Berger. "The Diet of *Australopithecus sediba*." *Nature* 487 (2012): 90–93.

Henshilwood, Christopher S. "Fully Symbolic *sapiens* Behaviour: Innovation in the Middle Stone Age at Blombos Cave, South Africa." In *Rethinking the Human Revolution: New Behavioural and Biological Perspectives on the Origin and Dispersal of Modern Humans*, edited by Paul Mellars, Katie Boyle, Ofer Bar-Yosef, and Chris Stringer, 123–132. Cambridge: McDonald Institute for Archaeological Research, Cambridge University, 2007.

Henshilwood, Christopher S., Francesco d'Errico, Curtis W. Marean, Richard G. Milo, and Royden Yates. "An Early Bone Tool Industry from the Middle Stone Age at Blombos Cave, South Africa: Implications for the Origins of Modern Human Behaviour, Symbolism and Language." *Journal of Human Evolution* 41 (2001): 631–678.

Henshilwood, Christopher S., Francesco d'Errico, Karen L. van Niekerk, Yvan Coquinot, Zenobia Jacobs, Stein-Erik Lauritzen, Michel Menu, and Renata García-Moreno. "A 100,000-Year-Old Ochre-Processing Workshop at Blombos Cave, South Africa." *Science* 334 (2011): 219–222.

Henshilwood, Christopher S., Francesco d'Errico, Royden Yates, Zenobia Jacobs, Chantal Tribolo, Geoff A. T. Duller, Norbert Mercier, Judith C. Sealy, Helene Valladas, Ian Watts, et al. "Emergence of Modern Human Behavior: Middle Stone Age Engravings from South Africa." *Science* 295 (2002): 1278–1280.

Herculano-Houzel, Suzana. "The Human Brain in Numbers: A Linearly Scaled-Up Primate Brain." *Frontiers in Human Neuroscience* 3 (2009): 1–11.

Hernando-Herraez, Irene, Javier Prado-Martinez, Paras Garg, Marcos Fernandez-Callejo, Holger Heyn, Christina Hvilsom, Arcadi Navarro, Manel Esteller, Andrew J. Sharp, and Tomas Marques-Bonet. "Dynamics of DNA Methylation in Recent Human and Great Ape Evolution." *PLoS Genetics* 9 (2013): e1003763.

Hershkovitz, Israel, Ofer Marder, Avner Ayalon, Miryam Bar-Matthews, Gal Yasur, Elisabetta Boaretto, Valentina Caracuta, Bridget Alex, Amos Frumkin, Mae Goder-Goldberger, et al. "Levantine Cranium from Manot Cave (Israel) Foreshadows the First European Modern Humans." *Nature* 520 (2015): 216–219.

Heupink, Tim H., Sankar Subramanian, Joanne L. Wright, Phillip Endicott, Michael Carrington Westaway, Leon Huynen, Walther Parson, Craig D. Millar, Eske Willerslev, and David M. Lambert. "Ancient mtDNA Sequences from the First Australians Revisited." *Proceedings of the National Academy of Sciences* 113 (2016): 6892–6897.

Hewes, Gordon W. "A History of Speculation on the Relation Between Tools and Language." In *Tools, Language and Cognition in Human Evolution*, edited by Kathleen R. Gibson and Tim Ingold, 20–31. Cambridge: Cambridge University Press, 1993.

Heylighen, Francis. "The Growth of Structural and Functional Complexity During Evolution." In *The Evolution of Complexity: The Violet Book of "Einstein Meets Magritte,"* edited by Francis Heylighen, Johan Bollen, and Alexander Riegler, 17–44. Dordrecht: Kluwer, 1999.

Higham, Thomas, Laura Basell, Roger Jacobi, Rachel Wood, Christopher Bronk Ramsey, and Nicholas J. Conard. "Testing Models for the Beginnings of the Aurignacian and the Advent of Figurative Art and Music: The Radiocarbon Chronology of Geißenklösterle." *Journal of Human Evolution* 62 (2012): 664–676.

Higham, Tom, Tim Compton, Chris Stringer, Roger Jacobi, Beth Shapiro, Erik Trinkaus, Barry Chandler, Flora Gröning, Chris Collins, Simon Hillson, et al. "The Earliest Evidence for Anatomically Modern Humans in Northwestern Europe." *Nature* 479 (2011): 521–524.

Hill, Kim R., Robert S. Walker, Miran Božičević, James Eder, Thomas Headland, Barry Hewlett, A. Magdalena Hurtado, Frank Marlowe, Polly Wiessner, and Brian Wood. "Co-residence Patterns in Hunter-Gatherer Societies Show Unique Human Social Structure." *Science* 331 (2011): 1286–1289.

Hiscock, Peter. *Archaeology of Ancient Australia*. London: Routledge, 2008.

Hiscock, Peter. "Cultural Diversification and the Global Dispersion of *Homo sapiens*." In *Emergence and Diversity of Modern Human Behavior in Paleolithic Asia*, edited by Yousuke Kaifu, Masami Izuho, Ted Goebel, Hiroyuki Sato, and Akira Ono, 225–236. College Station: Texas A&M University Press, 2015.

Hiscock, Peter. "Occupying New Lands: Global Migrations and Cultural Diversification with Particular Reference to Australia." In *Paleoamerican Odyssey*, edited by Kelly E. Graf, Caroline V. Ketron, and Michael R. Waters, 3–11. College Station: Texas A&M University Press, 2013.

Hiscock, Peter, Sue O'Connor, Jane Balme, and Tim Maloney. "World's Earliest Ground-Edge Axe Production Coincides with Human Colonisation of Australia." *Australian Archaeology* 82 (2016): 2–11.

Hobaiter, C., and R. W. Byrne. "The Gestural Repertoire of the Wild Chimpanzee." *Animal Cognition* 14 (2011): 745–767.

Hoffecker, John F. "Assemblage Variability in Beringia: The Mesa Factor." In *From the Yenisei to the Yukon: Interpreting Lithic Assemblage Variability in Late Pleistocene/ Early Holocene Beringia*, edited by Ted Goebel and Ian Buvit, 165–178. College Station: Texas A&M University Press, 2011.

Hoffecker, John F. *Desolate Landscapes: Ice-Age Settlement of Eastern Europe*. New Brunswick, N.J.: Rutgers University Press, 2002.

Hoffecker, John F. "The Early Upper Paleolithic of Eastern Europe Reconsidered." *Evolutionary Anthropology* 20 (2011): 24–39.

Hoffecker, John F. "The Eastern Gravettian 'Kostenki Culture' as an Arctic Adaptation." *Anthropological Papers of the University of Alaska*, n.s., 2 (2002): 115–136.

Hoffecker, John F. "The Information Animal and the Super-Brain." *Journal of Archaeological Method and Theory* 20 (2013): 18–41.

Hoffecker, John F. "Innovation and Technological Knowledge in the Upper Paleolithic of Northern Eurasia." *Evolutionary Anthropology* 14 (2005): 186–198.

Hoffecker, John F. *Landscape of the Mind: Human Evolution and the Archaeology of Thought*. New York: Columbia University Press, 2011.

Hoffecker, John F. *A Prehistory of the North: Human Settlement of the Higher Latitudes*. New Brunswick, N.J.: Rutgers University Press, 2005.

Hoffecker, John F. "Representation and Recursion in the Archaeological Record." *Journal of Archaeological Method and Theory* 14 (2007): 370–375.

Hoffecker, John F. "The Spread of Modern Humans in Europe." *Proceedings of the National Academy of Sciences* 106 (2009): 16040–16045.

Hoffecker, John F., and Scott A. Elias. *Human Ecology of Beringia*. New York: Columbia University Press, 2007.

Hoffecker, John F., Scott A. Elias, and Dennis H. O'Rourke. "Out of Beringia?" *Science* 343 (2014): 979–980.

Hoffecker, John F., Scott A. Elias, Dennis H. O'Rourke, G. Richard Scott, and Nancy H. Bigelow. "Beringia and the Global Dispersal of Modern Humans." *Evolutionary Anthropology* 25 (2016): 64–78.

Hoffecker, John F., Vance T. Holliday, M. V. Anikovich, A. E. Dudin, N. I. Platonova, V. V. Popov, G. M. Levkovskaya, I. E. Kuz'mina, E. V. Syromyatnikova, N. D. Burova, et al. "Kostenki 1 and the Early Upper Paleolithic of Eastern Europe." *Journal of Archaeological Science: Reports* 5 (2016): 307–326.

Hoffecker, John F., Vance T. Holliday, M. V. Anikovich, A. A. Sinitsyn, V. V. Popov, S. N. Lisitsyn, G. M. Levkovskaya, G. A. Pospelova, Steven L. Forman, and Biagio Giaccio. "From the Bay of Naples to the River Don: The Campanian Ignimbrite Eruption and the Middle to Upper Paleolithic Transition in Eastern Europe." *Journal of Human Evolution* 55 (2008): 858–870.

Hoffecker, John F., Vance T. Holliday, Vadim N. Stepanchuk, Alexis Brugère, Steven L. Forman, Paul Goldberg, Oleg Tubolzev, and Igor Pisarev. "Geoarchaeological and Bioarchaeological Studies at Mira, an Early Upper Paleolithic Site in the Lower Dnepr Valley, Ukraine." *Geoarchaeology: An International Journal* 29 (2014): 61–77.

Hoffecker, John F., I. E. Kuz'mina, E. V. Syromyatnikova, M. V. Anikovich, A. A. Sinitsyn, V. V. Popov, and Vance T. Holliday. "Evidence for Kill-Butchery Events of Early Upper Paleolithic Age at Kostenki, Russia." *Journal of Archaeological Science* 37 (2010): 1073–1089.

Hoffecker, John F., W. Roger Powers, and Ted Goebel. "The Colonization of Beringia and the Peopling of the New World." *Science* 259 (1993): 46–53.

Holland, John H. *Adaptation in Natural and Artificial Systems: An Introductory Analysis with Applications in Biology, Control, and Artificial Intelligence.* Ann Arbor: University of Michigan Press, 1975.

Hölldobler, Bert, and E. O. Wilson. *The Super-Organism: The Beauty, Elegance, and Strangeness of Insect Societies.* New York: Norton, 2009.

Holliday, Vance T. *Paleoindian Geoarchaeology of the Southern High Plains.* Austin: University of Texas Press, 1997.

Holliday, Vance T., John F. Hoffecker, Paul Goldberg, Richard I. Macphail, Steven L. Forman, Mikhail Anikovich, and Andrei Sinitsyn. "Geoarchaeology of the Kostenki-Borshchevo Sites, Don River Valley, Russia." *Geoarchaeology: An International Journal* 22 (2007): 181–228.

Holliday, Vance T., and D. Shane Miller. "The Clovis Landscape." In *Paleoamerican Odyssey*, edited by Kelly E. Graf, Caroline V. Ketron, and Michael R. Waters, 221–245. College Station: Texas A&M University Press, 2013.

Holloway, Ralph. "Evolution of the Human Brain." In *Handbook of Human Symbolic Evolution*, edited by Andrew Lock and Charles R. Peters, 74–125. Oxford: Clarendon Press, 1995.

Holloway, Ralph L., Chet S. Sherwood, Patrick R. Hof, and James K. Rilling. "Evolution of the Brain in Humans—Paleoneurology." In *Encyclopedia of Neuroscience*, edited by Marc D. Binder, Nobutaka Hirokawa, and Uwe Windhorst, 1326–1334. Berlin: Springer, 2009.

Holmes, Charles E. "Tanana River Valley Archaeology Circa 14,000 to 9000 B.P." *Arctic Anthropology* 38 (2001): 154–170.

Hooton, Earnest Albert. *Up from the Ape*. New York: Macmillan, 1937.

Hopkins, David M. "Cenozoic History of the Bering Land Bridge." *Science* 129 (1959): 1519–1528.

Hopkins, David M. "Introduction." In *The Bering Land Bridge*, edited by David M. Hopkins, 1–6. Stanford, Calif.: Stanford University Press, 1967.

Hopkins, David M. "Preface." In *The Bering Land Bridge*, edited by David M. Hopkins, vii–ix. Stanford, Calif.: Stanford University Press, 1967.

Hopkins, David M., P. A. Smith, and J. V. Matthews Jr. "Dated Wood from Alaska and the Yukon: Implications for Forest Refugia in Beringia." *Quaternary Research* 15 (1981): 217–249.

Horai, Satoshi, Kenji Hayasaka, Rumi Kondo, Kazuo Tsugane, and Naoyuki Takahata. "Recent African Origin of Modern Humans Revealed by Complete Sequences of Hominoid Mitochondrial DNAs." *Proceedings of the National Academy of Sciences* 92 (1995): 532–536.

Horn, Henry S. "The Adaptive Significance of Colonial Nesting in the Brewer's Blackbird (*Euphagus cyanocephalus*)." *Ecology* 49 (1968): 682–694.

Hrdlička, Aleš. "Shovel-Shaped Teeth." *American Journal of Physical Anthropology* 3 (1920): 429–465.

Hsieh, PingHsun, August E. Woerner, Jeffrey D. Wall, Joseph Lachance, Sarah A. Tishkoff, Ryan N. Gutenkunst, and Michael F. Hammer. "Model-Based Analyses of Whole-Genome Data Reveal a Complex Evolutionary History Involving Archaic Introgression in Central African Pygmies." *Genome Research* 26 (2016): 291–300.

Hsu, T. C. "A Possible Function of Constitutive Heterochromatin: The Bodyguard Hypothesis." *Genetics* 79, suppl. 2 (1975): 137–150.

Hublin, Jean-Jacques, Abdelouahed Ben-Ncer, Shara E. Bailey, Sarah E. Freidline, Simon Neubauer, Matthew M. Skinner, Inga Bergmann, Adeline Le Cabec, Stefano Benazzi, Katerina Harvati, et al. "New Fossils from Jebel Irhoud, Morocco and the Pan-African Origin of *Homo sapiens*." *Nature* 546 (2017): 289–292.

Hudjashov, Georgi, Toomas Kivisild, Peter A. Underhill, Phillip Endicott, Juan J. Sanchez, Alice A. Lin, Peidong Shen, Peter Oefner, Colin Renfrew, Richard Villems, et al. "Revealing the Prehistoric Settlement of Australia by Y Chromosome and mtDNA Analysis." *Proceedings of the National Academy of Sciences* 104 (2007): 8726–8730.

Huerta-Sánchez, Emilia, Xin Jin, Asan, Zhuoma Bianba, Benjamin M. Peter, Nicolas Vinckenbosch, Yu Liang, Xin Yi, Mingze He, Mehmet Somel, et al. "Altitude Adaptation in Tibetans Caused by Introgression of Denisovan-like DNA." *Nature* 512 (2014): 194–197.

Hultén, Eric. *Outline of the History of Arctic and Boreal Biota During the Quaternary Period*. New York: Cramer, 1937.

Hunt, Gavin R. "Manufacture and Use of Hook-Tools by New Caledonian Crows." *Nature* 379 (1996): 249–251.

Hunt, Gavin R., Jawad Abdelkrim, Michael G. Anderson, Jennifer C. Holzhaider, Amy J. Marshall, Neil J. Gemmell, and Russell D. Gray. "Innovative Pandanus-Tool Folding by New Caledonian Crows." *Australian Journal of Zoology* 55 (2007): 291–298.

Huxley, Thomas H. *Man's Place in Nature*. London: Williams and Norgate, 1863.

Hyman, Anthony. *Charles Babbage: Pioneer of the Computer*. Princeton, N.J.: Princeton University Press, 1982.

Ifrah, Georges. *The Universal History of Computing: From the Abacus to the Quantum Computer.* Translated by E. F. Harding. New York: Wiley, 2001.

Ingman, Max, Henrik Kaessmann, Svante Pääbo, and Ulf Gyllensten. "Mitochondrial Genome Variation and the Origin of Modern Humans." *Nature* 408 (2000): 708–713.

Irish, Joel D. "Afridonty: The 'Sub-Saharan Dental Complex' Revisited." In *Anthropological Perspectives on Tooth Morphology: Genetics, Evolution, Variation*, edited by G. Richard Scott and Joel D. Irish, 278–295. Cambridge: Cambridge University Press, 2013.

Irish, Joel D. "Ancestral Dental Traits in Recent Sub-Saharan Africans and the Origins of Modern Humans." *Journal of Human Evolution* 34 (1998): 81–98.

Irish, Joel D. "Biological Affinities of Late Pleistocene Through Modern African Aboriginal Populations: The Dental Evidence." Ph.D. diss., Arizona State University, 1993.

Isaac, Glynn. "The Food-Sharing Behavior of Protohuman Hominids." *Scientific American*, April 1978, 90–108.

Isler, Karin, and Carel P. van Schaik. "How Humans Evolved Large Brains: Comparative Evidence." *Evolutionary Anthropology* 23 (2014): 65–75.

Itard, Jean Marc. *The Memorandum and Report on Victor de l'Aveyron*. 1806.

Ives, John W., Duane Froese, Kisha Supernant, and Gabriel Yanicki. "Vectors, Vestiges, and Valhallas—Rethinking the Corridor." In *Paleoamerican Odyssey*, edited by Kelly E. Graf, Caroline V. Ketron, and Michael R. Waters, 149–169. College Station: Texas A&M University Press, 2013.

Izuho, Masami, and Yousuke Kaifu. "The Appearance and Characteristics of the Early Upper Paleolithic in the Japanese Archipelago." In *Emergence and Diversity of Modern Human Behavior in Paleolithic Asia*, edited by Yousuke Kaifu, Masami Izuho, Ted Goebel, Hiroyuki Sato, and Akira Ono, 289–313. College Station: Texas A&M University Press, 2015.

Jackendoff, Ray. *Foundations of Language: Brain, Meaning, Grammar, Evolution.* Oxford: Oxford University Press, 2002.

Jackendoff, Ray, and Eva Wittenberg. "What You Can Say Without Syntax: A Hierarchy of Grammatical Complexity." In *Measuring Grammatical Complexity*, edited by Frederick J. Newmeyer and Laurel B. Preston, 65–82. Oxford: Oxford University Press, 2014.

Jackson, Donald, César Méndez, Roxana Seguel, Antonio Maldonado, and Gabriel Vargas. "Initial Occupation of the Pacific Coast of Chile During Late Pleistocene Times." *Current Anthropology* 48 (2007): 725–731.

Jackson, Philip C., Jr. *Introduction to Artificial Intelligence.* 2nd ed. New York: Dover, 1985.

Jacobi, R. M. and T. F. G. Higham. "The 'Red Lady' Ages Gracefully: New Ultrafiltration AMS Determinations from Paviland." *Journal of Human Evolution* 55 (2008): 898–907.

Jacobs, Zenobia, Michael C. Meyer, Richard G. Roberts, Vera Aldeias, Harold Dibble, and Mohammed Abdeljalil El Hajraoui. "Single-Grain OSL Dating at La Grotte des Contrebandiers ('Smugglers Cave'), Morocco: Improved Age Constraints for the Middle Paleolithic Levels." *Journal of Archaeological Science* 38 (2011): 3611–3643.

Jacobs, Zenobia, Richard G. Roberts, Roland Nespoulet, Mohammed Abdeljalil El Hajraoui, and André Debénath. "Single-Grain OSL Chronologies for Middle

Palaeolithic Deposits at El Mnasra and El Harhoura 2, Morocco: Implications for Late Pleistocene Human-Environment Interactions Along the Atlantic Coast of Northwest Africa." *Journal of Human Evolution* 62 (2012): 377–394.

James, Hannah V. A. "The Emergence of Modern Human Behavior in South Asia: A Review of the Current Evidence and Discussion of Its Possible Implications." In *The Evolution and History of Human Populations in South Asia: Inter-disciplinary Studies in Archaeology, Biological Anthropology, Linguistics and Genetics*, edited by Michael D. Petraglia and Bridget Allchin, 201–227. Dordrecht: Springer, 2007.

Janeway, Charles A., Paul Travers, Mark Wolport, and Mark J. Shlomchik. *Immunobiology: The Immune System in Health and Disease*. 5th ed. New York: Garland, 2001.

Jaubert, Jacques. "The Paleolithic Peopling of Mongolia: An Updated Assessment." In *Emergence and Diversity of Modern Human Behavior in Paleolithic Asia*, edited by Yousuke Kaifu, Masami Izuho, Ted Goebel, Hiroyuki Sato, and Akira Ono, 453–469. College Station: Texas A&M University Press, 2015.

Jelinek, Arthur J. "The Lower Paleolithic: Current Evidence and Interpretation." *Annual Review of Anthropology* 6 (1977): 11–32.

Jenkins, Dennis L., Loren G. Davis, Thomas W. Stafford Jr., Paula F. Campos, Thomas J. Connolly, Linda Scott Cummings, Michael Hofreiter, Bryan Hockett, Katelyn McDonough, Ian Luthe, et al. "Geochronology, Archaeological Context, and DNA at the Paisley Caves." In *Paleoamerican Odyssey*, edited by Kelly E. Graf, Caroline V. Ketron, and Michael R. Waters, 485–510. College Station: Texas A&M University Press, 2013.

Jenkins, Dennis L., Loren G. Davis, Thomas W. Stafford Jr., Paula F. Campos, Bryan Hockett, George T. Jones, Linda Scott Cummings, Chad Yost, Thomas J. Connolly, Robert M. Yohe II, et al. "Clovis Age Western Stemmed Projectile Points and Human Coprolites at the Paisley Caves." *Science* 337 (2012): 223–228.

Johnson, Eileen. "Along the Ice Margin—The Cultural Taphonomy of Late Pleistocene Mammoth in Southeastern Wisconsin (USA)." *Quaternary International* 169–170 (2007): 64–83.

Johnson, Eileen. "Cultural Activities and Interactions." In *Lubbock Lake: Late Quaternary Studies on the Southern High Plains*, edited by Eileen Johnson, 120–158. College Station: Texas A&M University Press, 1987.

Johnson, Eileen. "The Taphonomy of Mammoth Localities in Southeastern Wisconsin (USA)." *Quaternary International* 142–143 (2006): 58–78.

Josenhans, Heiner, Daryl Fedje, Reinhard Pienitz, and John Southon. "Early Humans and Rapidly Changing Holocene Sea Levels in the Queen Charlotte Islands–Hecate Strait, British Columbia, Canada." *Science* 277 (1997): 71–74.

Joyce, Daniel J. "Pre-Clovis Megafauna Butchery Sites in the Western Great Lakes Region, USA." In *Paleoamerican Odyssey*, edited by Kelly E. Graf, Caroline V. Ketron, and Michael R. Waters, 467–483. College Station: Texas A&M University Press, 2013.

Kaestle, Frederika A., and David Glenn Smith. "Ancient Mitochondrial DNA Evidence for Prehistoric Population Movement: The Numic Expansion." *American Journal of Physical Anthropology* 115 (2001): 1–12.

Kaplan, Jed O. "Geophysical Applications of Vegetation Modeling." Ph.D. diss., Lund University, 2001.

Kayser, Manfred. "The Human Genetic History of Oceania: Near and Remote Views of Dispersal." *Current Biology* 20 (2010): R194–R201.

Keith, Sir Arthur. *The Antiquity of Man*. London: Williams and Norgate, 1915.

Kelly, Robert L. "Colonization of New Land by Hunter-Gatherers." In *Colonization of Unfamiliar Landscapes: The Archaeology of Adaptation*, edited by Marcy Rockman and James Steele, 44–58. London: Routledge, 2003.

Kelly, Robert L. *The Lifeways of Hunter-Gatherers: The Foraging Spectrum*. Cambridge: Cambridge University Press, 2013.

Kelly, Robert L., and Lawrence C. Todd. "Coming into the Country: Early Paleoindian Hunting and Mobility." *American Antiquity* 53 (1988): 231–244.

Kemp, Brian M., Ripan S. Malhi, John McDonough, Deborah A. Bolnick, Jason A. Eshleman, Olga Rickards, Cristina Martinez-Labarga, John R. Johnson, Joseph G. Lorenz, E. James Dixon, et al. "Genetic Analysis of Early Holocene Skeletal Remains from Alaska and Its Implications for the Settlement of the Americas." *American Journal of Physical Anthropology* 132 (2007): 605–621.

Kennedy, G. E. *Paleo-Anthropology*. New York: McGraw-Hill, 1980.

Kennedy, Kenneth A. R. "The Deep Skull of Niah: An Assessment of Twenty Years of Speculation Concerning Its Evolutionary Significance." *Asian Perspectives* 20 (1977): 32–50.

Kennedy, Kenneth A. R., and Siran U. Deraniyagala. "Fossil Remains of 28,000-Year-Old Hominids from Sri Lanka." *Current Anthropology* 30 (1989): 394–399.

Kennedy, Kenneth A. R., Siran U. Deraniyagala, William J. Roertgen, John Chiment, and Todd Disotell. "Upper Pleistocene Fossil Hominids from Sri Lanka." *American Journal of Physical Anthropology* 72 (1987): 441–461.

Kharitonov, V. M. "Iskopaemye Gominidy Severnogo Kavkaza." In *Drevneishii Kavkaz: Perekrestok Azii i Evropy*, edited by S. A. Vasil'ev and A. V. Larionova, 79–80. St. Petersburg: IIMK Russian Academy of Sciences, 2013.

Kimura, Motoo. "Evolutionary Rate at the Molecular Level." *Nature* 217 (1968): 624–626.

Kimura, Ryosuke. "Human Migrations and Adaptations in Asia Inferred from Genomic Diversity." In *Emergence and Diversity of Modern Human Behavior in Paleolithic Asia*, edited by Yousuke Kaifu, Masami Izuho, Ted Goebel, Hiroyuki Sato, and Akira Ono, 34–50. College Station: Texas A&M University Press, 2015.

Kitchen, Andrew, Michael M. Miyamoto, and Connie J. Mulligan. "A Three-Stage Colonization Model for the Peopling of the Americas." *PLoS ONE* 3 (2008): e1596.

Kittler, Ralf, Manfred Kayser, and Mark Stoneking. "Molecular Evolution of *Pediculus humanus* and the Origin of Clothing." *Current Biology* 13 (2003): 1414–1417.

Kivisild, Toomas. "Maternal Ancestry and Population History from Whole Mitochondrial Genomes." *Investigative Genetics* 6 (2015).

Klein, Richard G. "Anatomy, Behavior, and Modern Human Origins." *Journal of World Prehistory* 9 (1995): 167–198.

Klein, Richard G. *F. Clark Howell, 1925–2007: A Biographical Memoir*. Washington, D.C.: National Academy of Sciences, 2013.

Klein, Richard G. "Geological Antiquity of Rhodesian Man." *Nature* 244 (1973): 311–312.

Klein, Richard G. *The Human Career: Human Biological and Cultural Origins*. Chicago: University of Chicago Press, 1989.

Klein, Richard G. *The Human Career: Human Biological and Cultural Origins*. 3rd ed. Chicago: University of Chicago Press, 2009.

Klein, Richard G. "Stone Age Predation on Small African Bovids." *South African Archaeological Bulletin* 36 (1981): 55–65.

Klein, Richard G., Graham Avery, Kathryn Cruz-Uribe, David Halkett, John E. Parkington, Teresa Steele, Thomas P. Volman, and Royden Yates. "The Ysterfontein 1 Middle Stone Age Site, South Africa, and Early Human Exploitation of Coastal Resources." *Proceedings of the National Academy of Sciences* 101 (2004): 5708–5715.

Klein, Richard G., and Teresa E. Steele. "Archaeological Shellfish Size and Later Human Evolution in Africa." *Proceedings of the National Academy of Sciences* 110 (2013): 10910–10915.

Knoll, Andrew H. *Life on a Young Planet: The First Three Billion Years of Evolution on Earth*. Princeton, N.J.: Princeton University Press, 2003.

Kohn, Marek, and Steven J. Mithen, "Handaxes: Products of Sexual Selection?" *Antiquity* 73 (1999): 518–526.

Kolmogorov, A. N. "On Tables of Random Numbers." *Sankhyā: The Indian Journal of Statistics*, ser. A, 25 (1963): 369–375.

Korisettar, Ravi. "Antiquity of Modern Humans and Behavioral Modernity in the Indian Subcontinent." In *Emergence and Diversity of Modern Human Behavior in Paleolithic Asia*, edited by Yousuke Kaifu, Masami Izuho, Ted Goebel, Hiroyuki Sato, and Akira Ono, 80–93. College Station: Texas A&M University Press, 2015.

Kozlowski, Janusz K. "The Middle and Early Upper Paleolithic Around the Black Sea." In *Neandertals and Modern Humans in Western Asia*, edited by Takeru Akazawa, Kenichi Aoki, and Ofer Bar-Yosef, 461–482. New York: Plenum, 1998.

Kozlowski, Janusz K. "The Significance of Blade Technologies in the Period 50–35 kya BP for the Middle–Upper Palaeolithic Transition in Central and Eastern Europe." In *Rethinking the Human Revolution: New Behavioural and Biological Perspectives on the Origin and Dispersal of Modern Humans*, edited by Paul Mellars, Katie Boyle, Ofer Bar-Yosef, and Chris Stringer, 317–328. Cambridge: McDonald Institute for Archaeological Research, Cambridge University, 2007.

Kozlowski, Janusz K., A. Dagnam-Ginter, I. Gatsov, and S. Sirakova. "Upper Paleolithic Assemblages." In *Excavation in the Bacho Kiro Cave (Bulgaria): Final Report*, edited by Janusz K. Kozlowski, 119–162. Warsaw: Państwowe Wydawnictwo Naukowe, 1982.

Krasinski, Kathryn E. "Broken Bones and Cut Marks: Taphonomic Analyses and Implications for the Peopling of North America." Ph.D. diss., University of Nevada, Reno, 2010.

Krause, Johannes, Adrian W. Briggs, Martin Kircher, Tomislav Maricic, Nicolas Zwyns, Anatoli Derevianko, and Svante Pääbo. "A Complete mtDNA Genome of an Early Modern Human from Kostenki, Russia." *Current Biology* 20 (2010): 231–236.

Krause, Johannes, Carles Lalueza-Fox, Ludovic Orlando, Wolfgang Enard, Richard E. Green, Hernán A. Burbano, Jean-Jacques Hublin, Catherine Hänni, Javier Fortea, Marco de la Rasilla, et al. "The Derived FOXP2 Variant of Modern Humans Was Shared with Neandertals." *Current Biology* 17 (2007): 1908–1912.

Krings, Matthias, Anne Stone, Ralf W. Schmitz, Heike Krainitzki, Mark Stoneking, and Svante Pääbo "Neanderthal DNA Sequences and the Origin of Modern Humans." *Cell* 90 (1997): 19–30.

Krivoshapkin, Andrei I., A. A. Anoikan, and P. Jeffrey Brantingham. "The Lithic Industry of Obi-Rakhmat Grotto, Uzbekistan." *Bulletin of the Indo-Pacific Association* 26 (2007): 5–19.

Krivoshapkin, Andrei I., and P. Jeffrey Brantingham. "The Lithic Industry of Obi-Rakhmat Grotto, Uzbekistan." In *Actes du XIV Congres UISPP, 2–8 Septembre 2001*, 203–214. BAR International Series 1240. Oxford: Archaeopress, 2004.

Krivoshapkin, Andrei I., Yaroslav V. Kuzmin, and A. J. Timothy Jull. "Chronology of the Obi-Rakhmat Grotto (Uzbekistan): First Results on the Dating and Problems of the Paleolithic Key Site in Central Asia." *Radiocarbon* 52 (2010): 549–554.

Kuhlwilm, Martin, Ilan Gronau, Melissa J. Hubisz, Cesare de Filippo, Javier Prado-Martinez, Martin Kircher, Qiaomei Fu, Hernán A. Burbano, Carles Lalueza-Fox, Marco de la Rasilla, et al. "Ancient Gene Flow from Early Modern Humans into Eastern Neanderthals." *Nature* 530 (2016): 429–433.

Kuhn, Steven L., and Amilcare Bietti. "The Late Middle and Early Upper Paleolithic in Italy." In *The Geography of Neandertals and Modern Humans in Europe and the Greater Mediterranean*, edited by Ofer Bar-Yosef and David Pilbeam, 49–76. Cambridge, Mass.: Peabody Museum of Archaeology and Ethnology, 2000.

Kuhn, Steven L., Mary C. Stiner, and Erksin Güleç. "Initial Upper Palaeolithic in South-Central Turkey and Its Regional Context: A Preliminary Report." *Antiquity* 73 (1999): 505–517.

Kuhn, Steven L., Mary C. Stiner, Erksin Güleç, Ismail Özer, Hakan Yılmaz, Ismail Baykara, Ayşen Açıkkol, Paul Goldberg, Kenneth Martínez Molina, Engin Ünay, et al. "The Early Upper Paleolithic Occupations at Üçağızlı Cave (Hatay, Turkey)." *Journal of Human Evolution* 56 (2009): 87–113.

Kuhn, Steven L., Mary C. Stiner, David S. Reese, and Erksin Güleç. "Ornaments of the Earliest Upper Paleolithic: New Insights from the Levant." *Proceedings of the National Academy of Sciences* 98 (2001): 7641–7646.

Kuhn, Steven L., and Nicolas Zwyns. "Rethinking the Initial Upper Paleolithic." *Quaternary International* 347 (2014): 29–38.

Kumar, Vikrant, Arimanda N. S. Reddy, Jagedeesh P. Babu, Tipirisetti N. Rao, Banrida T. Langstieh, Kumarasamy Thangaraj, Alla G. Reddy, Lalji Singh, and Battini M. Reddy. "Y-Chromosome Evidence Suggests a Common Paternal Heritage of Austro-Asiatic Populations." *BMC Evolutionary Biology* 7 (2007): 47.

Kuper, Adam. *Culture: The Anthropologist's Account*. Cambridge, Mass.: Harvard University Press, 1999.

Kurzweil, Ray. *The Singularity Is Near: When Humans Transcend Biology*. New York: Viking, 2005.

Laden, Greg, and Richard Wrangham. "The Rise of the Hominids as an Adaptive Shift in Fallback Foods: Plant Underground Storage Organs (USOs) and Australopith Origins." *Journal of Human Evolution* 49 (2005): 482–498.

Laitman, Jeffrey. "The Anatomy of Human Speech." *Natural History*, August 1984, 20–27.

Lalueza-Fox, Carles, Holger Römpler, David Caramelli, Claudia Stäubert, Giulio Catalano, David Hughes, Nadin Rohland, Elena Pilli, Laura Longo, Silvana Condemi, et al. "A Melanocortin 1 Receptor Allele Suggests Varying Pigmentation Among Neanderthals." *Science* 318 (2007): 1453–1455.

Lalueza-Fox, Carles, Antonio Rosasb, Almudena Estalrrich, Elena Gigli, Paula F. Campos, Antonio García-Tabernero, Samuel García-Vargas, Federico Sánchez-Quinto, Oscar Ramírez, Sergi Civit, et al. "Genetic Evidence for Patrilocal Mating Behavior Among Neandertal Groups." *Proceedings of the National Academy of Sciences* 108 (2011): 250–253.

Lambeck, Kurt, Hélène Rouby, Anthony Purcell, Yiying Sun, and Malcolm Sambridge. "Sea Level and Global Ice Volumes from the Last Glacial Maximum to the Holocene." *Proceedings of the National Academy of Sciences* 111 (2014): 15296–15303.

Lambeck, Kurt, Yusuke Yokoyama, and Tony Purcell. "Into and Out of the Last Glacial Maximum: Sea-Level Change During Oxygen-Isotope Stages 3 and 2." *Quaternary Science Reviews* 21 (2002): 343–360.

Landauer, Rolfe. "Information Is Physical." *Physics Today* 44 (1991): 23–29.

Lander, Eric S., Lauren M. Linton, Bruce Birren, Chad Nusbaum, Michael C. Zody, Jennifer Baldwin, Keri Devon, Ken Dewar, Michael Doyle, William FitzHugh, et al. "Initial Sequencing and Analysis of the Human Genome." *Nature* 409 (2001): 860–921.

Landes, David S. *Revolution in Time: Clocks and the Making of the Modern World.* Cambridge, Mass.: Harvard University Press, 1983.

Landweber, Laura F., and Erik Winfree, eds. *Evolution as Computation: DIMACS Workshop, Princeton, January 1999.* Berlin: Springer, 2002.

Lane, Nick. *The Vital Question: Energy, Evolution, and the Origins of Complex Life.* New York: Norton, 2015.

Lane, Nick, and William Martin. "The Energetics of Genome Complexity." *Nature* 467 (2010): 929–934.

Langley, Michelle C., Sue O'Connor, and Elena Piotto. "42,000-Year-Old Worked and Pigment-Stained *Nautilus* Shell from Jerimalai (Timor-Leste): Evidence for an Early Coastal Adaptation in ISEA." *Journal of Human Evolution* 97 (2016): 1–16.

Latorre, Claudio, Calogero M. Santoro, Paula C. Ugalde, Eugenia M. Gayo, Daniela Osorio, Carolina Salas-Engaña, Ricardo De Pol-Holz, Delphine Joly, and Jason A. Rech. "Late Pleistocene Human Occupation of the Hyperarid Core in the Atacama Desert, Northern Chile." *Quaternary Science Reviews* 77 (2013): 19–30.

Lawler, Andrew. "In Search of Green Arabia." *Science* 345 (2014): 994–997.

Lazaridis, Iosif, Nick Patterson, Alissa Mittnik, Gabriel Renaud, Swapan Mallick, Karola Kirsanow, Peter H. Sudmant, Joshua G. Schraiber, Sergi Castellano, Mark Lipson, et al. "Ancient Human Genomes Suggest Three Ancestral Populations for Present-Day Europeans." *Nature* 513 (2014): 409–413.

Lbova, L. V. *Paleolit Severnoi Zoni Zapadnogo Zabaikal'ya.* Ulan-Ude: Buryat Science Center, 2000.

Leakey, L. S. B., P. V. Tobias, and J. R. Napier. "A New Species of the Genus *Homo* from Olduvai Gorge." *Nature* 202 (1964): 7–9.

Leakey, Meave G., Fred Spoor, M. Christopher Dean, Craig S. Feibel, Susan G. Anton, Christopher Kiarie, and Louise N. Leakey. "New Fossils from Koobi Fora in Northern Kenya Confirm Taxonomic Diversity in Early *Homo*." *Nature* 488 (2012): 201–204.

Leakey, Richard E. *One Life: An Autobiography.* Salem, Mass.: Salem House, 1983.

Lee, Richard B. *The !Kung San: Men, Women, and Work in a Foraging Society.* Cambridge: Cambridge University Press, 1979.

Lee, Richard B. "!Kung Spatial Organization: An Ecological and Historical Perspective." *Human Ecology* 1 (1972): 125–147.

Lepre, Christopher J., Helene Roche, Dennis V. Kent, Sonia Harmand, Rhonda Quinn, Jean-Philip Brugal, Pierre-Jean Texier, Arnaud Lenoble, and Craig S. Feibel. "An Earlier Origin for the Acheulian." *Nature* 477 (2011): 82–85.

Lewin, Roger. *Bones of Contention: Controversies in the Search for Human Origins*. New York: Simon and Schuster, 1987.

Lewis, Harry R., and Christos H. Papadimitriou. *Elements of the Theory of Computation*. 2nd ed. Upper Saddle River, N.J.: Prentice Hall, 1998.

Lewontin, Richard C. "The Apportionment of Human Diversity." *Evolutionary Biology* 6 (1972): 381–398.

Lewontin, Richard C. "Evolution and the Theory of Games." *Journal of Theoretical Biology* 1 (1961): 382–403.

Lewontin, Richard. *Human Diversity*. New York: Scientific American Books, 1982.

Lewontin, Richard C. "Theoretical Population Genetics in the Evolutionary Synthesis." In *The Evolutionary Synthesis: Perspectives on the Unification of Biology*, edited by Ernst Mayr and William B. Provine, 58–68. Cambridge, Mass.: Harvard University Press, 1980.

Lewontin, Richard C., and John L. Hubby. "A Molecular Approach to the Study of Genic Heterozygosity in Natural Populations of *Drosophila pseudoobscura*." *Genetics* 54 (1966): 595–609.

Li, Feng, Xing Gaoa, Fuyou Chena, Shuwen Peia, Yue Zhanga, Xiaoling Zhanga, Decheng Liua, Shuangquan Zhanga, Ying Guana, Huimin Wang, et al. "The Development of Upper Palaeolithic China: New Results from the Shuidonggou Site." *Antiquity* 87 (2013): 368–383.

Lieberman, Daniel E. *The Evolution of the Human Head*. Cambridge, Mass.: Belknap Press of Harvard University Press, 2011.

Lieberman, Daniel E. "Speculations About the Selective Basis for Modern Human Craniofacial Form." *Evolutionary Anthropology* 17 (2008): 55–68.

Lieberman, Daniel E., Brandeis M. McBratney, and Gail Krovitz. "The Evolution and Development of Cranial Form in *Homo sapiens*." *Proceedings of the National Academy of Sciences* 99 (2002): 1134–1139.

Lieberman, Philip. *The Biology and Evolution of Language*. Cambridge, Mass.: Harvard University Press, 1984.

Lieberman, Philip. *Eve Spoke: Human Language and Human Evolution*. New York: Norton, 1998.

Lieth, Helmut. "Primary Production of Major Vegetation Units of the World." In *Primary Productivity of the Biosphere*, edited by Helmut Lieth and Robert H. Whittaker, 203–215. New York: Springer, 1975.

Lister, Ryan, Eran A. Mukamel, Joseph R. Nery, Mark Urich, Clare A. Puddifoot, Nicholas D. Johnson, Jacinta Lucero, Yun Huang, Andrew J. Dwork, Matthew D. Schultz, et al. "Global Epigenomic Reconfiguration During Mammalian Brain Development." *Science* 341 (2013): 1237905.

Liu, Hua, Franck Prugnolle, Andrea Manica, and François Balloux. "A Geographically Explicit Genetic Model of Worldwide Human-Settlement History." *American Journal of Human Genetics* 79 (2006): 230–237.

Liu, Wu, Chang-zhu Jin, Ying-qi Zhang, Yan-jun Cai, Song Xing, Xiu-jie Wu, Hai Cheng, R. Lawrence Edwards, Wen-shi Pan, Da-gong Qin, et al. "Human Remains from Zhirendong, South China, and Modern Human Emergence in East Asia." *Proceedings of the National Academy of Sciences* 107 (2010): 19201–19206.

Liu, Wu, María Martinón-Torres, Yan-jun Cai, Song Xing, Hao-wen Tong, Shu-wen Pei, Mark Jan Sier, Xiao-hong Wu, R. Lawrence Edwards, Hai Cheng, et al. "The

Earliest Unequivocally Modern Humans in Southern China." *Nature* 526 (2015): 696–700.

Liu, Wu, Xianzhu Wu, Shuwen Pei, Xiujie Wu, and Christopher J. Norton. "Huanglong Cave: A Late Pleistocene Human Fossil Site in Hubei Province, China." *Quaternary International* 211 (2010): 29–41.

Llamas, Bastien, Lars Fehren-Schmitz, Guido Valverde, Julien Soubrier, Swapan Mallick, Nadin Rohland, Susanne Nordenfelt, Cristina Valdiosera, Stephen M. Richards, Adam Rohrlach, et al. "Ancient Mitochondrial DNA Provides High-Resolution Time Scale of the Peopling of the Americas." *Science Advances* 2 (2016): e1501385.

Lombard, Marlize, and Miriam Noël Haidle. "Thinking a Bow-and-Arrow Set: Cognitive Implications of Middle Stone Age Bow and Stone-Tipped Arrow Technology." *Cambridge Archaeological Journal* 22 (2012): 237–264.

Loogväli, Eva-Liis, Toomas Kivisild, Tõnu Margus, and Richard Villems. "Explaining the Imperfection of the Molecular Clock of Hominid Mitochondria." *PLoS ONE* 4 (2009): e8260.

Louie, Kenway, and Matthew A. Wilson. "Temporally Structured Replay of Awake Hippocampal Ensemble Activity During Rapid Eye Movement Sleep." *Neuron* 29 (2001): 145–156.

Lowe, J. John, and Michael J. C. Walker. *Reconstructing Quaternary Environments*. 3rd ed. Abingdon: Routledge, 2014.

Luce, R. Duncan, and Howard Raiffa. *Games and Decisions*. New York: Wiley, 1957.

Luck, Gary W. "The Relationships Between Net Primary Productivity, Human Population Density and Species Conservation." *Journal of Biogeography* 34 (2007): 201–212.

Lyman, R. Lee. *Vertebrate Taphonomy*. Cambridge: Cambridge University Press, 1994.

Macaulay, Vincent, Catherine Hill, Alessandro Achilli, Chiara Rengo, Douglas Clarke, William Meehan, James Blackburn, Ornella Semino, Rosaria Scozzari, Fulvio Cruciani, et al. "Single, Rapid Coastal Settlement of Asia Revealed by Analysis of Complete Mitochondrial Genomes." *Science* 308 (2005): 1034–1036.

Mackay, Alex, Brian A. Stewart, and Brian M. Chase. "Coalescence and Fragmentation in the Late Pleistocene Archaeology of Southernmost Africa." *Journal of Human Evolution* 72 (2014): 26–51.

Mackie, Quentin, Loren Davis, Daryl Fedje, Duncan McLaren, and Amy Gusick. "Locating Pleistocene-Age Submerged Archaeological Sites on the Northwest Coast: Current Status of Research and Future Directions." In *Paleoamerican Odyssey*, edited by Kelly E. Graf, Caroline V. Ketron, and Michael R. Waters, 133–147. College Station: Texas A&M University Press, 2013.

MacLarnon, A. M., and G. P. Hewitt. "The Evolution of Human Speech: The Role of Enhanced Breathing Control." *American Journal of Physical Anthropology* 109 (1999): 341–363.

Majumder, Partha P. "The Human Genetic History of South Asia." *Current Biology* 20 (2010): R184–R187.

Malyarchuk, Boris, Miroslava Derenko, Tomasz Grzybowski, Maria Perkova, Urzula Rogalla, Tomas Vanecek, and Iosif Tsybovsky. "The Peopling of Europe from the Mitochondrial Haplogroup U5 Perspective." *PLoS ONE* 5 (2010): e10285.

Mandryk, Carol A. S., Heiner Josenhans, Daryl W. Fedje, and Rolf W. Mathewes. "Late Quaternary Paleoenvironments of Northwestern North America: Implications for

Inland Versus Coastal Migration Routes." *Quaternary Science Reviews* 20 (2001): 301–314.

Marangoni, Aurelio, David Caramelli, and Giorgio Manzi. "*Homo sapiens* in the Americas: Overview of the Earliest Human Expansion in the New World." *Journal of Anthropological Sciences* 92 (2014): 79–97.

Marean, Curtis, Miryam Bar-Matthews, Jocelyn Bernatchez, Erich Fisher, Paul Goldberg, Andy I. R. Herries, Zenobia Jacobs, Antonieta Jerardino, Panagiotis Karkanas, Tom Minichillo, et al. "Early Use of Marine Resources and Pigment in South Africa During the Middle Pleistocene." *Nature* 449 (2007): 905–908.

Marks, Anthony E. "The Middle to Upper Paleolithic Transition in the Levant." In *Advances in World Archaeology*, edited by Fred Wendorf and Angela E. Close, 2:51–98. New York: Academic Press, 1983.

Marks, Anthony E., and C. Reid Ferring. "The Early Upper Paleolithic of the Levant." In *The Early Upper Paleolithic: Evidence from Europe and the Near East*, edited by J. F. Hoffecker and C. A. Wolf, 43–72. BAR International Series 437. Oxford: British Archaeological Reports, 1988.

Marlowe, Frank W. *The Hadza: Hunter-Gatherers of Tanzania*. Berkeley: University of California Press, 2010.

Marom, Anat, James S. O. McCullagh, Thomas F. G. Higham, Andrey A. Sinitsyn, and Robert E. M. Hedges. "Single Amino Acid Radiocarbon Dating of Upper Paleolithic Modern Humans." *Proceedings of the National Academy of Sciences* 109 (2012): 6878–6881.

Marr, David. *Vision: A Computational Investigation into the Human Representation and Processing of Visual Information*. San Francisco: Freeman, 1982.

Marzke, Mary W. "Joint Function and Grips of the *Australopithecus afarensis* Hand, with Special Reference to the Region of the Capitate." *Journal of Human Evolution* 12 (1983): 197–211.

Mattick, John S. "A New Paradigm for Developmental Biology." *Journal of Experimental Biology* 210 (2007): 1526–1547.

Mattick, John S., Paolo P. Amaral, Marcel E. Dinger, Tim R. Mercer, and Mark F. Mehler. "RNA Regulation of Epigenetic Processes." *BioEssays* 31 (2009): 51–59.

Matyukhin, A. E. "Mnogosloinye Paleoliticheskie Pamyatniki v Ust'e Severskogo Dontsa." In *Rannyaya Pora Verkhnego Paleolita Evrazii: Obshchee i Lokal'noe*, edited by M. V. Anikovich, 157–182. St. Petersburg: Russian Academy of Sciences, 2006.

Mayer-Oakes, William J. "Early Man in the Andes." *Scientific American*, May 1963, 117–128.

Mayfield, John E. *The Engine of Complexity: Evolution as Computation*. New York: Columbia University Press, 2013.

Maynard Smith, John. "The Concept of Information in Biology." *Philosophy of Science* 67 (2000): 177–194.

Maynard Smith, John. *Evolution and the Theory of Games*. Cambridge: Cambridge University Press, 1982.

Maynard Smith, John, and Eörs Szathmáry. *The Major Transitions in Evolution*. Oxford: Oxford University Press, 1995.

Maynard Smith, John, and Eörs Szathmáry. *The Origins of Life: From the Birth of Life to the Origin of Language*. Oxford: Oxford University Press, 1999.

Mayr, Ernst. *The Growth of Biological Thought: Diversity, Evolution, and Inheritance.* Cambridge, Mass.: Belknap Press of Harvard University Press, 1982.

McBrearty, Sally, and Alison S. Brooks. "The Revolution That Wasn't: A New Interpretation of the Origin of Modern Human Behavior." *Journal of Human Evolution* 39 (2000): 453–563.

McGrew, William. *Chimpanzee Material Culture: Implications for Human Evolution.* Cambridge: Cambridge University Press, 1992.

McGrew, William. *The Cultured Chimpanzee: Reflections on Cultural Primatology.* Cambridge: Cambridge University Press, 2004.

McGuinness, Evelynn, Dave Sivertsen, and John M. Allman. "Organization of the Face Representation in Macaque Motor Cortex." *Journal of Comparative Neurology* 193 (1980): 591–608.

McNabb, John, Francesca Binyon, and Lee Hazelwood. "The Large Cutting Tools from the South African Acheulean and the Question of Social Traditions." *Current Anthropology* 45 (2004): 653–677.

McNaughton, S. J., M. Oesterheld, D. A. Frank, and K. J. Williams. "Ecosystem-Level Patterns of Primary Productivity and Herbivory in Terrestrial Habitats." *Nature* 341 (1989): 142–144.

McShea, Daniel W. "Metazoan Complexity and Evolution: Is There a Trend?" *Evolution: International Journal of Organic Evolution* 50 (1996): 477–492.

Mealy, George H. "A Method for Synthesizing Sequential Circuits." *Bell System Technical Journal*, September 1955, 1045–1079.

Meignen, L., J.-M. Geneste, L. Koulakovskaia, and A. Sytnik. "Koulichivka and Its Place in the Middle-Upper Paleolithic Transition in Eastern Europe." In *The Early Upper Paleolithic Beyond Western Europe*, edited by P. Jeffrey Brantingham, Steven L. Kuhn, and Kristopher W. Kerry, 50–63. Berkeley: University of California Press, 2004.

Mellars, Paul. "Archeology and the Dispersal of Modern Humans in Europe: Deconstructing the 'Aurignacian.'" *Evolutionary Anthropology* 15 (2006): 167–182.

Mellars, Paul. *The Neanderthal Legacy: An Archaeological Perspective from Western Europe.* Princeton, N.J.: Princeton University Press, 1996.

Mellars, Paul. "Why Did Modern Human Populations Disperse from Africa ca. 60,000 Years Ago? A New Model." *Proceedings of the National Academy of Sciences* 103 (2006): 9381–9386.

Mellars, Paul, and Jennifer C. French. "Tenfold Population Increase in Western Europe at the Neanderthal-to-Modern Human Transition." *Nature* 333 (2011): 623–627.

Mellars, Paul, Kevin C. Gori, Martin Carr, Pedro A. Soares, and Martin B. Richards. "Genetic and Archaeological Perspectives on the Initial Modern Human Colonization of Southern Asia." *Proceedings of the National Academy of Sciences* 110 (2013): 10699–10704.

Mendez, Fernando L., G. David Poznik, Sergi Castellano, and Carlos D. Bustamente. "The Divergence of Neandertal and Modern Human Y Chromosomes." *American Journal of Human Genetics* 98 (2016): 728–734.

Menzel, R., and M. Giurfa. "Cognitive Architecture of a Mini-Brain: The Honeybee." *Trends in Cognitive Science* 5 (2001): 62–71.

Metni, M. "A Re-Examination of a Proposed Neandertal Maxilla from Ksar 'Akil Rock Shelter, Antelias, Lebanon." *American Journal of Physical Anthropology*, suppl. 28 (1999): 202.

Metspalu, Mait, Toomas Kivisild, Ene Metspalu, Jüri Parik, Georgi Hudjashov, Katrin Kaldma, Piia Serk, Monika Karmin, Doron M Behar, M. Thomas P. Gilbert, et al. "Most of the Extant mtDNA Boundaries in South and Southwest Asia Were Likely Shaped During the Initial Settlement of Eurasia by Anatomically Modern Humans." *BMC Genetics* 5 (2004): 26.

Meyer, Matthias, Juan-Luis Arsuaga, Cesare de Filippo, Sarah Nagel, Ayinuer Aximu-Petri, Birgit Nickel, Ignacio Martinez, Ana Gracia, José María Bermúdez de Castro, Eudald Carbonell, et al. "Nuclear DNA Sequences from the Middle Pleistocene Sima de los Huesos Hominins." *Nature* 531 (2016): 504–507.

Meyer, Matthias, Qiaomei Fu, Ayinuer Aximu-Petri, Isabelle Glocke, Birgit Nickel, Juan-Luis Arsuaga, Ignacio Martínez, Ana Gracia, José María Bermúdez de Castro, Eudald Carbonell, et al. "A Mitochondrial Genome Sequence of a Hominin from Sima de los Huesos." *Nature* 505 (2014): 403–406.

Meyer, Matthias, Martin Kircher, Marie-Theres Gansauge, Heng Li, Fernando Racimo, Swapan Mallick, Joshua G. Schraiber, Flora Jay, Kay Prüfer, Cesare de Filippo, et al. "A High-Coverage Genome Sequence from an Archaic Denisovan Individual." *Science* 338 (2012): 222–226.

Mijares, Armand Salvador. "Human Emergence and Adaptation to an Island Environment in the Philippine Paleolithic." In *Emergence and Diversity of Modern Human Behavior in Paleolithic Asia*, edited by Yousuke Kaifu, Masami Izuho, Ted Goebel, Hiroyuki Sato, and Akira Ono, 171–181. College Station: Texas A&M University Press, 2015.

Mijares, Armand Salvador, Florent Détroit, Philip Piper, Rainer Grün, Peter Bellwood, Maxime Aubert, Guillaume Champion, Nida Cuevas, Alexandra De Leon, and Eusebio Dizon. "New Evidence for a 67,000-Year-Old Human Presence at Callao Cave, Luzon, Philippines." *Journal of Human Evolution* 59 (2010): 123–132.

Miller, Gifford H., Marilyn L. Fogel, John W. Magee, Michael K. Gagan, Simon J. Clarke, and Beverly J. Johnson. "Ecosystem Collapse in Pleistocene Australia and a Human Role in Megafaunal Extinction." *Science* 309 (2005): 287–290.

Miller, Gifford, John Magee, Mike Smith, Nigel Spooner, Alexander Baynes, Scott Lehman, Marilyn Fogel, Harvey Johnston, Doug Williams, Peter Clark, et al. "Human Predation Contributed to the Extinction of the Australian Megafaunal Bird *Genyornis newtoni* ~47 ka." *Nature Communications* 7 (2016): 10496.

Minsky, Marvin L. *Computation: Finite and Infinite Machines.* Englewood Cliffs, N.J.: Prentice Hall, 1967.

Mishra, Sheila, Naveen Chauhan, and Ashkok K. Singhvi. "Continuity of Microblade Technology in the Indian Subcontinent Since 45 ka: Implications for the Dispersal of Modern Humans." *PLoS ONE* 8 (2013): e69280.

Mitani, John C., Josep Call, Peter M. Kappeler, Ryne A. Palombit, and Joan B. Silk, eds. *The Evolution of Primate Societies.* Chicago: University of Chicago Press, 2012.

Mitchell, Melanie. *Complexity: A Guided Tour.* Oxford: Oxford University Press, 2009.

Mithen, Steven J. *Thoughtful Foragers: A Study of Prehistoric Decision Making.* Cambridge: Cambridge University Press, 1990.

Mochanov, Y. A. *Drevneishie Etapy Zaseleniya Chelovekom Severo-Vostochnoi Azii.* Novosibirsk: Nauka, 1977.

Monod, Jacques. *Chance and Necessity: An Essay on the Natural Philosophy of Modern Biology.* Translated by Austryn Wainhouse. New York: Knopf, 1971.

Moore, Edward F. "Gedanken-Experiments on Sequential Machines." In *Automata Studies*, edited by C. E. Shannon and J. McCarthy, 129–153. Princeton, N.J.: Princeton University Press, 1956.

Morlan, Richard E. "Technological Characteristics of Some Wedge-Shaped Cores in Northwestern North America and Northeast Asia." *Asian Perspectives* 19 (1976): 96–106.

Morris, Simon Conway. *The Crucible of Creation: The Burgess Shale and the Rise of Animals*. New York: Oxford University Press, 1998.

Morse, Kate. "Shell Beads from Mandu Mandu Creek Rock-Shelter, Cape Range Peninsula, Western Australia, Dated Before 30,000 B.P." *Antiquity* 67 (1993): 877–883.

Mountcastle, Vernon B. *Perceptual Neuroscience: The Cerebral Cortex*. Cambridge, Mass.: Harvard University Press, 1998.

Mountcastle, Vernon B. *The Sensory Hand: Neural Mechanisms of Somatic Sensation*. Cambridge, Mass.: Harvard University Press, 2005.

Movius, Hallam L, Jr. "The Lower Palaeolithic Cultures of Southern and Eastern Asia." *Transactions of the American Philosophical Society* 38 (1948): 329–420.

Moyà-Solà, Salvador, and Meike Köhler. "A *Dryopithecus* Skeleton and the Origin of Great-Ape Locomotion." *Nature* 379 (1996): 156–159.

Mulligan, Connie J., and Andrew Kitchen. "Three-Stage Colonization Model for the Peopling of the Americas." In *Paleoamerican Odyssey*, edited by Kelly E. Graf, Caroline V. Ketron, and Michael R. Waters, 171–181. College Station: Texas A&M University Press, 2013.

Mulligan, R. Michael, Joanne Chory, and Joseph R. Ecker. "Signaling in Plants." *Proceedings of the National Academy of Sciences* 94 (1997): 2793–2795.

Murdoch, Bruce E. "The Cerebellum and Language: Historical Perspective and Review." *Cortex* 46 (2009): 858–868.

Naito, Yuichi I., Yoshito Chikarashi, Dorothée G. Drucker, Naohiko Ohkouchi, Patrick Semal, Christoph Wißing, and Hervé Bocherens. "Ecological Niche of Neanderthals from Spy Cave Revealed by Nitrogen Isotopes of Individual Amino Acids in Collagen." *Journal of Human Evolution* 93 (2016): 82–90.

Napier, John. "Fossil Hand Bones from Olduvai Gorge." *Nature* 196 (1962): 409–411.

Napier, John. *Hands*. Revised by Russell H. Tuttle. Princeton, N.J.: Princeton University Press, 1993.

Nash, David J., Sheila Coulson, Sigrid Staurset, J. Stewart Ullyott, Mosarwa Babutsi, and Martin P. Smith. "Going the Distance: Mapping Mobility in the Kalahari Desert During the Middle Stone Age Through Multi-Site Geochemical Provenancing of Silcrete Artefacts." *Journal of Human Evolution* 96 (2016): 113–133.

Nehoroshev, P. E. "Rezul'taty Datirovaniya Stoyanki Shlyakh." *Rossiiskaya arkheologiya* 3 (2006): 21–30.

Nehoroshev, P. E. *Tekhnologicheskii metod izucheniya pervobytnogo rasshchepleniya kamnya srednego paleolita*. St. Petersburg: Evropeiskii Dom, 1999.

Nelson, Richard K. *Hunters of the Northern Forest: Designs for Survival Among the Alaskan Kutchin*. Chicago: University of Chicago Press, 1973.

Newman, Mark E. J. "The Structure and Function of Complex Networks." *SIAM Review* 45 (2003): 167–256.

Nichols, Joanna. "Linguistic Diversity and the First Settlement of the New World." *Language* 66 (1990): 475–521.

Nigst, Philip R., Paul Haesaerts, Freddy Damblon, Christa Frank-Fellner, Carolina Mallol, Bence Viola, Michael Götzinger, Laura Niven, Gerhard Trnka, and Jean-Jacques Hublin. "Early Modern Human Settlement of Europe North of the Alps Occurred 43,500 Years Ago in a Cold Steppe-Type Environment." *Proceedings of the National Academy of Sciences* 111 (2014): 14394–14399.

Noad, Michael J., Douglas H. Cato, M. M. Bryden, Micheline N. Jenner, and K. Curt S. Jenner. "Cultural Revolution in Whale Songs." *Nature* 408 (2000): 537.

Noonan, James P. "Neanderthal Genomics and the Evolution of Modern Humans." *Genome Research* 20 (2010): 547–553.

Nowell, April. "Childhood, Play and the Evolution of Cultural Capacity in Neanderthals and Modern Humans," In *The Nature of Culture*, edited by Miriam N. Haidle, Nicolas J. Conard, and Michael Bolus, 87–98. Dordrecht: Springer, 2016.

Nowell, April, and Melanie Lee Chang. "The Case Against Sexual Selection as an Explanation of Handaxe Morphology." *PaleoAnthropology* (2009): 77–88.

Nyquist, Harry. "Certain Factors Affecting Telegraph Speed." *Bell System Technical Journal*, April 1924, 324.

O'Connell, James F., and Jim Allen. "Dating the Colonization of Sahul (Pleistocene Australia–New Guinea): A Review of Recent Research." *Journal of Archaeological Science* 31 (2004): 835–853.

O'Connell, James F., and Jim Allen. "The Process, Biotic Impact, and Global Implications of the Human Colonization of Sahul About 47,000 years ago." *Journal of Archaeological Science* 56 (2015): 73–84.

O'Connell, James F., Jim Allen, and Kristen Hawkes. "Pleistocene Sahul and the Origins of Seafaring." In *The Global Origins and Development of Seafaring*, edited by Atholl Anderson, James H. Barrett, and Katherine V. Boyle, 57–68. Cambridge: McDonald Institute for Archaeological Research, Cambridge University, 2010.

O'Connor, Sue. "Carpenter's Gap Rockshelter 1: 40,000 Years of Aboriginal Occupation in the Napier Ranges, Kimberley, WA." *Australian Archaeology* 40 (1995): 58–59.

O'Connor, Sue. "Crossing the Wallace Line: The Maritime Skills of the Earliest Colonists in the Wallacean Archipelago." In *The Emergence and Diversity of Modern Human Behavior in Paleolithic Asia*, edited by Yousuke Kaifu, Masami Izuho, Ted Goebel, Hiroyuki Sato, and Akira Ono, 214–224. College Station: Texas A&M University Press, 2015.

O'Connor, Sue, Rintaro Ono, and Chris Clarkson. "Pelagic Fishing at 42,000 BP and the Maritime Skills of Modern Humans." *Science* 334 (2011): 1117–1121.

O'Connor, Sue, Gail Robertson, and K. P. Aplin. "Are Osseous Artefacts a Window to Perishable Material Culture? Implications of an Unusually Complex Bone Tool from the Late Pleistocene of Timor." *Journal of Human Evolution* 67 (2014): 108–119.

Ojeda, Almerindo E. *A Computational Introduction to Linguistics: Describing Language in Plain Prolog.* Stanford, Calif.: CSLI Publications, 2013.

Ono, Rintaro, Santoso Soegondho, and Minoru Yoneda. "Changing Marine Exploitation During Late Pleistocene in Northern Wallacea: Shell Remains from Leang Sarru Rockshelter in Talaud Islands." *Asian Perspectives* 48 (2009): 318–341.

Oppenheimer, Stephen. *The Real Eve: Modern Man's Journey Out of Africa.* New York: Basic Books, 2003.

Orlando, Ludovic, and Eske Willerslev. "An Epigenetic Window into the Past?" *Science* 345 (2014): 511–512.

O'Rourke, Dennis H. "Why Do We Migrate? A Retrospective." In *Causes and Consequences of Human Migration: An Evolutionary Perspective*, edited by Michael H. Crawford and Benjamin C. Campbell, 527–536. Cambridge: Cambridge University Press, 2014.

O'Rourke, Dennis H., and Jennifer A. Raff. "Human Genetic History of the Americas." *Current Biology* 20 (2010): R202–R207.

Osborn, Henry Fairfield. *Man Rises to Parnassus: Critical Epochs in the Prehistory of Man.* 2nd ed. Princeton, N.J.: Princeton University Press, 1928.

Osgood, Cornelius. *The Ethnography of the Tanaina.* New Haven, Conn.: Yale University Press, 1937.

Osgood, Cornelius. *Ingalik Material Culture.* New Haven, Conn.: Yale University Press, 1940.

Oswalt, Wendell H. *An Anthropological Analysis of Food-Getting Technology.* New York: Wiley, 1976.

Oswalt, Wendell H. *Habitat and Technology: The Evolution of Hunting.* New York: Holt, Rinehart and Winston, 1973.

Oswalt, Wendell H. "Technological Complexity: The Polar Eskimos and the Tareumiut." *Arctic Anthropology* 24 (1987): 82–98.

Otte, Marcel, A. E. Matyukhin, and Damien Flas. "La chronologie de Biryuchya Balka (Région de Rostov, Russie)." In *Rannyaya Pora Verkhnego Paleolita Evrazii: Obshchee i Lokal'noe*, edited by M. V. Anikovich, 183–192. St. Petersburg: Russian Academy of Sciences, 2006.

Ovchinnikov, Igor V., and William Goodwin. "Ancient Human DNA from Sungir?" *Journal of Human Evolution* 44 (2003): 389–392.

Overstreet, David F. "Late-Glacial Ice-Marginal Adaptation in Southeastern Wisconsin." In *Paleoamerican Origins: Beyond Clovis*, edited by Robson Bonnichsen, Bradley T. Lepper, Dennis Stanford, and Michael R. Waters, 183–195. College Station: Texas A&M University Press, 2005.

Ovodov, Nikolai D., Susan J. Crockford, Yaroslav V. Kuzmin, Thomas F. G. Higham, Gregory W. L. Hodgins, and Johannes van der Plicht. "A 33,000-Year-Old Incipient Dog from the Altai Mountains of Siberia: Evidence of the Earliest Domestication Disrupted by the Last Glacial Maximum." *PLoS ONE* 6 (2011): e22821.

Owsley, Douglas W., Margaret A. Jodry, Thomas W. Stafford Jr., C. Vance Haynes Jr., and Dennis J. Stafford. *Arch Lake Woman: Physical Anthropology and Geoarchaeology.* College Station: Texas A&M University Press, 2010.

Pääbo, Svante. *Neanderthal Man: In Search of Lost Genomes.* New York: Basic Books, 2014.

Pääbo, Svante "The Y Chromosome and the Origin of All of Us (Men)." *Science* 268 (1995): 1141–1142.

Pagani, Luca, Stephan Schiffels, Deepti Gurdasani, Petr Danecek, Aylwyn Scally, Yuan Chen, Yali Xue, Marc Haber, Rosemary Ekong, Tamiru Oljira, et al. "Tracing the Route of Modern Humans Out of Africa by Using 225 Human Genome Sequences from Ethiopians and Egyptians." *American Journal of Human Genetics* 96 (2015): 986–991.

Pakendorf, Brigitte, Koen Bostoen, and Cesare de Filippo. "Molecular Perspectives on the Bantu Expansion: A Synthesis." *Language Dynamics and Change* 1 (2011): 50–88.

Papadimitriou, Christos H. *Computational Complexity*. Reading, Mass.: Addison-Wesley, 1994.

Papadimitriou, Christos H., and Kenneth Steiglitz. *Combinatorial Optimization: Algorithms and Complexity*. Mineola, N.Y.: Dover, 1998.

Paunero, Rafael S. "The Presence of a Pleistocene Colonizing Culture in La Maria Archaeological Locality: Casa del Minero 1, Argentina." In *Where the South Winds Blow: Ancient Evidence for Paleo South Americans*, edited by Laura Miotti, Monica Salemme, and Nora Flegenheimer, 127–132. College Station: Texas A&M University Press, 2003.

Pavlov, Pavel, John Inge Svendsen, and Svein Indrelid. "Human Presence in the European Arctic Nearly 40,000 Years Ago." *Nature* 413 (2001): 64–67.

Payne, Roger. *Among Whales*. New York: Delta, 1995.

Pearson, Georges A. "Early Occupation and Cultural Sequence at Moose Creek: A Late Pleistocene Site in Central Alaska." *Arctic* 52 (1999): 332–345.

Pearson, Osbjorn M. "Integration of the Genetic, Anatomical and Archaeological Data for the African Origin of Modern Humans: Problems and Prospects." In *African Genesis: Perspectives on Hominin Evolution*, edited by Sally C. Reynolds and Andrew Gallagher, 423–448. Cambridge: Cambridge University Press, 2012.

Pearson, Osbjorn M. "Statistical and Biological Definitions of 'Anatomically Modern' Humans: Suggestions for a Unified Approach to Modern Morphology." *Evolutionary Anthropology* 17 (2008): 38–48.

Pedersen, Mikkel W., Anthony Ruter, Charles Schweger, Harvey Friebe, Richard A. Staff, Kristian K. Kjeldsen, Marie L. Z. Mendoza, Alwynne B. Beaudoin, Cynthia Zutter, Nicolaj K. Larsen, et al. "Postglacial Viability and Colonization in North America's Ice-Free Corridor." *Nature* 537 (2016): 45–49.

Pelegrin, Jacques. "Cognition and the Emergence of Language: A Contribution from Lithic Technology." In *Cognitive Archaeology and Human Evolution*, edited by Sophie A. de Beaune, Frederick L. Coolidge, and Thomas Wynn, 95–108. Cambridge: Cambridge University Press, 2009.

Perego, Ugo A., Alessandro Achilli, Norman Angerhofer, Matteo Accetturo, Maria Pala, Anna Olivieri, Baharak Hooshiar Kashani, Kathleen H. Ritchie, Rosaria Scozzari, Qing-peng Kong, et al. "Distinctive Paleo-Indian Migration Routes from Beringia Marked by Two Rare mtDNA Haplogroups." *Current Biology* 19 (2009): 1–8.

Perera, Nimal, Nikos Kourampas, Ian A. Simpson, Siran U. Deraniyagala, David Bulbeck, Johan Kamminga, Jude Perera, Dorian Q. Fuller, Katherine Szabó, and Nuno V. Oliveira. "People of the Ancient Rainforest: Late Pleistocene Foragers at the Batadomba-lena Rockshelter, Sri Lanka." *Journal of Human Evolution* 61 (2011): 254–269.

Peresani, Marco, Emanuela Cristiani, and Matteo Romandini. "The Uluzzian Technology of Grotta di Fumane and Its Implication for Reconstructing Cultural Dynamics in the Middle–Upper Palaeolithic Transition of Western Eurasia." *Journal of Human Evolution* 91 (2016): 36–56.

Perreault, Charles, P. Jeffrey Brantingham, Steven L. Kuhn, Sarah Wurz, and Xing Gao. "Measuring the Complexity of Lithic Technology." *Current Anthropology* 54 (2013): S397–S406.

Petraglia, Michael, Ravi Korisettar, Nicole Boivin, Christopher Clarkson, Peter Ditchfield, Sacha Jones, Jinu Koshy, Marta Mirazón Lahr, Clive Oppenheimer, David

Pyle, et al. "Middle Paleolithic Assemblages from the Indian Subcontinent Before and After the Toba Super-Eruption." *Science* 317 (2007): 114–116.

Petraglia, Michael D., and Jeffrey I. Rose, eds. *The Evolution of Human Populations in Arabia: Paleoenvironments, Prehistory, and Genetics.* Dordrecht: Springer, 2009.

Piaget, Jean. *The Language and Thought of the Child.* New York: New American Library, 1974.

Piaget, Jean. *The Origins of Intelligence in Children.* Translated by Margaret Cook. New York: International Universities Press, 1952.

Pianka, Eric R. *Evolutionary Ecology.* 2nd ed. New York: Harper & Row, 1978.

Pierce, John R. *An Introduction to Information Theory: Symbols, Signals and Noise.* 2nd ed. New York: Dover, 1980.

Pilbeam, David. *The Ascent of Man: An Introduction to Human Evolution.* New York: Macmillan, 1972.

Pinker, Steven. *The Language Instinct: How the Mind Creates Language.* New York: Morrow, 1994.

Pitts, Michael, and Mark Roberts. *Fairweather Eden: Life in Britain Half a Million Years Ago as Revealed by the Excavations at Boxgrove.* London: Century, 1998.

Pitulko, Vladimir V., Aleksandr E. Basilyan, and Elena Y. Pavlova. "The Berelekh 'Graveyard': New Chronological and Stratigraphical Data from the 2009 Field Season." *Geoarchaeology: An International Journal* 29 (2014): 277–299.

Pitulko, Vladimir, Pavel Nikolskiy, Aleksandr Basilyan, and Elena Pavlova. "Human Habitation in Arctic Western Beringia Prior to the LGM." In *Paleoamerican Odyssey*, edited by Kelly E. Graf, Caroline V. Ketron, and Michael R. Waters, 13–44. College Station: Texas A&M University Press, 2013.

Pitulko, V. V., P. A. Nikolsky, E. Yu. Girya, A. E. Basilyan, V. E. Tumskoy, S. A. Koulakov, S. N. Astakhov, E. Yu. Pavlova, and M. A. Anisimov. "The Yana RHS Site: Humans in the Arctic Before the Last Glacial Maximum." *Science* 303 (2004): 52–56.

Pitulko, V. V., E. Y. Pavlova, P. A. Nikolskiy, and V. V. Ivanova. "The Oldest Art of the Eurasian Arctic: Personal Ornaments and Symbolic Objects from Yana RHS, Arctic Siberia." *Antiquity* 86 (2012): 642–659.

Pitulko, Vladimir V., Alexei N. Tikhonov, Elena Y. Pavlova, Pavel A. Nikolskiy, Konstantin E. Kuper, and Roman N. Polozov. "Early Human Presence in the Arctic: Evidence from 45,000-Year-Old Mammoth Remains." *Science* 351 (2016): 260–263.

Plagnol, Vincent, and Jeffrey D. Wall. "Possible Ancestral Structure in Human Populations." *PLoS Genetics* 2 (2006): e105.

Plavcan, J. Michael. "Understanding Dimorphism as a Function of Changes in Male and Female Traits." *Evolutionary Anthropology* 20 (2011): 143–155.

Plummer, Thomas. "Flaked Stones and Old Bones: Biological and Cultural Evolution at the Dawn of Technology." *Yearbook of Physical Anthropology* 47 (2004): 118–164.

Plummer, Thomas W., Peter W. Ditchfield, Laura C. Bishop, John D. Kingston, Joseph V. Ferraro, David R. Braun, Fritz Hertel, and Richard Potts. "Oldest Evidence of Toolmaking Hominins in a Grassland-Dominated Ecosystem." *PLoS ONE* 4 (2009): e7199.

Pollard, Katherine S., Sofie R. Salama, Bryan King, Andrew D. Kern, Tim Dreszer, Sol Katzman, Adam Siepel, Jakob S. Pedersen, Gill Bejerano, Robert Baertsch, et al. "Forces Shaping the Fastest Evolving Regions in the Human Genome." *PLoS Genetics* 2 (2006): e168.

Poltoraus, A. B., E. E. Kulikov, and I. A. Lebedeva. "Molekulyarnyi Analiz DNK iz Ostatkov Trekh Individuumov so Stoyanki Sungir' (Predvaritel'nye Itogi)." In *Homo Sungirensis: Verkhnepaleoliticheskii Chelovek: Ekologicheskie i Evolyutsion-nye Aspekty Issledovaniya*, 351–358. Moscow: Nauchnyi Mir, 2000.

Pope, Matt I., and Mark B. Roberts. "Observations on the Relationship Between Palaeolithic Individuals and Artifact Scatters at the Middle Pleistocene Site of Box-grove, UK." In *The Hominid Individual in Context: Archaeological Excavations of Lower and Middle Palaeolithic Landscapes, Locales and Artefacts*, edited by Clive Gamble and Martin Porr, 81–97. London: Routledge, 2005.

Posth, Cosimo, Gabriel Renaud, Alissa Mittnik, Dorothée G. Drucker, Hélène Rougier, Christophe Cupillard, Frédérique Valentin, Corinne Thevenet, Anja Furtwängler, Christoph Wißing, et al. "Pleistocene Mitochondrial Genomes Suggest a Single Major Dispersal of Non-Africans and a Late Glacial Population Turnover in Europe." *Current Biology* 26 (2016): 827–833.

Potter, Ben A., Charles E. Holmes, and David R. Yesner. "Technology and Economy Among the Earliest Prehistoric Foragers in Interior Eastern Beringia." In *Paleo-american Odyssey*, edited by Kelly E. Graf, Caroline V. Ketron, and Michael R. Waters, 81–103. College Station: Texas A&M University Press, 2013.

Potter, Ben A., Joel D. Irish, Josh D. Reuther, and Holly J. McKinney. "New Insights into Eastern Beringian Mortuary Behavior: A Terminal Pleistocene Double Infant Burial at Upward Sun River." *Proceedings of the National Academy of Sciences* 111 (2014): 17060–17065.

Poundstone, William. *Prisoner's Dilemma: John von Neumann, Game Theory, and the Puzzle of the Bomb*. New York: Anchor Books, 1992.

Poundstone, William. *The Recursive Universe: Cosmic Complexity and the Limits of Sci-entific Knowledge*. Rev. ed. Mineola, N.Y.:: Dover, 2013.

Powers, William R., and John F. Hoffecker. "Late Pleistocene Settlement in the Nenana Valley, Central Alaska." *American Antiquity* 54 (1989): 263–287.

Poznik, G. David, Brenna M. Henn, Muh-ching Yee, Elzbieta Sliwerska, Ghia M. Euskirchen, Alice A. Lin, Michael Snyder, Lluis Quintana-Murci, Jeffrey M. Kidd, Peter A. Underhill, et al. "Sequencing Y Chromosomes Resolves Discrepancy in Time to Common Ancestor of Males versus Females." *Science* 341 (2013): 562–569.

Poznik, G. David, Yali Xue, Fernando L. Mendez, Thomas F. Willems, Andrea Massaia, Melissa A. Wilson Sayres, Qasim Ayub, Shane A. McCarthy, Apurva Narechania, Seva Kashin, et al. "Punctuated Bursts in Human Male Demography Inferred from 1,244 Worldwide Y-Chromosome Sequences." *Nature Genetics* 48 (2016): 593–599.

Prat, Sandrine, Stéphane C. Péan, Laurent Crépin, Dorothée G. Drucker, Simon J. Puaud, Hélène Valladas, Martina Lázničková-Galetová, Johannes van der Plicht, and Alexander Yanevich. "The Oldest Anatomically Modern Humans from Far Southeast Europe: Direct Dating, Culture, and Behavior." *PLoS ONE* 6 (2011): e20834.

Qiu, Jane. "The Forgotten Continent." *Nature* 535 (2016): 218–220.

Quintana-Murci, Lluís, Raphaëlle Chaix, R. Spencer Wells, Doron M. Behar, Hamid Sayar, Rosaria Scozzari, Chiara Rengo, Nadia Al-Zahery, Ornella Semino, A. Sil-vana Santachiara-Benerecetti, et al. "Where West Meets East: The Complex mtDNA Landscape of the Southwest and Central Asian Corridor." *American Journal of Human Genetics* 74 (2004): 827–845.

Radcliffe-Brown, A. R. *The Andaman Islanders*. Cambridge: Cambridge University Press, 1922.

Radcliffe-Brown, A. R. "On Social Structure." *Journal of the Royal Anthropological Institute* 70 (1940): 1–12.

Raff, Jennifer A., and Deborah A. Bolnick. "Does Mitochondrial Haplogroup X Indicate Ancient Trans-Atlantic Migration to the Americas? A Critical Re-evaluation." *PaleoAmerica* 1 (2015): 297–304.

Raff, Jennifer A., Deborah A. Bolnick, Justin Tackney, and Dennis H. O'Rourke. "Ancient DNA Perspectives on American Colonization and Population History." *American Journal of Physical Anthropology* 146 (2011): 503–514.

Raghavan, Maanasa, Pontus Skoglund, Kelly E. Graf, Mait Metspalu, Anders Albrechtsen, Ida Moltke, Simon Rasmussen, Thomas W. Stafford Jr., Ludovic Orlando, Ene Metspalu, et al. "Upper Palaeolithic Siberian Genome Reveals Dual Ancestry of Native Americans." *Nature* 505 (2013): 87–91.

Raghavan, Maanasa, Matthias Steinrücken, Kelley Harris, Stephan Schiffels, Simon Rasmussen, Michael DeGiorgio, Anders Albrechtsen, Cristina Valdiosera, María C. Ávila-Arcos, Anna-Sapfo Malaspinas, et al. "Genomic Evidence for the Pleistocene and Recent Population History of Native Americans." *Science* 349 (2015): aab3884.

Ramirez Rozzi, Fernando V., and José Maria Bermudez de Castro. "Surprisingly Rapid Growth in Neanderthals." *Nature* 428 (2004): 936–939.

Ranov, V. A., and S. A. Laukhin. "Stoyanka na Puti Migratsii Srednepaleoliticheskogo Cheloveka iz Levanta v Sibir." *Priroda* 9 (2000): 52–60.

Rasmussen, Morten, Sarah L. Anzick, Michael R. Waters, Pontus Skoglund, Michael DeGiorgio, Thomas W. Stafford Jr., Simon Rasmussen, Ida Moltke, Anders Albrechtsen, Shane M. Doyle, et al. "The Genome of a Late Pleistocene Human from a Clovis Burial Site in Western Montana." *Nature* 506 (2014): 225–229.

Rasmussen, Morten, Xiaosen Guo, Yong Wang, Kirk E. Lohmueller, Simon Rasmussen, Anders Albrechtsen, Line Skotte, Stinus Lindgreen, Mait Metspalu, Thibaut Jombart, et al. "An Aboriginal Australian Genome Reveals Separate Human Dispersals into Asia." *Science* 334 (2011): 94–98.

Rasmussen, Morten, Yingrui Li, Stinus Lindgreen, Jakob Skou Pedersen, Anders Albrechtsen, Ida Moltke, Mait Metspalu, Ene Metspalu, Toomas Kivisild, Ramneek Gupta, et al. "Ancient Human Genome Sequence of an Extinct Palaeo-Eskimo." *Nature* 463 (2010): 757–762.

Rasmussen, Morten, Martin Sikora, Anders Albrechtsen, Thorfinn Sand Korneliussen, J. Víctor Moreno-Mayar, G. David Poznik, Christoph P. E. Zollikofer, Marcia S. Ponce de León, Morten E. Allentoft, Ida Moltke, et al. "The Ancestry and Affiliations of Kennewick Man." *Nature* 523 (2015): 455–458.

Redmond, Brian G., H. Gregory McDonald, Haskel J. Greenfield, and Matthew L. Burr. "New Evidence for Late Pleistocene Human Exploitation of Jefferson's Ground Sloth (*Megalonyx jeffersonii*) from Northern Ohio, USA." *World Archaeology* 44 (2012): 75–101.

Reich, David, Richard E. Green, Martin Kircher, Johannes Krause, Nick Patterson, Eric Y. Durand, Bence Viola, Adrian W. Briggs, Udo Stenzel, Philip L. F. Johnson, et al. "Genetic History of an Archaic Hominin Group from Denisova Cave in Siberia." *Nature* 468 (2010): 1053–1060.

Reich, David, Nick Patterson, Desmond Campbell, Arti Tandon, Stéphane Mazieres, Nicolas Ray, Maria V. Parra, Winston Rojas, Constanza Duque, Natalia Mesa, et al. "Reconstructing Native American Population History." *Nature* 488 (2012): 370–374.

Reich, David, Nick Patterson, Martin Kircher, Frederick Delfin, Madhusudan R. Nandineni, Irina Pugach, Albert Min-shan Ko, Ying-chin Ko, Timothy A. Jinam, Maude E. Phipps, et al. "Denisova Admixture and the First Modern Human Dispersals into Southeast Asia and Oceania." *American Journal of Human Genetics* 89 (2011): 516–528.

Reich, David, Kumarasamy Thangaraj, Nick Patterson, Alkes L. Price, and Lalji Singh. "Reconstructing Indian Population History." *Nature* 461 (2009): 489–494.

Relethford, John H. "Genetics and Modern Human Origins." *Evolutionary Anthropology* 4 (1995): 53–63.

Relethford, John H., and Henry C. Harpending. "Craniometric Variation, Genetic Theory, and Modern Human Origins." *American Journal of Physical Anthropology* 95 (1994): 249–270.

Renfrew, Colin. "Archaeogenetics—Towards a 'New Synthesis'?" *Current Biology* 20 (2010): R162–R165.

Reuleaux, Franz. *The Kinematics of Machinery: Outlines of a Theory of Machines.* Translated by Alexander B. W. Kennedy. New York: Dover, 1963.

Reynolds, Tim, and Graeme Barker. "Reconstructing Late Pleistocene Climates, Landscapes, and Human Activities in Northern Borneo from Excavations in the Niah Caves." In *Emergence and Diversity of Modern Human Behavior in Paleolithic Asia*, edited by Yousuke Kaifu, Masami Izuho, Ted Goebel, Hiroyuki Sato, and Akira Ono, 140–157. College Station: Texas A&M University Press, 2015.

Rich, Elaine. *Automata, Computability and Complexity: Theory and Applications.* Upper Saddle River, N.J.: Pearson Prentice Hall, 2008.

Richards, Michael P., Paul B. Pettitt, Mary C. Stiner, and Erik Trinkaus. "Stable Isotope Evidence for Increasing Dietary Breadth in the European Mid-Upper Paleolithic." *Proceedings of the National Academy of Sciences* 98 (2001): 6528–6532.

Richter, Daniel, Rainer Grün, Renaud Joannes-Boyau, Teresa E. Steele, Fethi Amani, Mathieu Rué, Paul Fernandes, Jean-Paul Raynal, Denis Geraads, Abdelouahed Ben-Ncer, et al. "The Age of the Hominin Fossils from Jebel Irhoud, Morocco, and the Origins of the Middle Stone Age." *Nature* 546 (2017): 293–296.

Richter, D., G. Tostevin, and P. Škrdla. "Bohunician Technology and Thermoluminescence Dating of the Type Locality of Brno-Bohunice (Czech Republic)." *Journal of Human Evolution* 55 (2008): 871–885.

Ridley, Matt. *The Red Queen: Sex and the Evolution of Human Nature.* New York: HarperCollins, 1993.

Rieux, Adrien, Anders Eriksson, Mingkun Li, Benjamin Sobkowiak, Lucy A. Weinert, Vera Warmuth, Andres Ruiz-Linares, Andrea Manica, and François Balloux. "Improved Calibration of the Human Mitochondrial Clock Using Ancient Genomes." *Molecular Biology and Evolution* 31 (2014): 2780–2792.

Rightmire, G. Philip. *The Evolution of* Homo erectus: *Comparative Anatomical Studies of an Extinct Human Species.* Cambridge: Cambridge University Press, 1990.

Rightmire, G. Philip. "Human Evolution in the Middle Pleistocene: The Role of *Homo heidelbergensis*." *Evolutionary Anthropology* 6 (1998): 218–227.

Roberts, Alice, *The Incredible Human Journey: The Story of How We Colonised the Planet.* London: Bloomsbury, 2009.

Roberts, Richard G., Timothy F. Flannery, Linda K. Ayliffe, Hiroyuki Yoshida, Jon M. Olley, Gavin J. Prideaux, Geoff M. Laslett, Alexander Baynes, M. A. Smith, Rhys Jones, et al. "New Ages for the Last Australian Megafauna: Continent-Wide Extinction About 46,000 Years Ago." *Science* 292 (2001): 1888–1892.

Roberts, Richard G., Rhys Jones, and M. A. Smith. "Thermoluminescence Dating of a 50,000-Year-Old Human Occupation Site in Northern Australia." *Nature* 345 (1990): 94–97.

Roberts, Richard G., Rhys Jones, Nigel A. Spooner, M. J. Head, Andrew S. Murray, and M. A. Smith. "The Human Colonisation of Australia: Optical Dates of 53,000 and 60,000 Years Bracket Human Arrival at Deaf Adder Gorge, Northern Territory." *Quaternary Geochronology* 13 (1994): 575–583.

Rogachev, A. N. *Mnogosloinye Stoyanki Kostenkovsko-Borshevskogo Raiona na Donu i Problema Razvitiya Kul'tury v Epokhy Verkhnego Paleolita na Russkoi Ravnine.* Materialy i Issledovaniya po Arkheologii SSSR 59. Moscow: Nauka, 1957.

Rogers, Jeffrey, and Richard A. Gibbs. "Comparative Primate Genomics: Emerging Patterns of Genome Content and Dynamics." *Nature Reviews Genetics* 15 (2014): 347–359.

Rose, Jeffrey I. "New Light on Human Prehistory in the Arabo-Persian Gulf Oasis." *Current Anthropology* 51 (2010): 849–883.

Rose, Jeffrey I., Vitaly I. Usik, Anthony E. Marks, Yamandu H. Hilbert, Christopher S. Galletti, Ash Parton, Jean Marie Geiling, Viktor Černý, Mike W. Morley, and Richard G. Roberts. "The Nubian Complex of Dhofar, Oman: An African Middle Stone Age Industry in Southern Arabia." *PLoS ONE* 6 (2011): e28239.

Rosenberg, Noah A., Jonathan K. Pritchard, James L. Weber, Howard M. Cann, Kenneth K. Kidd, Lev A. Zhivotovsky, and Marcus W. Feldman. "Genetic Structure of Human Populations." *Science* 298 (2002): 2381–2385.

Rule, Susan, Barry W. Brook, Simon G. Haberle, Chris S. M. Turney, A. Peter Kershaw, and Christopher N. Johnson. "The Aftermath of Megafaunal Extinction: Ecosystem Transformation in Pleistocene Australia." *Science* 335 (2012): 1483–1486.

Rybin, Evgeny P. "Middle and Upper Paleolithic Interactions and the Emergence of 'Modern Behavior' in Southern Siberia and Mongolia." In *Emergence and Diversity of Modern Human Behavior in Paleolithic Asia,* edited by Yousuke Kaifu, Masami Izuho, Ted Goebel, Hiroyuki Sato, and Akira Ono, 470–489. College Station: Texas A&M University Press, 2015.

Rymer, Russ. *Genie: A Scientific Tragedy.* New York: Harper Perennial, 1993.

Sahlins, Marshall. *Culture and Practical Reason.* Chicago: University of Chicago Press, 1976.

Sahoo, Sanghamitra, Anamika Singh, G. Himabindu, Jheelam Banerjee, T. Sitalaximi, Sonali Gaikwad, R. Trivedi, Phillip Endicott, Toomas Kivisild, Mait Metspalu, et al. "A Prehistory of Indian Y Chromosomes: Evaluating Demic Diffusion Scenarios." *Proceedings of the National Academy of Sciences* 103 (2006): 843–848.

Sali, S. A. *The Upper Palaeolithic and Mesolithic Cultures of Maharashtra.* Pune: Deccan College Post Graduate and Research Institute, 1989.

Sanchez, Guadalupe, Vance T. Holliday, Edmund P. Gaines, Joaquín Arroyo-Cabrales, Natalia Martínez-Tagüeña, Andrew Kowler, Todd Lange, Gregory W. L. Hodgins, Susan M. Mentzer, and Ismael Sanchez-Morales. "Human (Clovis)–Gomphothere (*Cuvieronius* sp.) Association ~13,390 Calibrated yBP in Sonora, Mexico." *Proceedings of the National Academy of Sciences* 111 (2014): 10972–10977.

Sánchez-Quinto, Federico, and Carles Lalueza-Fox. "Almost 20 Years of Neanderthal Pal-aeogenetics: Adaptation, Admixture, Diversity, Demography, and Extinction." *Philosophical Transactions of the Royal Society B: Biological Sciences* 370 (2015): 20130374.

Sandgathe, Dennis M., Harrold L. Dibble, Paul Goldberg, Shannon P. McPherron, Alain Turq, Laura Niven, and Jamie Hodgkins. "On the Role of Fire in Neandertal Adaptations in Western Europe: Evidence from Pech de l'Aze IV and Roc de Marsal, France." *PaleoAnthropology* 2011 (2011): 216–242.

Sandweiss, Daniel H., Heather McInnis, Richard L. Burger, Asunción Cano, Bernadino Ojeda, Rolando Paredes, María del Carmen Sandweiss, and Michael D. Glascock. "Quebrada Jaguay: Early South American Maritime Adaptations." *Science* 281 (1998): 1830–1832.

Sankararaman, Sriram, Swapan Mallick, Nick Patterson, and David Reich. "The Combined Landscape of Denisovan and Neanderthal Ancestry in Present-Day Humans." *Current Biology* 26 (2016): 1241–1247.

Sano, Katsuhiro. "Evidence for the Use of Bow-and-Arrow Technology by the First Modern Humans in the Japanese Islands." *Journal of Archaeological Science: Reports* 10 (2016): 130–141.

Sarich, Vincent M., and Allan C. Wilson. "Immunological Time Scale for Hominid Evolution." *Science* 158 (1967): 1200–1203.

Saunders, Jeffrey J. "Processing Marks on Remains of *Mammuthus columbi* from the Dent Site, Colorado, in Light of Those from Clovis, New Mexico." In *Frontiers in Colorado Paleoindian Archaeology: From the Dent Site to the Rocky Mountains*, edited by Robert H. Brunswig and Bonnie L. Pitblado, 155–184. Boulder: University Press of Colorado, 2007.

Savage-Rumbaugh, Sue, and Roger Lewin. *Kanzi: The Ape at the Brink of the Human Mind.* New York: Wiley, 1994.

Savage-Rumbaugh, Sue, Stuart G. Shanker, and Talbot J. Taylor. *Apes, Language, and the Human Mind.* New York: Oxford University Press, 1998.

Sawyer, Susanna, Gabriel Renauda, Bence Viola, Jean-Jacques Hublin, Marie-Theres Gansauge, Michael V. Shunkov, Anatoly P. Derevianko, Kay Prüfer, Janet Kelso, and Svante Pääbo. "Nuclear and Mitochondrial DNA Sequences from Two Denisovan Individuals." *Proceedings of the National Academy of Sciences* 112 (2015): 15696–15700.

Schick, Kathy, and Nicholas Toth. *Making Silent Stones Speak: Human Evolution and the Dawn of Technology.* New York: Simon and Schuster, 1993.

Schick, Kathy, and Nicholas Toth. "An Overview of the Oldowan Industrial Complex: The Sites and the Nature of Their Evidence." In *The Oldowan: Case Studies into the Earliest Stone Age*, edited by Nicholas Toth and Kathy Schick, 3–42. Gosport, Ind.: Stone Age Institute Press, 2006.

Schlanger, Nathan. "The Chaîne Opératoire." In *Archaeology: The Key Concepts*, edited by Colin Renfrew and Paul Bahn, 25–31. Abington: Routledge, 2005.

Schlanger, Nathan. "Understanding Levallois: Lithic Technology and Cognitive Archaeology." *Cambridge Archaeological Journal* 6 (1996): 231–254

Schneider, Thomas D. "Evolution of Biological Information." *Nucleic Acids Research* 28 (2000): 2794–2799.

Schroeder, K. B., T. G. Schurr, J. C. Long, N. A. Rosenberg, M. H. Crawford, L. A. Tarskaia, L. P. Osipova, S. I. Zhadanov, and D. G. Smith. "A Private Allele Ubiquitous in the Americas." *Biology Letters* 3 (2007): 218–223.

Scott, G. Richard, and Joel D. Irish. *Tooth Crown and Root Morphology: The Arizona State University Dental Anthropology System.* Cambridge: Cambridge University Press, 2017.

Scott, G. Richard, Kirk Schmitz, Kelly N. Heim, Kathleen S. Paul, Roman Schomberg, and Marin A. Pilloud. "Sinodonty, Sundadonty, and the Beringian Standstill Model: Issues of Timing and Migrations into the New World." *Quaternary International* (2016): in press.

Scott, G. Richard, and Christy G. Turner II. *The Anthropology of Modern Human Teeth: Dental Morphology and Its Variation in Recent Human Populations.* Cambridge: Cambridge University Press, 1997.

Seeley, Thomas D. *Honeybee Democracy.* Princeton, N.J.: Princeton University Press, 2010.

Seeley, Thomas D. *Honeybee Ecology: A Study of Adaptation in Social Life.* Princeton, N.J.: Princeton University Press, 1985.

Seguin-Orlando, Andaine, Thorfinn S. Korneliussen, Martin Sikora, Anna-Sapfo Malaspinas, Andrea Manica, Ida Moltke, Anders Albrechtsen, Amy Ko, Ashot Margaryan, Vyacheslav Moiseyev, et al. "Genomic Structure in Europeans Dating Back at Least 36,200 Years." *Science* 346 (2014): 1113–1118.

Selfridge, Oliver G. "Pandemonium: A Paradigm for Learning." In *Mechanisation of Thought Processes: Proceedings of a Symposium Held at the National Physical Laboratory on 24th, 25th, 26th and 27th November 1958,* 511–530. London: Her Majesty's Stationary Office, 1959.

Serre, David, André Langaney, Mario Chech, Maria Teschler-Nicola, Maja Paunovic, Philippe Mennecier, Michael Hofreiter, Göran Possnert, Svante Pääbo. "No Evidence of Neandertal mtDNA Contribution to Early Modern Humans." *PLoS Biology* 2 (2004): 0313–0317.

Seyfarth, Robert, Dorothy Cheney, and Peter Marler. "Vervet Monkey Alarm Calls: Semantic Communication in a Free-Ranging Primate." *Animal Behaviour* 28 (1980): 1070–1094.

Shang, Hong, Haowen Tong, Shuangquan Zhang, Foyou Chen, and Erik Trinkaus. "An Early Modern Human from Tianyuan Cave, Zhoukoudian, China." *Proceedings of the National Academy of Sciences* 104 (2007): 6573–6578.

Shang, Hong, and Erik Trinkaus. *The Early Modern Human from Tianyuan Cave, China.* College Station: Texas A&M University Press, 2010.

Shannon, Claude E. "A Mathematical Theory of Communication." *Bell System Technical Journal* July 1948, 379–423.

Shannon, Claude E., and Warren Weaver. *The Mathematical Theory of Communication.* Urbana: University of Illinois Press, 1949.

Shapiro, Harry Lionel. *Peking Man.* New York: Simon and Schuster, 1974.

Shapiro, James A. "Genome System Architecture and Natural Genetic Engineering." In *Evolution as Computation: DIMACS Workshop, Princeton, January 1999,* edited by Laura F. Landweber and Eric Winfree, 1–14. Berlin: Springer, 2002.

Shea, John J. "*Homo sapiens* Is as *Homo sapiens* Was." *Current Anthropology* 52 (2011): 1–15.

Shea, John J., and Matthew L. Sisk. "Complex Projectile Technology and *Homo sapiens* Dispersal into Western Eurasia." *PaleoAnthropology* 2010 (2010): 100–122.

Shen, Guanjun, Wei Wang, Qian Wang, Jianxin Zhao, Kenneth Collerson, Chunlin Zhou, and Phillip V. Tobias. "U-Series Dating of Liujiang Hominid Site in Guangxi, Southern China." *Journal of Human Evolution* 43 (2002): 817–829.

Shen, Guanjun, Xianzhu Wu, Qian Wang, Hua Tu, Yue-xing Feng, and Jian-xin Zhao. "Mass Spectrometric U-Series Dating of Huanglong Cave in Hubei Province, Central China: Evidence for Early Presence of Modern Humans in Eastern Asia." *Journal of Human Evolution* 65 (2013): 162–167.

Shipman, Pat. *The Invaders: How Humans and Their Dogs Drove Neanderthals to Extinction.* Cambridge, Mass.: Harvard University Press, 2015.

Shulha, Hennady P., Jessica L. Crisci, Denis Reshetov, Jogender S. Tushir, Iris Cheung, Rahul Bharadwaj, Hsin-jung Chou, Isaac B. Houston, Cyril J. Peter, Amanda C. Mitchell, et al. "Human-Specific Histone Methylation Signatures at Transcription Start Sites in Prefrontal Neurons." *PLoS Biology* 10 (2012): e1001427.

Sicoli, Mark A. and Gary Holton. "Linguistic Phylogenies Support Back-Migration from Beringia to Asia." *PLoS ONE* 9 (2014): e91722.

Simon, Herbert A. "The Architecture of Complexity." *Proceedings of the American Philosophical Society* 106 (1962): 467–482.

Simonti, Corinne N., Benjamin Vernot, Lisa Bastarache, Erwin Bottinger, David S. Carrell, Rex L. Chisholm, David R. Crosslin, Scott J. Hebbring, Gail P. Jarvik, Iftikhar J. Kullo, et al. "The Phenotypic Legacy of Admixture Between Modern Humans and Neandertals." *Science* 351 (2016): 737–741.

Singh, Simon. *The Code Book: The Science of Secrecy from Ancient Egypt to Quantum Cryptography.* New York: Anchor Books, 1999.

Sinitsyn, A. A. "Nizhnie Kul'turnye Sloi Kostenok 14 (Markina Gora) (Raskopki 1998–2001 gg.)." In *Kostenki v Kontekste Paleolita Evrazii*, edited by A. A. Sinitsyn, V. Ya. Sergin, and J. F. Hoffecker, 219–236. St. Petersburg: Russian Academy of Sciences, 2002.

Skoglund, Pontus, Erik Ersmark, Eleftheria Palkopoulou, and Love Dalén. "Ancient Wolf Genome Reveals an Early Divergence of Domestic Dog Ancestors and Admixture into High-Latitude Breeds." *Cell* 25 (2015): 1515–1519.

Slimak, Ludovic, John Inge Svendsen, Jan Mangerud, Hughes Plisson, Herbjørn Presthus Heggen, Alexis Brugère, and Pavel Yurievich Pavlov. "Late Mousterian Persistence near the Arctic Circle." *Science* 332 (2011): 841–845.

Smith, Colin I., Andrew T. Chamberlain, Michael S. Riley, Chris Stringer, and Matthew J. Collins. "The Thermal History of Human Fossils and the Likelihood of Successful DNA Amplification." *Journal of Human Evolution* 45 (2003): 203–217.

Smith, David Glenn, Ripan S. Malhi, Jason A. Eshleman, Frederika A. Kaestle, and Brian M. Kemp. "Mitochondrial DNA Haplogroups of Paleoamericans in North America." In *Paleoamerican Origins: Beyond Clovis*, edited by Robson Bonnichsen, Bradley T. Lepper, Dennis Stanford, and Michael R. Waters, 243–254. College Station: Texas A&M University Press, 2005.

Smith, M. A. "Characterizing Late Pleistocene and Holocene Stone Artefact Assemblages from Puritjarra Rock Shelter: A Long Sequence from the Australian Desert." *Records of the Australian Museum* 58 (2006): 371–410.

Smith, Tanya M., Paul Tafforeau, Donald J. Reid, Rainer Grün, Stephen Eggins, Mohamed Boutakiourt, and Jean-Jacques Hublin. "Earliest Evidence of Modern Human Life History in North African Early *Homo sapiens.*" *Proceedings of the National Academy of Sciences* 104 (2007): 6128–6133.

Smith, Tanya M., Paul Tafforeauc, Donald J. Reid, Joane Pouech, Vincent Lazzari, John P. Zermeno, Debbie Guatelli-Steinberg, Anthony J. Olejniczak, Almut Hoffman,

Jakov Radovčić, et al. "Dental Evidence for Ontogenetic Differences Between Modern Humans and Neanderthals." *Proceedings of the National Academy of Sciences* 107 (2010): 20923–20928.

Soares, Pedro, Alessandro Achilli, Omella Semino, William Davies, Vincent Macaulay, Hans-Jürgen Bandelt, Antonio Torroni, and Martin B. Richards. "The Archaeogenetics of Europe." *Current Biology* 20 (2010): R174–R183.

Sollas, W. J. *Ancient Hunters and Their Modern Representatives*. London: Macmillan, 1911.

Soriano, Sylvain, Paola Villa, and Lyn Wadley. "Blade Technology and Tool Forms in the Middle Stone Age of South Africa: The Howiesons Poort and Post-Howiesons Poort at Rose Cottage Cave." *Journal of Archaeological Science* 34 (2007): 681–703.

Sponheimer, Matt, and Darna L. Dufour. "Increased Dietary Breadth in Early Hominin Evolution: Revisiting Arguments and Evidence with a Focus on Biogeochemical Contributions." In *The Evolution of Hominid Diets: Integrating Approaches to the Study of Palaeolithic Subsistence*, edited by Jean-Jacques Hublin and Michael P. Richards, 229–240. Berlin: Springer, 2009.

Sponheimer, Matt, and Julia A. Lee-Thorp. "Isotopic Evidence for the Diet of an Early Hominid *Australopithecus africanus*." *Science* 283 (1999): 368–370.

Stanford, Dennis J., and Bruce A. Bradley. *Across Atlantic Ice: The Origin of America's Clovis Culture*. Berkeley: University of California Press, 2012.

Steele, James, and Gustavo Politis. "AMS ^{14}C Dating of Early Human Occupation of Southern South America." *Journal of Archaeological Science* 36 (2009): 419–429.

Stegman, Ulrich E. "The Arbitrariness of the Genetic Code." *Biology and Philosophy* 19 (2004): 205–222.

Steiper, M. E., and N. M. Young. "Timing Primate Evolution: Lessons from the Discordance between Molecular and Paleontological Estimates." *Evolutionary Anthropology* 17 (2008): 179–188.

Stepanchuk, V. N. "The Archaic to True Upper Paleolithic Interface: The Case of Mira in the Middle Dnieper Area." *Eurasian Prehistory* 3 (2005): 23–41.

Sterelny, Kim, Kelly C. Smith, and Michael Dickison. "The Extended Replicator." *Biology and Philosophy* 19 (1996): 205–222.

Stevens, Nancy J., Erik R. Seiffert, Patrick M. O'Connor, Eric M. Roberts, Mark D. Schmitz, Cornelia Krause, Eric Gorscak, Sifa Ngasala, Tobin L. Hieronymus, and Joseph Temu. "Palaeontological Evidence for an Oligocene Divergence Between Old World Monkeys and Apes." *Nature* 497 (2013): 611–614.

Stewart, Alexander J., and Joshua B. Plotkin, "Collapse of Cooperation in Evolving Games," *Proceedings of the National Academy of Sciences* 111 (2014): 17558–17563.

Stewart, J. R., and C. B. Stringer. "Human Evolution Out of Africa: The Role of Refugia and Climate Change." *Science* 335 (2012): 1317–1321.

Stiner, Mary C., Natalie D. Munro, Todd A. Surovell, Eitan Tchernov, and Ofer Bar-Yosef. "Paleolithic Population Growth Pulses Evidenced by Small Animal Exploitation." *Science* 283 (1999): 190–194.

Stock, Jat T. "The Skeletal Phenotype of 'Negritos' from the Andaman Islands and Philippines Relative to Global Variation Among Hunter-Gatherers." *Human Biology* 85 (2013): 67–94.

Stocking, George W., Jr. *Race, Culture, and Evolution: Essays in the History of Anthropology*. Chicago: University of Chicago Press, 1968.

Stokes, Chris R., Lev Tarasov, and Arthur S. Dyke. "Dynamics of the North American Ice Sheet Complex During Its Inception and Build-up to the Last Glacial Maximum." *Quaternary Science Reviews* 50 (2012): 86–104.

Stone, Anne C., and Mark Stoneking. "Genetic Analyses of an 8000-Year-Old Native American Skeleton." *Ancient Biomolecules* 1 (1996): 83–87.

Stoneking, Mark, "Genetic Evidence Concerning the Origins and Dispersals of Modern Humans." In *Causes and Consequences of Human Migration: An Evolutionary Perspective*, edited by Michael H. Crawford and Benjamin C. Campbell, 11–20. Cambridge: Cambridge University Press, 2012.

Streenathan, M., V. R. Rao, and R. G. Bednarik. "Palaeolithic Cognitive Inheritance in Aesthetic Behavior of the Jarawas of the Andaman Islands." *Anthropos* 103 (2008): 367–392.

Strieder, Georg F. *Principles of Brain Evolution.* Sunderland, Mass.: Sinauer, 2005.

Stringer, Chris. *Lone Survivors: How We Came to Be the Only Humans on Earth.* New York: Times Books, 2012.

Stringer, Chris. "The Origin of Early Modern Humans: A Comparison of the European and Non-European Evidence." In *The Human Revolution: Behavioural and Biological Perspectives on the Origins of Modern Humans*, edited by Paul Mellars and Chris Stringer, 232–244. Princeton, N.J.: Princeton University Press, 1989.

Stringer, Chris. "The Status of *Homo heidelbergensis* (Schoetensack 1908)." *Evolutionary Anthropology* 21 (2012): 101–107.

Stringer, Chris, and Peter Andrews. *The Complete World of Human Evolution.* 2nd ed. London: Thames and Hudson, 2011.

Stringer, Christopher, and Clive Gamble. *In Search of the Neanderthals: Solving the Puzzle of Human Origins.* New York: Thames and Hudson, 1993.

Stringer, Christopher, and Robin McKie. *African Exodus: The Origins of Modern Humanity.* New York: Holt, 1996.

Suleimanov, R. Kh. *Statisticheskoe Izuchenie Kul'tury Grota Obi-Rakhmat.* Tashkent: Fan, 1972.

Summerhayes, Glenn R., Matthew Leavesley, Andrew Fairbairn, Herman Mandui, Judith Field, Anne Ford, and Richard Fullagar. "Human Adaptation and Plant Use in Highland New Guinea 49,000 to 44,000 Years Ago." *Science* 330 (2010): 78–81.

Susman, Randall L. "Hand of *Paranthropus robustus* from Member 1, Swartkrans: Fossil Evidence for Tool Behavior." *Science* 240 (1988): 781–784.

Svoboda, Jiří A. "On Modern Human Penetration into Northern Eurasia: The Multiple Advances Hypothesis." In *Rethinking the Human Revolution: New Behavioural and Biological Perspectives on the Origin and Dispersal of Modern Humans*, edited by Paul Mellars, Katie Boyle, Ofer Bar-Yosef, and Chris Stringer, 329–339. Cambridge: McDonald Institute for Archaeological Research, Cambridge University, 2007.

Svoboda, Jiří A., and Ofer Bar-Yosef, eds. *Stránská skála: Origins of the Upper Paleolithic in the Brno Basin, Moravia, Czech Republic.* Cambridge, Mass.: Peabody Museum of Archaeology and Ethnology, 2003.

Swanson, Larry W. *Brain Architecture: Understanding the Basic Plan.*2nd ed. New York: Oxford University Press, 2012.

Swisher, Carl C., III, Garniss H. Curtis, and Roger Lewin. *Java Man: How Two Geologists' Dramatic Discoveries Changed Our Understanding of the Evolutionary Path to Modern Humans.* New York: Scribner, 2000.

Swisher, C. C., III, W. J. Rink, S. C. Antón, H. P. Schwarcz, G. H. Curtis, and A. Suprijo Widiasmoro. "Latest *Homo erectus* of Java: Potential Contemporaneity with *Homo sapiens* in Southeast Asia." *Science* 274 (1996): 1870–1874.

Szathmary, Emöke J. E. "Genetics of Aboriginal North Americans." *Evolutionary Anthropology* 1 (1993): 202–220.

Szathmary, Emöke J. E., and Nancy S. Ossenberg. "Are the Biological Differences Between North American Indians and Eskimos Truly Profound?" *Current Anthropology* 19 (1978): 673–701.

Szathmáry, Eörs. "Toward Major Evolutionary Transitions Theory 2.0." *Proceedings of the National Academy of Sciences* 112 (2015): 10104–10111.

Szathmáry, Eörs, and Chrisantha Fernando. "Concluding Remarks." In *The Major Transitions in Evolution Revisited*, edited by Brett Calcott and Kim Sterelny, 301–310. Cambridge, Mass.: MIT Press, 2011.

Szathmáry, Eörs, and John Maynard Smith. "The Major Evolutionary Transitions." *Nature* 374 (1995): 227–232.

Szilard, Leo. "Über die Entropieverminderung in Einem Thermodynamischen System bei Eingriffen Intelligenter Wesen." *Zeitschrift für Physik* 53 (1929): 840–856.

Tackney Justin C., Ben A. Potter, Jennifer Raff, Michael Powers, W. Scott Watkins, Derek Warner, Joshua D. Reuther, Joel D. Irish, and Dennis H. O'Rourke. "Two Contemporaneous Mitogenomes from Terminal Pleistocene Burials in Eastern Beringia." *Proceedings of the National Academy of Sciences* 112 (2015): 13833–13838.

Tamm, Erika, Toomas Kivisild, Maere Reidla, Mait Metspalu, David Glenn Smith, Connie J. Mulligan, Claudio M. Bravi, Olga Rickards, Cristina Martinez-Labarga, Elsa K. Khusnutdinova, et al. "Beringian Standstill and Spread of Native American Founders." *PLoS ONE* 9 (2007): e829.

Tattersall, Ian. "Human Origins: Out of Africa." *Proceedings of the National Academy of Sciences* 106 (2009): 16018–16021.

Tattersall, Ian, and Rob DeSalle. *Race? Debunking a Scientific Myth*. College Station: Texas A&M University Press, 2011.

Tattersall, Ian, and Jeffrey H. Schwartz. "The Morphological Distinctiveness of *Homo sapiens* and Its Recognition in the Fossil Record: Clarifying the Problem." *Evolutionary Anthropology* 17 (2008): 49–54.

Templeton, Alan R. "The 'Eve' Hypothesis: A Genetic Critique and Re-analysis." *American Anthropologist* 95 (1993): 51–72.

Templeton, Alan R. "Human Origins and the Analysis of Mitochondrial DNA Sequences." *Science* 255 (1992): 737.

Templeton, Alan R. "Human Races: A Genetic and Evolutionary Perspective." *American Anthropologist* 100 (1998): 632–650.

Texier, Pierre-Jean, Guillaume Porraz, John Parkington, Jean-Philippe Rigaud, Cedric Poggenpoel, Christopher Miller, Chantal Tribolo, Caroline Cartwright, Aude Coudenneau, Richard Klein, et al. "A Howiesons Poort Tradition of Engraving Ostrich Eggshell Containers Dated to 60,000 Years Ago at Diepkloof Rock Shelter, South Africa." *Proceedings of the National Academy of Sciences* 107 (2010): 6180–6185.

Thalmann, O., B. Shapiro, P. Cui, V. J. Schuenemann, S. K. Sawyer, D. L. Greenfield, M. B. Germonpré, M. V. Sablin, F. López-Giráldez, X. Domingo-Roura, et al. "Complete Mitochondrial Genomes of Ancient Canids Suggest a European Origin of Domestic Dogs." *Science* 342 (2013): 871–874.

Thayer, Zaneta M., and Seth D. Dobson. "Sexual Dimorphism in Chin Shape: Implications for Adaptive Hypotheses." *American Journal of Physical Anthropology* 143 (2010): 417–425.

Theodosopoulos, Patricia K., and Theodore V. Theodosopoulos. "Evolution at the Edge of Chaos: A Paradigm for the Maturation of the Humoral Immune Response." In *Evolution as Computation: DIMACS Workshop, Princeton, January 1999*, edited by Laura F. Landweber and Eric Winfree, 41–66. Berlin: Springer, 2002.

Théry-Parisot, Isabelle, Sadrine Costamagno, Jean-Philip Brugal, Philippe Fosse, and Raphaële Guilbert. "The Use of Bone as Fuel During the Palaeolithic, Experimental Study of Bone Combustible Properties." In *The Zooarchaeology of Fats, Oils, Milk and Dairying*, edited by J. Mulville and A. K. Outram, 50–59. Oxford: Oxbow Books, 2005.

Tishkoff, Sarah A., Floyd A. Reed, Françoise R. Friedlaender, Christopher Ehret, Alessia Ranciaro, Alain Froment, Jibril B. Hirbo, Agnes A. Awomoyi, Jean-Marie Bodo, Ogobara Doumbo, et al. "The Genetic Structure and History of Africans and African Americans." *Science* 324 (2009): 1035–1043.

Tobias, Phillip V. "The Brain of *Homo habilis*: A New Level of Organization in Cerebral Evolution." *Journal of Human Evolution* 16 (1987): 741–762.

Tomasello, Michael, and Josep Call. *Primate Cognition*. New York: Oxford University Press, 1997.

Torrence, Robin. "Time Budgeting and Hunter-Gatherer Technology." In *Hunter-Gatherer Economy in Prehistory: A European Perspective*, edited by Geoff Bailey, 11–22. Cambridge: Cambridge University Press, 1983.

Toth, Nicholas. "The Oldowan Reassessed: A Close Look at Early Stone Artifacts." *Journal of Archaeological Science* 12 (1985): 101–120.

Toth, N., K. Schick, E. S. Savage-Rumbaugh, R. A. Sevick, and D. M. Rumbaugh. "Pan the Toolmaker: Investigations into the Stone-Tool Making and Tool-Using Capabilities of a Bonobo (*Pan paniscus*)." *Journal of Archaeological Science* 20 (1993): 81–91.

Traversi, R., S. Becagli, S. Poluianov, M. Severi, S. K. Solanki, I. G. Usoskin, and R. Udisti. "The Laschamp Geomagnetic Excursion Featured in Nitrate Record from EPICA-Dome C Ice Core." *Scientific Reports* 6 (2016): 20235.

Trigger, Bruce G. *A History of Archaeological Thought*. Cambridge: Cambridge University Press, 1989.

Trinkaus, Erik. "Early Modern Humans." *Annual Review of Anthropology* 34 (2005): 207–230.

Trinkaus, Erik, Oana Moldovan, Ştefan Milota, Adrian Bîlgăr, Laurenţiu Sarcina, Sheela Athreya, Shara E. Bailey, Ricardo Rodrigo, Gherase Mircea, Thomas Higham, et al. "An Early Modern Human from the Peştera cu Oase, Romania." *Proceedings of the National Academy of Sciences* 100 (2003): 11231–11236.

Trinkaus, Erik, and Pat Shipman. *The Neandertals: Of Skeletons, Scientists, and Scandal*. New York: Vintage Books, 1994.

Trivers, Robert L. "The Evolution of Reciprocal Altruism." *Quarterly Review of Biology* 46 (1971): 35–57.

Tsanova, Tsenka, and Jean-Guillaume Bordes. "Contribution au débat sur l'origine de l'Aurignacien: Principaux résultats d'une étude technologique de l'industrie lithique de la couche 11 de Bacho Kiro." In *The Humanized Mineral World: Towards Social and Symbolic Evaluation of Prehistoric Technologies in Southeastern Europe*, edited by Tsoni Tsonev and Emmanuela Montagnari Kokelj, 41–50. ERAUL 103. Liege: University of Liege, 2003.

Turing, Alan M. "On Computable Numbers, with an Application to the *Entscheidungsproblem.*" *Proceedings of the London Mathematical Society* 42 (1936): 230–265.

Turner, Christy G., II. "Advances in the Dental Search for Native American Origins." *Acta Anthropogenetica* 8 (1984): 23–78.

Turner, Christy G., II. "Dental Evidence for the Peopling of the Americas." *National Geographic Society Research Reports* 19 (1985): 573–596.

Turner, Christy G., II. "Dental Evidence on the Origins of the Ainu and Japanese." *Science* 193 (1976): 911–913.

Turner, Christy G., II. "The First Americans: The Dental Evidence." *National Geographic Research* 2 (1986): 37–46.

Turner, Christy G., II. "Major Features of Sundadonty and Sinodonty, Including Suggestions About East Asian Microevolution, Population History, and Late Pleistocene Relationships with Australian Aboriginals." *American Journal of Physical Anthropology* 82 (1990): 295–317.

Turner, Christy G., II. "Teeth and Prehistory in Asia." *Scientific American*, February 1989, 88–96.

Turner, Christy G., II, Christian R. Nichol, and G. Richard Scott. "Scoring Procedures for Key Morphological Traits of the Permanent Dentition: The Arizona State University Dental Anthropology System." In *Advances in Dental Anthropology*, edited by Mark A. Kelley and Clark Spencer Larsen, 13–31. New York: Wiley-Liss, 1990.

Turney, Chris S. M., Michael I. Bird, L. Keith Fifield, Richard G. Roberts, Mike Smith, Charles E. Dortch, Rainer Grün, Ewan Lawson, Linda K. Ayliffe, Gifford H. Miller, et al. "Early Occupation at Devil's Lair, Southwestern Australia 50,000 Years Ago." *Quaternary Research* 55 (2001): 3–13.

Tylor, E. B. *Primitive Culture.* London: Murray, 1871.

Underhill, Peter A., Giuseppe Passarino, Alice A. Lin, Peidong Shen, Marta Mirazón Lahr, Robert A. Foley, Peter J. Oefner, and Luigi L. Cavalli-Sforza. "The Phylogeography of Y Chromosome Binary Haplotypes and the Origins of Modern Human Populations." *Annals of Human Genetics* 65 (2001): 43–62.

Underhill, Peter A., Peidong Shen, Alice A. Lin, Li Jin, Giuseppe Passarino, Wei H. Yang, Erin Kauffman, Batsheva Bonné-Tamir, Jaume Bertranpetit, Paolo Francalacci, et al. "Y Chromosome Sequence Variation and the History of Human Populations." *Nature Genetics* 26 (2000): 358–361.

Urrego, D. H., M. F. Sánchez Goñi, A.-L. Daniau, S. Lechevrel, and V. Hanquiez. "Increased Aridity in Southwestern Africa During the Warmest Periods of the Last Interglacial." *Climate of the Past* 11 (2015): 1417–1431.

Val, Aurore, Paloma de la Peña, and Lyn Wadley. "Direct Evidence for Human Exploitation of Birds in the Middle Stone Age of South Africa: The Example of Sibudu Cave, KwaZulu-Natal." *Journal of Human Evolution* 99 (2016): 1–17.

Vandermeersch, Bernard. "The First *Homo sapiens sapiens* in the Near East." In *The Transition from Lower to Middle Palaeolithic and the Origin of Modern Man*, edited by Avraham Ronen, 297–299. BAR International Series S151. Oxford: British Archaeological Reports, 1982.

van Geel, Bas, André Aptroot, Claudia Baittinger, Hilary H. Birks, Ian D. Bull, Hugh B. Cross, Richard P. Evershed, Barbara Gravendeel, Erwin J. O. Kompanje, Peter

Kuperus, et al. "The Ecological Implications of a Yakutian Mammoth's Last Meal." *Quaternary Research* 69 (2008): 361–376.

Vanhaeran, Marian, Francesco d'Errico, Chris Stringer, Sarah L. James, Jonathan A. Todd, and Henk K. Mienis. "Middle Paleolithic Shell Beads in Israel and Algeria." *Science* 312 (2006): 1785–1788.

Van Peer, Philip. "The Nile Corridor and the Out-of-Africa Model: An Examination of the Archaeological Record." *Current Anthropology* 39 (1998): S115–S140.

Van Valen, Leigh. "A New Evolutionary Law." *Evolutionary Theory* 1 (1973): 1–30.

Vereshchagin, N. K., and I. E. Kuz'mina. "Ostatki Mlekopitayushchikh iz Paleoliticheskikh Stoyanok na Donu i Verkhnei Desne." *Trudy Zoologicheskogo Instituta AN SSSR* 72 (1977): 77–110.

Vermeersch, P. M., E. Paulissen, S. Stokes, C. Charlier, P. Van Peer, C. Stringer, and W. Lindsay. "A Middle Palaeolithic Burial of a Modern Human at Taramsa Hill, Egypt." *Antiquity* 72 (1998): 475–484.

Vernot, Benjamin, Serena Tucci, Janet Kelso, Joshua G. Schraiber, Aaron B. Wolf, Rachel M. Gittelman, Michael Dannemann, Steffi Grote, Rajiv C. McCoy, Heather Norton, et al. "Excavating Neandertal and Denisovan DNA from the Genomes of Melanesian Individuals." *Science* 352 (2016): 235–239.

Veth, Peter, Mike Smith, Jim Bowdler, Kathryn Fitzsimmons, Alan Williams, and Peter Hiscock. "Excavations at Parnkupirti, Lake Gregory, Great Sandy Desert: OSL Ages for Occupation Before the Last Glacial Maximum." *Australian Archaeology* 69 (2009): 1–10.

Vigilant, L., M. Stoneking, H. Harpending, K. Hawkes, and A. C. Wilson. "African Populations and the Evolution of Human Mitochondrial DNA." *Science* 253 (1991): 1503–1507.

Villa, Paola, and Wil Roebroeks, "Neandertal Demise: An Archaeological Analysis of the Modern Human Superiority Complex," *PLoS ONE* 9 (2014): e96424.

Villa, Paola, Sylvain Soriano, Nicolas Teyssandier, and Sarah Wurz. "The Howiesons Poort and MSA III at Klasies River Main Site, Cave 1A." *Journal of Archaeological Science* 37 (2010): 630–635.

Villa, Paola, Sylvain Soriano, Tsenka Tsanova, Ilaria Degano, Thomas F. G. Higham, Francesco d'Errico, Lucinda Backwell, Jeannette J. Lucejko, Maria Perla Colombini, and Peter B. Beaumont. "Border Cave and the Beginning of the Later Stone Age in South Africa." *Proceedings of the National Academy of Sciences* 109 (2012): 13208–13213.

Villmoare, Brian, William H. Kimbel, Chalachew Seyoum, Christopher J. Campisano, Erin N. DiMaggio, John Rowan, David R. Braun, J Ramón Arrowsmith, and Kaye E. Reed. "Early *Homo* at 2.8 Ma from Ledi-Geraru, Afar, Ethiopia." *Science* 347 (2015): 1352–1355.

Vincent, Jean-Didier, and Pierre-Marie Lledo. *The Custom-Made Brain: Cerebral Plasticity, Regeneration, and Enhancement.* Translated by Laurence Garey. New York: Columbia University Press, 2014.

Voelker, A. H. L., P. M. Grootes, M. J. Nadeau, and M. Sarnthein. "Radiocarbon Levels in the Iceland Sea from 25–53 kyr and Their Link to the Earth's Magnetic Field Intensity." *Radiocarbon* 42 (2000): 437–452.

von Neumann, John. *The Computer and the Brain.* 3rd ed.. New Haven, Conn.: Yale University Press, 2012.

von Neumann, John. *Theory of Self-Reproducing Automata*. Edited and completed by Arthur W. Burks. Urbana: University of Illinois Press, 1966.

von Neumann, John, and Oskar Morgenstern. *Theory of Games and Economic Behavior*. Princeton, N.J.: Princeton University Press, 1944.

Vygotsky, L. S. *Mind in Society: The Development of Higher Psychological Processes*. Cambridge, Mass.: Harvard University Press, 1978.

Vygotsky, L. S. *Thought and Language*. Revised and edited by Alex Kozulin. Cambridge, Mass.: MIT Press, 1986.

Wade, Nicholas. *A Troublesome Inheritance: Genes, Race, and Human History*. New York: Penguin Press, 2014.

Wadley, Lyn. "Recognizing Complex Cognition Through Innovative Technology in Stone Age and Palaeolithic Sites." *Cambridge Archaeological Journal* 23 (2013): 163–183.

Wadley, Lyn. "Were Snares and Traps Used in the Middle Stone Age and Does It Matter? A Review and a Case Study from Sibudu, South Africa." *Journal of Human Evolution* 58 (2010): 179–192.

Waguespack, Nicole M. "Pleistocene Extinctions: The State of Evidence and the Structure of Debate." In *Paleoamerican Odyssey*, edited by Kelly E. Graf, Caroline V. Ketron, and Michael R. Waters, 311–319. College Station: Texas A&M University Press, 2013.

Walker, Alan, and Richard Leakey. "The Postcranial Bones." In *The Nariokotome Homo erectus Skeleton*, edited by Alan Walker and Richard Leakey, 136–138. Berlin: Springer, 1993.

Walker, Alan, and Pat Shipman. *The Ape in the Tree: An Intellectual and Natural History of Proconsul*. Cambridge, Mass.: Belknap Press of Harvard University Press, 2005.

Walker, Alan, and Pat Shipman. *The Wisdom of the Bones: In Search of Human Origins*. New York: Knopf, 1996.

Wall, Jeffrey D. "Detecting Ancient Admixture in Humans Using Sequence Polymorphism Data." *Genetics* 154 (2000): 1271–1279.

Wallace, Anthony F. C. *The Social Context of Innovation: Bureaucrats, Families, and Heroes in the Early Industrial Revolution, as Foreseen in Bacon's* New Atlantis. Princeton, N.J.: Princeton University Press, 1982.

Wang, J. "Estimation of Effective Population Sizes from Data on Genetic Markers." *Philosophical Transactions of the Royal Society B: Biological Sciences* 360 (2005): 1395–1409.

Wang, Sija, Cecil M. Lewis Jr., Mattias Jakobsson, Sohini Ramachandran, Nicolas Ray, Gabriel Bedoya, Winston Rojas, Maria V. Parra, Julio A. Molina, Carla Gallo, et al. "Genetic Variation and Population Structure in Native Americans." *PLoS Genetics* 3 (2007): 2049–2067.

Waters, Michael R., Steven L. Forman, Thomas A. Jennings, Lee C. Nordt, Steven G. Driese, Joshual M. Feinberg, Joshua L. Keene, Jessi Halligan, Anna Lindquist, James Pierson, et al. "The Buttermilk Creek Complex and the Origins of Clovis at the Debra L. Friedkin Site, Texas." *Science* 331 (2011): 1599–1603.

Waters, Michael R., and Thomas Wier Stafford Jr. "The First Americans: A Review of the Evidence for the Late-Pleistocene Peopling of the Americas." In *Paleoamerican Odyssey*, edited by Kelly E. Graf, Caroline V. Ketron, and Michael R. Waters, 541–560. College Station: Texas A&M University Press, 2013.

Waters, Michael R., Thomas W. Stafford Jr., Brian Kooyman, and L. V. Hills. "Late Pleistocene Horse and Camel Hunting at the Southern Margin of the Ice-Free Corridor: Reassessing the Age of Wally's Beach, Canada." *Proceedings of the National Academy of Sciences* 112 (2015): 4263–4267.

Waters, Michael R., Thomas W. Stafford Jr., H. Gregory McDonald, Carl Gustafson, Morten Rasmussen, Enrico Cappellini, Jesper V. Olsen, Damian Szklarczyk, Lars Juhl Jensen, M. Thomas P. Gilbert, et al. "Pre-Clovis Mastodon Hunting 13,800 Years Ago at the Manis Site, Washington." *Science* 334 (2011): 351–353.

Watson, E., P. Forster, M. Richards, and H.-J. Bandelt. "Mitochondrial Footprints of Human Expansion in Africa." *American Journal of Human Genetics* 61 (1997): 691–704.

Watts, Ian. "The Origin of Symbolic Culture." In *The Evolution of Culture*, edited by RobinDunbar, Chris Knight, and Camilla Power, 113–146. New Brunswick, N.J.: Rutgers University Press, 1999.

Weaver, Anne H. "Reciprocal Evolution of the Cerebellum and Neocortex in Fossil Humans." *Proceedings of the National Academy of Sciences* 102 (2005): 3576–3580.

Weaver, Timothy D. "Did a Discrete Event 200,000–100,000 Years Ago Produce Modern Humans?" *Journal of Human Evolution* 63 (2012): 121–126.

Weaver, Timothy D. "The Meaning of Neandertal Skeletal Morphology." *Proceedings of the National Academy of Sciences* 106 (2009): 16028–16033.

Weidenreich, Franz. *Apes, Giants, and Man.* Chicago: University of Chicago Press, 1946.

Weiner, J. S., K. P. Oakley, and W. E. Le Gros Clark. "The Solution of the Piltdown Problem." *Bulletin of the British Museum (Natural History), Geology* 2 (1953): 139–146.

Weiner, Jürgen, and Harald Floss. "Eine Schwefelkiesknolle aus dem Aurignacien vom Vogelherd, Baden-Württemberg—Zu den Anfängen der Feuernutzung im europäischen Paläolithikum." *Archäologische Informationen* 27 (2004): 59–78.

Wei Wei, Qasim Ayub, Yuan Chen, Shane McCarthy, Yiping Hou, Ignazio Carbone, Yali Xue, and Chris Tyler-Smith. "A Calibrated Human Y-Chromosomal Phylogeny Based on Resequencing." *Genome Research* 23 (2013): 388–395.

Wells, Spencer. *Deep Ancestry: Inside the Genographic Project.* Washington, D.C.: National Geographic Society, 2007.

Wells, Spencer. *The Journey of Man: A Genetic Odyssey.* New York: Random House, 2002.

Weninger, Bernhard, and Olaf Jöris. "A ^{14}C Age Calibration Curve for the Last 60 ka: The Greenland-Hulu U/Th Timescale and Its Impact on Understanding the Middle to Upper Paleolithic Transition in Western Eurasia." *Journal of Human Evolution* 55 (2008): 772–781.

West, Frederick Hadleigh. *The Archaeology of Beringia.* New York: Columbia University Press, 1981.

Westbrook, Rachel, Sarah Fowell, Nancy Bigelow, and Sam Vanlaningham. "Evidence for Glacial Refugium in Central Beringia." Poster presented at the annual meeting of the Geological Society of America, Charlotte, N.C., November 4–7, 2012.

Whallon, Robert, William A. Lovis, and Robert K. Hitchcock, eds. *Information and Its Role in Hunter-Gatherer Bands.* Los Angeles: Cotsen Institute of Archaeology Press, 2011.

Whitfield, Charles W., Susanta K. Behura, Stewart H. Berlocher, Andrew G. Clark, J. Spencer Johnston, Walter S. Sheppard, Deborah R. Smith, Andrew V. Suarez, Daniel Weaver, and Neil D. Tsutsui. "Thrice Out of Africa: Ancient and Recent Expansions of the Honey Bee *Apis mellifera*." *Science* 314 (2006): 642–645.

Whittaker, Robert H. *Communities and Ecosystems.* 2nd ed. New York: Macmillan, 1975.

Wiessner, Polly. "Risk, Reciprocity and Social Influences on !Kung San Economics." In *Politics and History in Band Societies,* edited by Eleanor Leacock and Richard Lee, 61–84. Cambridge: Cambridge University Press, 1982.

Wild, Eva M., Maria Teschler-Nicola, Walter Kutschera, Peter Steier, Erik Trinkaus, and Wolfgang Wanek. "Direct Dating of Early Upper Palaeolithic Human Remains from Mladeč." *Nature* 435 (2005): 332–335.

Wilkins, Jayne, Benjamin J. Schoville, Kyle S. Brown, and Michael Chazan. "Evidence for Early Hafted Hunting Technology." *Science* 338 (2012): 942–946.

Willerslev, Eske, Anders J. Hansen, Jonas Binladen, Tina B. Brand, M. Thomas P. Gilbert, Beth Shapiro, Michael Bunce, Carsten Wiuf, David A. Gilichinsky, and Alan Cooper. "Diverse Plant and Animal Genetic Records from Holocene and Pleistocene Sediments." *Science* 300 (2003): 791–795.

Williams, Alan, Sean Ulm, Andrew R. Cook, Michelle C. Langley, and Mark Collard. "Human Refugia in Australia During the Last Glacial Maximum and Terminal Pleistocene: A Geospatial Analysis of the 25–12 ka Australian Archaeological Record." *Journal of Archaeological Science* 40 (2013): 4612–4625.

Williams, George C. *Adaptation and Natural Selection: A Critique of Some Current Evolutionary Thought.* Princeton, N.J.: Princeton University Press, 1966.

Williams, George C. *Natural Selection: Domains, Levels, and Challenges.* New York: Oxford University Press, 1992.

Williams, George C. *Sex and Evolution.* Princeton, N.J.: Princeton University Press, 1975.

Willoughby, Pamela R. *The Evolution of Modern Humans in Africa: A Comprehensive Guide.* Lanham, Md.: Altamira Press, 2007.

Witt, Kelsey E., Kathleen Judd, Andrew Kitchen, Colin Grier, Timothy A. Kohler, Scott G. Ortman, Brian M. Kemp, and Ripan S. Malhi. "DNA Analysis of Ancient Dogs of the Americas: Identifying Possible Founding Haplotypes and Reconstructing Population Histories." *Journal of Human Evolution* 79 (2014): 105–118.

Wolpoff, Milford H. "Multiregional Evolution: The Fossil Alternative to Eden." In *The Human Revolution: Behavioural and Biological Perspectives on the Origins of Modern Humans,* edited by Paul Mellars and Chris Stringer, 62–108. Princeton, N.J.: Princeton University Press, 1989.

Wolpoff, Milford H. *Paleo-Anthropology.* New York: Knopf, 1980.

Wolpoff, Milford H., Wu Xin Zhi, and Alan G. Thorne. "Modern *Homo sapiens* Origins: A General Theory of Hominid Evolution Involving the Fossil Evidence from East Asia." In *The Origins of Modern Humans: A World Survey of the Fossil Evidence,* ed. Fred H. Smith and Frank Spencer, 411–483. New York: Liss, 1984.

Wrangham, Richard. *Catching Fire: How Cooking Made Us Human.* New York: Basic Books, 2009.

Wright, Sewall. "Evolution of Mendelian Populations." *Genetics* 16 (1931): 97–159.

Wroe, Stephen, Judith H. Field, Michael Archer, Donald K. Grayson, Gilbert J. Price, Julien Louys, J. Tyler Faith, Gregory E. Webb, Iain Davidson, and Scott D. Mooney. "Climate Change Frames Debate over the Extinction of Megafauna in Sahul (Pleistocene Australia-New Guinea)." *Proceedings of the National Academy of Sciences* 110 (2013): 8777–8781.

Wurz, Sarah. "Technological Trends in the Middle Stone Age of South Africa Between MIS 7 and MIS 3." *Current Anthropology* 54 (2013): S305–S319.

Wynn, Jonathan G., Matt Sponheimer, William H. Kimbell, Zeresenay Alemseged, Kaye Reed, Zalalem K. Bedaso, and Jessica N. Wilson. "Diet of *Australopithecus afarensis* from the Pliocene Hadar Formation, Ethiopia." *Proceedings of the National Academy of Sciences* 110 (2013): 10495–10500.

Wynn, Thomas and Frederick L. Coolidge. "Archeological Insights into Hominin Cognitive Evolution." *Evolutionary Anthropology* 25 (2016): 200–213.

Yellen, John E., Alison S. Brooks, Els Cornelissen, Michael J. Mehlman, and Kathlyn Stewart. "A Middle Stone Age Worked Bone Industry from Katanda, Upper Semliki Valley, Zaire." *Science* 268 (1995): 553–556.

Yellen, John, and Henry Harpending. "Hunter-Gatherer Populations and Archaeological Inference." *World Archaeology* 4 (1972): 244–252.

Yotova, Vania, Jean-Francois Lefebvre, Claudia Moreau, Elias Gbeha, Kristine Hovhannesyan, Stephane Bourgeois, Sandra Bédaridal, Luisa Azevedo, Antonio Amorim, Tamara Sarkisian, et al. "An X-Linked Haplotype of Neandertal Origin Is Present Among All Non-African Populations." *Molecular Biology and Evolution* 28 (2011): 1957–1962.

Yunis, J. J., and W. G. Yasmineh. "Heterochromatin, Satellite DNA, and Cell Function." *Science* 174 (1971): 1200–1209.

Yurtsev, Boris A. "The Pleistocene Tundra-Steppe and the Productivity Paradox: The Landscape Approach." *Quaternary Science Reviews* 20 (2001): 165–174.

Yurtsev, Boris A. "Problems of the Late Cenozoic Paleogeography of Beringia in Light of Phytogeographic Evidence." In *Beringia in the Cenozoic Era*, edited by V. L. Kontrimavichus, 129–153. New Delhi: Amerind, 1984.

Zegura, Stephen L., Tatiana M. Karafet, Lev A. Zhirotovsky, and Michael F. Hammer. "High-Resolution SNPs and Microsatellite Haplotypes Point to a Single, Recent Entry of Native American Y Chromosomes into the Americas." *Molecular Biology and Evolution* 21 (2004): 164–175.

Ziesemer, Kirsten, Menno L. P. Hoogland, Corinne L. Hofman, Christina Warinner, and Hannes Schroeder. "Challenging Environments: Ancient DNA Research in the Circum-Caribbean." In *Abstracts of the SAA 81st Annual Meeting* (Orlando, Fla.), 496. Washington, D.C.: Society for American Archaeology, 2016.

Zilhão, João, Erik Trinkaus, Silviu Constantin, Ştefan Milota, Mircea Gherase, Laurenţiu Sarcina, Adrian Danciu, Hélène Rougier, Jérôme Quilès, and Ricardo Rodrigo. "The Peştera cu Oase People, Europe's Earliest Modern Humans." In *Rethinking the Human Revolution: New Behavioural and Biological Perspectives on the Origin and Dispersal of Modern Humans*, edited by Paul Mellars, Katie Boyle, Ofer Bar-Yosef, and Chris Stringer, 249–262. Cambridge: McDonald Institute for Archaeological Research, Cambridge University, 2007.

ACKNOWLEDGMENTS

Looking through my files, I see that I signed a contract with Columbia University Press (CUP) in June 2011 to produce a draft of this book by October 1, 2012. The complete draft was, in fact, not submitted until July 2015, and it was revised significantly during the year and a half that followed. I mention these grisly details to underscore my debt to the staff at CUP for their guidance and support throughout the writing of the book. Patrick Fitzgerald, publisher for the life sciences, championed the original concept of the book in 2011 and provided the ideal balance of exhortation and encouragement as I wrote and rewrote the draft. Editorial assistant Ryan Groendyk and his predecessors (Bridget Flannery-McCoy and Kathryn Schell) were always attentive and helpful, while the art department patiently redrew my homemade maps and figures. I am especially indebted to Irene Pavitt, who painstakingly edited the final manuscript with care and precision. The jacket and interior were designed by Milenda Lee.

A number of colleagues reviewed some or all of the chapters, and I am most grateful for their helpful comments and suggestions. Dennis H. O'Rourke (University of Kansas) and James F. O'Connell (University of Utah) reviewed the complete draft. G. Richard Scott (University of Nevada, Reno) reviewed chapter 2 (synthesis) and contributed the box and figures on dental anthropology in the same chapter. Teresa E. Steele (University of California, Davis) reviewed chapter 4 (Africa), Sheela Athreya (Texas A&M University) reviewed chapter 5 (South Asia), and Vance T. Holliday

(University of Arizona) reviewed chapter 7 (Americas). Several colleagues at the University of Colorado, Boulder (Cathy Cameron, Steve Lekson, Scott Ortman, Payson Sheets, and Paola Villa), also reviewed an earlier draft of chapter 7. Nancy H. Bigelow and Amy Hendricks (University of Alaska, Fairbanks) provided the BIOME4 model output on Beringian vegetation during the Last Glacial Maximum (figure 7.4). Ian T. Hoffecker (Karolinska Institutet) made the original drawings for several figures, including figures 2.7, 3.5, 7.3, and 7.5, and box figures 3.1 and 3.2, and contributed valuable ideas to the discussions of technological complexity and its measurement.

I see clearly now that the genesis of this book lies in my arrival in the autumn of 1979 at the University of Chicago to complete my graduate study. I have a vivid memory of sitting in Karl W. Butzer's office at some point in 1980 as he talked to me about the relatively early dating of modern human remains in Africa. As recounted in the book, Professor Butzer played a major role in establishing the antiquity of modern humans in Africa and thus providing original basis for the Recent African Origin model. My Ph.D. adviser, Richard G. Klein, had described the contrast between modern humans and their predecessors in *Ice-Age Hunters of the Ukraine* (1973) as a "quantum advance," and he eventually became the most prominent advocate of this view. Most of our colleagues in paleoanthropology probably disagree with him, especially now, but, as I have argued in this book, I believe that Professor Klein is correct. If anything, I think he is understating it somewhat.

INDEX

"major transitions" in evolution, 2, 12, 17, 21, 38–40, 136, 139, 144–145, 162, 200, 349n.4; and ribozyme, 2. *See also* complexity; eukaryotes: origin of; Maynard Smith, John; modern humans: as "major transition" in evolution; Szathmáry, Eörs
Makarovo-4 (Siberia), 295, 298
Malakunanja II (Australia), 242
Malaya Syia (Siberia), 295
Mamanwa, 237
Mammuthus jeffersoni, 337. *See also* Hebior
Mamontovaya kurya (Russia), 255, 282, 291
Man, E. H., 221
Mandu Mandu Creek Rock Shelter (Australia), 243
Manis (Washington), 335–336, 420n.120
Manot Cave (Israel), 82, 252, 266–267, 270, 272, 274, 406n.61; cranium from, 266–267
Marine Isotope Stages (MIS), 355n.94; MIS 2 (Last Glacial Maximum), 214–215; MIS 3, 214–216; MIS 4, 82, 210, 214; MIS 5, 82, 214; MIS 5c, 216; MIS 5e, 216
Marks, Anthony E., 271, 274. *See also* Initial Upper Paleolithic
Marler, Peter, 119
marriage. *See* pair-bonding: in humans, long-term
massively parallel signature sequencing (MPSS). *See* DNA: sequencing techniques for
massive parallel processing, of metazoan brain, 111, 117. *See also* brain or central nervous system
mass-kill techniques (hunting of animal herds), 161, 265, 288–289, 400n.128. *See also* Kostenki: Kostenki 12 at
materials, transportation of, over large distances, 62–63, 65, 80, 108, 205, 265, 277, 297; fossil shells, 108, 290; silcrete, 195. *See also* Howiesons Poort; networks
Matja Kuru (Timor), 231

mattocks (digging tools), of antler, 290. *See also* Kostenki: Kostenki 14 at
Mauer (Germany), 155, 158, 354n.84; jaw from, 22–23. See also *Homo heidelbergensis*
Maxwell, James Clerk, 33
"Maxwell's Demon," 33, 373n.23
Mayfield, John, 153, 373n.17. *See also* computation: evolution as
Maynard Smith, John, 2, 12, 17, 32, 38, 111, 140, 144, 197, 200, 349n.4, 357n.117. *See also* game theory; information; "major transitions" in evolution; reciprocity: and altruism
Mayr, Ernst, 18, 353n.59
McBrearty, Sally, 173
Mead (Alaska), 324
Mead, Margaret, 44. *See also* culture, concept of
Meadowcroft Rockshelter (Pennsylvania), 420n.125
"Megaberingia." *See* Beringia: definition of; Yurtsev, Boris
Mehkakheri (India), 227–230
Melanesians, 56–57, 59, 239–240
Mellars, Paul, 172, 227, 292
metazoa, origin of, 39–40
Mezmaiskaya Cave (Russia), 282, 288, 364n.60, 403n.29, 409n.98
Middle Paleolithic (Eurasia), 267, 273–274, 278, 281, 284, 287
Middle Stone Age (MSA; African), 50–52, 265
Miocene epoch, 85, 121
Mira (Ukraine), 283
Mishra, Sheila, 227
"mitochondrial Adam." *See* "African Adam"
mitochondrial DNA (mtDNA), 37, 49–52, 54, 58, 71, 73, 82, 89–93, 102, 276, 308, 310, 313, 320–321; decreasing genetic diversity of, as function of distance from Africa, 89, 308; haplogroup A, 90, 92–93, 309, 328, 340; haplogroup B, 91–93, 293, 298, 309, 328–329, 340; haplogroup C, 93, 309, 311, 328–329; haplogroup D, 54,